D0416942

THE BIOCHEMISTRY AND PHYSIOLOGY OF GIBBERELLINS

Volume I

Edited by

Alan Crozier

PRAEGER SPECIAL STUDIES • PRAEGER SCIENTIFIC

Library of Congress Cataloging in Publication Data
Main entry under title:

The Biochemistry and physiology of gibberellins.

   Includes bibliographies and indexes.
   1. Gibberellins. 2. Plants, Effect of gibberellins
on. I. Crozier, Alan.
QK898.G45B56   1983   581.19'27   83-13862
ISBN 0-03-059054-X (v. 1:alk. paper)
ISBN 0-03-059056-6 (v. 2:alk. paper)

Published in 1983 by Praeger Publishers
CBS Educational and Professional Publishing
a Division of CBS Inc.
521 Fifth Avenue, New York, NY 10175 USA
© 1983 by Praeger Publishers

3456789 052 987654321

Printed in the United States of America
on acid-free paper

CONTENTS

PREFACE

This is the first of two volumes dealing with the biochemistry and physiology of gibberellins (GAs). The discovery of GAs can be traced back to studies by Kurosawa on the "bakanae" disease of rice that were reported in *The Transactions of the Natural History Society of Formosa* in 1926. Kurosawa demonstrated that "bakanae" seedlings were infected with a fungal pathogen, *Gibberella fujikuroi*, which secreted a factor that markedly enhanced the rate of shoot elongation. He also noted that the active factor promoted the growth of seedlings of maize, sesame, millet, and oats. In the 1930s scientists from the Department of Agricultural Chemistry at the University of Tokyo succeeded in crystallizing the growth inducing factor which they named gibberellin. The structure of the first fungal GA was elucidated in the mid-1950s by investigators in the United States and Great Britain. Shortly thereafter GAs were shown to be natural constituents of higher plant tissues, and it became apparent that they functioned as endogenous regulators of many aspects of growth and development.

Although GAs have been subjected to extensive investigation during the past twenty-five years, it is still possible to keep abreast with general developments in the field. However, many aspects of GA research have become technically complex and unless one is actively involved in a specific area it is difficult to evaluate findings critically and place them in proper perspective. For these reasons it would be very difficult for one person to write an authoritative account of anything other than a very restricted area of GA research. The aim of this and the accompanying volume is to overcome this problem by bringing together authors with a specialist knowledge of topics that collectively encompass most of the main areas of GA biochemistry and physiology. It is hoped that the compendium will be of value to researchers in the field and also provide a useful resume of the current status of our knowledge of GAs for those with more general interests in either plant growth regulation or plant physiology and biochemistry.

Some overlap of material between chapters is inevitable in a multiauthor volume. Attempts have been made to reduce this to a minimum, and what remains is intended to be of help to the reader insofar as continuity of the text is maintained rather than being interrupted by cross references to other contributions.

I would like to express my gratitude to the authors, who painstakingly prepared their manuscripts and responded so promptly to editorial comment. I am also indebted to Dr. J. D. Connelly, Department of Chemistry, University of Glasgow, who read and helpfully commented on selected manuscripts, advised on the style to be used when illustrating chemical structures, and provided me with ready access to his "Dictionary of Organic Compounds." I wish to thank Professor J. MacMillan, FRS, School of Chemistry, University of Bristol, for providing prepublication details of the more recently discovered GAs.

I would also like to thank Professor M. B. Wilkins, Botany Department, University of Glasgow, for his advice and interest in the project. Finally, I am especially grateful to Mrs. A. Sutcliffe and Mr. T. N. Tait, Department of Botany, University of Glasgow, who burnt much midnight oil, respectively drafting and photographing the figures. Without their valient efforts publication would have been delayed by many months.

Alan Crozier

LIST OF CONTRIBUTORS

John R. Bearder
Biosciences Laboratory
Shell Research Ltd.
Sittingbourne Research Centre
Kent ME9 8AG
England

Ronald C. Coolbaugh
Department of Botany
Iowa State University
Ames, Iowa 50011
USA

Alan Crozier
Department of Botany
The University
Glasgow G12 8QQ
Scotland

Richard C. Durley
Department of Forest Science
Forest Research Laboratory
Oregon State University
Corvallis, Oregon 97331
USA

Peter Hedden
Pflanzenphysiologisches Institut
    der Universitat
D-3400 Gottingen
German Federal Republic

Bernard O. Phinney
Department of Biology
University of California, Los Angeles
Los Angeles, California 90024
USA

Gernot Schneider
Institut fur Biochemie der Pflanzen
Forschungszentrum fur Molekular-
    biologie und Medizin
Akademie der Wissenschaften der DDR
Halle (Saale)
DDR

Valerie M. Sponsel
Department of Organic Chemistry
University of Bristol
Bristol BS8 1TS
England

Nobutaka Takahashi
Department of Agricultural Chemistry
University of Tokyo
Bunkyo-Ku, Tokyo 113
Japan

Isomaro Yamaguchi
Department of Agricultural Chemistry
University of Tokyo
Bunkyo-Ku, Tokyo 113
Japan

Introduction:
The Occurrence and Structure of
Gibberellins

Since the first gibberellin (GA) was structurally elucidated in the mid-1950s (Curtis & Cross, 1954; Stodola et al., 1955), GAs have been subjected to extensive investigation and at the same time there have been radical advances in the analytical sciences. One of the consequences of these developments is that 66 naturally occurring GAs have now been characterized (Fig. 1). Eleven GAs have been found only in *Gibberella fujikuroi* cultures, 41 are exclusive to higher plants, while 14 are ubiquitous, having been detected in extracts from the fungus and higher plant tissues (Table 1). Many more potential permutations of the GA structure exist, and there will undoubtedly continue to be additions to this list for some time to come. In order to avoid confusion the trivial nomenclature GA_1–GA_{66} has been adopted and there will be a sequential allocation of numbers GA_{67} . . . GA_n as further GAs are characterized (MacMillan & Takahashi, 1968).

In the earlier literature the systematic nomenclature of the GAs was based on the gibbane skeleton (*1*). However, as this was unrelated to other diterpenoids, it has been superseded and the systematic nomenclature for both GAs and their kaurenoid precursors is now in accordance with the accepted rules for tetracyclic diterpenes being based on gibberellane (*2*) and kaurane (*3*) (see Rowe, 1968; McCrindle & Overton, 1969). Since all known GAs and their kaurenoid precursors have an absolute stereochemistry which is enantiomeric to (*2*) and (*3*), their systematic nomenclature is based on *ent*-gibberellane (*4*) and *ent*-kaurane (*5*). The operator *ent* reverses the stereochemical

1

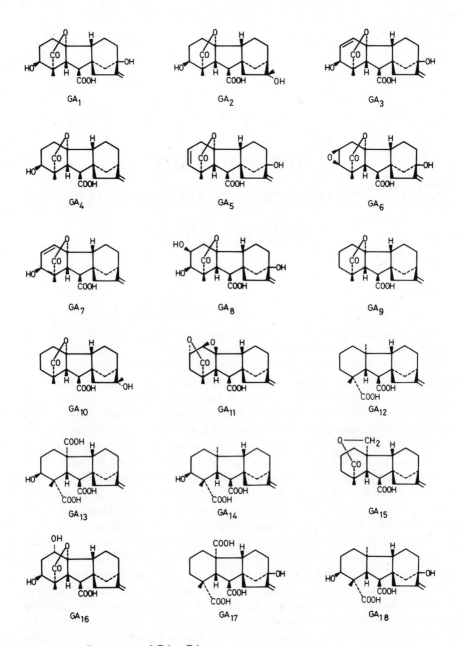

Figure 1   Structures of GA_1–GA_{66}.

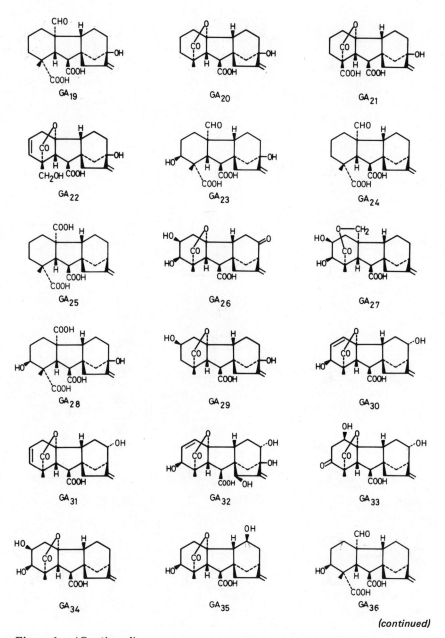

Figure 1   (*Continued*)

(continued)

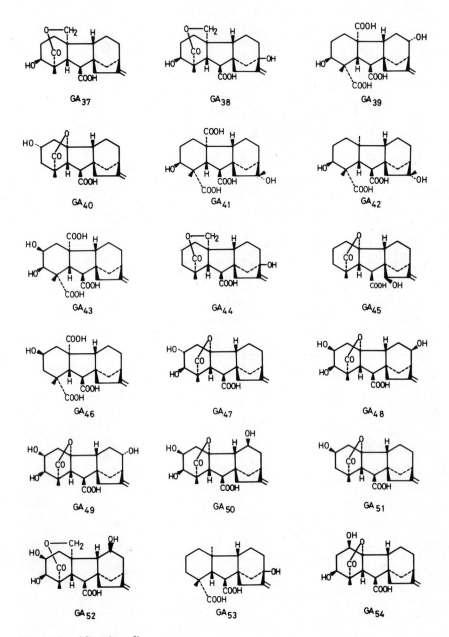

Figure 1 *(Continued)*

4

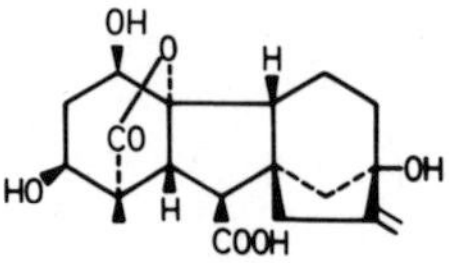

GA_{55}

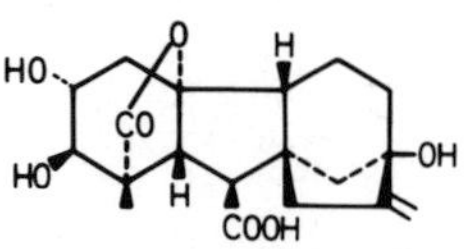

GA_{56}

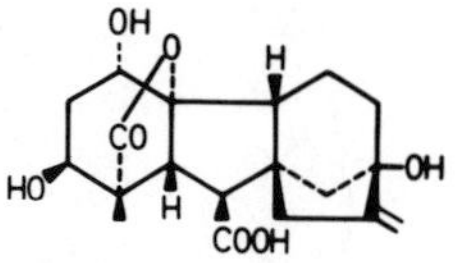

GA_{57}

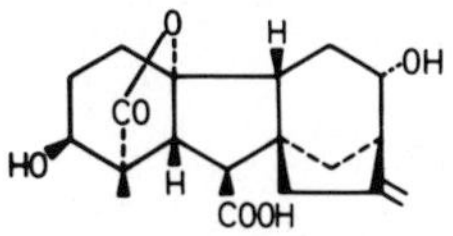

GA_{58}

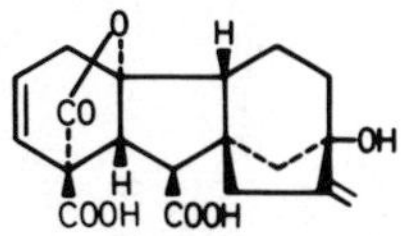

GA_{59}

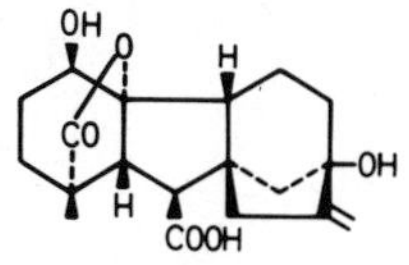

GA_{60}

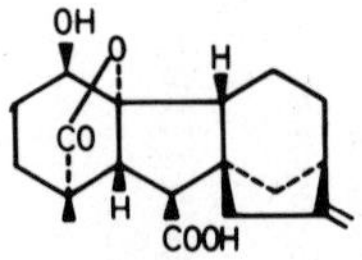

GA_{61}

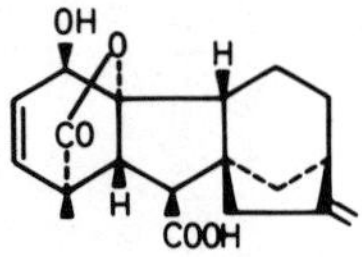

GA_{62}

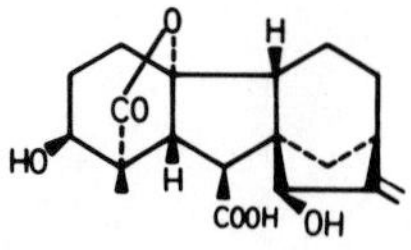

GA_{63}

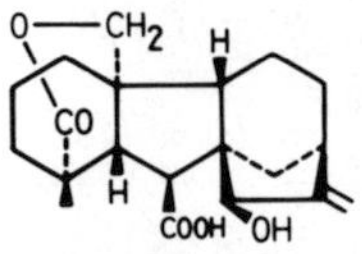

GA_{64}

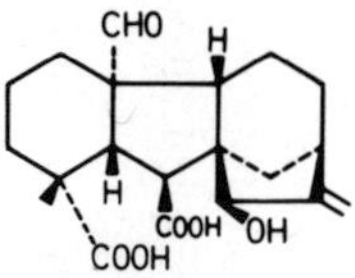

GA_{65}

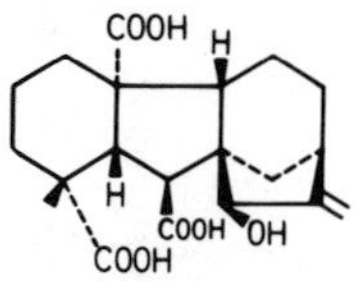

GA_{66}

Figure 1 (*Continued*)

designation at all chiral centers, and this can be a source of confusion because substituents that are α or β with reference to traditionally drawn structural formulas (Fig. 1) become *ent*-β or *ent*-α, respectively, in the systematic nomenclature. Thus GA_{39} is *ent*-3α,12β-dihydroxy-gibberell-16-ene-7,19,20 trioic acid. More frequently, however, GAs are referred to by their trivial names $GA_{1 \ldots n}$, and in these circumstances the conventional method of designating substituents the way they are actually drawn is applied. Similarly, it is common practice for trivial names to be used for kaurenolides. Thus, structure (*6*) is 7β-hydroxykaurenolide. In contrast, other kaurane derivatives are usually referred to by their systematic or semi-systematic names. Structure (*7*) is, therefore, *ent*-7α-hydroxykaur-16-en-19-oic acid or, less formally, *ent*-7α-hydroxykaurenoic acid, and the hydroxyl group at C-7 is referred to as having either a 7β or an *ent*-7α orientation.

Table 1. Discovery and origin of naturally occurring gibberellins

Gibberellin	Occurrence (G:*G. fujikuroi* H:higher plant)	Material from which the gibberellin was originally isolated	Reference
GA$_1$	G, H	*Gibberella fujikuroi* mycelial filtrate	Takahashi et al. (1955)
GA$_2$	G	*Gibberella fujikuroi* mycelial filtrate	Takahashi et al. (1955)
GA$_3$	G, H	*Gibberella fujikuroi* mycelial filtrate	Curtis & Cross (1954); Stodola et al. (1955)
GA$_4$	G, H	*Gibberella fujikuroi* mycelial filtrate	Takahashi et al. (1957)
GA$_5$	H	*Phaseolus vulgaris* immature seed	West & Phinney (1959); MacMillan et al. (1959)
GA$_6$	H	*Phaseolus coccineus*[a] immature seed	MacMillan et al. (1962)
GA$_7$	G, H	*Gibberella fujikuroi* mycelial filtrate	Cross et al. (1962)
GA$_8$	H	*Phaseolus coccineus*[a] immature seed	MacMillan et al. (1962)
GA$_9$	G, H	*Gibberella fujikuroi* mycelial filtrate	Cross et al. (1962)
GA$_{10}$	G	*Gibberella fujikuroi* mycelial filtrate	Hanson (1966)
GA$_{11}$	G	*Gibberella fujikuroi* mycelial filtrate	Brown et al. (1967)
GA$_{12}$	G, H	*Gibberella fujikuroi* mycelial filtrate	Cross & Norton (1965)
GA$_{13}$	G, H	*Gibberella fujikuroi* mycelial filtrate	Galt (1965)

GA_{14}	G	*Gibberella fujikuroi* mycelial filtrate	Cross (1966)
GA_{15}	G, H	*Gibberella fujikuroi* mycelial filtrate	Hanson (1967)
GA_{16}	G, H	*Gibberella fujikuroi* mycelial filtrate	Galt (1968)
GA_{17}	H	*Phaseolus coccineus*[a] immature seed	Pryce & MacMillan (1967)
GA_{18}	H	*Lupinus luteus* immature seed	Koshimizu et al. (1966)
GA_{19}	H	*Phyllostachys edulis* shoots	Murofushi et al. (1966)
GA_{20}	H	*Pharbitis nil* immature seed	Takahashi et al. (1967a)
GA_{21}	H	*Canavalia gladiata* immature seed	Takahashi et al. (1967b)
GA_{22}	H	*Canavalia gladiata* immature seed	Takahashi et al. (1967b)
GA_{23}	H	*Lupinus luteus* immature seed	Koshimizu et al. (1968)
GA_{24}	G, H	*Gibberella fujikuroi* mycelial filtrate	Harrison et al. (1968)
GA_{25}	G, H	*Gibberella fujikuroi* mycelial filtrate	Harrison & MacMillan (1971)
GA_{26}	H	*Pharbitis nil* immature seed	Yokota et al. (1969)
GA_{27}	H	*Pharbitis nil* immature seed	Yokota et al. (1969)
GA_{28}	H	*Lupinus luteus* immature seed	Fukui et al. (1971)
GA_{29}	H	*Calonyction aculeatum* immature seed	Murofushi et al. (1970)
GA_{30}	H	*Calonyction aculeatum* immature seed	Murofushi et al. (1970)
GA_{31}	H	*Calonyction aculeatum* immature seed	Murofushi et al. (1970)
GA_{32}	H	*Prunus armenica* immature seed	Coombe (1971)

[a]Previously referred to as *Phaseolus multiflorus*.

(continued)

7

Table 1. *(Continued)*

GA$_{33}$	H	*Calonyction aculeatum* immature seed	Murofushi et al. (1971)
GA$_{34}$	H	*Calonyction aculeatum* immature seed	Murofushi et al. (1971)
GA$_{35}$	H	*Cytisus scoparius* immature seed	Yamane et al. (1971)
GA$_{36}$	G	*Gibberella fujikuroi* mycelial filtrate	Bearder & MacMillan (1972)
GA$_{37}$	G, H	*Phaseolus vulgaris* immature seed	Hiraga et al. (1972)
GA$_{38}$	H	*Phaseolus vulgaris* immature seed	Hiraga et al. (1972)
GA$_{39}$	H	*Cucurbita pepo* immature seed	Fukui et al. (1977)
GA$_{40}$	G	*Gibberella fujikuroi* mycelial filtrate	Yamaguchi et al. (1973)
GA$_{41}$	G	*Gibberella fujikuroi* mycelial filtrate	Bearder & MacMillan (1973)
GA$_{42}$	G	*Gibberella fujikuroi* mycelial filtrate	Bearder & MacMillan (1973)
GA$_{43}$	H	*Cucurbita maxima* immature seed	Graebe et al. (1974)
GA$_{44}$	H	*Pisum sativum* seed	Frydman et al. (1974)
GA$_{45}$	H	*Pyrus communis* immature seed	Bearder et al. (1975)
GA$_{46}$	H	*Marah macrocarpus*[b] immature seed	Beeley & MacMillan (1976)
GA$_{47}$	G	*Gibberella fujikuroi* mycelial filtrate	Beeley & MacMillan (1976)
GA$_{48}$	H	*Cucurbita pepo* immature seed	Fukui et al. (1977)-
GA$_{49}$	H	*Cucurbita pepo* immature seed	Fukui et al. (1977)
GA$_{50}$	H	*Lagenaria leucantha* immature seed	Fukui et al. (1978)

8

GA$_{51}$	H	*Pisum sativum* immature seed	Sponsel & MacMillan (1977)
GA$_{52}$	H	*Lagenaria leucantha* immature seed	Fukui et al. (1978)
GA$_{53}$	H	*Vicia faba* seed	Sponsel et al. (1979)
GA$_{54}$	G, H	*Gibberella fujikuroi* mycelial filtrate	Murofushi et al. (1979)
GA$_{55}$	G, H	*Gibberella fujikuroi* mycelial filtrate	Murofushi et al. (1979)
GA$_{56}$	G	*Gibberella fujikuroi* mycelial filtrate	Murofushi et al. (1979)
GA$_{57}$	G	*Gibberella fujikuroi* mycelial filtrate	Murofushi et al. (1980)
GA$_{58}$	H	*Cucurbita maxima* immature seed	Beale et al. (in preparation)
GA$_{59}$	H	*Canavalia gladiata* immature seed	Yokota & Takahashi (1981)
GA$_{60}$	H	*Triticum aestivum* immature seed	Kirkwood & MacMillan (1982)
GA$_{61}$	H	*Triticum aestivum* immature seed	Kirkwood & MacMillan (1982)
GA$_{62}$	H	*Triticum aestivum* immature seed	Kirkwood & MacMillan (1982)
GA$_{63}$	H	*Prunus malus* immature seed	Dolan, Hutchinson, & MacMillan (unpublished data)
		Prunus communis immature seed	
GA$_{64}$	H	*Helianthus annus* immature seed	Hutchison (1983)
GA$_{65}$	H	*Helianthus annus* immature seed	Hutchison (1983)
GA$_{66}$	H	*Helianthus annus* immature seed	Hutchison (1983)

[b] Previously referred to as *Echinocystis maccocarpa*.

9

Handwritten annotations:

Hutchison et al. (1988)

9A60 = Apple

9A67
9A72

15β-6H 9A17
15β-6H 9A19
15β-0H 9A44
15β-6H 9A53

Kirkwood + MacMillan
J. Chem Soc Perkin
Trans I, 689-697

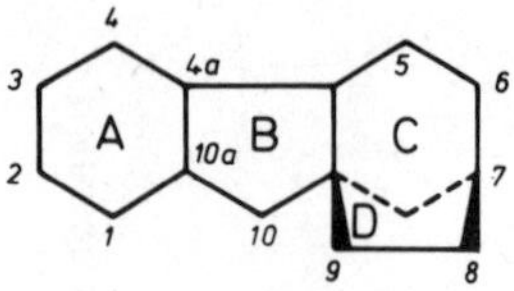

(1) gibbane

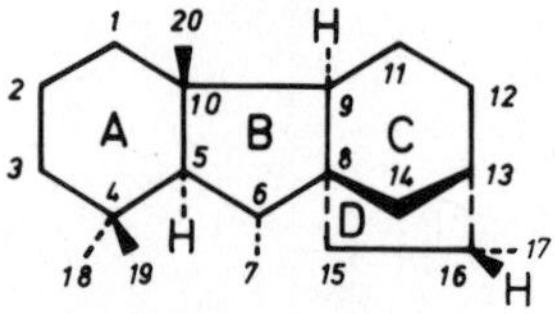

(2) gibberellane

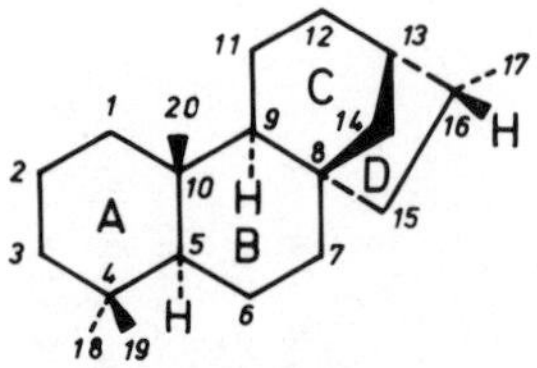

(3) kaurane

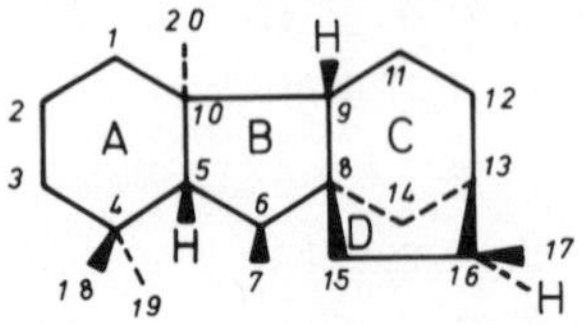

(4) ent–gibberellane

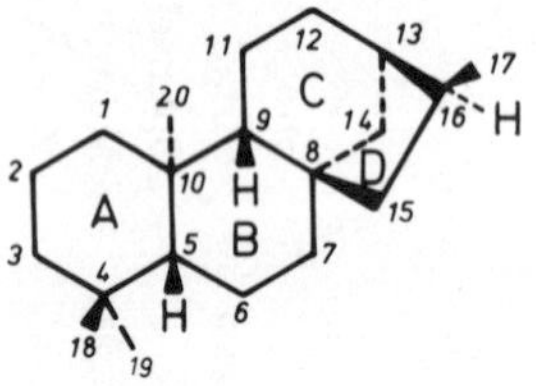

(5) ent –kaurane

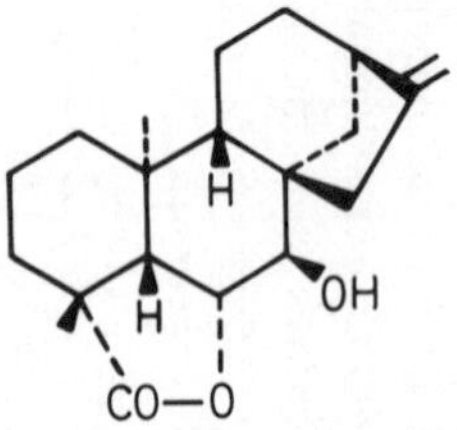

(6)

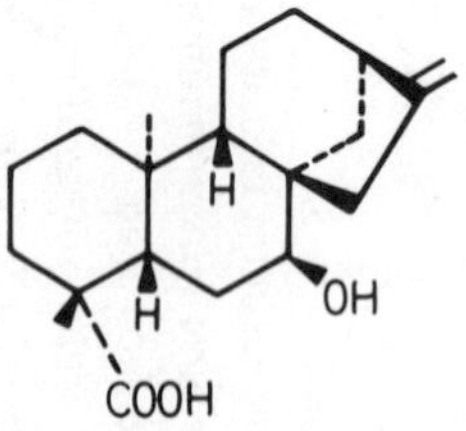

(7)

Structures (1)–(7)

While the large number of endogenous GAs presents intriguing problems to the chemist, it can be a daunting prospect to those with training in the plant sciences and interests in growth regulation rather than GA biochemistry per se. Fortunately, familiarization with the structures illustrated in Fig. 1 is not as difficult a task as first impressions might suggest. As mentioned above, all the GAs possess an *ent*-gibberellane skeleton (4) and can be divided into two groups by virtue of the possession of either 19 or 20 carbon atoms. The C_{20}-

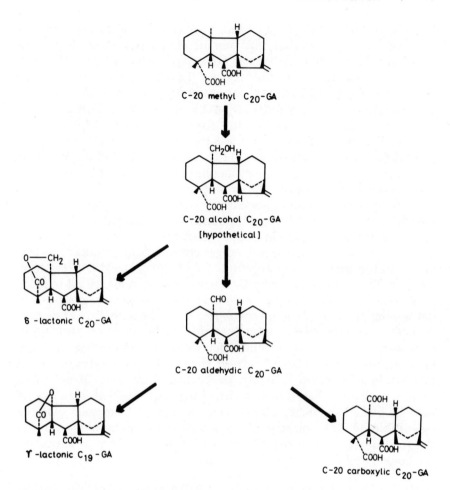

Figure 2 Basic GA structures and probable biosynthetic relationships of C_{19}-GAs and C_{20}-GAs. Biosynthetic information based on data obtained with cell-free preparations from liquid endosperm of *Cucurbita maxima* seed (Graebe, Hedden, & Rademacher, 1980).

GAs are characterized by the presence of carbon-20 which can exist as either a CH_3, CH_2OH, CHO, or COOH function, and there is evidence that these structures are related biosynthetically by the pathway illustrated in Fig. 2. On extraction from plant tissues the C-20 alcohol C_{20}-GAs are believed to lactonize to produce δ-lactonic C_{20}-GAs. δ-Lactonic C_{20}-GAs such as GA_{15}, GA_{37}, GA_{38}, and GA_{44} are, thus, likely to be artifacts, although the possibility that they are also native products cannot be ruled out. C-20 aldehydic GAs give rise to C-20 carboxylic C_{20}-GAs and also appear to act as the immediate precursor of C_{19}-GAs, all but one of which possess a $19 \rightarrow 10$ γ-lactonic bridge. The exception is GA_{11}, which has a $19 \rightarrow 2$ linkage. The variations in the oxidation state at C-20 and the presence or absence of 3β- or 13-hydroxyl groups account for 20 GAs (Table 2). The remaining GAs are represented by additional modifications to these basic configurations in the form of 2,3 and 1,10 epoxide groups, C-3 and C-12 keto groups, β-hydroxylation at C-1, C-2, C-12, and C-15, α-hydroxylations at C-1, C-2, C-12, and C-16, oxidation of the 18 methyl group to primary alcohol and carboxyl functions, and the introduction of 1,2 and 2,3 double bonds.

A number of naturally-occurring GA conjugates have also been identified (Table 3). These compounds are discussed in detail and their structures illustrated in Chapter 6 of this volume.

The application of combined gas chromatography-mass spectrometry to the analysis of endogenous GAs (Binks, MacMillan, & Pryce, 1969) is the main reason why GAs have now been identified in more than 45 species of higher plants. Fourteen GAs have been identified in developing grain of *Triticum aestivum*, 13 in the immature *Phaseolus coccineus* seed, and 11 in immature seed of *Calonyction aculeatum*. Immature seed has proved to be a rich source of GAs and can contain, in total, up to 50 mg GA kg^{-1} fresh weight. This is several orders of magnitude more than is likely to be found in other higher plant tissues, and it is therefore not surprising that the vast majority of GAs identified in higher plants have originated from seed material (see Table 1). GAs present in immature seed can be conjugated as the seed develops, and seeds have also been the source of almost all the conjugated GAs that have been identified to date (Table 3).

Table 2. GA structures based on variation in the oxidation state at C-20 and the presence or absence of hydroxyl groups at C-3 and C-13

		Hydroxylation			
Oxidation at C-20		None	3β	13	3β, 13
C_{20}-GAs	CH_3	GA_{12}	GA_{14}	GA_{53}	GA_{18}
	δ-lactone	GA_{15}	GA_{37}	GA_{44}	GA_{38}
	CHO	GA_{24}	GA_{36}	GA_{19}	GA_{23}
	COOH	GA_{25}	GA_{13}	GA_{17}	GA_{28}
C_{19}-GAs	γ-lactone	GA_9	GA_4	GA_{20}	GA_1

Table 3. Discovery and origin of naturally occurring gibberellin conjugates

Gibberellin conjugate	Materials from which GA conjugate was originally isolated	Reference
Glucosides		
GA_1-3-O-β-D-glucosyl ether	*Dolichos lablab* seed	Yokota et al. (1978)
GA_3-3-O-β-D-glucosyl ether	*Pharbitis nil* immature seed	Tamura et al. (1968)
GA_8-2-O-β-D-glucosyl ether	*Phaseolus coccineus* seed	Schreiber et al. (1967)
GA_{26}-2-O-β-D-glucosyl ether	*Pharbitis nil* immature seed	Yokota et al. (1969)
GA_{27}-2-O-β-D-glucosyl ether	*Pharbitis nil* immature seed	Yokota et al. (1969)
GA_{29}-2-O-β-D-glucosyl ether	*Pharbitis nil* immature seed	Yokota et al. (1970)
GA_{35}-11-O-β-D-glucosyl ether	*Cytisus scoparius* immature seed	Yamane et al. (1971)
Glucosyl esters		
GA_1-β-D-glucosyl ester	*Phaseolus vulgaris* mature seed	Hiraga et al. (1974)
GA_4-β-D-glucosyl ester	*Phaseolus vulgaris* mature seed	Hiraga et al. (1972)
GA_5-β-D-glucosyl ester	*Pharbitis purpurea* immature seed	Yamaguchi et al. (1980)
GA_9-β-D-glucosyl ester	*Picea sitchensis* needles	Lorenzi et al. (1976)
GA_{37}-β-D-glucosyl ester	*Phaseolus vulgaris* mature seed	Hiraga et al. (1972)
GA_{38}-β-D-glucosyl ester	*Phaseolus vulgaris* mature seed	Hiraga et al. (1972)
GA_{44}-β-D-glucosyl ester	*Pharbitis purpurea* immature seed	Yamaguchi et al. (1980)
Other conjugates		
GA_1 n-propyl ester	*Cucumus sativus* seed	Hemphill et al. (1973)
GA_3 n-propyl ester	*Cucumus sativus* seed	Hemphill et al. (1973)
3-O-β-acetyl GA_3	*Gibberella fujikuroi* mycelial filtrate	Schreiber et al. (1966)
GA_9 methyl ester	*Lygodium japonicum* prothallus	Yamane et al. (1979)
Gibberethione	*Pharbitis nil* immature seed	Yokota et al. (1974)

REFERENCES

Beale, M. H., Bearder, J. R., Hedden, P., Graebe, J. E., & MacMillan, J. (in preparation). Gibberellin A_{58} and *ent*-$6\alpha,7\alpha,13$-trihydroxykaur-16-en-19-oic acid from seeds of *Curcurbita maxima*.

Bearder, J. R., Dennis, F. G., MacMillan, J., Martin, G. C., & Phinney, B. O. (1975). A new gibberellin (A_{45}) from seed of *Pyrus communis* L. Tetrahedron Lett., 669–670.

Bearder, J. R., & MacMillan, J. (1972). Gibberellin A_{36}, isolation from *Gibberella fujikuroi*, structure and conversion to gibberellin A_{37}. Agr. Biol. Chem. *36*, 342–344.

Bearder, J. R., & MacMillan, J. (1973). Fungal products. Part IX. Gibberellins A_{16}, A_{36}, A_{37}, A_{41} from *Gibberella fujikuroi*. J. Chem. Soc., Perkin Trans. I, 2824–2830.

Beeley, L. J., & MacMillan, J. (1976). Partial syntheses of 2-hydroxy gibberellins: Characterisation of two new gibberellins, A_{46} and A_{47}. J. Chem. Soc., Perkin Trans. I, 1022–1028.

Binks, R., MacMillan, J., & Pryce, R. J. (1969). Plant hormones VII. Combined gas chromatography-mass spectrometry of the methyl esters of gibberellins A_1 to A_{24} and their trimethylsilyl ethers. Phytochem. *8*, 271–284.

Brown, J. C., Cross, B. E., & Hanson, J. R. (1967). Two gibbane $1\rightarrow3$-lactones. Tetrahedron *23*, 4095–4103.

Coombe, B. G. (1971). GA_{32}: A polar gibberellin with high biological potency. Science *172*, 856–857.

Cross, B. E. (1966). Gibberellin A_{14}. J. Chem. Soc., 501–504.

Cross, B. E., Galt, R. H. B., & Hanson, J. R. (1962). Gibberellin A_7 and gibberellin A_9. Tetrahedron *18*, 451–459.

Cross, B. E., & Norton, K. (1965). Gibberellin A_{12}. J. Chem. Soc., 1570–1572.

Curtis, P. J., & Cross, P. E. (1954). Gibberellic acid: A new metabolite from the culture filtrates of *Gibberella fujikuroi*. Chem. Ind., 1066.

Frydman, V. M., Gaskin, P., & MacMillan, J. (1974). Qualitative and quantitative analysis of gibberellins throughout seed maturation in *Pisum sativum*. cv. Progress No. 9. Planta *118*, 123–132.

Fukui, H., Koshimizu, K., & Mitsui, T. (1971). Gibberellin A_{28} in the fruits of *Lupinus luteus*. Phytochem. *10*, 671-673.

Fukui, H., Koshimizu, K., & Nemori, R. (1978). Two new gibberellins A_{50} and A_{52} in seeds of *Lagenaria leucantha*. Agr. Biol. Chem. *42*, 1571-1576.

Fukui, H., Koshimuzu, K., Usuda, S., & Yamazaki, Y. (1977). Isolation of plant growth regulators from seeds of *Cucurbita pepo* L. Agr. Biol. Chem. *41*, 175-180.

Galt, R. H. B. (1965). Gibberellin A_{13}. J. Chem. Soc., 3143-3151.

Galt, R. H. B. (1968). Gibberellin A_{16} methyl ester. Tetrahedron *24*, 1337-1333.

Graebe, J. E., Hedden, P., Gaskin, P., & MacMillan, J. (1974). The biosynthesis of a C_{19}-gibberellin from mevalonic acid in a cell-free system from a higher plant. Planta *120*, 307-309.

Graebe, J. R., Hedden, P., & Rademacher, W. (1980). Gibberellin biosynthesis. In: Gibberellins: Chemistry, Physiology and Use. British Plant Growth Regulator Group Monograph No. 5. pp. 31-47, Lenton, J. R., ed. British Plant Growth Regulator Group, Wantage.

Hanson, J. R. (1966). Gibberellin A_{10}. Tetrahedron *22*, 701-703.

Hanson, J. R. (1967). Gibberellin A_{15}. Tetrahedron *23*, 733-735.

Harrison, D. M., & MacMillan, J. (1971). Two new gibberellins, A_{24} and A_{25}, from *Gibberella fujikuroi*: Their isolation, structure and correlation with gibberellins A_{13} and A_{15}. J. Chem. Soc. (C), 631-636.

Harrison, D. M., MacMillan, J., & Galt, R. H. B. (1968). Gibberellin A_{24}, an aldehydic gibberellin from *Gibberella fujikuroi*. Tetrahedron Lett., 3137-3139.

Hemphill, D. D., Baker, L. R., & Sell, H. M. (1973). Isolation of novel conjugated gibberellins from *Cucumis sativus* seed. Can. J. Biochem. *51*, 1647-1653.

Hiraga, K., Kawake, S., Yokota, T., Murofushi, N., & Takahashi, N. (1974). Isolation and characterisation of gibberellins in mature seeds of *Phaseolus vulgaris*. Agr. Biol. Chem. *38*, 2511-2520.

Hiraga, K., Yokota, T., Murofushi, N., & Takahashi, N. (1972). Isolation and characterisation of a free gibberellin and glucosyl esters of gibberellins in mature seed of *Phaseolus vulgaris.* Agr. Biol. Chem. *36*, 345–347.

Hutchison, M. (1983). Studies on hydroxylated gibberellins. Ph.D. Thesis, University of Bristol.

Hutchison, M. et al. Phytochemistry 27: 2695-2701/1988

Kirkwood, P. S., & MacMillan, J. (1982). Gibberellins A_{60}, A_{61} and A_{62}: Partial synthesis and natural occurrence. J. Chem. Soc., Perkin Trans. I, 689–697.

Koshimizu, K., Fukui, H., Inui, M., Ogawa, Y., & Mitsui, T. (1968). Gibberellin A_{23} in immature seeds of *Lupinus luteus.* Tetrahedron Lett., 1143–1147.

Koshimizu, K., Fukui, H., Kusaki, T., Mitsui, T., & Ogawa, Y. (1966). A new C_{20} gibberellin in immature seeds of *Lupinus luteus.* Tetrehedron Lett., 2453–2463.

Lorenzi, R., Horgan, R., & Heald, J. K. (1976). Gibberellin A_9 glucosyl ester in needles of *Picea sitchensis.* Phytochem. *15*, 789–790.

MacMillan, J., Seaton, J. C., & Suter, P. J. (1959). A new plant-growth promoting acid—gibberellin A_5 from the seed of *Phaseolus multiflorus.* Proc. Chem. Soc., 325.

MacMillan, J., Seaton, J. C., & Suter, P. J. (1962). Plant hormones—II. Isolation and structures of gibberellin A_6 and gibberellin A_8. Tetrahedron *18*, 349–355.

MacMillan, J., & Takahashi, N. (1968). Proposed procedure for the allocation of trivial names to the gibberellins. Nature *217*, 170–171.

McCrindle, R., & Overton, K. H. (1969). The diterpenoids, sesterterpenoids and triterpenoids. In: Rodd's Chemistry of Carbon Compounds, Vol. II, Pt. C, 2nd edition, pp. 369–482, Coffey, S., ed. Elsevier, Amsterdam.

Murofushi, N., Iriuchijima, Takashi, N., Tamura, S., Kato, J., Wada, Y., Watanabe, E., & Aoyama, T. (1966). Isolation and structures of a novel C_{20} gibberellin in bamboo shoots. Agr. Biol. Chem. *30*, 317–324.

Murofushi, N., Sugimoto, M., Itoh, K., & Takahashi, N. (1979). Three novel gibberellins produced by *Gibberella fujikuroi.* Agr. Biol. Chem. *43*, 2179–2185.

Murofushi, N., Sugimoto, M., Itah, K., & Takahashi, N. (1980). A novel gibberellin, GA_{57}, produced by *Gibberella fujikuroi.* Agr. Biol. Chem. *44*, 1583–1587.

Murofushi, N., Yokota, T., & Takahashi, N. (1970). Isolation and structure of gibberellins from immature seed of *Calonyction aculeatum.* Agr. Biol. Chem. *34*, 1436-1438.

Murofushi, N., Yokota, T., & Takahashi, N. (1971). Structures of gibberellins A_{33} and A_{34} from immature seeds of *Calonyction aculeatum.* Agr. Biol. Chem. *35*, 441-443.

Pryce, R. J., & MacMillan, J. (1967). A new gibberellin in the seed of *Phaseolus multiflorus.* Tetrahedron Lett., 4173-4175.

Rowe, J. W. (1968). The common and systematic nomenclature of cyclic diterpenes, 3rd revision. Forest Products Laboratory, U.S. Department of Agriculture, Madison, Wisconsin.

Schreiber, K., Schneider, G., Sembdner, G., & Folke, I. (1966). Isolierung von *O*(2)-acetyl-gibberellinsäure als stoffwechselproduckt von *Fusarium moniliforme* sheld. Phytochem. *5*, 1221-1225.

Schreiber, K., Weiland, J., & Sembdner, G. (1967). Isolierung und struktur eines gibberellinglucosids. Tetrahedron Lett., 4285-4288.

Sponsel, V. M., Gaskin, P., & MacMillan, J. (1979). The identification of gibberellins in immature seeds of *Vicia faba*, and some chemotaxonomic considerations. Planta *146*, 101-105.

Sponsel, V. M., & MacMillan, J. (1977). Further studies on the metabolism of gibberellins (GAs) A_9, A_{20} and A_{29} in immature seeds of *Pisum sativum.* cv. Progress No. 9. Planta *135*, 129-136.

Takahashi, N., Murofushi, N., Yokota, T., & Tamura, S. (1967a). Gibberellin in immature seeds of *Pharbitis nil.* Tetrahedron Lett. 1065-1068.

Takahashi, N., Kitamura, H., Kawarada, A., Seta, Y., Takai, M., Tamura, S., & Sumiki, Y. (1955). Isolation of gibberellins and their properties. Bull. Agr. Chem. Soc. Japan *19*, 267-277.

Takahashi, N., Murofushi, N., Yokota, T., & Tamura, S. (1967b). Structures of new gibberellins in immature seeds of *Canavalia gladiata.* Tetrahedron Lett., 4861-4865.

Takahashi, N., Seta, Y., Kitamura, H., & Sumiki, Y. (1957). A new gibberellin, gibberellin A_4. Bull. Agr. Chem. Soc. Japan *21*, 396.

Tamura, S., Takahashi, N., Yokota, T., Murofushi, N., & Ogawa, Y. (1968). Isolation of water-soluble gibberellins from immature seeds of *Pharbitis nil.* Planta *78*, 208-212.

West, C. A., & Phinney, B. O. (1959). Gibberellins from flowering plants. I. Isolation and properties of a gibberellin from *Phaseolus vulgaris*. L. J. Amer. Chem. Soc. *81*, 2424-2427.

Yamaguchi, I., Kobayashi, M., & Takahashi, N. (1980). Isolation and characterization of glucosyl esters of gibberellin A_5 and A_{44} from immature seeds of *Pharbitis purpurea*. Agr. Biol. Chem. *44*, 1975-1977.

Yamaguchi, I., Miyamoto, M., Yamane, H., Takahashi, N., Fujita, K., & Imanari, M. (1973). Structure of gibberellin A_{40}. Agr. Biol. Chem. *37*, 2453-2454.

Yamane, H., Takahashi, N., Takeno, K., & Furuya, M. (1979). Identification of gibberellin A_9 methyl ester as a natural substance regulating formation of reproductive organs in *Lygodium japonicum*. Planta *147*, 251-256.

Yamane, H., Yamaguchi, I., Murofushi, N., & Takahashi, N. (1971). Isolation and structure of gibberellin A_{35} and its glucoside from immature seed of *Cytisus scoparius*. Agr. Biol. Chem. *35*, 1144-1146.

Yokota, T., Kobayashi, S., Yamane, H., & Takahashi, N. (1978). Isolation of a novel gibberellin glucoside, 3-*O*-β-D-glucopyranosylgibberellin A_1 from *Dolichos lablab* seed. Agr. Biol. Chem. *42*, 1811-1812.

Yokota, T., Murofushi, N., & Takahashi, N. (1970). Structure of a new gibberellin glucoside in immature seeds of *Pharbitis nil*. Tetrahedron Lett., 1489-1491.

Yokota, T., Murofushi, N., Takahashi, N., & Tamura, S. (1971). Gibberellins in immature seeds of *Pharbitis nil*. Part III. Isolation and structures of gibberellin glucosides. Agr. Biol. Chem. *35*, 583-595.

Yokota, T., Takahashi, N., Murofushi, N., & Tamura, S. (1969). Isolation of gibberellins A_{26} and A_{27} and their glucosides from immature seeds of *Pharbitis nil*. Planta *87*, 180-184.

Yokota, T., & Takahashi, N. (1981). Gibberellin A_{59}: A new gibberellin from *Canavalia gladiata*. Agr. Biol. Chem. *45*, 1251-1254.

Yokota, T., Yamazaki, S., Takahashi, N., & Iitaka, Y. (1974). Structure of pharbitic acid, a new gibberellin-related diterpenoid. Tetrahedron Lett., 2957-2960.

1

The History of Gibberellins

Bernard O. Phinney

The discovery of gibberellins in higher plants is an intriguing example of serendipity in science. The fact that gibberellins are now accepted as plant hormones is the result of a long series of steps, some of which were quite unrelated to the final product. These steps were initiated by the motivation of Japanese plant pathologists to control a fungal infection, the "bakanae" disease, which sometimes had devastating effects on the yield of rice in the Orient. Their research was to lead ultimately to the identification from the fungus of a new class of natural products, the gibberellins.*

In the early stages of this fascinating story the Japanese investigated not only the effects of gibberellin on rice, but also the gibberellin-induced growth of other economic plants. However, it was in the Western world, notably Great Britain and the United States, where it was appreciated that the enhancement of stem elongation by fungal gibberellins could indicate that gibberellins were also native constituents of higher plant tissues. This interpretation led to studies that established gibberellins as a new class of hormones endogenous to higher plants.

It is the purpose of this chapter to present and evaluate selected events involved in the history of gibberellin research. The period to be

---

*Interestingly, the identification of gibberellins as the causal agent of the "bakanae" disease had nothing to do with its ultimate control, as the pathogen was virtually eliminated by treating seed with mercuric fungicides prior to planting.

covered begins with the first published accounts of the "bakanae" disease at the end of the nineteenth century and terminates with the identification of gibberellins from higher plants in the late 1950s. The account is by no means complete or detailed, but rather is a presentation of selected events that give a chronological continuity to information as it unfolded over the years. Likewise, the account does not cover the effects of gibberellins on the numerous economic plants that were originally reported by the Japanese and extensively exploited in the United States in the 1950s and 1960s.

A personal touch is given to the history by first-hand experiences in the details that led up to the isolation and identification of gibberellins in higher plants. These later events occurred so rapidly that the literature provides few clues as to their chronology. In addition, I have, over the years, had the opportunity to discuss the history of "gibberellinology" with Frank H. Stodola and John E. Mitchell from the United States; Teijiro Yabuta (albeit by translators and sign language), Yusuke Sumiki, Saburo Tamura, and Hidefumi Asuyama from Japan; and George Elson, Jake MacMillan, and Margaret Radley from Great Britain. Liberal use has also been made of secondary reference material, especially Stodola's *Source Book on Gibberellins (1828-1957)* published in 1958, Bruce Stowe and Toshio Yamaki's review in 1957 on the history and physiology of the gibberellins, and review chapters by Saburo Tamura published in 1969 and 1977 on the history and physiological action of the gibberellins. British accounts on the early development of gibberellins in the UK have been written by Percy Brian (1959) and by George Mees and George Elson (1978). Figure 1.1 shows the "carbon skeleton" for some of the early gibberellin workers.

The early research that led to the first isolation and identification of gibberellins was the exclusive product of the efforts of Japanese scientists. Their studies began with the description of the disease symptoms of the rice plant, which were later shown to be due to gibberellins produced by the fungus *Gibberella fujikuroi*. This direction of research led to chemical investigations by the Japanese that defined the gibberellin molecule as containing a fluorene ring system.

Although lay descriptions of the rice disease appeared in the literature in the early nineteenth century (see Konishi, 1828, cited by Ito & Kamura, 1931), scientific description awaited the publication in 1898 of an article by Shotaro Hori of the Mycological Laboratory at the Imperial Agricultural Experiment Station located at Nishigahara, Tokyo. This paper (Hori, 1898) reported the symptoms of the disease and clearly demonstrated that it could be induced by infecting a healthy seedling with the "bakanae" fungus.

Table 1.1. Names used by Japanese farmers in different localities to describe the disease of rice resulting from infection by the fungus *Gibberella fujikuroi*

Name	Translation	District (prefecture)
"Bakanae"	silly seedling	Tokyo, Aichi
"Bakanae"	silly rice crop	Miyagi
"Ahoine"	stupid rice crop	Shiga
"Yurei"	ghost	Shimane
"Somen nae"	thin noodle seedling	Shimane, Fukushima
"Naganae"	long seedling	Osaka
"Sasanae"	bamboo seedling	Mie
"Yarinae"	spear seedling	Miyagi
"Yarikatsugi"	spear warrier	Mie
"Oyakata"	boss	Chiba
"Otokonae"	male seedling	Ishikawa, Miyagi Akita, Shimane, Shizuoka Niigata
"Onnanae"	female seedling	Niigata

Information in this table is taken from Tamura (1977).

In 1903, Hori listed a series of common names used by farmers to describe the disease (Table 1.1). These names—the translations of which included male seedling, thin noodle seedling, foolish seedling, and stupid rice crop—often reflected the sense of humor of the farmer (see Tamura, 1977).* In recent times, especially in the Western literature, the term "bakanae" has become widely accepted as referring to the seedling elongation resulting from infection by the fungus. Originally, the term was associated with a combination of responses—excessive elongation of the seedlings combined with elongated mature rice plants that lacked fruit. Clearly, this is why such plants were called "foolish seedlings."

---

*In 1966 and 1967 I isolated 1500 strains of *Fusarium moniliforme* in Japan from rice seedlings infected with the "bakanae" disease. The seedlings were collected from paddy fields located on the islands of Hokkaido, Honshu, Shikoku, and Kyushu. Whenever farmers were questioned using the expression "ine-bakanae-byo . . . doko-deska . . . kudasai . . ." (rice-foolish seedling-disease . . . where it is . . . please . . .), they would respond with a broad smile, bow, and use words and sign language equivalent to "follow me."

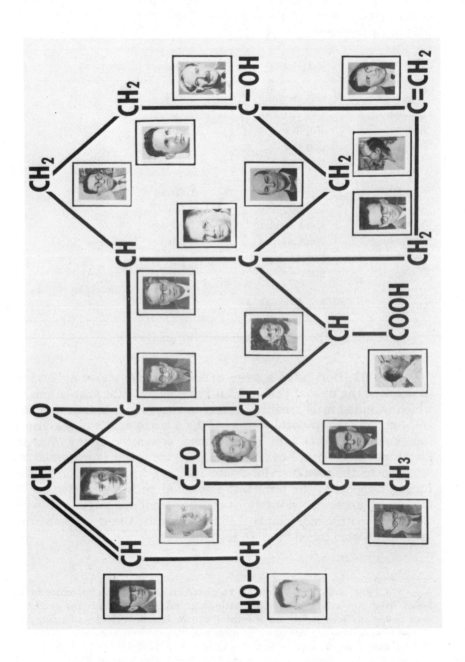

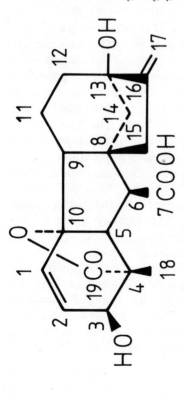

1. Hayashi—National Institute of Agricultural Science, Tokyo, Japan
2. MacMillan—Akers Research Laboratories, Imperial Chemical Industries, Welwyn, Herts, England
3. Lang—Department of Botany, University of California, Los Angeles, USA
4. Mulholland—Akers Research Laboratories, ICI, Welwyn, Herts, England
5. Fennell—Northern Regional Research Laboratory, Peoria, Illinois, USA

6. Kurosawa—Imperial Research Institute, Department of Agriculture, Formosa (Taiwan)
7. Marth—USDA Plant Industry Station, Beltsville, Maryland, USA
8. Raper—USDA Northern Regional Research Laboratory, Peoria, Illinois, USA
9. Sumiki—Department of Agricultural Chemistry, University of Tokyo, Japan
10. Yabuta—Department of Agricultural Chemistry, University of Tokyo, Japan
11. Shimada—Hokkaido Imperial University, Sapporo, Japan
12. Wittwer—Department of Horticulture, Michigan State University, East Lansing, USA
13. Brian—Akers Research Laboratories, ICI, Welwyn, Herts, England
14. Phinney—Department of Botany, University of California, Los Angeles, USA
15. Cross—Akers Research Laboratories, ICI, Welwyn, Herts, England
16. Mitchell—USDA Plant Industry Station, Beltsville, Maryland, USA
17. West—Department of Chemistry, University of California, Los Angeles, USA
18. Grove—Akers Research Laboratories, ICI, Welwyn, Herts, England
19. Stodola—USDA Northern Regional Research Laboratory, Peoria, Illinois, USA

Figure 1.1 The "carbon skeleton" for some of the chemists and biologists associated with the early history of gibberellins. (Prepared by Dr. F. H. Stodola, NRRL 1959.)

There are several symptoms associated with the disease (Hori, 1898). At the time of germination infected plants show reduced root growth and elongated narrow leaves that are pale yellow in color. Plants that survive often have elongated stems with aerial adventitious roots and sterile glumes. Heavy infections result in pigmentation at the base of the plant and mycelial growth and pink coloration on aerial parts. Mature infected plants are sometimes taller than their healthy counterparts, and fruit is either absent or poorly developed. The "bakanae" disease had been the subject of discussion in Japanese experiment stations for an extended period of time before Hori's publication (see Tamura, 1969). The damage to rice plants could be extensive, and 40% reductions in the yield of rice were often recorded (Sawada, 1912). While the disease was often widespread in the paddy field, it could cause a disaster in the seedling beds where rice plants are grown prior to transplanting to the field, the seedling loss being such that multiple replantings were frequently required. The result was late maturity and lowered yield, since production of rice, as for many crops, is dependent on the time of planting.

Up to the 1930s, the taxonomic position of the fungus responsible for the "bakanae" symptoms was ill defined. As a result there was much controversy concerning the nomenclature, which of course led to confusion in relating the symptoms of the disease to a specific species of fungus. The problem arose in part from the dual nomenclature of fungi, and in part from the lack of critical data necessary to develop a systematic nomenclature for the genus. Thus, the imperfect or vegetative stage had been called *Fusarium heterosporum* and *Fusarium moniliforme*, while the perfect or sexual stage was referred to as either *Gibberella fujikuroi* or *Lisea fujikuroi*. The problem was resolved in 1931 by Hans W. Wollenweber in his publication on the revision of the taxonomy of the *Fusaria*. In this monograph, and in a subsequent joint publication in 1935 with Otto A. Reinking, the imperfect stage of the fungus responsible for the "bakanae" disease was named *Fusarium moniliforme* (Sheldon) and the perfect stage named *Gibberella fujikuroi* (Saw.) Wr. The priority given to the latter name had followed recommendations made in 1931 and 1932 by the Japanese plant pathologist Yutaka Nisikado.

The first clue that a substance produced by the fungus resulted in the "bakanae" effect came from a paper published in 1912 by Kenkichi Sawada, a plant pathologist at the Imperial Research Institute of the Department of Agriculture in Taipaei, Taiwan. At the turn of the century the disease was widespread in Taiwan, and as a result Sawada had been actively working on methods of

Figure 1.2 Dr. Eiichi Kurosawa in his laboratory at the experiment station in Nishigahara (*upper*), and in later years (*left*).

control. In 1912 he published a paper in *The Formosan Agricultural Review* in which he suggested that the elongation symptoms "... must be related to stimulus given by these mycelia. . . ." While this statement would appear to imply a chemical mechanism, Tamura (1969) points out that Sawada's use of the term "stimulus" was in an abstract sense. His philosophical approach to the subject would have precluded a mechanistic interpretation at that time.

The story that led to studies on the stimulus itself did not unfold until 1924, when Eiichi Kurosawa moved to Taipei to work with Sawada. Kurosawa (Fig. 1.2) had previously been an assistant at the Horticultural School of Chiba Prefecture in Japan following graduation from Chiba Horticultural "Higher School." During studies with Sawada on the prevention of the "bakanae" disease, Kurosawa became interested in the mechanism responsible for the elongation symptoms associated with the fungal infection. As a consequence, he initiated a program of his own in the summer of 1925 with the goal of isolating the "secretion" responsible for the "bakanae" symptoms. Within the short time of one year Kurosawa obtained definitive results, which were published in 1926 in *The Transactions of the Natural History Society of Formosa.*

Kurosawa's approach was simple and direct. He obtained a sterile filtrate from a culture that would mimic the "bakanae" symptoms produced by the fungus itself. Cultures were grown on a semi-solid rice medium as well as on a supplemented "Knopf" medium. After incubation, both heated and nonheated media were passed through a Chamberlain filter. Rice kernels were soaked in the extracts and allowed to germinate and grow. In other experiments the sterile filtrate was injected directly into the young plants. Kurosawa even separated the root inhibition/shoot elongation symptoms associated with the disease by removing roots and injecting the sterile "secretion solution" directly into the hollow stem of young plants. He concluded that the activity of his sterile extract ". . . was not due to enzyme action but rather to some kind of chemical" (Kurosawa, 1926). This landmark paper also included the first published picture of the effect of the disease on rice seedlings (Fig. 1.3). The secretion factor was also shown to stimulate elongation of seedlings of maize, sesame, millet, and oats. Interestingly, Kurosawa called the active material a toxin, clearly implying that the "bakanae" effect was a pathological symptom, unrelated to growth in a healthy plant. His summary, which reads as follows, supports this position:

1. The rice "bakanae" fungus secretes a type of toxin, which accelerates rice growth.
2. Besides accelerating rice growth, the toxin damages root growth and interferes with the formation of chlorophyll.
3. It is conceivable that the toxin exerts a similar action on other plants besides rice.
4. The toxin was scarcely altered, even after being kept at 100° for 4 hours.
5. It appears that resistivity of the rice plant to the toxin is dependent on the variety.
6. Apparently the rice plant does not form an antibody against the toxin.

(Trans. by Sasame; see Stodola, 1958.)

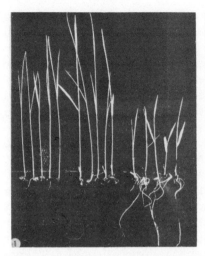

Figure 1.3   The first published photographs showing the growth response of rice seedlings to Kurosawa's "secretion factor" (Kurosawa, 1926).

Kurosawa published on a variety of "bakanae" effects, including descriptions of the symptoms of the disease, studies on variations in the appearance of the fungus under different culture conditions, and methods to control the disease (i.e., treatment of the seed with mercuric chloride and formaldehyde). His publication of 1930 reported the presence of not only a "growth-stimulating substance" in his "broth extracts" but also "another product" that suppresses the growth of plants. In 1932 he reported that elevated temperatures (35°C) were associated with growth suppression and low temperatures (20°C) with growth stimulation. He also subsequently found that incubation of the fungus at pH 3 suppressed "production of the inhibitor concomitant with a stimulation of the growth accumulation substance" (Kurosawa, 1934). The span of his publications covered the years 1924–34; they were 13 in number, of which 5 were concerned with studies on the culture filtrates from the fungus. Kurosawa died in 1953 at the age of 59.

After Kurosawa's 1926 report, a number of plant pathologists initiated studies on the fungal filtrate, with their interest centered on the isolation of the active principle and its biological properties. More than 50 publications appeared on the subject between 1927 and 1940, mainly from Kyoto University (T. Hemmi and F. Seto), Hokkaido University (S. Ito and S. Shimada), Mie University for Agricultural Research, Kurashike Agricultural college (T. Takahashi), and the Ohara Institute (Y. Nisikado, H. Matsumoto, and K. Yamauchi). It became apparent that the active principle could be adsorbed onto charcoal, could be passed through a semipermeable membrane, and was stable when boiled; also, low carbohydrate levels favored an enhanced production of the stimulant. Several unsuccessful attempts were made to isolate crystalline material.

The next critical step in the development of the story occurred in the 1930s with the entry into the field by the chemist Teijiro Yabuta (Figs. 1.4 and 1.5). By this time the Japanese plant pathologists had provided sound evidence for the presence of a substance produced by the fungus that had considerable biological significance. Yabuta was eminently qualified to attempt the isolation and identification of the active ingredient, as a result of his training as an organic chemist with interest in natural products. He was Professor of Agricultural Chemistry in the Faculty of Agriculture at the University of Tokyo, and he also held a joint appointment as an engineer at the Imperial Agricultural Experiment Station, Nishigahara, Tokyo. This latter appointment turned out to be highly significant since in 1933 Kurosawa moved from Taiwan to a new position at Nishigahara. It was fortunate for Yabuta that Kurosawa made this move since he was to become invaluable to Yabuta's future program. For instance, the original "bakanae" cultures were obtained from Kurosawa along with information on their maintenance and handling. Kurosawa also provided details of the early methods used for the extraction of the active principle.

By 1934 Yabuta's group had obtained crystalline material from their culture filtrates. However, they found it to be inhibitory rather than stimulatory when assayed on rice seedlings, even at high dilutions.

Figure 1.4   Yutaka Sumiki (*left*) and Teijiro Yabuta (*right*) in 1956.

Figure 1.5 Teijiro Yabuta, at the time he received the Order of Cultural Merits, Japan.

Obviously Yabuta's group had isolated the inhibitory "substance" originally reported by Kurosawa in 1930. The substance was given the trivial name fusaric acid and was identified as a picolinic acid (Yabuta, Kambe, & Hayashi, 1934). The compound was subsequently shown to be 5-*n*-butyl picolinic acid (Yabuta & Hayashi, 1940).

After discussing the inhibitory problem with Kurosawa, Yabuta changed the composition of the culture medium, and extracts were

Figure 1.6 The plant physiologist Takeshi Hayashi in 1956.

soon obtained that had high biological activity. These samples were purified to give a noncrystalline solid, which was called gibberellin. This was the first time the word gibberellin had been used as a name for the "bakanae" substance isolated from culture filtrates of the fungus. Yabuta's results were summarized in *Agriculture and Horticulture* in 1935, while details of the isolation procedures appeared in the spring of 1936 in a publication co-authored by Takeshi Hayashi (Fig. 1.6). Since the material was noncrystalline, they were unable to determine its chemical properties. The purified gibberellin preparation was shown to stimulate seedling elongation of a number of economic plants, including rice, barley, buckwheat, soybean, gourd, tomato, cucumber, and morning glory (Yabuta & Hayashi 1936, 1938).

The gates would thus appear to have been opened for a new approach to studies on the control of plant growth. However, this did not occur. The supply of gibberellin was not a limiting factor: Yabuta's samples were available and, in Japan, the fungus was always a source of crude gibberellin. Perhaps, at that time, it was too simple an idea to realize that a metabolite present in a fungus could also be found in a higher plant as a natural growth regulator. This possibility was further clouded by the accepted dogma that auxin was the only hormone present in plants (see Letham et al., 1978). Since the gibberellin growth responses were so different from those of auxin, it was apparently difficult for biologists to envisage that gibberellins might also act as natural regulators of growth in higher plants.

The long-awaited crystallization of gibberellin into gibberellin A and gibberellin B was first reported by Teijiro Yabuta and Yusuke

Sumiki in 1938 (see Fig. 1.4). Both materials were biologically active, yet they had different chemical properties. Also the names were reversed in later publications (see Yabuta et al., 1941b). Between 1938 and 1941 a series of papers appeared on the chemical properties of the two gibberellins. Unfortunately, both gibberellin A and gibberellin B were subsequently found to be impure (see Takahashi et al., 1955), and so the chemical studies were of limited value. During this same period several reports were made on the biological properties of gibberellin A (e.g., Yabuta et al., 1941a; Yabuta & Sumiki, 1944). Virtually no studies were conducted between 1941 and 1945 on the chemistry of the gibberellins, and the subsequent resumption of research was disappointingly slow owing to problems stemming from the war (see Tamura, 1969, 1977).

By the early 1950s gibberellin research had entered a new phase, becoming international in scope and activities. There appears to be no clear rationale to explain why scientists in the Western world did not become interested in the Japanese findings prior to 1950. Two examples will show by hindsight what could have been foresight. First, contrary to popular opinion, English translations of the Japanese work were available through *Chemical Abstracts* (U.S.A.) for the years 1935, 1939, and 1940–41. They were apparently not seen by "the right eyes." A second example comes from conversations with Vern Stoutemyer, then a Horticulturist in the Division of Plant Exploration and Introduction, U.S. Department of Agriculture (USDA), Beltsville, Maryland. His friend, Samuel Detweiler, then associated with the USDA through the Bureau of Agriculture and Industrial Chemistry, had a son who traveled to Japan in the late 1930s and whose interests in agricultural research led him to visit plant pathologists and chemists working on gibberellin. He returned with a written account of the disease, its effects on plants, a structure of the ring system then proposed by Japanese chemists, and the names of Japanese scientists involved in the studies. This account was circulated in the USDA in Beltsville. Presumably the report also was not seen by "the right eyes."

The events that *did* lead to a world-wide interest in gibberellins are likewise difficult to analyze. Some are as follows: After Yabuta's retirement in 1950, Yusuke Sumiki, with whom he is pictured in Fig. 1.4, became Professor of Agricultural Chemistry at The University of Tokyo, having been a former student and Assistant Professor in Yabuta's group. In 1951 Sumiki obtained permission to present his studies on the chemistry of gibberellins at the *Fifth International Microorganism Conference* held in Rio de Janeiro. This trip was followed by his attendance, also in 1951, at meetings of the International Congress of Pure and Applied Chemistry held in New York

City. There he met Frank H. Stodola, a chemist at the USDA Northern Regional Research Laboratories, Peoria, Illinois. In 1953 Sumiki presented a similar paper at the *VIth International Congress for Microbiology* in Rome. A British chemist, Jake MacMillan, was in attendance and invited Sumiki to visit the Akers Research Laboratories of the Imperial Chemical Industries (ICI) in the UK. Upon reaching England, Sumiki found that scientists at the Akers Laboratories were already pursuing research on the chemistry and biology of the fungal gibberellins. Obviously, the "word was out" prior to Sumiki's visits to the West.

The entrance of the British (Figs. 1.7 and 1.8) into the field was the result of an interesting source of communication. In 1950, *Chemical Abstracts* published a collection of reports on the early Japanese studies. This series of abstracts immediately caught the eye of W. A. (Seccy) Sexton, Research Director of ICI's Pharmaceutical Division, located at Alderley Edge, near Manchester. Sexton contacted Percy W. Brian, a botanist and mycologist in charge of basic research at the ICI Akers Laboratories in Welwyn, north of London, and suggested that Brian's group might be interested in studying this new class of natural products. Such an interchange of ideas and information was not uncommon at the ICI Research Laboratories. As a result, a screening program was set up in which a search was made for gibberellin-producing strains in the ICI collection of *Fusaria*. Several gibberellin producers were found, and one strain was selected for fermentation studies. Fortuitously, the selected strain turned out to be a wise choice since it was later found to produce essentially only one gibberellin, which greatly facilitated the purification steps necessary to obtain a single crystalline product. Purity of product was, of course, absolutely essential for identification studies in those times. Initially, the fungus was grown as small batches in still culture, then in submerged flask culture, and finally on a large scale in Hoover washing machines! From the initial extraction of the fungal filtrates, crude crystalline preparations were obtained directly which had high biological activity in their bioassays. The preparation was then given to John Frederick Grove, a natural products chemist in charge of the basic research group on organic chemistry at the Akers Laboratories. It was these ICI scientists that Sumiki met during his visit to the UK in 1953.

In the United States, the first research on gibberellins originated from a different line of communication. After the Second World War, studies were initiated at a research unit at Camp Dietrick, Maryland, an organization closely associated with the armed forces. In 1946 Axel Anderson, later a plant pathologist at Michigan State University,

Figure 1.7   Percy W. Brian and chemists at the ICI Akers Research Laboratories in 1956. *Left to right*: Jake MacMillan, Percy W. Brian, W. A. (Seccy) Sexton, D. G. Davies, and Alfred Spinks.

Figure 1.8   A period picture of gibberellin chemists at the ICI Akers Research Laboratories. *Left to right:* Jake MacMillan, Brian E. Cross, and T. P. C. (Paddy) Mulholland., *Grove*   *name missing*   *Grove?*

was involved in graduate work on *Fusarium* pathogens at Camp Dietrick. During these studies he isolated a *Fusarium* strain from field-grown wheat, which produced "bakanae" symptoms in his seedling test plots. As a result, the Camp Dietrick group became interested in the pathogen, taking the subject to the stage where the symptoms could be reproduced from filtrates of the fungus. Anderson received his Ph.D. degree and left the group in 1948. The project was then terminated. That same year John E. Mitchell accepted a position at Camp Dietrick as a plant pathologist, and he soon became interested in gibberellins after reading the

Japanese literature on the subject, although completely unaware of Anderson's studies. In 1950 Mitchell obtained a giberellin-producing strain from Yoshikazu Nishikado, Director of the Ohara Institute, Kurashike, Japan. He initiated fermentation studies and presented a preliminary report in December 1950 at the national meetings of the American Phytopathological Society held in Memphis, Tennessee (Mitchell & Angel, 1951). The report described optimal fermentation procedures for the fungus, as well as effects of the fungal extracts on the growth of seedlings of bean (*Vicia faba*). This was the first report of gibberellin studies in the United States.

As explained by Stodola (1958), Mitchell was well aware of the potential of gibberellins for agriculture. He, therefore, recommended that the facilities of the USDA be used to develop methods for the production of the fungal gibberellin, and as a result the microbiology group at the USDA Laboratories at Peoria became associated with gibberellin research. The unit was headed by Kenneth B. Raper, a mycologist in charge of the culture collection of the Fermentation Division. Raper's group began to investigate fermentation procedures in August 1951 with the Nisikado strain provided by Mitchell. Preliminary studies were carried out in small shaker flasks; later, 20-liter vessels were used; and finally, stainless steel 300-gallon fermentors were employed (Stodola et al., 1955). The purpose of the program was to produce pure gibberellin A for agricultural uses. However, in these initial fermentations the filtrates were found to be biologically inactive.

It was at this time that Sumiki visited the Peoria Laboratories, shortly after presenting his paper on gibberellin A at the chemistry meetings in New York City. Sumiki's visit to Peoria was apparently motivated by his strong interest in the industrial utilization of surplus agricultural products which, in turn, was the major justification for the existence of the Peoria group at that time. Consequent to this visit, Sumiki sent new cultures to the United States. However, in the hands of the Peoria workers these cultures were also found to be nonproducers. The difficulty was finally traced to a lack of magnesium in the culture medium. Whereas the Japanese used tap water in which this essential ingredient was present, the Americans had followed the usual practice of preparing media with distilled water. Good yields of gibberellin were obtained as soon as the Peoria workers supplemented the culture medium with magnesium sulphate.

With the fermentation problems resolved, direction of the work was turned over to Frank H. Stodola (Fig. 1.9), head of the chemistry section of the Fermentation Division. It was Stodola who was to isolate gibberellin from the fermentations. Stodola developed a con-

Figure 1.9 The U.S. gibberellin chemist Frank Stodola in 1956.

suming interest in the subject and became the moving force in the gibberellin program at Peoria. By July of 1952 he had isolated by recrystallization an almost pure gibberellin preparation that could be studied chemically. Surprisingly, he found the physical properties to be different from those reported by the Japanese for giberellin A. Stodola's gibberellin had the empirical formula of $C_{19}H_{22}O_6$ with an optical rotation of $[\alpha]_D^{20}$ +92. He named this new compound gibberellin-X and recorded his findings in a USDA station report in April 1953, as well as in an article in *Archives of Biochemistry and Biophysics* in 1955 which was submitted in July 1954. Additional data were published in 1957 (Stodola, Nelson, & Spence, 1957).

During the same period, beginning in 1951, the British were busily analyzing their gibberellin preparations. The chemists in the team associated with these studies were Philip Curtis, Brian Cross, John Grove, Jake MacMillan, and T. P. C. (Paddy) Mulholland. As previously mentioned, crude crystalline preparations were obtained directly from culture filtrates. On further purification by chromatography and recrystallization, Curtis and Cross obtained a gibberellin, which they called gibberellic acid, that had physical properties different from the Japanese gibberellin A. The British gibberellin had an optical rotation of $[\alpha]_D^{14}$ + 82 and a molecular formula of $C_{19}H_{22}O_6$.

Their results were submitted to *Chemistry and Industry* in July 1954, and the paper appeared in the August issue of that same year (Curtis & Cross, 1954). Further information on the new compound was published in the *Journal of the Chemical Society* (Cross, 1954). Following an exchange of samples between Stodola and Grove, the British and U.S. gibberellins were found to have identical chemical properties and the name gibberellic acid was accepted by both parties. Thus, by 1954, gibberellic acid had been defined chemically by the British as a tetracyclic-dihydroxy-lactonic acid with the molecular formula of $C_{19}H_{22}O_6$. A structure for gibberellic acid was proposed in 1956 by Cross, Grove, MacMillan, and Mulholland. Hanson (1968) and Grove (1961) have reviewed the work that led to the final accepted structure for gibberellic acid.

In the 1950s Nobutaka Takahashi, Saburo Tamura (Fig. 1.10), and colleagues at Tokyo University reinvestigated the original gibberellin A preparation. Three compounds were isolated from this material, albeit as methyl esters, and they were named gibberellin A_1, gibberellin A_2, and gibberellin A_3. Gibberellin A_3 was identical to gibberellic acid (Takahashi et al., 1955).

In the late 1950s there was an amazing explosion in the Western world in the number of publications on gibberellin responses in plants—3 in 1954, 6 in 1955, over 41 in 1956, and more than 150 in 1957. In addition, popular accounts reached the hundreds by 1956, and the number of publications increased exponentially for several years. It is difficult to unravel the factors responsible for the origin of this spray-and-pray approach. The number of papers that appeared in this short period of time was such that publication dates became meaningless in terms of who did what and when.

Some of the initial reports that probably caught the eye of U.S. biologists were the spectacular responses of both dwarf and rosetted plants to gibberellic acid. These included the elongation of a pea dwarf cultivar reported by Percy Brian (1954, 1955), the elongation of single-gene dwarf mutants of maize published by Bernard Phinney (1956a, 1956b) (Fig. 1.11), and the elongation of biennial rosette plants observed by Anton Lang (1956a, 1956b) (Fig. 1.12) and Sylvan Wittwer (1957). In September 1956 a symposium sponsored by the American Society of Plant Physiologists was held at the University of Connecticut. The subject was "Naturally Occurring Plant Growth Regulators other than Auxins," and close to one thousand people attended the meeting. As a part of this symposium, Stodola reviewed the chemistry of the gibberellins and Phinney spoke on their physiology. Six short reports were also given on gibberellins and plant responses, including a presentation by

Figure 1.10 The Japanese chemists Saburo Tamura (*left*) and Nobutaka Takahashi (*right*) in the late 1950s (*above*) and in 1982 in the Department of Agricultural Chemistry at the University of Tokyo (*below*).

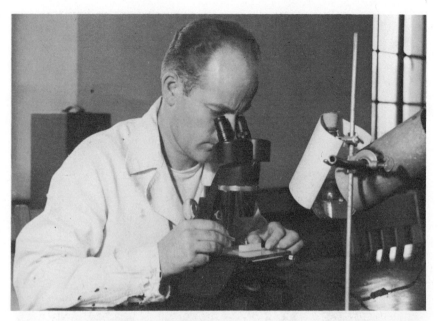

Figure 1.11 Bernard O. Phinney in pre-gibberellin days (1948) working on the genetics of *Neurospora* (*above*) and in more recent times (1982) taking a brief rest from administrative chores at the University of California, Los Angeles (*below*).

Figure 1.12   Anton Lang after receiving an honorary degree at the University of Glasgow in 1981.

Paul C. Marth, W. V. Audia, and John W. Mitchell (1956) of the USDA in Beltsville showing effects of gibberellic acid on a wide variety of horticultural, agronomic, and forest tree species. At these same meetings West and Phinney (1956) presented the first evidence for the presence of gibberellins in higher plants. By late 1956 semi-popular accounts on the effects of gibberellins on higher plants were beginning to appear in the literature, especially from Michigan State University at East Lansing and from the USDA in Beltsville.

The next step in the story of gibberellins was the discovery that extracts from higher plants contained material that induced biological responses identical to those elicited by the fungal gibberellin, gibberellic acid. These unidentified components were called gibberellin-like substances. This finding was a major advance, comparable to Kurosawa's report on "bakanae"-producing substances in the fungus. The time lapse between the two events was 30 years. While the presence of gibberellins in plants is now taken for granted, the idea that they could be naturally occurring substances in flowering plants did not appear in the literature until the mid-1950s. The evidence that gibberellin-like substances could be obtained from higher plants originated from two research centers, namely the ICI laboratories in Welwyn, England, and the University of California, Los Angeles.

In England the spectacular growth response of Brian's dwarf peas to gibberellic acid (Fig. 1.13) led Brian's group to speculate that dwarfism might be due to the absence of *endogenous* gibberellins. Since the response to *exogenous* gibberellin mimicked the growth of the tall cultivar of peas, it was believed that endogenous gibberellins would be present in nondwarf cultivars. As a result, a program was initiated seeking evidence for the presence of gibberellins in plants. The work was carried out by Margaret Radley (Fig. 1.14) under the supervision of P. W. Brian. Her methods were based on techniques

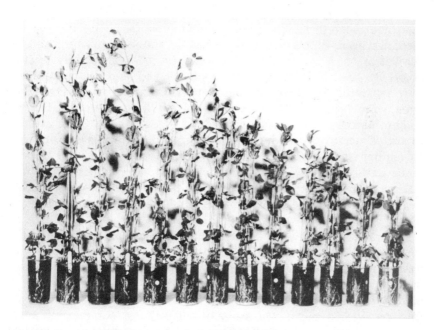

Figure 1.13   An early photograph from the ICI group showing the effects of gibberellic acid on Meteor pea seedlings. Careful inspection shows that the plants are kept erect by being tied to glass tubing (Brian & Hemming, 1955).

Figure 1.14 Margaret Radley at an ICI cocktail party in 1956.

that had been used to isolate gibberellins from the fungus. Pea seedlings were extracted with ethanol, the ethanol removed by distillation, and the water residue partitioned against ethyl acetate at pH 3. The ethyl acetate was concentrated, and the components were separated by paper chromatography. Eluates from the chromatogram were bioassayed, and a zone of activity was detected which co-chromatographed with gibberellic acid. Radley submitted a report of her work to *Nature* in August 1956, and this appeared in a November issue. The title of this most interesting paper was "Occurrence of Substances Similar to Gibberellic Acid in Higher Plants." The report was followed by a series of publications in which Radley demonstrated the presence of gibberellin-like substances in different plant organs and in several species of plants.

In the United States, the first evidence for the presence of gibberellin-like substances in plants also came from studies on dwarfism. In this case it was the response of single-gene dwarf mutants of maize to gibberellic acid (Fig. 1.15). In fact, these studies culminated a 20-year search by U.S. scientists for a hormonal basis for dwarfism in maize. The story of this subplot, summarized in the following account, is another example of serendipity in the history of gibberellins.

Since the 1930s, maize geneticists had recorded some 20 non-allelic dwarf mutants (Emerson, Beadle, & Frazer, 1935). These were simple recessive mutants that originated spontaneously in stocks used by geneticists. Geneticists, especially at the California Institute of Technology (Cal Tech), Pasadena, had regularly encouraged physiologists to use this material for growth studies. As a result, the plant physiologist Johannes van Overbeek began investigations at Cal Tech in the early 1930s into auxin levels in normal and dwarf genotypes of maize. In 1933 and 1938 he reported that several mutants had appreciably lower levels of auxin than did normals. His work was not further investigated at that time. In the early 1950s Robert Harris, a graduate student at the University of California, Los Angeles, working under Phinney's supervision, confirmed van Overbeek's results using improved methods of auxin analysis. He also found that the mutants would not respond to the auxin, indole-3-acetic acid. In further observations, he noted that normal coleoptile tips placed on decapitated dwarf coleoptiles would not stimulate elongation of the dwarf coleoptiles and also that dwarf tips would not inhibit the growth of decapitated normal coleoptiles. Harris concluded that some other factor than auxin must be limiting the

Figure 1.15   One of the original photographs demonstrating the effects of gibberellic acid on dwarf maize. *Left to right:* normal control, normal plus 10 µg gibberellic acid, dwarf-1 mutant control, and dwarf-1 plus 10 µg gibberellic acid (Phinney, 1956b).

growth of the dwarfs (Harris, 1953). A cause-and-effect interpretation could not be given to the low auxin level, since the dwarfs did not respond to exogenous auxin. Phinney (1946) had earlier studied one of the maize dwarfs, looking for clues to explain the dwarf habit of growth.

The answer to the problem came in October 1955 as the result of a seminar by Phinney given at Cal Tech on the physiology of dwarfism. The talk was highlighted by a picture of Brian's dwarf-pea responses that had just appeared in the August 1955 issue of *Physiologia Plantarum*, and the discussion ended with the intriguing question: "Do maize dwarfs also respond to gibberellic acid?" A sample of gibberellin was obtained that day from James Bonner, Professor of Biology at Cal Tech. Within weeks, gibberellin was found to promote shoot elongation for five of the maize mutants, and so the long-sought-for hormonal explanation for dwarfism in maize had been found. Subsequent samples of gibberellic acid came from Stodola, portions of which were given to Lang for his studies on bolting in rosetted biennials (Lang, 1956a).

Phinney's group, together with Charles West (Fig. 1.16) of the Chemistry Department at the University of California, Los Angeles, immediately embarked on a screening program looking for evidence for gibberellins in higher plants. Seedlings of dwarf mutants were used for bioassay since any response from an extract would be evidence for the presence of gibberellin-like substances. The term gibberellin-like substance was coined since the evidence was biological, not chemical. Within a few months gibberellin-like activity was obtained from diethyl ether extracts of the immature seed of corn, peas, lupin, and tobacco. The active material resembled gibberellic acid when subjected to paper chromatography yet differed in that it did not fluoresce under UV light when treated with aqueous sulfuric acid. The results made the March 1956 deadline for abstracts to be given at the September meetings of the American Society of Plant Physiologists at Storrs, Connecticut, and were also included in a symposium on "Plant Growth Regulators other than Auxins." That autumn, Faust Lona, Professor of Botany at the University of Parma, Italy, visited the University of California group, at which time he showed interest in the evidence for gibberellins in higher plants. In 1957 he reported the presence of gibberellin-like substances from young flowers of rape (Lona, 1957). In the same year, other papers reported that gibberellin-like substances from wild cucumber endosperm would induce bolting in *Hyoscyamus* (Lang, Sandoval, & Bedri, 1957), and that extracts of immature bean seed would stimulate the growth of rice seedlings (Murakami, 1957).

Figure 1.16 The U.S. biochemist Charles A. West in the early 1980s.

The details of the work from the California group in Los Angeles appeared in the May 1957 issue of the *Proceedings of the National Academy of Sciences*, with the title "Evidence for Gibberellin-like Substances from Flowering Plants" and authored by Phinney, West, Ritzel, and Neely. These substances were found in extracts from 9 genera representing 7 different families of flowering plants. Gibberellin-like substances became a popular subject for research, and it was soon clear that they were of widespread if not universal occurrence in the plant kingdom.

The stage was thus set for the final step in the story, the identification of endogenous gibberellins from higher plants. Before embarking on this episode, another example of foresight/hindsight should be described. While in the process of investigating the occurrence of gibberellin-like substances in plants, both Radley and Phinney were shown a 1951 publication from *Science* that, in retrospect, contained information for the presence of gibberellin-like components in an extract from a higher plant. In this paper, John D. Mitchell, Dorothy P. Skaggs, and W. Powell Anderson described ether-extractable substances from young bean seed which greatly stimulated shoot elongation in bean seedlings. The extract also induced curvature when applied unilaterally to the first internode of bean seedlings, and this auxin-like property was used to monitor levels of activity during seed maturation. It is now apparent that Mitchell's group was probably working with a mixture of gibberellins and auxins.

The detection in 1956 and 1957 of gibberellin-like substances immediately led to studies that resulted in the isolation and identification of gibberellins in higher plants. As plant tissues usually contain only about 0.001 to 1.0 mg of gibberellic acid equivalents per kilogram of fresh weight, kilogram quantities of material had to be analyzed in order to obtain sufficient gibberellin for chemical characterization. In general, the approach was to extract tissues with an organic solvent and to purify by solvent partitioning and assorted chromatographic procedures. The purified product was then repeatedly crystallized and its physical properties determined.

In 1958, MacMillan reported the identification of gibberellin A_1 from the immature seed of runner bean (*Phaseolus multiflorus**). This was the first identification of a gibberellin from a higher plant. The title of the paper, which appeared in *Die Naturwissenschaften*, was "The Occurrence of Gibberellin A_1 in Higher Plants: Isolation from the Seed of Runner Bean (*Phaseolus multiflorus*)" (MacMillan & Suter, 1958). A second paper, published in the October 1959 issue of *The Proceedings of the Chemical Society*, reported the identification of gibberellin A_5 from the same material (MacMillan, Seaton, & Suter, 1959). In 1959, West presented the results of his isolation studies at the *IVth International Conference of Plant Growth Regulation* held at the Boyce Thompson Institute for Plant Research, Yonkers, New York. He reported the isolation of two crystalline substances, Bean Factor I and Bean Factor II, from immature bean seed (*Phaseolus vulgaris*). The former was subsequently shown to be identical to gibberellin A_1, the latter to gibberellin A_5. Further information on the isolation was also presented in the *Journal of the American Chemical Society* by West and Phinney (1959). In addition, in 1959 Sumiki and Kawarada reported the identification of gibberellin A_1 from the elongated water sprouts of a cultivar of mandarin orange (*Citrus unshiu*) (see Sumiki & Kawarada 1961; also Kawarada & Sumiki, 1959). A spate of publications then appeared on the identification of gibberellins from the fungus *Gibberella fujikuroi* as well as from higher plants. To date, a total of 66 gibberellins have been characterized (MacMillan, 1980, 1983), of which 56 have been isolated from plants and 26 from the fungus.

---

*Now referred to as *Phaseolus coccineus*.

As a final episode in this brief history of gibberellins, it is appropriate to remind readers of an important agreement that prevented the gibberellin nomenclature from drowning in a mass of trivial names. Over the years, the fungal gibberellins had been assigned names by numbers, i.e., gibberellin A_{1-x}, a system that had been practiced on an informal basis and without problems. However, with the identification of new gibberellins from higher plants, trivial names began to appear in the literature based on their plant origin— i.e., bamboo gibberellin, Lupinus I gibberellin, Lupinus II gibberellin. This potential nomenclatural problem was privately discussed in 1967 at the *VIth International Conference on Plant Growth Substances* held in Ottawa, Canada, and it was agreed that the gibberellin A_{1-x} terminology should be extended to higher plant gibberellins. A recommendation to this effect appeared in a January 1968 issue of *Nature* (MacMillan & Takahashi, 1968), and MacMillan and Takahashi are currently responsible for the bookkeeping task of assigning numbers to new gibberellins as they are identified. As a result, nomenclaturial chaos was bypassed by the use of a chronological numbering system.

At this point, I shall bring this historical account to an end, being well aware that the closer one approaches the present, the more controversial becomes the subject, and the greater becomes the number of people who may feel they have been misquoted, not quoted, or maligned.

REFERENCES

Brian, P. W. (1959). Effects of gibberellins on plant growth and development. Biol. Rev. *34*, 37–84.

Brian, P. W., Elson, G. W., Hemming, H. G., & Radley, M. (1954). The plant growth promoting properties of gibberellic acid, a metabolic product of the fungus, *Gibberella fujikuroi*. J. Sci. Food Agr. *5*, 602–612.

Brian, P. W., & Hemming, H. G. (1955). The effect of gibberellic acid on shoot growth and pea seedlings. Physiol. Plantarum *8*, 669–681.

Cross, B. E. (1954). Gibberellic acid. Part I. J. Chem. Soc. (London), 4670–4676.

Cross, B. E., Grove, J. F., MacMillan, J., & Mulholland, T. P. C. (1956). Gibberellic acid, Part IV. The structures of gibberic and allogibberic acids and possible structures for gibberellic acid. Chem. Ind. (London), 954-955.

Curtis, P. J., & Cross, B. E. (1954). Gibberellic acid. A new metabolite from the culture filtrates of *Gibberella fujikuroi.* Chem. Ind. 1066.

Emerson, R. A., Beadle, G. W., & Fraser, A. C. (1935). A summary of linkage studies on maize. N.Y. State Agr. Exp. Sta. Mem. *39,* 1-83.

Grove, J. F. (1961). The gibberellins. Quart. Rev. (Chem. Soc. London) *15,* 46-70.

Hanson, J. R. (1968). The gibberellins. In: The Tetracyclic Diterpenes, pp. 41-59, Barton, D. H. R., & Raphael, R. A., eds. Pergamon Press, New York.

Harris, R. M. (1953). Auxin relations in a dwarf-1 allele of *Zea mays* L. Ph.D. Thesis, University of California, Los Angeles.

Hori, S. (1898). Some observations on "Bakanae" disease of the rice plant. Mem. Agr. Res. Sta. (Tokyo) *12*(1), 110-119.

Hori, S. (1903). Bakanae disease of rice. Lectures on plant disease, Seibido, Tokyo. First edition, 114-121.

Ito, S., & Kimura, J. (1931). Studies on the "bakanae" disease of the rice plant. Hokkaido Agr. Exp. Sta. Rept. 27, 1-99.

Kawarada, A., & Sumiki, Y. (1959). The occurrence of gibberellin A_1 in water sprouts of Citrus. Bull. Agr. Chem. Soc. (Japan) *23,* 343-344.

Konishi, T. A. (1828). Nogyo-yowa, Part 1, Subsection Nawashiro, Section kome, Edo (Tokyo), publisher unknown. Reprinted in Takimoto, S. (1929), Nippon Keizai Taiten, *29,* 549-628.

Kurosawa, E. (1926). Experimental studies on the nature of the substance excreted by the "bakanae" fungus. Trans. Nat. Hist. Soc. Formosa *16,* 213-227.

Kurosawa, E. (1930). On the overgrowth phenomenon of rice seedlings related to the excretion of the cultures of *Lisea fujikuroi* × Sawada and related organisms. Trans. Nat. Hist. Soc. Formosa *20,* 218-239.

Kurosawa, E. (1932). On certain experimental results concerning the over-elongation phenomenon of rice plants which owe to the filtrate got from the culture solution of 'bakanae' fungus. Trans. Nat. Hist. Soc. Formosa *22,* 198-201.

Kurosawa, E. (1934). Concerning the results of transplanting rice seedlings affected with bakanae disease. Ann. Phytopath. Soc. Japan *4* (No. 1 and 2), 33–34.

Lang, A. (1956a). Bolting and flowering in biennial *Hyoscyamus niger*, induced by gibberellic acid. Plant Physiol. *31* (Suppl), xxxv.

Lang, A. (1956b). Gibberellin and flower formation. Naturwissenschaften *43*, 544.

Lang, A., Sandoval, J. A., & Bedri, A. (1957). Induction of bolting and flowering in *Hyoscyamus* and *Samolus* by a gibberellin-like material from a seed plant. Proc. Nat. Acad. Sci. (U.S.A.) *43*, 960–964.

Letham, D. S., Higgins, T. J. V., Goodwin, P. B., & Jacobsen, J. V. (1978). In: Phytohormones and Related Compounds: A Comprehensive Treatise, Vol. I. The Biochemistry of Phytohormones and Related Compounds, pp. 1–28, Letham, D. S., Goodwin, P. B., & Higgins, T. J. V., eds. Elsevier/North-Holland Biomedical Press, New York.

Lona, F. (1957). Gibberellin-like action of extracts obtained from young floral structures of Brassica napus L. var. Oleifera. (Preliminary Note.) 1. Atenes, parmense *28*, 111–115.

MacMillan, J. (1980). Introduction. In: Hormonal Regulation of Development I. Molecular Aspects of Plant Hormones, pp. 1–8, MacMillan, J., ed., Encyclopedia of Plant Physiology New Series, Vol. 9. Springer-Verlag, Berlin, Heidelberg, New York.

MacMillan, J. (1983). Gibberellins in higher plants. Biochem. Soc. Trans. (in press).

MacMillan, J., Seaton, J. C., & Suter, P. J. (1959). A new plant-growth promoting acid—gibberellin A_5 from the seed of *Phaseolus multiflorus*. Proc. Chem. Soc. 325.

MacMillan, J., & Suter, P. J. (1958). The occurrence of gibberellin A_1 in higher plants: isolation from the seed of runner bean (*Phaseolus multiflorus*). Naturwissenschaften *45*, 46–47.

MacMillan, J., & Takahashi, N. (1968). Proposed procedure for the allocation of trivial names to the gibberellins. Nature *217*(5124), 170–171.

Marth, P. C., Audia, W. V., & Mitchell, J. W. (1956). Effects of gibberellic acid on growth and development of various species of plants. Plant Physiol. *31* (Suppl.), xliii.

Mees, G. C., & Elson, G. W. (1978). The gibberellins. In: Jealott's Hill, Fifty

Years of Agricultural Research, 1928-1978, pp. 55-60, Peacock, F. C., ed. Kynoch Press, Birmingham.

Mitchell, J. E., & Angel, C. R. (1951). The growth-stimulating properties of a metabolic product of *Fusarium moniliforme*. Phytopath. *41*, 26-27.

Mitchell, J. W., Skaggs, D. P., & Anderson, W. P. (1951). Plant growth stimulating hormones in immature bean seeds. Science *114*, 159-161.

Murakami, Y. (1957). The effect of the extract of immature bean seeds on the growth of the coleoptile and leaf of rice plant. Bot. Mag. Tokyo *70*, 376-382.

Nisikado, Y. (1931). Comparative investigations on the diseases of Gramineae caused by *Lisea fujikuroi* Saw. and *Gibberella moniliformis* (Sh.) Winel. (Preliminary note.) Ber. des Ohara Inst. für landwirt. Forsch. in Kurashiki Provinz Okayama, Japan *5*, 87-106.

Nisikado, Y. (1932). Naming of the organism causing the bakanae disease and plants affected by the disease. Jap. J. Plant Protection *19*, 491-497.

Phinney, B. O. (1946). Gene action in the development of the leaf in *Zea mays* L. Ph.D. Thesis, University of Minnesota, Minneapolis.

Phinney, B. O. (1956a). Biochemical mutants in maize: Dwarfism and its reversal with gibberellins. Plant Physiol. *31* (Suppl.), xx.

Phinney, B. O. (1956b). Growth response of single gene mutants of maize to gibberellic acid. Proc. Nat. Acad. Sci. (U.S.A.) *42*, 185-189.

Phinney, B. O., West, C. A., Ritzel, M. B., & Neely, P. M. (1957). Evidence for gibberellin-like substances from flowering plants. Proc. Nat. Acad. Sci. (U.S.A.) *43*, 398-404.

Radley, M. (1956). Occurrences of substances similar to gibberellic acid in higher plants. Nature *178*, 1070-1071.

Sawada, K. (1912). Diseases of agricultural products in Japan. Formosan Agr. Rev. No. 63, 10 and 16.

Stodola, F. H. (1958). Source book on gibberellins 1828-1957. Agricultural Research Service. U.S. Department of Agriculture, 1-421.

Stodola, F. H., Nelson, G. E. N., & Spence, D. J. (1957). The separation of gibberellin A and gibberellic acid on buffered partition columns. Arch. Biochem. Biophys. *66*, 438-443.

Stodola, F. H., Raper, K. B., Fennell, D. I., Conway, H. F., Johns, V. E., Langford, C. T., & Jackson, R. W. (1955). The microbial production of gibberellins A and X. Arch. Biochem. Biophys. *54*, 240-245.

Stowe, B. B., & Yamaki, T. (1957). The history of the physiological action of the gibberellins. Ann. Rev. Plant Physiol. *8*, 181-216.

Sumiki, Y., & Kawarada, A. (1961). Occurrence of gibberellin A_1 in the water sprouts of Citrus. In: IVth International Conference on Plant Growth Regulation (1959), pp. 483-487, Klein, R. M., ed. Iowa State University Press, Ames, Iowa.

Takahashi, N., Kitamura, H., Kawarada, A., Seta, Y., Takai, M., Tamura, S., & Sumiki, Y. (1955). Biochemical studies on "bakanae" fungus. Part XXXIV. Isolation of gibberellins and their properties. Bull. Agr. Chem. Soc. Japan *19*, 267-277.

Tamura, S. (1969). The history of research on gibberellin. In: Gibberellins, Chemistry, Biochemistry and Physiology, pp. 3-27, Tamura, S., ed. Tokyo Daigakushuppankai, Tokyo.

Tamura, S. (1977). The history of plant hormone gibberellin. In: Plant Hormones, pp. 18-50. Dai nippon tosho Co. Ltd., Tokyo.

van Overbeek, J. (1935). The growth hormone and the dwarf type of growth in corn. Proc. Nat. Acad. Sci. (U.S.A.) *21*, 292-299.

van Overbeek, J. (1938). Auxin production in seedlings of dwarf maize. Plant Physiol. *13*, 587-598.

West, C. A. (1961). The chemistry of gibberellins from flowering plants. In: Plant Growth Regulation, pp. 473-481, Klein, R. M., ed. Iowa University Press, Ames.

West, C. A., & Phinney, B. O. (1956). Properties of gibberellin-like factors from extracts of higher plants. Plant Physiol. *31* (Suppl.), xx.

West, C. A., & Phinney, B. O. (1959). Gibberellins from flowering plants. I. Isolation and properties of a gibberellin from *Phaseolus vulgaris* L. J. Amer. Chem. Soc. *81*, 2424-2427.

Wittwer, S. H., & Bukovac, M. J. (1957). Gibberellins—new chemicals for crop production. M.S.U. Agric. Exp. Sta. Quart. Bull. *39*, 469-494.

Wollenweber, H. W. (1931). Fusarium monograph. Parasitic and saprophytic fungi. Zeit. Parasit. *3*, 269-516.

Wollenweber, H. W., & Reinking, O. A. (1935). The Fusaria: Their Description, Injurious Effects, and Control, Vol. VIII, pp. 1-135. P. Parey, Berlin.

Yabuta, T. (1935). Biochemistry of the "bakanae" fungus of rice. Agr. Hort. *10*, 17-22.

Yabuta, T., & Hayashi, T. (1936). Biochemistry of the "bakanae" fungus of rice. Agr. Hort. *11*, 27-33.

Yabuta, T., & Hayashi, T. (1938). Biochemistry of the "bakanae" fungus of rice (additional report). Agr. Hort. *13*, 21-25.

Yabuta, T., & Hayashi, T. (1940). Biochemical studies of "bakanae" fungus of rice. J. Imp. Agr. Exp. Sta. (Japan) *3*, 365-400.

Yabuta, T., Kambe, K., & Hayashi, T. (1934). Biochemistry of the "bakanae" fungus of rice. Part 1. Fusaric acid, a new product of the bakanae fungus. J. Agr. Chem. Soc. Japan *10*, 1059-1068.

Yabuta, T., & Sumiki, T. (1938). Communication to the editor. J. Agr. Chem. Soc. Japan *14*, 1526.

Yabuta, T., & Sumiki, Y. (1944). Biochemistry of the "bakanae" fungus of rice. Part 18. Action of gibberellin on the growth of buds of *Paulownia tomentosa*. J. Agr. Chem. Soc. Japan *20*, 52.

Yabuta, T., Sumiki, Y., Aso, K., & Hayashi, T. (1941a). Biochemistry of the "bakanae" fungus of rice. Part 13. The effects of gibberellin on special components and special tissues of plants. (4). The action of gibberellin on tobacco seedlings. J. Agr. Chem. Soc. Japan *17*, 1001-1004.

Yabuta, T., Sumiki, Y., Aso, K., Tamura, T., Igarashi, H., & Tamari, K. (1941b). Biochemistry of the "bakanae" fungus of rice. Part 10. The chemical constitution of gibberellin. J. Agr. Chem. Soc. Japan *17*, 721-730.

2

Early Stages of Gibberellin Biosynthesis

Ronald C. Coolbaugh

2.1 INTRODUCTION

An understanding of gibberellin (GA) biochemistry and physiology requires information on a number of topics, including (1) the sites, reactions, and conditions controlling the biosynthesis of GAs; (2) the modes and rates of transport of GAs within plants under different environmental conditions and at different stages of development; (3) the mechanism(s) of action, fate, and rates of turnover of GAs; and (4) the identities, quantities, localization, and physiological state of native GAs. This chapter will deal primarily with the first of these areas of study—GA biosynthesis—and will be further limited to the early stages in the GA biosynthetic pathway, the formation of *ent*-kaurene from mevalonic acid (MVA) (see Fig. 2.1).

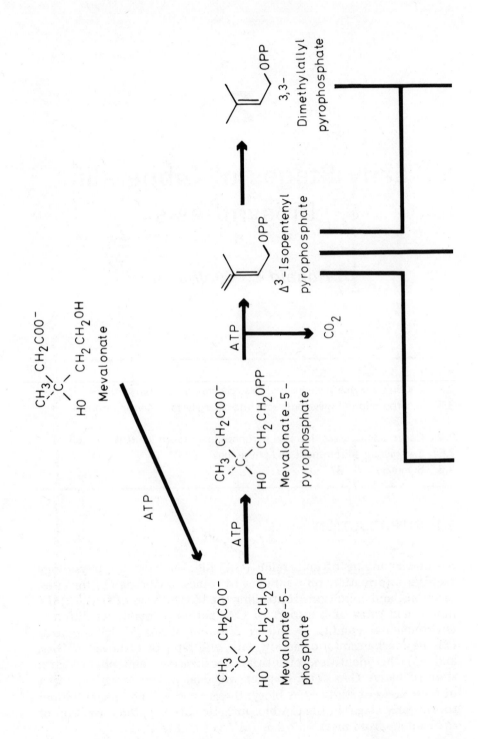

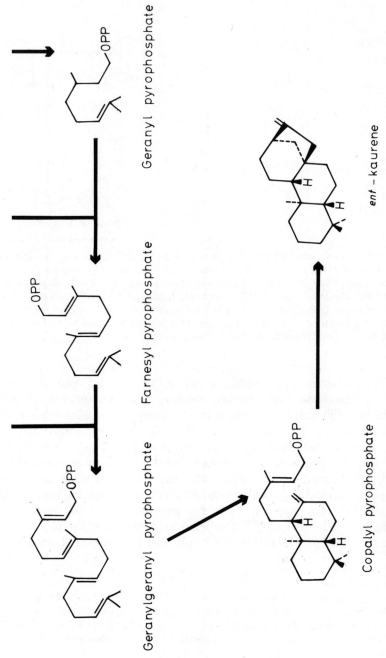

Geranyl pyrophosphate

Farnesyl pyrophosphate

Geranylgeranyl pyrophosphate

Copalyl pyrophosphate

ent – kaurene

Figure 2.1 Biosynthetic pathway from mevalonic acid to *ent*-kaurene.

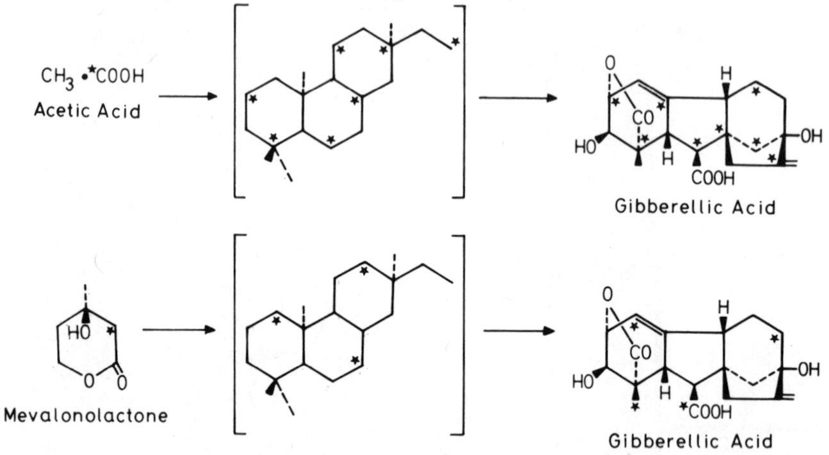

Figure 2.2 Incorporation of [^{14}C]acetic acid and [^{14}C]MVA into gibberellic acid (GA$_3$). Labeled GA$_3$, obtained by culturing *G. fujikuroi* in liquid medium containing either CH$_3 \cdot ^{14}$CO$_2$H or [2-^{14}C]mevalonic lactone, was chemically degraded to demonstrate the locations of ^{14}C, which are indicated by asterisks. (Redrawn, with permission, from Birch et al., 1958.) The structure of GA$_3$ illustrated is as indicated by Birch et al. (1958). The correct positions of the 4,10-lactone ring and 1,2-double bond were determined subsequently.

The isolation and crystallization of GA, which is the causative agent of "bakanae," or "foolish seedling" disease, by Yabuta and Sumiki in 1935 and the subsequent discovery of GA-like substances in higher plants (Radley, 1956, West & Phinney, 1956) led to investigations related to the biosynthetic origin of GAs. The initial work on this pathway was carried out by Birch and co-workers (1958, 1959) who demonstrated that the GAs, which are branched and cyclic C$_{19}$ and C$_{20}$ compounds, are formed through the isoprenoid biosynthetic pathway. This finding was based on the results of feeding experiments using ^{14}C-labeled acetate and MVA in cultures of the fungus *Gibberella fujikuroi*,* followed by chemical degradation

---

*The names *Gibberella fujikuroi* and *Fusarium moniliforme* are used interchangeably in the literature. This organism was assigned to the class Fungi Imperfecti and the genus *Fusarium* until its sexual reproductive (perfect) stage was discovered. At that time it was reclassified as an ascomycete in the genus *Gibberella*.

of the resultant $[^{14}C]GA_3$ (Fig. 2.2). *Ent*-kaurene was later identified as both a product of MVA metabolism (Cross et al., 1962) and a precursor of GAs (Cross, Galt, & Hansen, 1964; see also Chapter 5, this volume) in fermentations of *G. fujikuroi*.

As diterpenes, the GAs share many early biosynthetic intermediates and enzymes with other isoprenoids, and several points of divergence occur in this pathway which lead to other important groups of compounds (Fig. 2.3). For instance, isopentenyl pyrophosphate (IPP) is a substrate for side chains of cytokinins; geranyl pyrophosphate (GPP) is metabolized to form monoterpenes; farnesyl pyrophosphate (FPP) is converted in some systems to abscisic acid; and two FPP molecules may also be condensed to form squalene and subsequently converted into sterols. Geranylgeranyl pyrophosphate (GGPP) appears to be an important pivotal intermediate. In addition to being a more-or-less direct substrate for the formation of *ent*-kaurene, which is the first committed intermediate in GA formation, GGPP can also be converted into chlorophyll esters, macrocyclic diterpenes, and plastoquinone and phytoene, the immediate precursor of the carotenes. Some excellent reviews on various aspects of isoprenoid biosynthesis have appeared recently (Beytia & Porter, 1976; Loomis & Croteau, 1980; Moore, 1979; Lang, 1970; Hedden, MacMillan, & Phinney, 1978; Hanson, 1977; Spurgeon & Porter, 1980).

A number of techniques have been brought to bear on the questions of GA biosynthesis. Pharmacological tests have been conducted with many synthetic and natural compounds in order to ascertain their effects on selected aspects of plant growth and development. Native GAs and their precursors from plant tissues grown under a variety of environmental conditions have been isolated, identified, and quantified in attempts to correlate quantities of specific GAs with physiological activities. Radioactive compounds have been applied to whole plants, excised plant parts, cell cultures, and cell-free preparations and the subsequent isolation and identification of products has yielded much information about the enzyme capacities of a variety of plants. The application of pure organic chemistry has helped to further clarify the mechanisms of some of these reactions. This chapter will focus primarily on the cell-free enzyme systems that have been used to investigate the initial stages of GA biosynthesis. Reference to other methods or other reactions is included where it enhances our understanding of this section of the biosynthetic pathway.

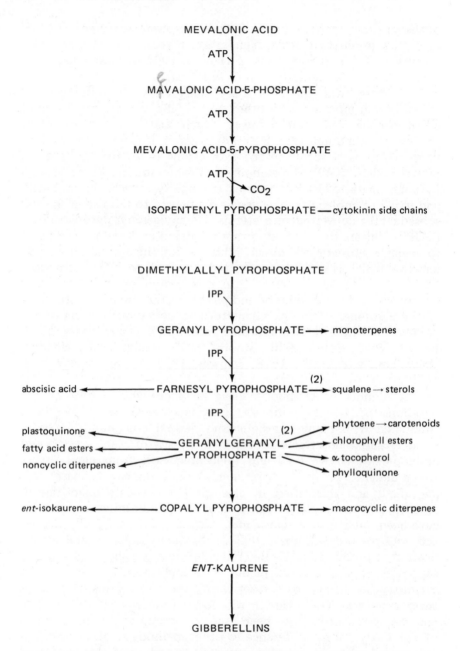

Figure 2.3 Biosynthetic relationships among isoprenoid pathways.

2.2 CELL-FREE ENZYME SYSTEMS USED TO STUDY *ENT*-KAURENE BIOSYNTHESIS

The development of cell-free enzyme systems capable of catalyzing reactions leading to the formation of *ent*-kaurene has facilitated the study of selected biochemical reactions. The first cell-free enzyme system capable of incorporating MVA into *ent*-kaurene and other intermediates was derived from the liquid endosperm of wild cucumber (*Marah macrocarpus*) seeds as described by Graebe et al. (1965). Reaction mixtures contained 0.021 μmol [^{14}C]MVA, 1.0 μmol adenosine triphosphate (ATP), 1.0 μmol $MgCl_2$, and 1.0 ml of the gently mixed acellular liquid endosperm of *M. macrocarpus* seeds, in a total volume of 1.2 ml. Reactions were incubated for 60 min at 33°C and stopped either by boiling or by the addition of acetone. The diterpene hydrocarbon, *ent*-kaurene, was extracted from the reaction mixtures with benzene and separated from other products by thin-layer chromatography (TLC) in hexane (Fig. 2.4).

Since that time, other cell-free enzyme systems have been developed which catalyze, to varying degrees, the conversion of intermediates in this pathway. The major cell-free systems that synthesize *ent*-kaurene are briefly summarized below in order to establish some perspective of the apparent assets and limitations of each system. However, one feature common to all of the systems

Figure 2.4 Resolution of the ^{14}C-lipid fraction produced from [2-^{14}C]-mevalonic acid in endosperm of *Marah macrocarpus*. *Ent*-kaurene is separated from more polar compounds by TLC using a hexane mobile phase (System A). After removal of silica gel containing *ent*-kaurene (*Rf* 0.8–1.0), the plate was rechromatographed in benzene-ethyl acetate (9:1) to separate other products (System B). Fraction 1 is *ent*-kaurene; fractions 2, 3, and 4 contain *ent*-kaurenoic acid, *ent*-kaurenol, and *ent*-kaurenal, respectively. (Redrawn, with permission, from Graebe et al., 1965.)

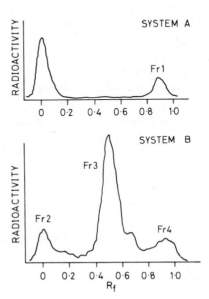

is the presence of the enzymes catalyzing the incorporation of MVA into *ent*-kaurene in the high-speed supernatant fraction of plant extracts after ultracentrifugation. This is in contrast to the enzymes involved in several subsequent reactions and has led to the common notion that these enzymes are "soluble." This implies that they are not associated with a membrane fraction, but makes no assumption regarding the *in vivo* localization of the enzyme within either the free cytoplasm or membrane-bound organelles. In a few cases such data are available, though procedures for tissue homogenization and enzyme isolation usually disturb the membranes to such an extent that the original compartmentation of the enzyme is lost. Thus, "soluble" enzymes derived from crude tissue homogenates may well have come from organelles broken in the homogenization process. Likewise, preparations of isolated organelles may carry with them certain enzyme activities which are cytoplasmic in origin, but which bind to the outer membranes of the organelles during the isolation process. The assignment of specific localizations is especially equivocal where enzyme activity associated with cellular fractions is low and other supporting evidence is lacking.

2.2.1 Gibberella fujikuroi

The fungus *G. fijukuroi* represents perhaps the simplest known organism that will readily incorporate [^{14}C]MVA into [^{14}C]GA$_3$ *in vivo*. Cell-free enzyme extracts of *G. fujikuroi* will also incorporate MVA into *ent*-kaurene (Evans & Hanson, 1972) and subsequently, under appropriate conditions, into GAs. Among the several advantages of using this organism as a test system are the very high enzyme activities and the apparent comparability of early intermediates with those of the same pathway in higher plants. Other assets include: (1) easy preparation of large quantities of fungal cultures and enzymes; (2) easy incorporation of intermediates and inhibitors into the growing organism; and (3) isolation of mutants blocked at one or more steps in the GA biosynthetic pathway.

It is important to note, however, that *G. fujikuroi* is an unusual organism in that it produces far larger quantities of GAs than any higher plant, yet the presence of GAs does not appear to be an essential requirement for any known function in the fungus. It is apparent from the wealth of literature on the subject that the early stages of GA biosynthesis in the fungus and in the higher plants are identical. However, this does not necessarily mean that the enzymes in these two plant groups are either identical or subject to the same controls. Indeed, the extraordinary rate of GA production in *G. fujikuroi* may be an indication that the pathway in the fungus lacks some of the controls present in higher plants. Nevertheless, the fungus has been an excellent model organism in which to study various aspects of the GA biosynthetic pathway. The pathways do diverge at later points in the metabolic sequence, and thus some differences occur in the endogenous GA spectra of the fungus and higher plants (see Chapters 3, 4, and 5). Similarly, the GA profile can vary from one species of higher plant to another, in different tissues of the same plant, and within the same tissue at different stages of development. *Gibberella fujikuroi* will continue to be used in important studies related to this pathway, but we must be cautious about extrapolating results obtained to higher plants wherein, ultimately, lies our primary interest.

2.2.2   *Marah macrocarpus* and *Cucurbita maxima*

The liquid endosperm of *M. macrocarpus* seeds is acellular and contains large amounts of GAs. It is easy to work with as it can be obtained in relatively large quantities (0.5–1.0 ml per seed), has a gelatinous texture, and can be used as an enzyme source after either gentle homogenization or partial purification. The *M. macrocarpus* system has been used for detailed study of several of the early steps in the biosynthesis and oxidation of *ent*-kaurene (see Upper & West, 1967; Oster & West, 1968; Shechter & West, 1969; Frost & West, 1977; Duncan & West, 1981). The related system from the liquid endosperm of pumpkin (*Cucurbita maxima*) seed incorporates MVA into *ent*-kaurene with exceptional efficiency and has yielded much of our knowledge on the later steps in the GA biosynthesis pathway including the formation of C_{19}-GAs (Graebe, 1969; Graebe et al., 1974; see also Chapter 3, this volume). However, liquid endosperm is atypical when compared to the cellular tissues of other plant parts including seeds. We do not know, except in some of the earliest reactions, whether enzymes in the endosperm are subject to the same controls as those in other plant parts. For example, if GA biosynthesis in shoot tips is controlled, to some degree, by light intensity and photoperiod, is the same true of the pathway in endosperm tissue? It should be noted that *Marah macrocarpus* has a limited distribution and that the extremely large root system of both *C. maxima* and *M. macrocarpus* make them difficult to grow under controlled environmental conditions.

2.2.3 Ricinus communis

Cell-free preparations from castor bean (*Ricinus communis*) seedlings are especially interesting (Robinson & West, 1970a, 1970b). The *Ricinus* system was the first to be derived from a vegetative tissue and, unlike other *in vitro* systems studied to date, it yields five diterpenes in addition to *ent*-kaurene. Details on the products of MVA metabolism in the *Ricinus* system will be described in Section 2.3.

2.2.4 Pisum sativum

The garden pea (*Pisum sativum*) has also been extensively used for studies on the cell-free biosynthesis of *ent*-kaurene. Initial studies on *P. sativum* centered on seeds rather than seedlings because they contain and synthesize large amounts of GAs in comparison to the vegetative parts. This system has been useful in confirming some of the early steps and in investigating potential differences between varieties with different phenotypic characteristics or physiological requirements. The enzyme extracts are, however, complicated by the nature of the seed itself as the *ent*-kaurene–synthesizing enzymes are localized in the highly proteinaceous, green cotyledons (Coolbaugh & Moore, 1971). In addition, competing sterol synthesizing enzymes are present and enzyme activities are low relative to the *Marah* and *Cucurbita* preparations. More recently, *P. sativum* seed systems have been developed to investigate oxidative reactions in *ent*-kaurene metabolism (Ropers et al., 1978) and interconversions of GAs during seed development (see Frydman & MacMillan, 1973, 1975, 1976; see also Chapters 3 and 4, this volume).

The cell-free enzyme systems from *P. sativum* shoot tips are also of potential value for studying the GA biosynthetic pathway, especially as there are a number of genetic varieties which exhibit different physiological characteristics (Coolbaugh et al., 1973; Ecklund & Moore, 1974; Choinski & Moore, 1980). However, enzyme preparations from *P. sativum* shoot tips have very low activities—only about 2% of that of *P. sativum* seeds—which makes extensive investigations on specific enzymes prohibitive under these conditions. Nevertheless, a number of interesting investigations into the overall incorporation of MVA into *ent*-kaurene have been carried out.

2.2.5 Other plants

Enzyme preparations from etiolated shoots of corn (*Zea mays*) have also proved to be useful (Hedden & Phinney, 1976, 1979). This system provides an alternative vegetative material with several interesting dwarf mutants. Although not yet extensively used, *Z. mays* has great promise for expanding our knowledge of GA biosynthesis. Recently described systems from tomato (*Lycopersicum esculatum*) and tobacco (*Nicotiana tabacum*) cell cultures and germinating *Lycopersicum esculatum* seeds (Yafin & Schechter, 1975), runner bean (*Phaseolus coccineus*) suspensors (Ceccarelli, Lorenzi, & Alpi, 1979, 1981), and sunflower shoots (*Helianthus annuus*) (Shen-Miller & West, 1982) also show potential for this field of study.

Some investigations on the early stages of GA biosynthesis have focused on specific enzymes and reactions, whereas others have dealt with a number of biochemical and physiological aspects of the overall sequence of reactions that yield *ent*-kaurene. In the sections that follow the incorporation of MVA into GGPP is discussed first, followed by the conversion of GGPP to *ent*-kaurene, and finally the overall synthesis of *ent*-kaurene from MVA.

2.3 INCORPORATION OF MEVALONIC ACID INTO GERANYLGERANYL PYROPHOSPHATE

2.3.1 Enzymology

Early stages in the isoprenoid pathway include the activation of MVA and its conversion to C_5 isoprene units which can be subsequently condensed to form mono-, sequestra-, di-, tri-, and polyterpenes. MVA itself is generated from acetyl Co A derived from the Krebs cycle, fat catabolism, and other pathways. Recent reviews by Beytia and Porter (1976) and Spurgeon and Porter (1980) include the steps from acetate to MVA. Although biosynthesis of MVA from acetate has not been demonstrated in higher plants, enzymes catalyzing specific reactions involved in the incorporation of acetyl Co A into MVA have been isolated from a number of higher plant sources. Acetate also serves as a substrate for *ent*-kaurene and carotenoid biosynthesis. The labeling patterns of carotenoids are consistent with the formation of MVA from three acetate molecules.

MVA kinase (ATP: mevalonate-5-phosphotransferase, EC 2.7.1.36) catalyzes the first reaction in this sequence. Cell-free enzyme systems capable of incorporating MVA into mevalonate-5-phosphate (MVAP) have been prepared from a number of plant sources including *Ricinus communis* (Shewry & Stobart, 1973), green leaves and etiolated cotyledons of French bean (*Phaseolus vulgaris*) (Gray & Kekwick, 1973a), *Marah macrocarpus* endosperm (Knotz, Coolbaugh, & West, 1977), *Cucurbita pepo* seedlings (Loomis & Battaile, 1963), latex serum from rubber (*Hevea brasiliensis*) (Williamson & Kekwick, 1965), *Pinus pinaster* seedlings, extracts of *Agave americana* (Garcia-Peregrin, Suarez, & Mayor, 1973), and *P. sativum* seeds (Moore & Coolbaugh, 1976). Downing and Mitchell (1974) have obtained 95% conversion of active [^{14}C]MVA to phosphorylated intermediates using an enzyme preparation from leaves of mint (*Nepeta cataria*).

Gray and Kekwick (1973a, 1973b) reported that six different plant species gave enzymes with similar molecular characteristics: mol wt, 94,800–103,500 daltons; diffusion coefficients, 5.39–5.62 × 10^{-7} cm^2 sec^{-1}; and sedimentation coefficients, 5.85–6.00 S. They also reported that MVA kinase from green leaves and etiolated cotyledons of *Ph. vulgaris* had indistinguishable characteristics with regard to their affinities for MVA and ATP and that the pH optimum of both preparations was 7.0. This result contrasted with previous publications by Rogers, Shah, and Goodwin (1966a, 1966b) reporting the presence of two isoenzymes in *Ph. vulgaris*, one located in the chloroplasts and the other in the soluble fraction. A divalent cation is required for enzyme activity. Most reports indicate that Mn^{2+} is more effective in activating the enzyme than Mg^{2+}, although a combination of the two elements appears to be superior in some systems. Several workers have reported that thiol compounds improve yields, suggesting the presence of one or more sulfhydryl groups at or near the active site of the enzymes.

The enzyme catalyzing the conversion of MVAP to mevalonate-5-pyrophosphate (MVAPP), namely 5-pyrophospho-MVA kinase (EC 2.7.4.2), has not been studied extensively due to the difficulties in preparing MVAP for use as a substrate. In addition, the isolation of MVAPP as a single product is difficult because it is often rapidly metabolized or attacked by phosphatases. This enzyme has, however, been partially purified from the latex of *Hevea brasiliensis* (Skilleter, Williamson, & Kekwick, 1966; Skilleter & Kekwick, 1971), as well as from yeast and several animal systems (see Beytia & Porter, 1976). Enzyme from *H. brasiliensis* had an apparent pH optimum of 7.2, a *Km* for 5-phosphomevalonate of 0.042 m*M*, and a *Km* for ATP of 0.19 m*M*. The presence of thiol compounds was required for activity. Mevalonate-5-pyrophosphate has been reported as a product in several other cell-free enzyme systems (see Spurgeon & Porter, 1980).

Mevalonate pyrophosphate decarboxylation, catalyzed by 5-pyrophosphomevalonate decarboxylase (EC 4.1.1.33), represents an interesting reaction in this sequence. As in the case of the previous two reactions, this step also requires ATP and a divalent cation. However, the reaction is further involved in the overall energy metabolism of the plant as it results in the loss of CO_2. As such, it can be considered a branch point of great importance. The decarboxylase has been partially purified from extracts of *H. brasiliensis* (Skilleter & Kekwick, 1971). It has a pH optimum of 5.5–6.5, a *Km* for 5-pyrophosphomevalonate of 0.004 mM, and a *Km* for ATP of 0.12 mM. Mg^{2+} stimulated the enzyme at a slightly lower concentration than did Mn^{2+}.

The sequence of reactions between IPP and GGPP has been studied extensively and thoroughly reviewed by Beytia and Porter (1976). The first reaction is an isomerization to dimethylallyl pyrophosphate (DMAPP). The enzyme catalyzing this reaction, IPP isomerase (EC 5.3.3.2), was first isolated from yeast autolysates by Agranoff et al. (1960) and from pumpkin fruit by Ogura, Nishino, and Seto (1968). The enzyme catalyzes the isomerization of the double bond from the 3-position of IPP to the 2-position of DMAPP.

Further reactions in this part of the pathway involve the sequential addition of IPP to DMAPP, GPP, and FPP. The enzymes involved in these head-to-tail condensation reactions are called prenyl transferases. Based upon evidence of copurification of enzyme activities in the plant and animal systems studied, it is believed that specific enzymes may catalyze the sequential reactions in C_{15}-, C_{20}-, and longer-chain–specific systems.

A C_{15}-specific prenyl transferase, farnesyl pyrophosphate synthetase (EC 2.5.1.1), from yeast has been purified to homogeneity (Eberhardt & Rilling, 1975). Partially purified enzymes have been obtained from pumpkin fruit (Ogura et al., 1968) and germinating *R. communis* (Green & West, 1974). The latter reported isolation and purification of two forms of prenyl transferase from *Ricinus*. At higher protein concentrations (40–60 ug ml^{-1}) the enzymes showed almost identical molecular weights (72,500 ± 3,000 daltons), while at lower enzyme concentrations (20 ug ml^{-1}) the molecular weights of transferases I and II were 56,000 ± 2,000 and 60,000 ± 2,000, respectively. The pH optimum of both enzymes was 6.8 and Mg^{2+} was a better activator than Mn^{2+}. Mn^{2+} was, however, more effective than Mg^{2+} at concentrations below about 0.06 mM. Sulfhydryl inhibitors were much less effective on these enzymes than on animal transferases. On the basis of these data, Green and West (1974) proposed interconversions between the two forms of transferase as a possible regulatory mechanism.

C_{20}-specific prenyl transferases have been obtained from many plant materials including the fungus *Neurospora crassa* (Grob, Kirschner, & Lynen, 1961), carrot root (Nandi & Porter, 1964), and pumpkin fruit (Ogura, Shinka, & Seto, 1972). Enzyme systems that form geranylgeranyl pyrophosphate have also been reported from *M. macrocarpus* endosperm (Oster & West, 1968), barley (*Hordeum vulgare*) embryos (Davies, Reese, & Taylor, 1975), and *P. sativum* seeds (Moore & Coolbaugh, 1976). A time course with the pea system showed FPP disappearing as GGPP accumulated. Mn^{2+} is apparently a better activator in these systems than Mg^{2+}. The mechanisms of the prenyl transferase reactions are discussed by Beytia and Porter (1976) and Spurgeon and Porter (1980).

2.3.2 Compartmentation

Evidence for the localization of MVA activating enzymes and prenyl transferases has been reported from a variety of plant materials. Early attempts to incorporate MVA into carotenoids led Goodwin and Mercer (1963) to suggest that isoprenoid biosynthesis is compartmentalized and occurs independently in the cytoplasm and in plastids. There is now much evidence to support this hypothesis. The association of MVA kinase activity with known cytoplasmic enzymes in *Ph. vulgaris* (Gray & Kekwick, 1973a) is consistent with other data indicating the presence of these enzymes in the cytoplasm. On the other hand, the biosynthesis of phytoene via GGPP in extracts of higher plant plastids (Anderson & Porter, 1962) and later studies by Jungalwala and Porter (1967) provided evidence for the presence of these enzymes in plastids. The direct incorporation of MVA into isoprenoids in chloroplasts has, however, proved difficult due to the apparent impermeability of the chloroplast membrane to MVA. Wellburn and Hampp (1976) demonstrated that the envelopes of etioplasts became impermeable to MVA after 1–2 hr of illumination. Acetate penetrated the envelope after 4 hr of illumination, but not after exposure to light for 8–24 hr. In addition, the MVA-activating enzymes or some necessary cofactor were apparently lost from chloroplasts upon extraction with normal techniques. Direct evidence for MVA-activating enzymes in plastids came from the incorporation of [2-^{14}C]MVA into GGPP and phytoene using chloroplasts isolated by nonaqueous techniques (Charlton, Treharne, & Goodwin, 1967; Buggy, Britton, & Goodwin, 1974).

Although the direct incorporation of MVA into *ent*-kaurene in chloroplast preparations has not yet been reported, the presence of MVA-activating enzymes in wheat (*Triticum aestivum*) etioplasts (Cooke, 1977) and prenyl transferase in proplastids of *R. communis* endosperm (Green, Dennis, & West, 1975) as well as the reports of *ent*-kaurene synthetase in developing *R. communis* and *M. macrocarpus* seed endosperm (Simcox, Dennis, & West, 1975) and in both proplastids and chloroplasts of *Pisum sativum* shoot tips (Simcox, Dennis, & West, 1975; Moore & Coolbaugh, 1976) certainly suggest that the entire pathway exists in plastids. It is noteworthy that abscisic acid, which is a C_{15} isoprenoid, has also been isolated from *P. sativum* chloroplast preparations (Railton et al., 1974). The biosynthesis of abscisic acid from MVA in cell-free extracts of avocado (*Persea gratissima*) chloroplasts has also been reported (Milborrow, 1974).

2.3.3   Regulation

Inhibition of MVA kinase by GPP and FPP, but not by IPP and DMAPP, has been demonstrated in animal tissues (Dorsey & Porter, 1968; Flint, 1970). These results were confirmed and extended with MVA kinase from green leaves and etiolated cotyledons of *Ph. vulgaris* (Gray & Kekwick, 1973a). FPP and phytyl pyrophosphate showed the lowest inhibitor constants at 7.1-7.5 μM and 3.5-3.6 μM, respectively; while GPP and GGPP were somewhat less inhibitory (*Ki* values of 30.8-32.5 μM and 48.0-48.6 μM, respectively). Adenosine diphosphate (ADP) and IPP had no effect on MVA kinase in these studies.

MVAPP decarboxylase is subject to inhibition by its immediate products. Partially purified decarboxylase from *H. brasiliensis* was inhibited by both IPP and ADP (Skilleter & Kekwick, 1971). In addition, the GA biosynthetic pathway from MVA to *ent*-kaurene as measured in extracts of *Marah macrocarpus* was shown to be subject to energy charge control (Knotz et al., 1977). High reaction rates were possible only when the energy charge approached 1.0 (Fig. 2.5) and the reaction was specifically inhibited by ADP (Fig. 2.6). The effect appeared to be limited to the decarboxylase reaction.

GA_3 was shown to stimulate MVA kinase activity (specific and total) in extracts of hazel (*Corylus avellana*) embryonic axes (Shewry,

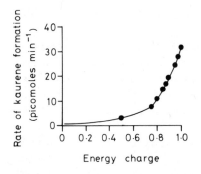

Figure 2.5 Effect of adenylate energy charge on rates of *ent*-kaurene biosynthesis. Reaction mixtures contained $11\,\mu M$ [^{14}C]mevalonic acid, $2.5\,mM$ Mn^{2+}, $5\,mM$ Mg^{2+}, $200\,\mu l$ enzyme extract, and combinations of AMP, ADP, and ATP (total concentration of $2\,mM$) at the adenylate energy charges indicated. The total volume of reactions was $0.5\,ml$. Incubation times were $10\,min$. (Redrawn, with permission, from Knotz et al., 1977.)

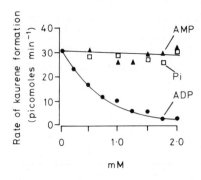

Figure 2.6 Rates of *ent*-kaurene synthesis in the presence of AMP, ADP, or Pi. Incubation conditions were as in Fig. 2.5 except $2\,mM$ ATP was used in the presence of the indicated concentrations of AMP (▲), ADP (●), or Pi (□). (Redrawn, with permission, from Knotz et al., 1977.)

Pinfield, & Stobart, 1974). GA_3 also caused a significant increase in decarboxylase activity in these extracts, but this was probably a result of increased MVA kinase activity in the enzyme preparation that used MVA as substrate. Based on a comparison of decarboxylase activity in sucklings and weaned rats on a low cholesterol diet, and weaned rats on a diet supplemented with cholesterol, it appears that rat liver decarboxylase is under feedback control by cholesterol (Ramachandran & Shaw, 1976).

2.3.4 Inhibitors

The value of inhibitors in investigations of biochemical pathways is well known. The utility of inhibitors is enhanced when they inhibit individual reactions in a pathway directed toward the synthesis of a single product or a group of closely related products. Conversely, an inhibitor is of limited use when it acts upon a reaction common to many pathways. In addition to the feedback inhibition of reactions between MVA and GGPP already noted, there are a variety of substances which inhibit general isoprenoid biosynthetic reactions. Some may be fairly specific, but most are general enzyme inhibitors. Obviously, sulfhydryl inhibitors are generally active on the MVA activating enzymes as well as on several other enzymes catalyzing subsequent steps in the pathway. Fimognari and Rodwell (1965) have reported on the inhibition of acetate incorporation into MVA in rat livers by bile salts including cholate and its derivatives. These authors considered this to be a feedback inhibition which blocked the MVA-nicotine-adenine dinucleotide phosphate (NADP) oxido-reductase reaction. Similar inhibition of acetate incorporation into MVA was obtained with AMO-1618 (2'-isopropyl-4'-[trimethyl-ammonium chloride]-5'-methylphenyl piperidine-1-carboxylate) which was previously thought to be more specific for *ent*-kaurene synthetase (Douglas & Paleg, 1972, 1978). Holmes and DiTullio (1962) had earlier tested a number of potential inhibitors in this pathway. Their results indicated that vanadium and a number of farnesoic acid derivatives inhibit MVA kinase. SK&F 525-A (2,2-diphenyl pentanoic acid 2-[dimethylamino]ethyl ester) inhibits MVA decarboxylase and IPP isomerase. SK&F 3301-A (tris-[N,N-diethylaminoethyl]phosphate trichloride), triparanol [1-((4-diethyl-aminoethoxy)phenyl)-1-(p-tolyl)-2-(p-chlorophenyl)-ethanol], and benzmalacene [N-(1-methyl-2,3-di-p-chlorophenylpropyl)maleamic acid] inhibit between IPP and squalene, but the precise site(s) have not been determined. All these substances except vanadium and the farnesoic acid derivatives also inhibit reactions subsequent to squalene in the cholesterol pathway.

2.4 CONVERSION OF GERANYLGERANYL PYROPHOSPHATE TO *ENT*-KAURENE

2.4.1 Enzymology

The cyclization of GGPP to *ent*-kaurene is a key point in the GA biosynthetic pathway. It is the step that commits GGPP to GA biosynthesis instead of carotenoids, phytol, or other products. The reactions are catalyzed by *ent*-kaurene synthetase in two stages (Fig. 2.7). The two stages, catalyzed by *ent*-kaurene synthetase activity A and activity B, were first predicted by Upper and West (1967) and later demonstrated by Shechter and West (1969) in extracts of *G. fujikuroi*, where cyclization can be stopped after the formation of copalyl phrophosphate (CPP). Small quantities of a product with similar chromatographic mobility have also been

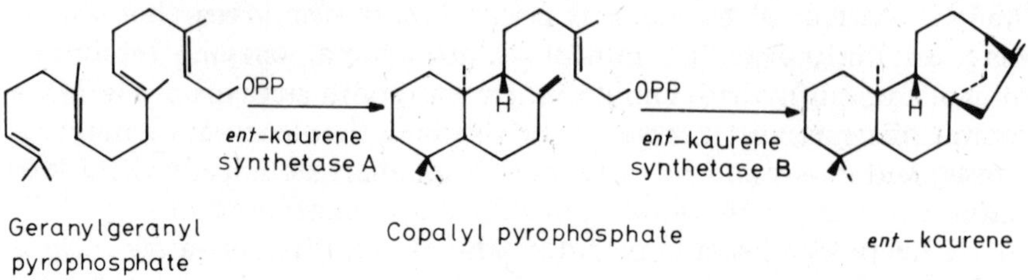

Geranylgeranyl pyrophosphate

Copalyl pyrophosphate

ent-kaurene

Figure 2.7   Cyclization of geranylgeranyl pyrophosphate to *ent*-kaurene.

observed in the *M. macrocarpus* system, and CPP is readily converted to *ent*-kaurene in extracts of *M. macrocarpus, R. communis* and *P. sativum* (Simcox et al., 1975), and *Helianthus annuus* (Shen-Miller & West, 1982), as well as cell cultures of *Lycopersicum esculatum* and *Nicotiana tobaccum* (Yafin & Shechter, 1975). It was not until quite recently, however, that activity A and activity B could be physically resolved (Duncan & West, 1981).

Ent-kaurene synthetase was purified 170-fold from cell-free extracts of *G. fujikuroi* (Fall & West, 1971). The molecular weight of the purified enzyme, which exhibited both A and B activities, was estimated at $4.3-4.9 \times 10^5$ daltons by gel filtration chromatography and sucrose density-gradient centrifugation. The pH optima of the A and B activities were 7.5 and 6.9, respectively. The presence of a divalent cation was also required for both activities; Mg^{2+} was the most effective cation tested. Ni^{2+} is capable of supporting only activity A, a feature that can be used to advantage in the accumulation of CPP. High levels of ATP were also useful in enhancing A activity over that of B. The addition of 20 mM 2-mercaptoethanol or 2.0 mM dithiothreitol gave a consistent enhancement of activity. The activity of the purified enzyme was stable in 25% glycerol at $-20°C$ for several months; however, repeated freezing and thawing caused loss of activity. The Km for GGPP in activity A was approximately 0.7 μM and that of CPP in activity B was approximately 1.0 μM. Kinetic studies with dual isotopes indicated preferential utilization of CPP generated from GGPP over that added exogenously (Fall & West, 1971).

Unlike the *G. fujikuroi* enzyme, *ent*-kaurene synthetase from *M. macrocarpus* could initially be purified only about 5-fold (Frost & West, 1977). The pH optima for both activities and the dependence on divalent cations were quite similar to those from *G. fujikuroi*. In the absence of substrate inhibition the average value of the apparent Km for GGPP was 1.6 μM. The Km for CPP was determined as

0.6 μM. Attempts to resolve the two activities from this source were also initially unsuccessful. Although no physical resolution of the A and B activities could be achieved with enzymes from either source, distinct differences in pH optima, divalent cation requirements, and sensitivity to sulfhydryl reagents were found for the two activities.

The recent separation of A and B activities from *M. macrocarpus* endosperm preparations by Duncan and West (1981) was achieved by ammonium sulfate fractionation followed by QAE Sephadex A-50 chromatography (Fig. 2.8) and polyacrylamide gel electrophoresis. The molecular weights of both enzymes as determined by gel filtration chromatography and sedimentation velocities were estimated at approximately 82,000 daltons. The authors confirmed the earlier finding that CPP derived from the active site of the A enzyme was preferentially channelled to the B enzyme rather than equilibrating with the free CPP in the system. Furthermore, a model was proposed for the efficient cyclization of GGPP to *ent*-kaurene by a noncovalently bonded complex of enzymes A and B which is in equilibrium with free A and B enzymes.

The third enzyme source from which *ent*-kaurene synthetase has been studied in some detail is *R. communis* seedlings (Robinson & West, 1970a, 1970b). This system is greatly complicated by the formation of five separate diterpene hydrocarbons from GGPP (Fig.

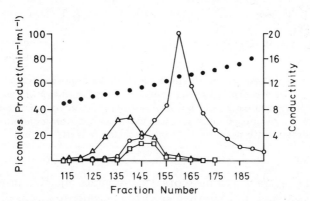

Figure 2.8 Resolution of *ent*-kaurene synthetase A and *ent*-kaurene synthetase B. A 45% ammonium sulfate fraction from the endosperm of *Marah macrocarpus* seeds containing 58 mg of protein was applied to a 400-ml bedvolume column of QAE Sephadex A-50. The column was eluted with a linear gradient of KCl in buffer, and 200 μl aliquots were assayed for 10 min for A activity ($\triangle$), B activity ($\circ$), and AB activity ($\square$). Conductivity in mMhos is indicated by ($\bullet$). (Redrawn, with permission, from Duncan & West, 1981.)

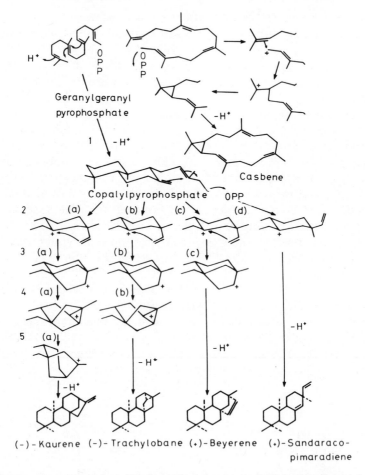

Figure 2.9 Biosynthesis of diterpene hydrocarbons from geranylgeranyl pyrophosphate. These schemes can account for the biosynthesis of casebene, *ent*-kaurene, (—)-trachylobane, (+)-beyerene, and (+)-sandaracopimaradine from geranylgeranyl pyrophosphate in cell-free extracts of *Ricinus communis* seedling shoots. (Redrawn, with permission, from Robinson & West, 1970b.)

2.9). A TLC system was developed to separate the five products on a single plate, and an extensive set of experiments was performed to characterize the components (Fig. 2.10). Primarily on the basis of the elution of separate activities from a DEAE-cellulose column, Robinson and West (1970b) suggested that *ent*-kaurene, (—)-trachylobane, (+)-beyerene, and (+)-sandaracopimaradiene were all derived from a common intermediate (CPP) by separate enzymic components.

RONALD C. COOLBAUGH   /   73

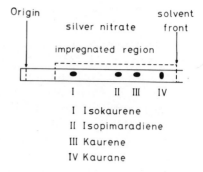

Figure 2.10 Separation of diterpene hydrocarbons by TLC. The top three-fourths of the thin-layer plate was impregnated with silver nitrate solution prior to chromatography in *n*-hexane/benzene (17:3). The positions of reference standards after chromatography are indicated. (Redrawn, with permission, from Robinson & West, 1970b.)

Casbene, which was later shown to be an antifungal agent (Sitton & West, 1975) whose production is induced by fungal attack (Dueber, Adolf, & West, 1978), is derived independently from GGPP.

According to the biogenetic theory of Ruzicka (1963), the cyclization of GGPP to *ent*-kaurene should involve the introduction of one hydrogen from the medium into the 3-position of CPP and *ent*-kaurene. The other 3-hydrogen of *ent*-kaurene is derived from the 4-pro-*R* hydrogen of MVA (Cornforth et al., 1966). Hanson and White (1969) have shown that the GA synthesized from $(4R)$-$[4$-^3H, 2-$^{14}C]$MVA does not contain ^3H at the 3-position whereas *ent*-7α-hydroxykaurenoic acid $19\rightarrow6$ lactone does. Using *ent*-$[3\beta$-^3H, 17-$^{14}C]$kaurene as a precursor to GA in cultures of *G. fujikuroi*, Dawson, Jefferies, and Knox (1975) have elucidated the stereochemistry of the 3-hydroxylation of *ent*-kaurene. This data combined with previously published information also permitted the assignment of the stereochemistry of the initial cyclization step. All available evidence suggests a proton-initiated cyclization with antiplanar addition to the terminal double bond with the polyene in an incipient chair conformation for the A-ring (Fig. 2.11). The stereochemistry of the conversion of copalyl pyrophosphate to *ent*-kaurene with respect to C-15 and C-17 has been determined in enzyme preparations from *M. macrocarpus* (Coates & Cavender, 1980).

Figure 2.11 A-ring stereochemistry in the cyclization of geranylgeranyl pyrophosphate derived from $[4R$-$^3H]$-mevalonic acid. (Redrawn, with permission, from Dawson et al., 1975.)

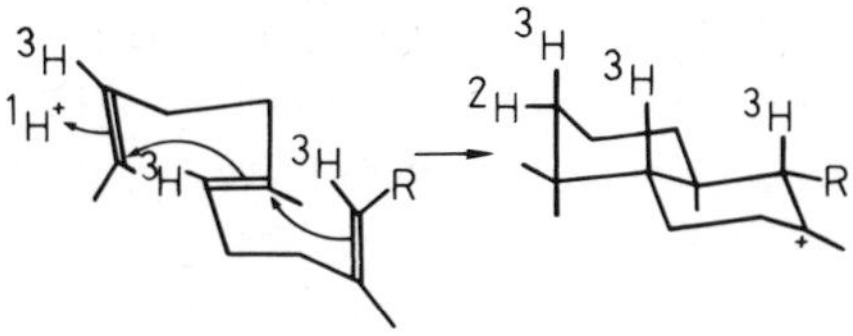

2.4.2   Compartmentation

Although it is well known that all enzymes involved in the incorporation of MVA into *ent*-kaurene can be obtained in the soluble fraction of tissue homogenates, the methods of preparation of these extracts normally do not exclude the possibility that these enzymes are derived partially or totally from cellular organelles. Stoddart (1968, 1969) was the first to suggest the involvement of chloroplast enzymes in GA biosynthesis when he reported the isolation of GA-like materials from chloroplast fractions and the incorporation of *ent*-kaurenoic acid into GA-like substances by sonicated chloroplasts of *Brassica oleracea* leaves. These findings coupled with the previously described data on the incorporation of MVA into GGPP in plastid extracts and additional studies on the localization of phytochrome in plastids and its apparent regulation of GA levels (Cooke & Saunders, 1975; Evans & Smith, 1976; Cooke, Saunders, & Kendrick, 1975; Railton & Reid, 1974) have led many to the conclusion that a significant portion of the GA biosynthesis occurs within proplastids or chloroplasts. Simcox et al. (1975) reported that *ent*-kaurene synthetase activity B is present in proplastids of *M. macrocarpus*, *P. sativum*, and *R. communis*. They considered the level of activity A to be insignificant. Moore and Coolbaugh (1976), on the other hand, demonstrated very low activity for the conversion of GGPP to *ent*-kaurene (AB activity) in sonicated chloroplasts of *Pisum* shoot tips. No activity was detected in intact chloroplast preparations. Quite recently West, Shen-Miller, and Railton (1982) have reported a stimulation of B activity by chloroplast membrane fractions, and Railton, Fellows, and West (personal communication) have extended these results using etioplasts of *Hordeum vulgare* and chloroplasts isolated from *Ph. coccineus*, *H. annuus*, and normal and *d-5* mutants of *Z. mays*. In these systems chloroplasts metabolized CPP, but not GGPP, to *ent*-kaurene.

2.4.3   Regulation

The data described above show clearly that *ent*-kaurene synthetase is the branch point in the pathway which commits the cell or organelle to the production of either GA or other alternative products. Depending upon the activity of *ent*-kaurene synthetase A, GGPP can be shunted to a variety of pathways leading, for instance, to carotenoids, quinones, phytyl esters, and macrocyclic diterpenes such as casbene. However, once GGPP is converted to CPP, the system is destined to produce *ent*-kaurene and/or one of a few very closely related diterpene hydrocarbons such as trachylobane, beyerene, and sandaracopimaradiene. The functions of these other diterpenes which are produced in *Ricinus* seedling preparations are not known, but these substances have been identified in other plants and are generally synthesized in lesser amounts than *ent*-kaurene. *Ent*-kaurene is a direct precursor of GAs in *G. fujikuroi* and in all higher plant systems investigated to date. The key position of *ent*-kaurene synthetase in the GA biosynthetic pathway suggests that it is likely to be subject to metabolic control. The paper by Duncan and West (1981) establishes a sound model for natural control at this site. *Ent*-kaurene synthetase B appears to be ubiquitous in the systems tested to date. AB activity, on the other hand, is present in only a few systems under the conditions tested. It is therefore quite feasible that the overall synthesis of GAs might be controlled by the relative activity of *ent*-kaurene synthetase A in a tissue or the equilibrium of A, B, and AB.

Another possible form of control in this pathway was suggested by Hedden and Phinney (1976, 1979). These authors reported that a cell-free extract from a *d-5* dwarf mutant of *Z. mays* synthesizes more *ent*-isokaurene than *ent*-kaurene (Figs. 2.12, 2.13). *Ent*-isokaurene is biologically inactive and is apparently not converted to GAs. In this system CPP serves as substrate for both *ent*-kaurene synthesis and *ent*-isokaurene synthesis. The regulation in this instance would appear to be of the form of *ent*-kaurene synthetase B. As noted above, this is also the site where CPP may be diverted to other diterpenes in the *R. communis* system. Small quantities of *ent*-isokaurene are also produced from MVA and GGPP in cell-free enzyme extracts of seedlings of *Z. mays* (Early Golden Giant Hybrid) infected with *Rhizopus stolonifer* (Mellon & West, 1979). *Ent*-isokaurene is not, however, among the products of GGPP or CPP metabolism in other cell-free systems tested to date. Indeed, *ent*-isokaurene was not found to be a product of CPP metabolism in chloroplast extracts of normal and *d-5* dwarf mutants of *Zea* (Railton, Fellows, and West, personal communication).

The results described below on more general cell-free enzyme systems that incorporate MVA into *ent*-kaurene are consistent with regulation of the pathway at the *ent*-kaurene synthetase level. Variations in the activities of the systems occur between dwarf and tall varieties of the same plant species, between tissues of the same plant, within a single tissue at different stages of development, and with different light treatments. Further investigations on these systems regarding the relative A, B, and AB activities are certainly needed. The discussion of inhibitors which follows this section is also pertinent to the regulation of this portion of the pathway.

Coates and co-workers (1976) have demonstrated that *ent*-kaurene synthetase is capable of functioning equally well with (*R,S*)-14,15-oxidogeranyl-geranyl pyrophosphate as with GGPP. Products of this reaction in soluble preparations of *M. macrocarpus* were primarily *ent*-3α- and *ent*-3β-hydroxykaurene. The initial rate of cyclization of the epoxy pyrophosphate was similar to that with GGPP. The *Km* values for the substrates were very similar, and GGPP was shown to be a competitive inhibitor of the cyclization of the epoxide. Thus, the same enzyme appears to be involved in both reactions. In the enzyme systems from *R. communis* and *G. fujikuroi*, 3-hydroxy products were also formed (Coates et al., 1976)

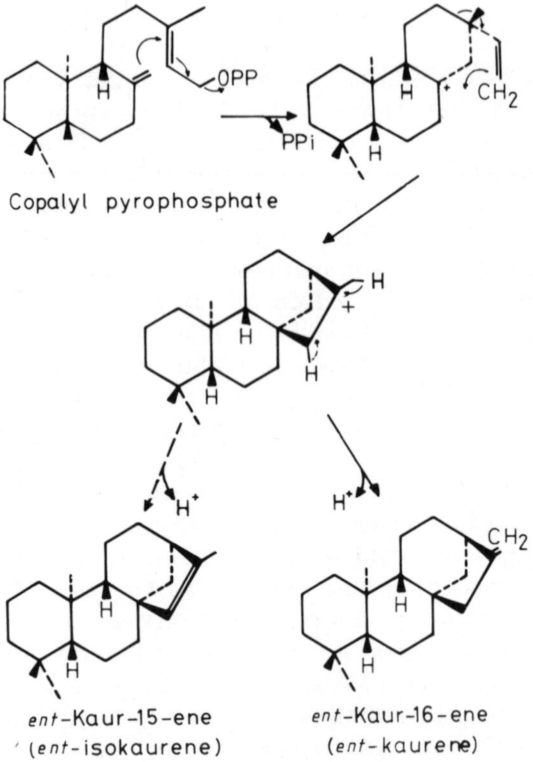

Copalyl pyrophosphate

ent-Kaur-15-ene
(*ent*-isokaurene)

ent-Kaur-16-ene
(*ent*-kaurene)

Figure 2.12 Cyclization of copalyl pyrophosphate to *ent*-kaurene or *ent*-isokaurene in cell-free extracts of normal and dwarf maize. (Courtesy of Hedden & Phinney, 1976; redrawn, with permission, from Moore, 1979.)

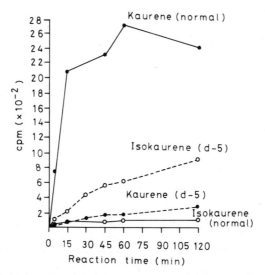

Figure 2.13   Comparative biosynthesis of *ent*-kaurene and *ent*-isokaurene from [³H]copalyl pyrophosphate in cell-free enzyme extracts prepared from normal and *d-5* dwarf mutants of *Zea mays*. (Courtesy of Hedden & Phinney, 1976; redrawn, with permission, from Moore, 1979.)

which illustrates the flexibility of the diterpene cyclase enzymes from a variety of sources. Furthermore, *ent*-3β-hydroxykaurene served as an effective substrate in preparations of *ent*-kaurene-oxidizing enzymes from *M. macrocarpus* (Hirano, 1976). These results suggested to Coates et al. (1976) the possibility of an alternative pathway to the GAs.

2.4.4   Inhibitors

A number of plant growth inhibitors affect *ent*-kaurene synthetase. One of the first instances of the identification of the mode of action of a growth regulator was the study by Dennis, Upper, and West (1965) which demonstrated the inhibition of *ent*-kaurene synthetase by AMO-1618, a quaternary ammonium compound (Fig. 2.14). It was later shown by Shechter and West (1969) that AMO-1618 and CCC (2-chloroethyltrimethylammonium chloride) block the initial cyclization of GGPP (activity A) in extracts of *G. fujikuroi*. Although CCC inhibited the incorporation of MVA into *ent*-kaurene in cell-free extracts and in cultures of *G. fujikuroi* (Shechter & West, 1969; Barnes, Light, & Lang, 1969), much higher

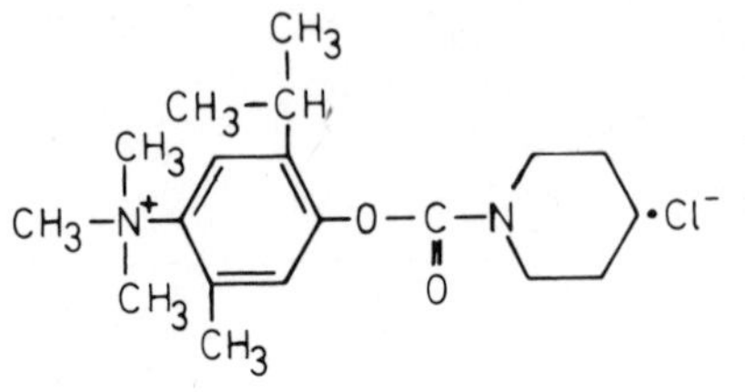

AMO-1618

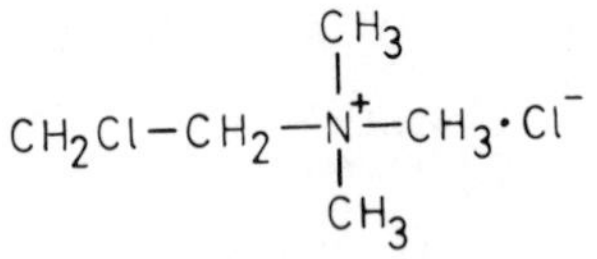

CCC

Figure 2.14 Structures of the quaternary ammonium plant growth inhibitors AMO-1618 and CCC (Cycocel).

concentrations of CCC than of AMO-1618 were required to inhibit activity in extracts of pea seeds (Anderson & Moore, 1967) and no inhibition was observed with CCC in extracts of *M. macrocarpus* (Dennis et al., 1965). Frost and West (1977), also working with *M. macrocarpus*, confirmed these observations with partially purified *ent*-kaurene synthetase. Phosphon D (tributyl-2,4-dichlorobenzyl-ammonium chloride) did not inhibit GA biosynthesis in *Gibberella* cultures (Kende, Ninnemann, & Lang, 1963), but it was later demonstrated that the fungus metabolizes this inhibitor (Harada & Lang, 1965). Neither AMO-1618 nor CCC significantly inhibit the conversion of CPP to *ent*-kaurene (*ent*-kaurene synthetase B). However, Phosphon D appears to be quite effective on synthetases A and B in extracts of *G. fujikuroi* (Shechter & West, 1969). Frost and West (1977) have carefully assessed the effects of a broad range of potential inhibitors, including GAs and abscisic acid, on the separate A and B activities in extracts of *M. macrocarpus* (Tables 2.1, 2.2).

Cho and co-workers (1979) have described the effects of a number of quaternary ammonium compounds on growth and GA biosynthesis of *G. fujikuroi*. Although all these compounds inhibited growth, they were quite variable in their effects on GA biosynthesis. The specificity of quaternary ammonium compounds and other plant growth inhibitors is not as great as once thought, since many of them have also been shown to inhibit sterol biosynthesis, and the growth inhibition in plants caused by these substances can be

Table 2.1.   Effects of inhibitors of *ent*-kaurene synthetase A activity

Inhibitor	Percent inhibition of activity at inhibitor concentration (M)				
	5×10^{-7}	1×10^{-6}	5×10^{-6}	5×10^{-5}	5×10^{-4}
SKF-525A	56	69	100	100	100
Phosphon D	20	44	100	100	100
Phosphon S	24	—	88	100	100
Q-64	17	—	84	100	100
Amo-1618	16	47	65	100	100
SKF-3301A	13	29	64	100	100
Q-58	0	30	60	100	100
Deoxycholate	5	—	35	75	100
Amchem	—	—	21	100	100
Carvadan	—	—	20	84	100
Delcosine	—	—	11	39	78
Q-53	—	—	0	10	79
Nicotine	—	—	0	11	65
CCC	—	—	0	0	4
Acetylcholine chloride	—	—	0	0	0
BCB	—	—	0	0	0
B995	—	—	3	4	1
SKF-7997	—	—	0	0	0
SKF-7732	—	—	3	0	3
Indole acetate	—	—	0	0	5
GA_1	—	—	0	0	2
GA_3	—	—	0	0	11
GA_4/GA_7	—	—	0	0	0
GA_{13}	—	—	0	0	12
Choline chloride	—	—	11	0	1
ABA	—	—	0	0	0

NOTE: Ent-kaurene synthetase A activity was measured by determining copalol and *ent*-kaurene formed after incubation of partially purified enzyme extracts from *Marah macrocarpus* endosperm with [^{14}C]geranylgeranyl pyrophosphate (1.0 μM), $MgCl_2$, and potential inhibitors at the concentrations indicated. (Frost & West, 1977)

Table 2.2. Effects of inhibitors on *ent*-kaurene synthetase B activity

Inhibitor	Percent inhibition of activity at inhibitor concentration (M)			
	5×10^{-6}	5×10^{-5}	5×10^{-4}	5×10^{-3}
Amchem	26	53	70	—
Deoxycholate	11	28	76	100
SKF-525A	8	28	63	—
SKF-3301A	2	38	52	—
Q-53	7	10	35	—
Amo-1618	11	12	—	—
Phosphon D	2	8	34	—
Acetylcholine chloride	12	6	9	9
ABA	0	13	24	—
Phosphon S	0	0	18	—
Indole acetate	0	5	0	—
GA$_1$	0	0	0	—
GA$_3$	0	0	12	—
GA$_4$/GA$_7$	0	0	7	—
GA$_{13}$	0	11	5	—
Q-64	6	0	0	—
Q-58	0	7	0	—
Carvadan	0	0	4	—
Delcosine	0	0	0	—
Nicotine	0	0	0	—
CCC	0	0	0	—
BCB	0	0	0	—
B995	2	0	0	0

NOTE: Ent-kaurene synthetase B activity was measured by determining *ent*-kaurene after incubation of partially purified enzyme extracts from *Marah macrocarpus* endosperm with [^{14}C] copalyl pyrophosphate (0.65 μM) MgCl$_2$, and potential inhibitors at the concentrations indicated. (Frost & West, 1977)

overcome by application of sterols in selected instances (see, for example, Newhall, 1969; Douglas & Paleg, 1974; 1981).

The presence of inhibitors of *ent*-kaurene synthetase A activity in crude extracts of plant tissues has been reported recently. Gafni and Shechter (1981a, 1981b) have isolated such a substance from extracts of *Ricinus* seedlings and cell suspension cultures, and Shen-Miller and West (1982) described an inhibitory activity from *Helianthus* seedlings. In both cases the AB activities were stimulated by dialysis of the extracts. Inhibitory activities in the extracts from *Ricinus* and *Helianthus* were also confirmed in these two studies by testing their effects on enzymes from *Gibberella* and *Marah*, respectively. The inhibitor from *Ricinus* was identified as diethylene glycol disulfide (Gafni & Shechter, 1981b). The possible roles of these factors in physiological regulation remain to be determined but, regardless of their physiological significance, they should be considered in the interpretation of data from cell-free enzyme systems.

2.5   *ENT*-KAURENE BIOSYNTHESIS AND PHYSIOLOGY

The preceding sections summarized data obtained from a number of plant systems on specific reactions in the pathway between MVA and *ent*-kaurene. Much research has also been done with cell-free enzyme systems to study the overall incorporation of MVA into *ent*-kaurene. The results of these investigations have provided additional information on the relationships between *ent*-kaurene biosynthesis and certain physiological activities.

2.5.1   *Ent*-kaurene biosynthesis and seed development

It is well known that seeds of many plants accumulate large quantities of GAs during development. Although the first GA to be identified from higher plants was GA_1 from *Ph. multiflorus* seeds (MacMillan & Suter, 1958), one of the first isolations of GA-like substances in higher plants was by West and Phinney (1956) using extracts of *M. macrocarpus* seeds. It was later shown by several investigators that the amounts of extractable GA-like substances in seeds increase during development and diminish as the seeds become mature (see Moore, 1979). Baldev, Lang, and Agatep (1967), in experiments that involved the injection of AMO-1618 into cultured pea fruits, showed that about 75% of the GA-like activity of the seeds was not needed for normal seed development (Table 2.3). Dennis and West (1967) reported that activity of *ent*-kaurene biosynthesis in *M. macrocarpus* seeds generally follows the same pattern, with maximum activity in the liquid endosperm prior to cotyledon

Table 2.3. Effects of AMO-1618 on growth and gibberellin content in excised developing *Pisum sativum* seeds

Conc. of AMO-1618 in medium (mg liter⁻¹)	Fresh weight of seed		GA content per seed	
	(g)	(%)	(μg)*	(%)
Before culture				
	0.0027		0.0015	
After 10 days of culture with AMO-1618				
	0.31	100	0.430	100
5	0.30	97	0.173	40
50	0.26	84	0.055	13
500	0.07	23	0.04	9

NOTE: Pods were excised, injected with solutions of AMO-1618 at the indicated concentrations, and grown in culture for 10 days. Ten seeds were used for each determination of weight and gibberellin-like activity, except before culture, when 35 seeds were used. (From B. Baldev et al., Science *147*, 155–156, 1965. Copyright 1965 by the American Association for the Advancement of Science)

*Results expressed as μg GA$_3$ equivalents.

expansion. Similar results were obtained by Coolbaugh and Moore (1969) for pea seeds in which the activity was present in the developing cotyledons; the activity reached a maximum when the seeds had achieved approximately half their maximum fresh weight (Fig. 2.15). The limiting character of *ent*-kaurene–synthesizing activity in comparison to activities of subsequent reactions in this pathway in pea seeds was further elucidated by Graebe (1980). The significance of the overabundance of GAs in developing seeds is not yet fully understood, but there is some evidence—albeit equivocal—that these GAs are conjugated and stored for release and activity during the early stages of seed germination (see Chapter 6).

2.5.2 *Ent*-kaurene biosynthesis and shoot development

The *ent*-kaurene biosynthesis capacity of tall peas, as measured in cell-free enzyme preparations from pea shoot tips, first appears after about 3 days of germination and reaches a maximum by the

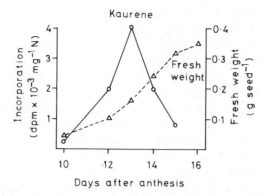

Figure 2.15 Changes in *ent*-kaurene biosynthesis during seed development in *Pisum sativum*. Net rates of *ent*-[^{14}C]kaurene biosynthesis were determined in extracts of seeds at different stages of development. (Redrawn, with permission, from Coolbaugh & Moore, 1969.)

eighth day (Moore & Ecklund, 1974). This is illustrated in Fig. 2.16. The lag period before the initiation of *ent*-kaurene synthesis could be correlated with the large quantities of GAs synthesized in developing seeds and the subsequent formation of conjugated GAs which may be released from the bound state early in germination.

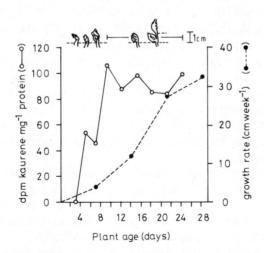

Figure 2.16 Ontogenic changes in growth rate and *ent*-kaurene-synthesizing capacity of *Pisum sativum* shoot tips, as measured in cell-free enzyme extracts. The drawings illustrate the shoot tips from which enzyme extracts were prepared. (Redrawn, with permission, from Moore & Ecklund, 1974.)

Cell-free enzyme systems from shoot tips of *P. sativum*, *R. communis*, and *Z. mays* have been described. In contrast to the highly active systems from developing seeds, shoot tips appear to be capable of synthesizing comparatively low quantities of *ent*-kaurene and GAs. Coolbaugh et al. (1973) reported that *P. sativum* shoot tips contain less than 1% of the synthesizing activity of developing seeds. Although this result is subject to limitations regarding the efficiency of extracting active enzymes from the two sources and the possible presence of inhibitory substances, it is generally consistent with the levels of endogenous GAs extracted from immature seed and seedling shoot tips.

Dwarfism in a number of plants can be overcome by addition of GAs. This fact implies that GA biosynthesis in dwarf varieties is impaired, although other factors such as changes in tissue sensitivities to GAs may also be involved. Studies of dwarfism in relation to *ent*-kaurene biosynthesis in *P. sativum* and *Z. mays* have produced different results. In *Z. mays*, where dwarfism is expressed in both light-grown and etiolated plants, Hedden and Phinney (1976, 1979) have reported that dwarf mutants shunt the major portion of *ent*-kaurene synthesis to *ent*-isokaurene (Fig. 2.13). The total activity for *ent*-kaurene plus *ent*-isokaurene in dwarf *Z. mays* is, however, considerably less than that found in the normal variety of *Zea mays*.

The situation is less clear cut in dwarf *P. sativum* which, in contrast to *Zea*, does not exhibit reduced stem elongation under etiolated conditions. Moore and Ecklund (1974) reported that enzyme extracts from a light-grown dwarf variety of *P. sativum* (Progress No. 9) shoots synthesize about half as much *ent*-kaurene as those from the light-grown tall variety (Alaska) (Table 2.4). A similar result was also obtained with extracts of developing *Pisum* seeds (Anderson & Moore, 1967). Under etiolated conditions the activity for *ent*-kaurene synthesis from shoot tips was reduced for both varieties, but a much greater reduction occurred in the tall cultivar. Thus, the etiolated dwarf actually exhibited more activity than the etiolated tall.

Investigations on the incorporation of [^{14}C]MVA into *ent*-kaurene in extracts of *P. sativum* seedlings have generally used the apical bud as a source of enzymes. Previously published data on extractable GAs and effects of applied GA$_3$ would suggest that this is a primary site of GA production. Quite recently, however, cell-free enzyme preparations from internodes and older leaves have been tested, and the rate of incorporation of MVA into *ent*-kaurene was actually higher in extracts of internodes immediately below the apical bud than in those from the bud itself (Coolbaugh, unpublished

Table 2.4. Comparison of *ent*-kaurene-synthesizing capacities of cell-free extracts from shoot tips of light- and dark-grown tall and dwarf *Pisum sativum*

Alaska		Progress No. 9	
Light-grown[a]	Etiolated[b]	Light-grown[c]	Etiolated[d]
dpm mg⁻¹ protein			
78	17	40	35
dpm g⁻¹ fresh wt			
944	170	558	350

NOTE: Seedlings from which enzyme extracts were prepared were 10–14 days old. Light-grown plants received 16 hr of light and 8 hr darkness per day. *Ent*-kaurene synthesis was measured after incuabtions of enzyme extracts with [^{14}C] mevalonic acid, ATP, $MgCl_2$, and $MnCl_2$ (Moore & Ecklund, 1974).
[a–d]Means of 5, 3, 4, and 1 independent experiments, respectively.

results). This was the case whether compared on a fresh weight basis or on a protein basis. Dialysis of the extracts stimulated the activity in all cases but did not change the relative levels of synthesis. Given the obvious limitations of using a crude cell-free enzyme extract as an indicator of *in vivo* activity, these data suggest that the rate of *ent*-kaurene biosynthesis in vegetative parts of peas are highest in tissues with the greatest immediate growth potential.

2.5.3   Effects of light on *ent*-kaurene biosynthesis

Some of the earliest reports on the isolation of GAs from higher plants indicated that light stimulates their formation and/or accumulation. Kohler (1965) reported that light-grown *P. sativum* seedlings contain 10 times more GA-like substances than etiolated plants. Although this result was not confirmed in an extensive investigation of both diffusible and extractable GAs from *P. sativum* seedlings (Jones & Lang, 1968), many other investigations point to the involvement of light in this pathway in higher plants (Kohler, 1966; Radley, 1963; Reid et al., 1972; Crozier & Audus, 1968; Zeevart, 1973; Kopcewicz, 1971; Reid, Clements, & Carr, 1968). Mertz (1970) has also reported light-enhanced incorporation of

L-leucine into GAs in cultures of *G. fujikuroi*. In addition, there are many papers describing evidence that parts of the GA biosynthetic pathway occur within the chloroplast, the primary light-capturing organelle. Ecklund and Moore (1974) using a cell-free enzyme system from *Pisum* seedlings which incorporates MVA into *ent*-kaurene have shown that *ent*-kaurene-synthesizing activity is induced to a maximum level with 12 hr of de-etiolation (Fig. 2.17). There was a 6.3-fold increase during the illumination period. Incubation of 10-day-old *Pisum* shoot tips in solutions of chloramphenicol and cycloheximide resulted in a reduction of the light-stimulated synthesis of *ent*-kaurene from MVA in subsequently prepared enzyme extracts (Gomez-Navarrete & Moore, 1978). This result suggests photoinduction of one or more of the enzymes involved in the biosynthesis of *ent*-kaurene from MVA. More recently, Choinski and Moore (1980) examined the action spectrum of the photo-induction of *ent*-kaurene biosynthesizing capacity using the same cell-free enzyme system. Their result is illustrated in Fig. 2.18. The action spectrum has distinct peaks in the red and the blue regions of the spectrum. Further experiments showed that far-red light was approximately half as effective as red light in inducing *ent*-kaurene-synthesizing activity. Red followed by far-red light gave the same effect as far-red light alone. Thus, the red light effect was partially photoreversible. The potential involvement of phytochrome in this

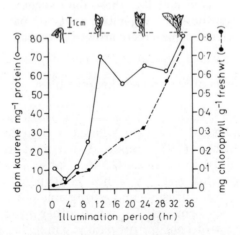

Figure 2.17 Changes in kaurene-synthesizing capacity and chlorophyll content during de-etiolation of 10-day-old seedlings of *Pisum sativum*. Drawings illustrate shoot tips from which enzyme and chlorophyll extracts were prepared. (Redrawn, with permission, from Moore & Ecklund, 1974.)

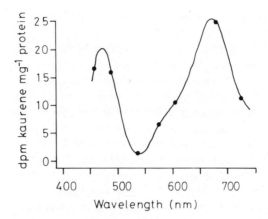

Figure 2.18 Action spectrum for photoinduction of *ent*-kaurene biosynthesis. *Ent*-kaurene synthesis was measured in extracts of etiolated *Pisum sativum* seedlings which were treated with 8 hr illumination $(2.4 \times 10^5 \, \text{ergs cm}^{-2} \text{s}^{-1})$ at the indicated wavelengths. (Redrawn, with permission, from Choinski & Moore, 1980.)

biosynthetic pathway has been predicted for some time. Evidence is slowly accumulating in support of this hypothesis, but further research is needed to clarify this important aspect of GA biosynthesis.

2.6 SUMMARY

The following features regarding the pathway of *ent*-kaurene bio-synthesis appear to be known with some degree of certainty.

1. In all systems tested to date, the enzymes that catalyze the incorporation of MVA into *ent*-kaurene are derived from the "soluble" (100,000 × g supernatant) fraction of homogenates.
2. The three initial enzymes in this pathway require ATP and a divalent cation for activity. At least one of these enzymes, probably MVAPP decarboxylase, is controlled by adenylate energy charge. It is severely inhibited by ADP.
3. *Ent*-kaurene biosynthesis may be a rate-limiting step in the production of GA. There is a much greater capacity for both *ent*-kaurene and GA biosynthesis in seed tissues than in vegetative tissues. In *P. sativum* seeds *ent*-kaurene-synthesizing activity is optimum when the seeds achieve approximately half their

maximum fresh weight; extractable GA follows a similar curve but delayed by about one day.

4. Dwarf and tall *Pisum* and *Zea* differ in their abilities to synthesize *ent*-kaurene. In light-grown *P. sativum* the rate of *ent*-kaurene biosynthesis is twice as great in cell-free extracts of the tall variety, Alaska, as in the dwarf variety, Progress No. 9. In *Z. mays*, there appears to be a different mechanism. *Ent*-kaurene is formed in the normal varieties, while *ent*-isokaurene, which is a dead-end product that is not converted into GAs, is preferentially synthesized in seedlings of the *d-5* dwarf mutant.

5. Enzymes involved in the conversion of MVA to *ent*-kaurene are localized in the liquid endosperm tissue of cucurbits and the cotyledons of *Pisum* seeds. There is also evidence that they are located in chloroplasts and etioplasts of the same plant materials.

6. *Ent*-kaurene synthesizing capacity in pea seedlings is greatly enhanced by light treatment. Full activity is achieved after 12 hr of de-etiolation. The action spectrum for this activity has peaks in the red and blue regions of the visible spectrum.

7. Several interesting inhibitors exist which will block one or more of the steps in this pathway. These have proved useful in studying many aspects of GA biosynthesis and physiology.

8. The recent separation of *ent*-kaurene synthetases A and B may lead to new information on the regulation of this pathway.

9. The oxidation of *ent*-kaurene and its further conversion to GAs are discussed in Chapter 3. The movement of reaction products from the "soluble" fraction to the "particulate" fraction and back again in successive reactions promises to be one of the more interesting aspects of the GA biosynthetic pathway.

ACKNOWLEDGMENTS

The author expresses appreciation to Drs. Alan Crozier, Peter Hedden, and Duane Isely for their constructive suggestions regarding this manuscript and to Dr. Charles A. West for providing a preprint of a paper.

REFERENCES

Anderson, D. G., & Porter, J. W. (1962). The biosynthesis of phytoene and other carotenes by enzymes of isolated higher plant plastids. Arch. Biochem. Biophys. *97*, 509–519.

Anderson, J. D., & Moore, T. C. (1967). Biosynthesis of (—)-kaurene in cell-free extracts of immature pea seeds. Plant Physiol. *42*, 1527–1534.

Agranoff, B. B., Eggerer, H., Henning, U., & Lynen, F. (1960). Biosynthesis of terpenes: VII. Isopentenyl pyrophosphate isomerase. J. Biol. Chem. *235*, 326–332.

Baldev, B., Lang, A., & Agatep, A. O. (1967). Gibberellin production in pea seeds developing in excised pods: Effect of growth retardant AMO-1618. Science *147*, 155–156.

Barnes, M. F., Light, E. N., & Lang, A. (1969). The action of plant growth retardants on terpenoid biosynthesis. Planta *88*, 172–182.

Beytia, E. D., & Porter, J. W. (1976). Biochemistry of polyisoprenoid biosynthesis. Ann. Rev. Biochem. *45*, 113–143.

Birch, A. J., Richards, R. W., & Smith, H. (1958). The biosynthesis of gibberellic acid. Proc. Chem. Soc. London, 192–193.

Birch, A. J., Richard, R. W., Smith, H., Harris, A., & Whalley, W. B. (1959). Studies in relation to biosynthesis: XII. Rosenonolactone and gibberellic acid. Tetrahedron 7, 241–251.

Buggy, M. J., Britton, G., & Goodwin, T. W. (1974). Terpenoid biosynthesis by chloroplasts isolated in organic solvents, Phytochem. *13*, 125–129.

Ceccarelli, N., Lorenzi, R., Alpi, A. (1979). Kaurene and kaurenol biosynthesis in cell-free systems of *Phaseolus coccineous* suspensor. Phytochem. *18*, 1657–1658.

Ceccarelli, N., Lorenzi, R., & Alpi, A. (1981). Kaurene metabolism in cell-free extracts of *Phaseolus coccineus* suspensor. Plant Sci. Lett. *21*, 325–332.

Charlton, J. M., Treharne, K. J., & Goodwin, T. W. (1967). Incorporation of [2-^{14}C] mevalonic acid into phytoene by isolated chloroplasts. Biochem. J. *105*, 205–212.

Cho, K. Y., Sakurai, A., Kamiya, Y., Takahashi, N., & Tamura, S. (1979). Effects of the new plant growth retardants of quaternary ammonium iodides on gibberellin biosynthesis in *Gibberella fujikuroi*. Plant Cell Physiol. *20*, 75–81.

Choinski, J. S., & Moore, T. C. (1980). Relationship between chloroplast development and *ent*-kaurene biosynthesis in peas. Plant Physiol. *65*, 1031–1035.

Coates, R. M., & Cavender, P. L. (1980). Stereochemistry of the enzymatic cyclization of copalyl pyrophosphate to kaurene in enzyme preparations from *Marah macrocarpus.* J. Amer. Chem. Soc. *102*, 6358-6359.

Coates, R. M., Conradi, R. A., Ley, D. A., Akeson, A., Harado, J., Lee, S. C., & West, C. A. (1976). Enzymic cyclization of (R,S)-14,15 oxidogeranyl-geranyl pyrophosphate to 3α- and 3β-hydroxykaurene. J. Amer. Chem. Soc. *98*, 4659-4661.

Cooke, R. J. (1977). Mevalonate-activating enzymes in wheat etioplasts. New Phytol. *78*, 91-94.

Cooke, R. J., & Saunders, P. F. (1975). Phytochrome mediated changes in extractable gibberellin activity in a cell-free system from etiolated wheat leaves. Planta *123*, 299-302.

Cooke, R. J., Saunders, P. F., & Kendrick, R. E. (1975). Red light induced production of gibberellin-like substances in homogenates of etiolated wheat leaves and in suspensions of intact etioplasts. Planta *124*, 319-328.

Coolbaugh, R. C., & Moore, T. C. (1969). Apparent changes in rate of kaurene biosynthesis during the development of pea seeds. Plant Physiol. *44*, 1364-1367.

Coolbaugh, R. C., & Moore, T. C. (1971). Localization of enzymes catalyzing kaurene biosynthesis in immature pea seeds. Phytochem. *10*, 2395-2400.

Coolbaugh, R. C., Moore, T. C., Barlow, S. A., & Ecklund, P. R. (1973). Biosynthesis of *ent*-kaurene in cell-free extracts of *Pisum sativum* shoot tips. Phytochem. *12*, 1613-1618.

Cornforth, J. W., Cornforth, R. H., Donninger, C., & Popjack, G. (1966). Studies on the biosynthesis of cholesterol: XIX. Steric course of hydrogen elimina-tions and of C—C bond formations in squalene biosynthesis. Proc. Royal Soc. (B) *163*, 492.

Cross, R. E., Galt, R. H. B., & Hansen, J. R. (1964). The biosynthesis of the gibberellins. Part I. (—)-kaurene as a precursor of gibberellic acid. J. Chem. Soc. 295-300.

Cross, B. E., Galt, R. H. B., Hanson, J. R., & Klyne, W. (1962). Some new meta-bolities of *Gibberella fujikuroi* and the stereochemistry of (—)-kaurene. Tetrahedron Lett. *4*, 145-150.

Crozier, A., & Audus, L. J. (1968). Distribution of gibberellin-like substances in light- and dark-grown seedlings of *Phaseolus multiflorus.* Planta *83*, 207-217.

Davies, B. H., Rees, A. F., & Taylor, R. F. (1975). Preparation of labelled terpenyl pyrophosphates using extracts of barley seed embryos. Phytochem. *14*, 719-722.

Dawson, R. M., Jefferies, P. R., & Knox, J. R. (1975). Cyclization and hydroxylation in the biosynthesis of gibberellic acid. Phytochem. *14*, 2593-2597.

Dennis, D. T., Upper, C. D., & West, C. A. (1965). An enzymic site of inhibition of gibberellin biosynthesis by AMO-1618 and other plant growth retardants. Plant Physiol. *40*, 948-952.

Dennis, D. T., & West, C. A. (1967). Biosynthesis of gibberellins. III. The conversion of (−)-kaurene-19-oic acid in endosperm of *Echinocystis macrocarpa* Greene. J. Biol. Chem. *212*, 3293-3300.

Dorsey, J. K., & Porter, J. W. (1968). The inhibition of mevalonic kinase by geranyl and farnesyl pyrophosphates. J. Biol. Chem. *243*, 4667-4670.

Douglas, T. J., & Paleg, L. G. (1972). Inhibition of sterol biosynthesis by 2-isopropyl-4-dimethylamino-5-methylphenyl-1-piperdine carboxylate methyl chloride in tobacco and rat liver preparations. Plant Physiol. *49*, 417-420.

Douglas, T. J., & Paleg, L. G. (1974). Plant growth retardants as inhibitors of sterol biosynthesis in tobacco seedlings. Plant Physiol. *54*, 238--245.

Douglas, T. J., & Paleg, L. G. (1978). Amo-1618 effects on incorporation of ^{14}C-MVA and ^{14}C-acetate into sterols in *Nicotiana* and *Digitalis* seedlings and cell-free preparations from *Nicotiana*. Phytochem. *17*, 713-718.

Douglas, T. J., & Paleg, L. G. (1981). Inhibition of sterol biosynthesis and stem elongation of tobacco seedlings induced by some hypocholesterolemic agents. J. Exp. Bot. *32*, 59-68.

Downing, M. R., & Mitchell, E. D. (1974). Metabolism of mevalonic acid to phosphorylated intermediates in a cell-free extract from *Nepeta cataria* leaves. Phytochem. *13*, 1419-1421.

Dueber, M. T., Adolf, W., & West, C. A. (1978). Biosynthesis of the diterpene phytoalexin casbene: Partial purification and characterization of casbene synthetase from *Ricinus communis*. Plant Physiol. *62*, 598-603.

Duncan, J. D., & West, C. A. (1981). Properties of kaurene synthetase from *Marah macrocarpus* endosperm: Evidence for the participation of separate but interacting enzymes. Plant Physiol. *68*, 1128-1134.

Eberhardt, N. L., & Rilling, H. C. (1975). Prenyl transferase from *Saccharomyces cerevisiae*: Purification to homogeneity and molecular properties. J. Biol. Chem. *250*, 863–866.

Ecklund, P. R., & Moore, T. C. (1974). Correlations of growth rate and de-etiolation with rate of *ent*-kaurene biosynthesis in pea (*Pisum sativum* L.). Plant Physiol. *53*, 5–10.

Evans, R., & Hanson, J. R. (1972). The formation of (—)-kaurene in a cell-free system from *Gibberella fujikuroi*. J. Chem. Soc., Perkin Trans. I, *18*, 2382–2385.

Evans, A., & Smith, H. (1976). Localization of phytochrome in etioplasts and its regulation *in vitro* of gibberellin levels. Proc. Nat. Acad. Sci. *73*, 138–142.

Fall, R. R., & West, C. A. (1971). Purification and properties of kaurene synthetase from *Fusarium moniliforme*. J. Biol. Chem. *246*, 6913–6928.

Fimognari, G. M., & Rodwell, V. W. (1965). Cholesterol biosynthesis: Mevalonate synthesis inhibited by bile salts. Science *147*, 1038.

Flint, A. P. F. (1970). The activity and kinetic properties of mevalonate kinase in super-ovulated rat ovary. Biochem. J. *120*, 145–150.

Frost, R. G., & West, C. A. (1977). Properties of kaurene synthetase from *Marah macrocarpus*. Plant Physiol. *59*, 22–29.

Frydman, V. M., & MacMillan, J. (1973). Identification of gibberellins A_{20} and A_{29} in seed of *P. sativum* cv. Progress No. 9 by combined gas chromatography-mass spectrometry. Planta *115*, 11–15.

Frydman, V. M., & MacMillan, J. (1975). The metabolism of gibberellins A_9, A_{20}, A_{29} in immature seeds of *Pisum sativum* cv. Progress No. 9. Planta *125*, 181–195.

Frydman, V. M., & MacMillan, J. (1976). Gibberellins in developing seed of *Pisum sativum* cv. Progress No. 9. ACTA Univer. Nicolai Copernici Biol. XVIII, Nauk. Matemat.-Prsyrod. Zesyt 37.

Gafni, Y., & Shechter, I. (1981a). Isolation of a kuarene synthetase inhibitor from castor bean seedlings and cell suspension cultures. Plant Physiol. *67*, 1169–1173.

Gafni, Y., & Shechter, I. (1981b). Diethylene glycol disulfide from castor bean cell suspension cultures. Phytochem. *20*, 2477–2479.

Garcia-Peregrin, E., Suarez, M. D., & Mayor, F. (1973). Isolation of two fractions with mevalonate kinase activity from *Pinus pinaster* and *Agave americana.* FEBS Lett. *30*, 15-17.

Gomez-Navarrete, G., & Moore, T. C. (1978). Effects of protein synthesis inhibitors on *ent*-kaurene biosynthesis during photomorphogenesis of etiolated pea seedlings. Plant Physiol. *61*, 889-892.

Goodwin, T. W., & Mercer, E. I. (1963). The regulation of sterol and carotenoid metabolism in germinating seedlings. Symp. Biochem. Soc. *24*, 37.

Graebe, J. E. (1969). Enzymic preparation of ^{14}C-kaurene. Planta. *85*, 171-174.

Graebe, J. E. (1980). GA-biosynthesis: development and application of cell-free systems for biosynthetic studies. In: Plant Growth Substances 1979, pp. 180-187, Skoog, F., ed. Springer-Verlag, Berlin, Heidelberg, New York.

Graebe, J. E., Dennis, D. T., Upper, C. D., & West, C. A. (1965). Biosynthesis of gibberellins: I. The biosynthesis of (−)-kaurene, (−)-kauren-19-ol, and *trans*-geranylgeraniol in endosperm nucellus of *Echinocystis macrocarpa* Greene. J. Biol. Chem. *240*, 1847-1854.

Graebe, J. E., Hedden, P., Gaskin, P., & MacMillan, J. (1974). The biosynthesis of a C_{19} gibberellin from mevalonic acid in a cell-free system from a higher plant. Planta. *120*, 307-309.

Gray, J. C., & Kekwick, R. G. O. (1973a). Mevalonate kinase in green leaves and etiolated cotyledons of the french bean *Phaseolus vulgaris*. Biochem. J. *133*, 335-347.

Gray, J. C., & Kekwick, R. G. O. (1973b). An assessment of some molecular parameters of mevalonate kinase from plant and animal sources. Arch. Biochem. Biophys. *159*, 458-462.

Green, T. R., Dennis, D. T., & West, C. A. (1975). Compartmentation of iso-pentenyl pyrophosphate isomerase and prenyl transferase in developing castor bean endosperm. Biochem. Biophys. Res. Commun. *64*, 976-982.

Green, T. R., & West, C. A. (1974). Purification and characterization of two forms of geranyl transferase from *Ricinus communis*. Biochem. *13*, 4720-4729.

Grob, E. C., Kirschner, K., & Lynen, F. (1961). Neues über die biosynthese der carotinoide. Chimia. *15*, 308-310.

Hanson, J. R. (1977). The biosynthesis of C_5-C_{20} terpenoid compounds. In: Biosynthesis, Vol 5, pp. 56-75. Specialist periodical reports, Chemical Society, London.

Hanson, J. R., & White, A. F. (1969). Studies in terpenoid biochemistry. Part IV. Biosynthesis of the kaurenolides and gibberellic acid. J. Chem. Soc. (C), 981–985.

Harada, H., & Lang, A. (1965). Effect of some 2-chloroethyltrimethylammonium chloride analogs and other growth retardants on gibberellin biosynthesis in *Fusarium moniliforme*. Plant Physiol. *40*, 176–183.

Hedden, P., MacMillan, J., & Phinney, B. O. (1978). The metabolism of the gibberellins. Ann. Rev. Plant Physiol. *29*, 149–192.

Hedden, P., & Phinney, B. O. (1976). The dwarf-5 mutant of *Zea mays:* A genetic lesion controlling the cyclization step (B activity) in kaurene biosynthesis. In: Collected Abstracts of the Paper—Demonstrations of the 9th International Conference on Plant Growth Substances, Lausanne, pp. 136.

Hedden, P., & Phinney, B. O. (1979). Comparison of *ent*-kaurene and *ent*-isokaurene synthesis in cell-free systems from etiolated shoots of normal and dwarf-5 maize seedlings. Phytochem. *18*, 1475–1479.

Hirano, S. S. (1976). Substrate specificity of the microsomal oxidases of *Echinocystis macrocarpa* endosperm. Ph.D. Thesis. University of California, Los Angeles.

Holmes, W. L., & DiTullio, N. W. (1962). Inhibitors of cholesterol biosynthesis which act at or beyond the mevalonic acid stage. Amer. J. Clin. Nutr. *10*, 310–322.

Jones, R. L., & Lang, A. (1968). Extractable and diffusable gibberellins from light- and dark-grown pea seedlings. Plant Physiol. *43*, 629–634.

Jungalwala, F. B., & Porter, J. W. (1967). Biosynthesis of phyloene from isopentenyl and farnesyl pyrophosphates by a partially purified tomato enzyme system. Arch. Biochem. Biophys. *119*, 209–219.

Kende, H., Ninnemann, H., & Lang, A. (1963). Inhibition of gibberellic acid biosynthesis in *Fusarium moniliforme* by AMO-1618 and CCC. Naturwissenschaften *18*, 599–600.

Knotz, J., Coolbaugh, R. C., & West, C. A. (1977). Regulation of the biosynthesis of *ent*-kaurene from mevalonate in the endosperm of immature *Marah macrocarpus* seeds by adenylate energy charge. Plant Physiol. *60*, 81–85.

Kohler, D. (1965). The effect of weak red light and chlorocholine/chloride on the gibberellin-content of normal pea seedlings, and the cause of the different sensibility of dwarf and normal pea seedlings toward the endogenous gibberellin. Planta *67*, 44-54.

Kohler, D. (1966). Die abhangigkeit der gibberellin-produktion von normalerbsen vom phytochrom-system. Planta *69*, 27-33.

Kopcewicz, J. (1971). Effect of white light irradiation on the endogenous growth regulators content in seeds and seedlings of pine (*Pinus silveestris* L.). Acta Soc. Botan. Palan. XL, 432-438.

Lang, A. (1970). Gibberellins: structure and metabolism. Ann. Rev. Plant Physiol. *21*, 537-570.

Loomis, W. D., & Battaile, J. (1963). Biosynthesis of terpenes. III. Mevalonic kinase from higher plants. Biochim. Biophys. Acta. *67*, 54-63.

Loomis, W. D., & Croteau, R. (1980). Biochemistry of terpenoids. In: The Biochemistry of Plants. A Comprehensive Treatise. Vol. 4, Lipids—Structure and Function. pp. 363-418, Stumpf, P. K., & Conn, E. E., ed. Academic Press, New York.

MacMillan, J., & Suter, P. J. (1958). The occurrence of gibberellin A_1 in higher plants: Isolation from the seed of runner bean (*Phaseolus multiflorus*). Naturwissenschaften *45*, 46–47.

Mellon, J. E., & West, C. A. (1979). Diterpene biosynthesis in maize seedlings in response to fungal infection. Plant Physiol. *64*, 406-410.

Mertz, D. (1970). Light stimulated incorporation of L-leucine into the gibberellin of *Gibberella fujikuroi.* Plant Cell Physiol. *11*, 273-279.

Milborrow, B. V. (1974). Biosynthesis of abscisic acid by a cell-free system. Phytochem. *13*, 131-136.

Moore, T. C. (1979). Biochemistry and Physiology of Plant Hormones. Springer-Verlag, Berlin, Heidelberg, New York.

Moore, T. C., & Coolbaugh, R. C. (1976). Conversion of geranylgeranyl pyrophosphate to *ent*-kaurene in enzyme extracts of sonicated chloroplasts. Phytochem. *15*, 1241-1247.

Moore, T. C., & Ecklund, P. R. (1974). Biosynthesis of *ent*-kaurene in cell-free extracts of pea shoots. In: Plant Growth Substances 1973, pp. 252-259. Hirokawa Publishing, Tokyo.

Nandi, D. L., & Porter, J. W. (1964). The enzymic synthesis of geranylgeranyl pyrophosphate by enzymes of carrot root and pig liver. Arch. Biochem. Biophys. *105*, 7-19.

Newhall, W. F. (1969). Correlation of pseudocholinesterase inhibition and plant growth by quaternary ammonium derivatives of (+)-limonene. Nature *223*, 965-966.

Ogura, K., Nishino, T., & Seto, S. (1968). The purification of prenyltransferase and isopentenyl pyrophosphate isomerase of pumpkin fruit and their some properties. J. Biochem. Tokyo *64*, 197-203.

Ogura, K., Shinka, T., & Seto, S. (1972). The purification and properties of geranylgeranyl pyrophosphate synthetase from pumpkin fruit. J. Biochem. Tokyo *72*, 1101-1110.

Oster, M. O., & West, C. A. (1968). Biosynthesis of trans-geranylgeranyl pyrophosphate in endosperm of *Echinocystis macrocarpa* Greene. Arch. Biochem. Biophys. *127*, 112-123.

Radley, M. (1956). Occurrence of substances similar to gibberellic acid in higher plants. Nature *178*, 1070-1071.

Radley, M. (1963). Gibberellin content of spinach in relation to photoperiod. Ann. Bot. *27*, 373-377.

Railton, I. D., & Reid, D. M. (1974). Studies on gibberellins in shoots of light grown peas: I. A re-evaluation of the data. Plant Sci. Lett. *2*, 157-163.

Railton, I. D., Reid, D. M., Gaskin, P., & MacMillan, J. (1974). Characterization of abscisic acid in chloroplasts of *Pisum sativum* L. cv. Alaska by combined gas chromatography-mass spectrometry. Planta *117*, 179.

Ramachandran, C. K., & Shaw, S. M. (1976). Decarboxylation of mevalonate pyrophosphate is one rate-limiting step in hepatic cholesterol synthesis in suckling and weaned rats. Biochem. Biophys. Res. Commun. *69*, 42-47.

Reid, D. M., Clements, J. B., Carr, J. D. (1968). Red light induction of gibberellin synthesis in leaves. Nature *217*, 580-582.

Reid, D. M., Tuing, M. S., Durley, R. C., & Railton, I. D. (1972). Red-light-enhanced conversion of tritiated gibberellin A₉ into other giberellin-like substances. Planta *108*, 67-75.

Robinson, D. R., & West, C. A. (1970a). Biosynthesis of cyclic diterpenes in extracts from seedlings of *Ricinus communis* L. I. Identification of diterpene hydrocarbons formed from mevalonate. Biochem. *9*, 70-79.

Robinson, D. R., & West, C. A. (1970b). Biosynthesis of cyclic diterpenes in extracts from seedlings of *Ricinus communis* L. II. Conversion of geranylgeranyl pyrophosphate into diterpene hydrocarbons and partial purification of the cyclization. Biochem. *9*, 80–89.

Rogers, L. J., Shah, S. P. J., & Goodwin, T. W. (1966a). Intracellular localization of mevalonate-activating enzyme in plant cells. Biochem. J. *100*, 14C–17C.

Rogers, L. J., Shah, S. P. J., Goodwin, T. W. (1966b). Mevalonate-kinase isoenzymes in plant cells. Biochem. J. *99*, 381–388.

Ropers, H. J., Graebe, J. E., Gaskin, P., & MacMillan, J. (1978). Gibberellin biosynthesis in a cell-free system from immature seeds of *Pisum sativum*. Biochem. Biophys. Res. Commun. *80*, 690–697.

Ruzicka, L. (1963). Perspektiven der biogenase und der chemie der terpene. Pure Appl. Chem. *6*, 493.

Shechter, I., & West, C. A. (1969). Biosynthesis of gibberellins. IV. Biosynthesis of cyclic diterpenes from trans-geranylgeranyl pyrophosphate. J. Biol. Chem. 3200–3209.

Shen-Miller, J., West, C. A. (1982). *Ent*-kaurene biosynthesis in extracts from *Helianthus annuus* seedlings. Plant Physiol. *69*, 637–641.

Shewry, P. R., Pinfield, N. J., & Stobart, A. K. (1974). Effect of gibberellic acid on mevalonate activation in germinating *Corylus avellana* seeds. Phytochem. *13*, 341–346.

Shewry, P. R., & Stobart, A. K. (1973). Properties of castor bean mevalonic acid kinase. Plant Sci. Lett. *1*, 473–477.

Simcox, P. D., Dennis, D. T., & West, C. A. (1975). Kaurene synthetase from plastids of developing plant tissues. Biochem. Biophys. Res. Commun. *66*, 166–172.

Sitton, D., & West, C. A. (1975). Casbene: An antifungal diterpene produced in cell-free extracts of *Ricinus communis* seedlings. Phytochem. *14*, 1921–1925.

Skilleter, D. M., & Kekwick, R. G. O. (1971). The enzymes forming isopentenyl pyrophosphate from 5-phosphomevalonate (mevalonate 5-phosphate) in the latex of *Hevea brasiliensis*. Biochem. J. *124*, 407–417.

Skilleter, D. N., Williamson, I. P., & Kekwick, R. G. O. (1966). Phosphomevalonate kinase from *Hevea brasiliensis* latex. Biochem. J. *98*, 27.

Spurgeon, S. L., & Porter, J. W. (1980). Carotenoids. In: The Biochemistry of Plants, Vol. 4, pp. 446-469, Stumpf, P. K., & Conn, E. E. eds. Academic Press, New York.

Stoddart, J. L. (1968). The association of gibberellin-like activity with the chloroplast fraction of leaf homogenates. Planta *81*, 106-112.

Stoddart, J. L. (1969). Incorporation of kaurenoic acid into gibberellins by chloroplast preparations of *Brassica oleracea*. Phytochem. *8*, 831-837.

Upper, C. D., & West, D. A. (1967). Biosynthesis of gibberellins. II. Enzymic cyclization of geranylgeranyl pyrophosphate to kaurene. J. Biol. Chem. *242*, 3285-3292.

Wellburn, A. R., & Hampp, R. (1976). Uptake of mevalonate and acetate during plastid development. Biochem. J. *158*, 231-233.

West, C. A., & Phinney, B. O. (1956). Properties of gibberellin-like factors from extracts of higher plants. Plant Physiol. Suppl. *31*, XX.

West, C. A., Shen-Miller, J., & Railton, I. D. (1982). Regulation of kaurene synthetase. In: Plant Growth Substances, 1982, pp. 81-90, Wareing, P. F. ed. Academic Press, London, New York.

Williamson, I. P., & Kekwick, R. G. O. (1965). The formation of 5-phospho-mevalonate by mevalonate kinase in *Hevea brasiliensis* latex. Biochem. J. *96*, 862-871.

Yafin, Y., & Shechter, I. (1975). Comparison between biosynthesis of *ent*-kaurene in germinating tomato seeds and cell suspension cultures of tomato and tobacco. Plant Physiol. *56*, 671-675.

Zeevart, J. A. D. (1973). Gibberellin A_{20} content of *Bryophyllum daigremontianum* under different photoperiodic conditions as determined by gas-liquid chromatography. Planta *114*, 285-288.

3

In Vitro Metabolism of Gibberellins

Peter Hedden

3.1 INTRODUCTION

The first report of a cell-free system that could catalyze steps in gibberellin (GA) biosynthesis appeared almost 20 years ago. This system, which was shown to convert mevalonic acid (MVA) into the GA precursors, *ent*-kaurene and *ent*-kaurenol, was prepared from the endosperm-nucellus of the Californian wild cucumber, *Marah macrocarpus* (initially classified as *Echinocystis macrocarpa*) (Graebe et al., 1965). It has subsequently been the source of a great deal of information on the early steps in GA biosynthesis. More recently, there has been a significant increase in the number of *in vitro* systems, the most successful from immature seeds. Complete pathways from MVA to endogenous GAs have now been demonstrated in at least four systems.

The use of cell-free systems offers several advantages over that of intact tissues for the study of metabolism. The difficulties associated with poor substrate uptake due to membrane impermeability or lack of translocation to the site of metabolism are obviated. The metabolism of different substrates can, therefore, be compared without regard to differences in uptake by the tissue under investigation. Cell-free systems are also more amenable to manipulation than are intact plants. Thus, by suitable changes in substrate or cofactor concentrations, it is often possible to accumulate intermediates that are normally present only in trace amounts. Furthermore, by the use of inhibitors or simple procedures for enzyme separation—for example, the separation of membrane-associated enzymes from soluble enzymes by differential centrifugation—the biosynthetic pathway can be interrupted at specific points. In this way, intermediates lying before the block can be obtained on a preparative scale. This technique has been used to obtain labeled intermediates that are difficult to synthesize chemically (see Chapter 7). An obvious attribute of *in vitro* systems is that they provide virtually the only opportunity to examine the properties of the enzymes involved in biosynthesis. Detailed information at the enzyme level is essential for a complete understanding of GA metabolism, in particular as a means to answer questions on the regulation of GA synthesis or the mechanism of action of inhibitors.

In vitro systems suffer from the disadvantage that they are "unnatural." Homogenization of plant tissue results in the breakdown of intercellular compartmentation and thus in the mixing of proteins, cofactors, and ions that are normally strictly segregated. This can lead to abnormal metabolism of the substrates, or to loss of activity due to poisoning of the enzymes by factors from other cell compartments. The situation can be even more serious when a mixture of tissues, such as in whole seeds or shoots, is used as a source of material. The results of metabolism studies *in vitro*, therefore, should be interpreted carefully. Probably the best evidence for the validity of a pathway *in vivo* that has been determined by the use of a cell-free system is that the end products can be detected as endogenous components of the tissue under investigation. Studies on biosynthesis *in vivo* are by no means free from the problem of unnatural metabolism. Depending upon the method and site of application, the substrate may, for instance, not reach its normal site of metabolism and/or be transported to other sites where it undergoes unusual reactions.

The GA-biosynthetic pathway can be divided into three stages. The first stage, the conversion of MVA to *ent*-kaurene by soluble enzymes, is covered in Chapter 2. In the second stage, *ent*-kaurene is oxidized by membrane-associated enzymes to the first structural GAs, which are then further oxidized in stage three by soluble enzymes. Stages two and three provide a convenient basis for dividing this chapter, which also includes a discussion of methodology. The topic of *in vitro* GA metabolism has been included in several recent reviews (Bearder & Sponsel, 1977; Hedden, MacMillan, & Phinney, 1978; Graebe & Ropers, 1978; MacMillan, 1978; Phinney, 1979; Sembdner et al., 1980).

3.2 METHODOLOGY

3.2.1 Preparation of cell-free systems

Cell-free systems for GA biosynthesis have been prepared from a variety of plant material, the methods employed depending on the type of tissue under investigation. GA biosynthesis has also been studied in intact organelles, but this subject will not be discussed here. The simplest and mildest procedures for cell disruption have been employed with liquid endosperm from *Marah macrocarpus* and *Cucurbita maxima*, for which the cell walls are sufficiently fragile to be ruptured in a glass homogenizer (Graebe et al., 1965; Graebe, 1969). Large cell debris were removed by filtration through glass wool or by centrifugation at low speed. Maximum activity for later steps in the pathway were obtained in the *C. maxima* system after dialysis (Graebe, 1972), which presumably served to remove inhibitors released during the homogenization. Dialysis of the endosperm homogenate against 0.05 M phosphate buffer at pH 8.0 containing 2.5 mM $MgCl_2$ raised the pH to around 7.5, which is optimal for most enzymes in the pathway.

Tougher material such as immature seeds or fruit of *Pisum sativum* were ground in the presence of buffer using a chilled mortar and pestle (Graebe, 1968; Coolbaugh & Moore, 1971; Ropers et al., 1978). Seed parts such as embryos of *M. macrocarpus* (Graebe et al., 1965) and *P. sativum* (Coolbaugh & Moore, 1971), and suspensors of *Phaseolus coccineus* (Ceccarelli, Lorenzi, & Alpi, 1979) were also homogenized by this method. Phosphate was used most commonly as homogenization medium, although the use of tris-HCl has also been described (Graebe, 1968). The homogenates were filtered through cheese-cloth and/or sintered glass and then centrifuged at low speed (2,000–10,000 g) before use. In contrast to the *C. maxima* endosperm system, the activity of the *P. sativum* seed homogenate was not increased by dialysis (Ropers, 1978).

Homogenates of germinating seeds or isolated parts thereof require more vigorous procedures. Cell-free extracts have been prepared from whole germinating seeds of *Ricinus communis* (Robinson & West, 1970) and *Phaseolus vulgaris* (Patterson & Rappaport, 1974), from embryos of *Hordeum vulgare* (Murphy & Briggs, 1973), and from cotyledons and embryonic axes of *Ph. vulgaris* (Patterson et al., 1975). In these cases the material was homogenized mechanically in the presence of buffer, usually tris buffer in the pH range 6.5–7.3. Polyvinylpyrrolidone (PVP), a reagent to remove phenols (Loomis & Battaile, 1966), and a reducing agent (2-mercaptoethanol or dithiothreitol) were included in the homogenization medium. Coolbaugh and Durst (1978) have found that both 2-mercaptoethanol and dithiothreitol form addition compounds with *ent*-kaurene when the reducing agents are present at relatively high concentrations. This should obviously be kept in mind when these reagents are used, especially as it has not been demonstrated that they significantly increase the activity of the homogenates.

A cell-free system from vegetative shoots of *P. sativum* was prepared by first grinding the frozen material to a fine powder in a mortar and then mixing with PVP and 0.01 *M* phosphate buffer at pH 6.8 containing chloramphenicol (an antibiotic) and dithiothreitol (Coolbaugh et al., 1973). This system catalyzed the conversion of MVA to *ent*-kaurene in low yield. Similar procedures were used to prepare *in vitro* systems from seedlings of *Lycopersicum esculentum* (tomato) (Yafin & Shechter, 1975) and *Zea mays* (Hedden & Phinney, 1979; Wurtele, Hedden, & Phinney, 1982). Seedlings of *Z. mays* have also been homogenized with a blender in the presence of phosphate buffer, PVP, and 2-mercaptoethanol (Mellon & West, 1979). In general, *in vitro* systems from vegetative tissues have shown low enzyme activity for GA biosynthesis. This low activity may simply reflect the low levels of GAs present in these tissues. It may thus be unrealistic to expect to observe more than one biosynthetic step at a time in such systems.

Mycelia from the fungus *Gibberella fujikuroi* have been the source for *in vitro* systems that catalyze *ent*-kaurene biosynthesis (Shechter & West, 1969, Evans & Hanson, 1972) and kaurenoid oxidation (West, 1973). The tough fungal cell wall was broken either by passing a suspension of the mycelia in phosphate buffer through a French press (Evans & Hanson, 1972) or by freezing such a suspension and crushing in a Sagers' press (Shechter & West, 1969). Alternatively, the mycelial mat was frozen as such, and, after crushing in the Sagers' press, the homogenate was suspended in tricine buffer (pH 8.0) containing sucrose and 2-mercaptoethanol (West, 1973).

Extensive purification of the enzymes for GA biosynthesis from tissue homogenates have seldom been undertaken. An exception is the purification of *ent*-kaurene synthetase from various sources by West and co-workers (see Chapter 2, this volume). Also, Patterson, Rappaport, and Breidenbach (1975) have described the partial purification of a GA-hydroxylase from *Ph. vulgaris* cotyledons by ammonium sulfate precipitation and sucrose density-gradient centrifugation, and Hoad et al. (1982) have recently reported a 500:1 purification of the same enzyme. Most commonly, however, enzyme purification has been limited to separating the homogenate into microsomal and soluble fractions by centrifugation at 100,000 g or above.

3.2.2 Identification of products

The conclusive identification of the products obtained from incubations with *in vitro* systems (and *in vivo* systems) is one of the most difficult aspects of metabolism studies. Without adequate identification of the products, the results obtained from such experiments are impossible to interpret. It is not only important that the products be unambiguously identified but also that they be shown to originate from the substrate. In order to trace their metabolic fate, substrates are usually isotopically labeled, most commonly with a radioactive isotope. However, the identity of the radio-labeled product—or, to put it another way, the presence of the label in the suspected product—must be established. To this end co-chromatography of the radio-labeled product with an authentic substance using thin-layer chromatography or, better, gas-liquid chromatography has been used as a criterion of identity. However, the limited resolution of these procedures renders this method unreliable. Co-crystallization of product and authentic substance is a better criterion of identity, but the possibility of different compounds co-crystallizing cannot be ignored. The use of combinations of these methods obviously improves the chances of a correct identification. These methods, however, require that the identities of the products are suspected and that authentic compounds are available, in the case of co-crystallization, in substantial quantities.

Mass spectrometry offers the best possibility for identifying unambiguously small quantities (10^{-3} to 10^{0} μg) of compounds.

However, it must still be shown that the identified compound is, in fact, a product and not an endogenous component of the system under investigation. This can be achieved if the substrate contains sufficiently high amounts of an isotope label for it to be detectable by mass spectrometry. The stable isotopes ^{2}H, ^{13}C, ^{15}N, and ^{18}O can be used for this purpose but, being nonradioactive, they are useless as a means of detecting the products during purification. The radioisotope ^{3}H, because of its high specific radioactivity (29 Ci matom^{-1}), is not usually used with a high enough degree of isotopic substitution to be detectable by mass spectrometry. However, combinations of ^{2}H and ^{3}H, substituted at the same position, have been used successfully (Sponsel & MacMillan, 1978). ^{14}C is an ideal label as it is both radioactive and has a sufficiently low specific radioactivity (63 mCi matom^{-1}) to be detectable by mass spectrometry without impractically high levels of radioactivity having to be used. The ^{14}C-label can be easily detected by mass spectrometry in molecules with a specific radioactivity greater than about 20 mCi mmol^{-1}. Since the degree of isotopic substitution can be measured, it is possible to determine the specific radioactivity of the products (Bowen, MacMillan, & Graebe, 1972) and hence the amount of dilution by endogenous metabolites that has occurred.

The presence of endogenous metabolites in a cell-free system can cause the label to become so dilute as to be no longer detectable by mass spectrometry. This is particularly true of end products of the metabolic pathways, but intermediates can also accumulate. Microsomes of *C. maxima* were found to contain the GA-precursor, *ent*-kaurene, which is converted to further intermediates when the microsomes are incubated (Hedden & Graebe, 1981). *Gibberella fujikuroi* can accumulate relatively high levels of *ent*-kaurene in its mycelium (Cross et al., 1963b). Therefore, it is often desirable to remove or reduce the concentration of endogenous compounds before feeding a substrate. This can be achieved by such methods as differential centrifugation, dialysis, or gel filtration. Alternatively, inhibitors can be employed to prevent the conversion of endogenous pools of intermediates. For example, the oxidation of *ent*-kaurene can be prevented by ancymidol, which is a sufficiently specific inhibitor not to affect later steps (subsequent to *ent*-kaurenoic acid) (Coolbaugh, Hirano, & West, 1978). The removal of all endogenous metabolites from the system may indeed render the need for isotopically labeled substrates no longer necessary if the products are formed in amounts large enough to be detected by nonisotope techniques.

For clarity in the following discussion, it will not be continuously stated whether isotopically labeled substrates were used or not. However, unless otherwise indicated, it can be assumed that the products were adequately identified and shown beyond reasonable doubt to result from conversions of the applied substrates.

3.3 REACTIONS CATALYZED BY MICROSOMAL ENZYMES

The conversion of *ent*-kaurene to more polar GA precursors by a series of oxidative steps has been observed in cell-free systems from a number of higher plants and from the fungus *G. fujikuroi*. A general scheme for the metabolism of *ent*-kaurene by plant and *G. fujikuroi* microsomes is shown in Fig. 3.1. The main pathway of GA biosynthesis as far as GA_{12} aldehyde, which is depicted with thick arrows, is common to all systems so far investigated and is probably universal. The evidence from experiments *in vitro* for the steps in Fig. 3.1 is listed in Table 3.1.

3.3.1 *Ent*-kaurene to *ent*-7α-hydroxykaurenoic acid

This part of the pathway (steps 1–4 in Fig. 3.1) has been studied in some detail by West and co-workers using microsomes from endosperm of *Marah macrocarpus* (Dennis & West, 1967; Murphy & West, 1969; Lew & West, 1971; Hasson & West, 1976a, 1976b). Each step requires O_2 and a reduced pyridine nucleotide, preferably NADPH, for activity, which are properties characteristic of mixed-function oxygenases. The sequential oxidation of C-19 of *ent*-kaurene from a methyl group to a carboxylic acid group can be considered as a series of successive hydroxylations:

$$-CH_3 \longrightarrow -CH_2OH \longrightarrow -CH(OH)_2 \longrightarrow -C(OH)_3$$
$$\Big\Updownarrow -H_2O \qquad\qquad \Big\Updownarrow -H_2O$$
$$-CHO \qquad \longrightarrow \qquad -CO_2H$$

It might be expected that such a series of hydroxylations on a single carbon atom would be catalyzed by a single active site, but evidence from Hirano, cited by West (1980), suggests otherwise. Thus, although it was shown that members of the series *ent*-kaurene to *ent*-kaurenal

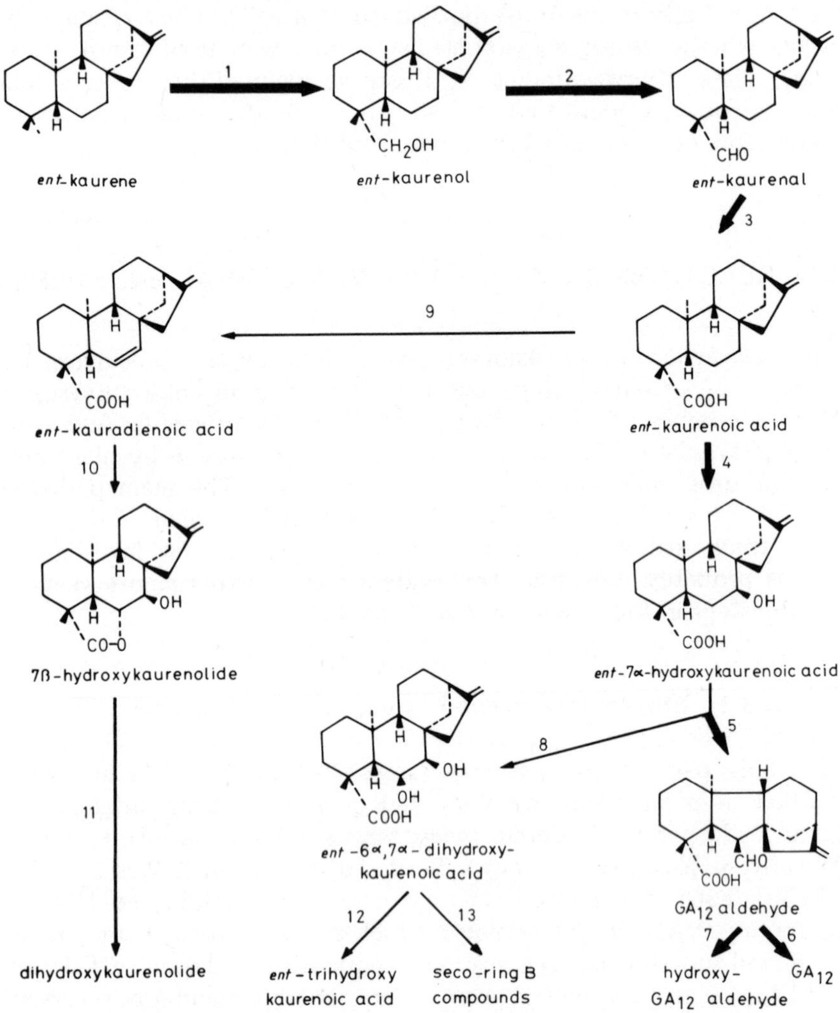

Figure 3.1 Reactions catalyzed by membrane-associated (microsomal) enzymes. The thick arrows chart the pathway to the GAs.

competitively inhibited the oxidation of the others, the concentration required for 50% inhibition was between 10 and 100 times higher than the Km value for the substrate. For example, the oxidation of ent-kaurenol ($Km = 0.5 \ \mu M$) was 50% inhibited by ent-kaurene at 50 μM. Furthermore, substrate analogues, which are not part of the biosynthetic sequence, were about equally effective as inhibitors. These data suggest separate sites for each substrate. Certainly, the

Table 3.1. Cell-free preparations in which the reactions illustrated in Fig. 3.1 have been observed

Reaction as indicated in Fig. 3.1	Plant material	Reference
1	*Marah macrocarpus* endosperm	Graebe et al. (1965)[a] Dennis & West (1967) Murphy & West (1969)
	Cucurbita maxima endosperm	Graebe (1972)[b]
	Marah oreganus endosperm	Coolbaugh & Hamilton (1976)[b]
	Pisum sativum immature seeds	Coolbaugh & Moore (1971)[a] Ropers et al (1978)
	Phaseolus coccineus suspensors	Ceccarelli et al. (1981)
2 and *3*	*M. macrocarpus* endosperm	Dennis & West (1967)
	C. maxima endosperm	Graebe (1972)[a,b]
	M. oreganus endosperm	Coolbaugh & Hamilton (1976)[a,b]
	P. sativum immature seeds	Coolbaugh & Moore (1971)[b] Ropers et al. (1978)
	P. coccineus suspensors	Ceccarelli et al. (1981a)[a]
	Hordeum distichon germinating embryos	Murphy & Briggs (1975)
	Zea mays coleoptiles	Wurtele et al. (1982)
4	*M. macrocarpus* endosperm	Lew & West (1971)
	C. maxima endosperm	Graebe et al. (1972)[a] Graebe et al. (1974)

(continued)

Table 3.1.   *(Continued)*

Reaction as indicated in Fig. 3.1	Plant material	Reference
	Gibberella fujikuroi mycelium	West (1973)
	H. distichon germinating embryos	Murphy & Briggs (1975)
	P. sativum immature seeds	Ropers et al. (1978)
	P. coccineus suspensors	Ceccarelli et al. (1981a)[a]
5	*C. maxima* endosperm	Graebe et al. (1972)[a] Graebe et al. (1974)
	M. macrocarpus endosperm	West (1973)
	G. fujikuroi mycelium	West (1973)
	P. sativum immature seeds	Ropers et al. (1978)
	P. coccineus suspensors	Ceccarelli et al. (1981b)[a]
6	*C. maxima* endosperm	Graebe et al. (1972)[a] Graebe et al. (1974)
	M. macrocarpus endosperm	West (1973)[b]
	G. fujikuroi mycelium	West (1973)[b]
	P. sativum immature seeds	Ropers et al. (1978)
7		
3β-hydroxylation	*G. fujikuroi* mycelium	West (1973)[a]
13-hydroxylation	*P. sativum* immature seeds	MacMillan (1978)

Table 3.1. (*Continued*)

Reaction as indicated in Fig. 3.1.	Plant material	Reference
12α-hydroxylation	*C. maxima* endosperm	Hedden & Graebe (1982)
8	*M. macrocarpus* endosperm	West (1973)[b]
	G. fujikuroi mycelium	West (1973)
	C. maxima endosperm	Graebe et al. (1974)
9 and *10*	*G. fujikuroi* mycelium	West (1973)
	C. maxima endosperm	Hedden & Graebe (1981)
11		
12α-hydroxylation	*C. maxima* endosperm	Hedden & Graebe (1981)
13	*C. maxima* endosperm	Graebe et al. (1974) Hedden & Graebe (unpublished)

[a]The reaction was not observed directly but is assumed to have occurred as part of a reaction sequence.

[b]Only chromatographic evidence available for the identity of the reaction product.

intermediates become readily detached from the active site(s), as evidenced by their accumulation when precursors are incubated with *in vitro* systems. Furthermore, the existence of the B1-41a mutant of *G. fujikuroi*, in which GA biosynthesis is blocked between *ent*-kaurenal and *ent*-kaurenoic acid (Bearder et al., 1974), lends additional support to separate binding sites in the C-19 oxidation series. However, the sites must be very similar since each step between *ent*-kaurene and *ent*-kaurenoic acid in *M. macrocarpus* microsomes is inhibited by ancymidol (Coolbaugh et al., 1978). This inhibitor is highly specific and has little effect on the oxidation of *ent*-kaurenoic acid and later steps in the pathway (see Section 3.3.3).

The involvement of a soluble, noncatalytic protein in the metabolism of *ent*-kaurene has been suggested on the basis of results obtained with homogenates of *P. sativum* seeds (Moore, Barlow, & Coolbaugh, 1972) and *G. fujikuroi* mycelium (Hanson, Willis, & Parry, 1980). It was suggested that this protein binds *ent*-kaurene produced by soluble enzymes and facilitates its oxidation by the microsomes. Analogous sterol carrier proteins have been reported to participate in the conversion of squalene to cholesterol in mammalian liver (Scallen et al., 1972). Moore et al. (1972) found that *ent*-[^{14}C]kaurene, either added to a 100,000-g supernatant fraction from *P. sativum* seeds or generated from [2-^{14}C]MVA in the same fraction, was bound to a high molecular-weight protein. This protein appeared to be formed from lower molecular-weight components in the presence of *ent*-kaurene. Evidence was also presented that *ent*-kaurene bound to the high molecular-weight protein was converted by *P. sativum* seed microsomes, whereas unbound *ent*-kaurene was not. However, the data were based on differences in the amount of extractable *ent*-[^{14}C]kaurene after incubation, and the products were not identified. Graebe and Ropers (1978) have argued that the *ent*-kaurene might in fact be binding to microsomal fragments that were not sedimented during centrifugation. This argument must also apply to the results of Hanson et al. (1980) who found that *ent*-[^{3}H]kaurene produced from [2-^{3}H]MVA in a cell-free system from *G. fujikuroi* mycelium became bound to a protein fraction. The protein-bound *ent*-kaurene could be displaced partially by *ent*-16β,17-epoxykaurene [(6) in Fig. 3.3 below], a compound that inhibited both the formation and the oxidation of *ent*-kaurene in fungal cultures. Hanson et al. (1980) suggested that the epoxide prevented *ent*-kaurene from binding to its "carrier protein" but did not show that the protein was involved in *ent*-kaurene metabolism. Hasson and West (1976a) reported a high molecular-weight, heat-labile (protein?) factor in the high-speed supernatant fraction, from a homogenate of *M. macrocarpus* endosperm, that stimulated the microsomal oxidation of *ent*-kaurene in this system. They also reported that *ent*-kaurene became associated with high molecular-weight material in the high-speed supernatant, but had no evidence that an *ent*-kaurene-protein complex was a preferential substrate for the microsomes in this system. Thus, despite some circumstantial evidence, the case for a lipid carrier protein in *ent*-kaurene oxidation has still to be proved.

3.3.2 Cytochrome P-450 and the microsomal oxidases

Murphy and West (1969) demonstrated the incorporation of ^{18}O into *ent*-kaurenol when *ent*-kaurene was hydroxylated in an atmosphere containing $^{18}O_2$. They also showed that the hydroxylation was inhibited by CO and that the inhibition was reversed by light with maximum effectiveness at 450 nm. This light-reversible CO inhibition was also demonstrated for the oxidation of *ent*-kaurenal to *ent*-kaurenoic acid. Thus, cytochrome P-450 was implicated at the active center of the microsomal oxidases. The oxidation of *ent*-kaurene by cell-free extracts of *P. sativum* cotyledons was reported to be inhibited by CO (Coolbaugh & Moore, 1971); but, as this system did not metabolize exogenous *ent*-kaurene, the results were based on differences in the amounts of *ent*-kaurene produced from MVA in the presence and absence of CO and are therefore inconclusive.

The occurrence of microsomal cytochrome P-450 and its involvement in mixed-function oxidation has been established for many plant systems (West, 1980). Reduced microsome preparations from *M. macrocarpus* endosperm (Murphy & West, 1969) and *P. sativum* cotyledons (Coolbaugh & Moore, 1971) gave CO-difference spectra with maxima at *ca.* 420 and 450 nm. The peak at 420 nm is thought to be due to inactivated cytochrome P-450. Hasson and West (1976a, 1976b) have studied the properties of the *M. macrocarpus* system for the oxidation of *ent*-kaurene in some detail and concluded that it is similar to the mammalian hepatic system. This latter system has been studied much more extensively than have plant systems (see, for example, White & Coon, 1980). Consideration of a model of the hepatic system may, therefore, help in interpreting results obtained with plant microsomes.

Hydroxylations catalyzed by mixed-function oxygenases result in the incorporation of one atom of molecular oxygen into the substrate and the reduction of the second atom to water. The reaction can be considered as the equation

$$RH + O_2 + 2H^+ + 2e^- \rightarrow ROH + H_2O$$

In cytochrome P-450 mixed-function oxygenases, the source of the two electrons are ultimately reduced pyridine nucleotides, which transfer their reducing equivalents via short chains to the final acceptor at the active site. The mechanistic scheme for cytochrome P-450-catalyzed hydroxylations shown in Fig. 3.2 is adapted from that proposed by White and Coon (1980). Three points of particular

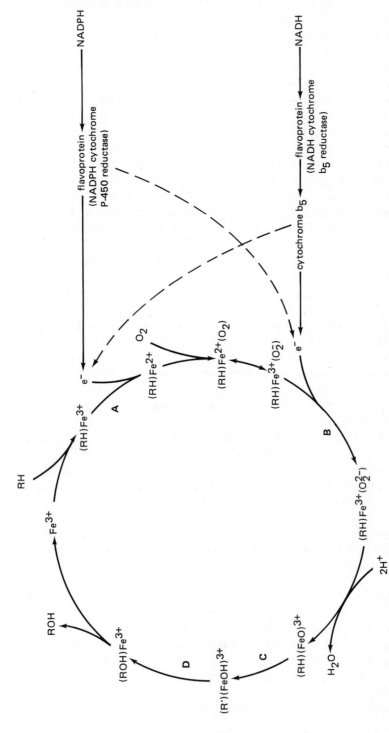

Figure 3.2 Mechanistic scheme for cytochrome P-450-mediated hydroxylation in liver microsomes. The catalytic cycle is that proposed by White and Coon (1980). (Reproduced, with permission, from the Annual Review of Biochemistry, Volume 49, © 1980 by Annual Reviews Inc.)

interest in this scheme are the reduction steps (A and B) and the hydrogen abstraction (or oxygen insertion) reaction (C and D). In the mammalian liver system, the first electron is supplied by NADPH and is transferred via NADPH cytochrome P-450 reductase, a flavoprotein that has been shown to contain both FAD and FMN (Vermilion & Coon, 1978; Vermilion et al., 1981). The second electron transfer is less well understood. There is evidence for the involvement of cytochrome b_5, which is reduced by NADH via NADH cytochrome b_5 reductase, a suggestion that is supported by the synergistic action of NADH and NADPH (Hilderbrandt & Estabrook, 1971). However, cytochrome b_5 is not an obligatory component of the oxygenase and may react other than as an electron donor in some cases (Coon et al., 1977). Recent results have indicated that both electrons can be supplied via NADPH cytochrome P-450 reductase or cytochrome b_5, the electron transfer pathway utilized depending on the reduced pyridine nucleotides available (Noshiro, Harada, & Omura, 1981). The actual mechanism of hydrogen abstraction and oxygen insertion (steps C and D) is not completely understood but is thought to involve an oxenoid intermediate (Hamilton, 1974). The available evidence supports a free-radical mechanism (Groves et al., 1978) in which a hydrogen atom is extracted from the substrate and the resulting radicals ($R^{\cdot} + HO^{\cdot}$) combine.

Working with the *M. macrocarpus* microsomal system, Hasson and West (1976a) found that the rate of *ent*-kaurene oxidation to *ent*-kaurenol, *ent*-kaurenal, *ent*-kaurenoic acid, and *ent*-7α-hydroxy-kaurenoic acid was supported much more effectively by NADPH than by NADH. They also found that both nucleotides together at low concentrations (100 μM NADH and 1 μM NADPH) were more effective than either alone at equivalent concentration. This synergistic effect is analogous to that found with the liver oxidase system. Hasson and West also found that a mixture of NADH, ATP, and NADP was more effective than NADH alone, and they proposed that the microsomes were capable of ATP-stimulated reduction of NADP. Low levels (0.5 μM) of FAD and FMN stimulated *ent*-kaurene oxidation rates, a finding consistent with the involvement of flavin nucleotides in the oxygenases. In a subsequent paper, Hasson and West (1976b) reported the following electron transfer components as being present in the microsomes: cytochrome P-450, cytochrome b_5, NADPH cytochrome c reductase, NADH cytochrome c reductase, and NADH cytochrome b_5 reductase. However, with the exception of cytochrome P-450, their involvement in the oxygenases has not been demonstrated directly. It was possible to recover some *ent*-kaurene oxidase activity after solubilization with 1% sodium deoxycholate and removal of the detergent, but all activity was lost after separation of the components and recombination.

3.3.3 Inhibitors of *ent*-kaurene oxidation

A number of inhibitors of GA biosynthesis have been found to block the oxidation of *ent*-kaurene. The most extensively studied is ancymidol [(*1*) in Fig. 3.3)], which inhibits the three oxidation steps in the conversion of *ent*-kaurene to *ent*-kaurenoic acid by *Marah* microsomes (Coolbaugh & Hamilton, 1976, Coolbaugh et al., 1978). It is highly effective ($Ki = 2 \times 10^{-9} M$) and specific. The oxidation of *ent*-kaurenoic acid in the same system was only slightly inhibited by ancymidol, and *trans*-cinnamic acid hydroxylase, a cytochrome P-450–dependent plant hydroxylase, was not affected by the substance. Ancymidol is not, therefore, a general inhibitor of plant cytochrome P-450 oxygenases. The widespread effectiveness of ancymidol as a plant growth retardant (Leopold, 1971) probably indicates that it is a general inhibitor of *ent*-kaurene oxidation in higher plants. Coolbaugh, Swanson, & West, (1982b) have compared ancymidol with a number of other substituted pyrimidines and found a good correlation between their ability to inhibit *ent*-kaurene oxidation by *Marah* microsomes and their effectiveness as growth retardants. Although ancymidol was one of the most effective compounds in these tests, it was considerably less efficient at inhibiting GA biosynthesis in *G. fujikuroi* than some of the other pyrimidines (Coolbaugh, Heil, & West, 1982a). In this case the degree of inhibition appeared to be correlated with fungicidal activity, although the reduction in GA production could not be completely accounted for by reduced fungal growth. Indeed, there was also a good correlation between fungicidal activity and ability to inhibit *in vitro ent*-kaurene oxidation in the fungus, the order of effectiveness of these substances being roughly opposite to that found for the inhibition of the same reaction in *Marah* microsomes. It is of interest that these substituted pyrimidines and other nitrogen-containing heterocyclic fungicides are thought to act by inhibiting 14-demethylation in the biosynthesis of ergosterol (Buchenauer, 1977). The inhibited enzyme is a mixed-function oxygenase that has much in common with *ent*-kaurene oxidase. There is some evidence that ancymidol may be binding directly to cytochrome P-450 (Coolbaugh et al., 1978). Addition of the inhibitor to a suspension of *Marah* microsomes produced a difference spectrum with a maximum at 427 nm, a characteristic of substrate- or inhibitor-induced difference spectra

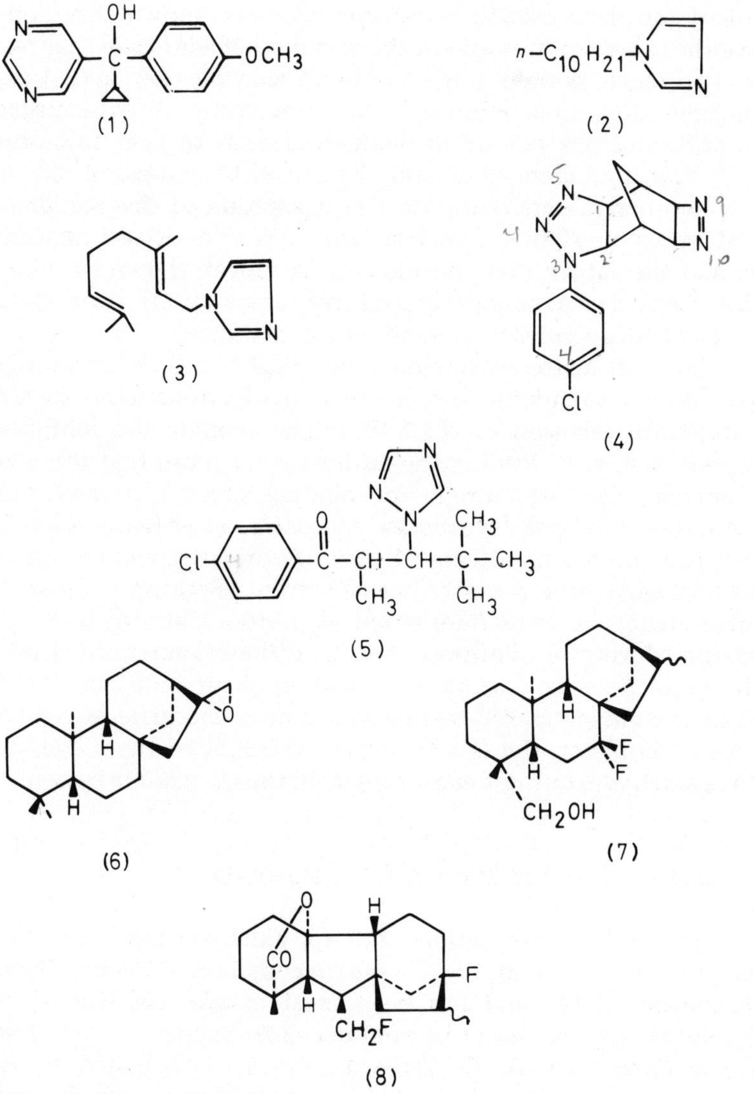

Figure 3.3 Inhibitors of GA biosynthesis that act after the formation of *ent*-kaurene.

observed in liver microsomes (Yoshida & Kumoaka, 1975). Coolbaugh et al. (1982a) have suggested that the fungicides may interact more strongly with the fungal cytochrome P-450 than does ancymidol. Conversely, ancymidol would have a greater affinity for the higher plant system.

(1) Ancymidol
(2) 1-n-decylimidazole
(3) 1-geranylimidazole
(4) Tetcyclacis
(5) Paclobutrazol (PP333)

Many plant growth retardants are now known that appear to have the same point of action as ancymidol. Wada (1978) has reported on two such compounds, 1-*n*-decylimidazole (*2*) and 1-geranyl-imidazole (*3*), that inhibited the conversion of *ent*-kaurene and *ent*-kaurenol, but not of *ent*-kaurenoic acid, to GA_3 in cultures of *G. fujikuroi*. Rademacher and Jung (1981) compared the effects of a number of retardants on the elongation of rice seedlings. The most efficient of these, such as the norbornenodiazetine derivative (*4*) and the substituted triazole (*5*), have been shown to inhibit the same steps as ancymidol in cell-free preparations from *Cucurbita maxima* endosperm (Hildebrandt et al., 1982).

The *ent*-kaurene-analogue, *ent*-16β,17-epoxykaurene (*6*), has been shown to inhibit the conversion of *ent*-kaurene to GA_3 by *G. fujikuroi* (Hanson et al., 1980). The epoxide also inhibited *ent*-kaurene synthesis, leading the authors to propose that the analogue competed with *ent*-kaurene for binding sites on a carrier protein that links *ent*-kaurene synthetase and oxidase (see Section 3.3.1).

The inhibitory effects of some fluorinated *ent*-kaurenoid and GA analogues have recently been reported (Boulton & Cross, 1981). These analogues were fluorinated at positions known to be sites of reaction during biosynthesis. *Ent*-7,7-difluorokaurenol (*7*) inhibited the steps between *ent*-kaurene and *ent*-kaurenoic acid in the *C. maxima* endosperm cell-free system, but it had little effect on later steps. In contrast, the GA analogue (*8*) inhibited the metabolism of GA_{12} aldehyde, but not earlier steps, in the *C. maxima* system.

3.3.4 The formation of GA_{12} aldehyde

The cell-free formation of GA_{12} aldehyde (step 5 in Fig. 3.1) was first observed in the *C. maxima* system (Graebe, Bowen, & MacMillan, 1972) and has been most extensively studied in this system. It has also been observed in cell-free systems from *M. macrocarpus* (West, 1973), *G. fujikuroi* (West, 1973), and *P. sativum* (Ropers et al., 1978). GA_{12} aldehyde is formed in a single step from *ent*-7α-hydroxykaurenoic acid in a reaction that requires NADPH and O_2 (Graebe & Hedden, 1974). The reaction is catalyzed by the microsomes and, thus, has properties similar to the preceding mixed-function oxygenase reactions, although the involvement of cytochrome P-450 has not been demonstrated. An interesting feature of the reaction is that it is accompanied by the formation of *ent*-6α,7α-dihydroxykaurenoic acid (an *ent*-6α-hydroxylation, step 8 in Fig. 3.1). This latter reaction has been observed in *M. macrocarpus* and *G. fujikuroi* (West, 1973) as well as in *C. maxima* (Graebe et al., 1974a). It has not been possible to separate the two activities since they have the same cofactor requirements, pH (7.0–7.4) and temperature (30°C) optima (Graebe & Hedden, 1974; Becker, 1974). The oxidation of *ent*-7α-hydroxykaurenoic acid to the two products shows Michaelis-Menten kinetics with an apparent Km of 9–11 μM (Graebe et al., in preparation).

GA$_{12}$ aldehyde is formed from *ent*-7α-hydroxykaurenoic acid by a reaction in which ring B contracts from a 6- to a 5-membered ring with the extrusion of C-7 (see Fig. 3.5 below). The reaction represents a single-step oxidation. The mechanism of this reaction has been studied in some detail with the *C. maxima* system by the use of specifically labeled *ent*-7α-hydroxykaurenoic acid. Doubly labeled substrate, consisting of a mixture of *ent*-[6-^{3}H$_2$]7α-hydroxy-kaurenoic acid and *ent*-[^{14}C]7α-hydroxykaurenoic acid, was found to lose half of the ^{3}H in the conversion to both GA$_{12}$ aldehyde and *ent*-6α,7α-dihydroxykaurenoic acid (Graebe, Hedden, & MacMillan, 1975). It has subsequently been shown, using doubly labeled *ent*-7α-hydroxykaurenoic acid in which the *ent*-6α-hydrogen was labeled as ^{3}H or ^{2}H, that the *ent*-6α-hydrogen was lost (Graebe et al., in preparation). The tritium was recovered in the water of the incubation medium. As expected, the *ent*-7β-^{3}H atom of *ent*-[7β-^{3}H]7α-hydroxy-kaurenoic acid was retained in the conversion to both products but was lost on the subsequent enzymatic oxidation of the aldehyde group of GA$_{12}$ aldehyde to a carboxylic acid group in GA$_{12}$. Time-course studies on the conversion of *ent*-[6α-^{3}H,^{14}C]7α-hydroxy-kaurenoic acid mixtures showed that the tritiated substrate was oxidized more slowly than the protium equivalent (Fig. 3.4). The apparent magnitude of this isotope effect (*7–9*) lies within the range typical for a primary tritium isotope effect. Thus, breakage of the C—H bond is at least partially rate-limiting in both reactions. The simplest interpretation of the data is that both products—GA$_{12}$ aldehyde and *ent*-6α,7α-dihydroxykaurenoic acid—are formed from a common intermediate, the formation of which involves breaking the *ent*-C-6-αH bond and is rate-limiting. If a carbon radical mechanism is assumed for the reaction, the scheme in Fig. 3.5 can be proposed. In this case, the ratio of the two products (almost 1:1 in *C. maxima*) would depend on the position of the equilibrium, its rate of achievement, and the rate at which the radicals combine with HO˙. There is evidence that the radical intermediates may exist long enough for some rearrangement to occur (Groves et al., 1978).

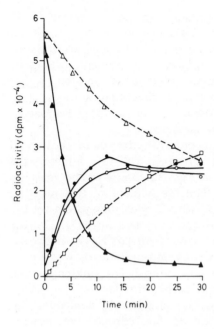

Figure 3.4 Comparison of the rates of oxidation of *ent*-[^{14}C]7α-hydroxy-kaurenoic acid (—▲—) and *ent*-[6α-^{3}H]7α-hydroxykaurenoic acid (- -△- -) by resuspended microsomes from *C. maxima* endosperm. Also shown is the formation of the ^{14}C-labeled products, GA$_{12}$ aldehyde (—●—) and *ent*-6α,7α-dihydroxykaurenoic acid (—○—), and the release of tritium as water into the incubation medium (- -□- -).

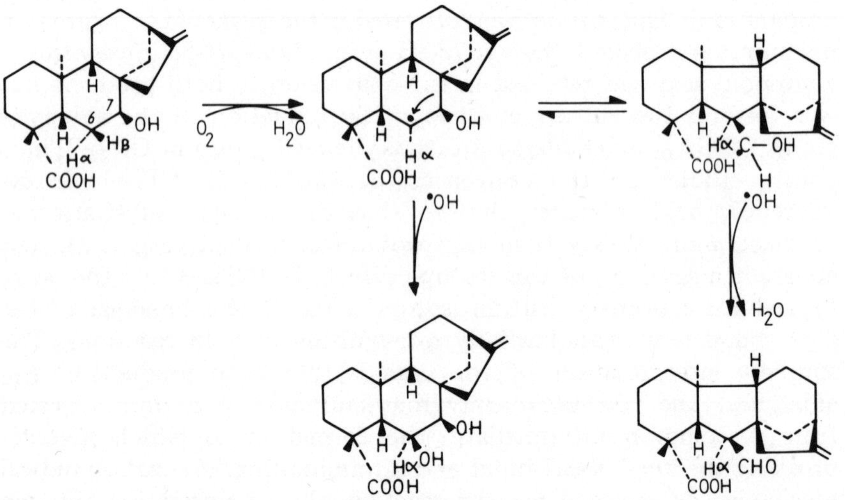

Figure 3.5 Scheme proposed for the conversion of *ent*-7α-hydroxykaurenoic acid to GA$_{12}$ aldehyde and *ent*-6α,7α-dihydroxykaurenoic acid based on a free-radical mechanism.

3.3.5 The oxidation of GA_{12} aldehyde

The final GA products produced from *ent*-kaurenoic acid by microsome preparations from the fungus *G. fujikuroi* were identified as GA_{14} and GA_{12}, although the identification of the latter was based only on chromatographic evidence (West, 1973). The same products were shown to be formed from GA_{12} aldehyde by enzymatic reactions that required NADPH and O_2 and were inhibited by CO. The oxidation of GA_{12} aldehyde in *G. fujikuroi* microsomes is therefore catalyzed by mixed-function oxygenases similar to those that oxidize *ent*-kaurene and its metabolites in *M. macrocarpus*. The formation of GA_{14} involves a 3β-hydroxylation for which the substrate appears to be GA_{12} aldehyde. The immediate product of this reaction, GA_{14} aldehyde, does not accumulate, although Dockerill, Evans, and Hanson (1977) identified this intermediate as a product after incubating radio-labeled GA_{12} aldehyde with a 10,000-g supernatant fraction. They obtained a second product, which, on tentative evidence, was suggested to be GA_{13} aldehyde. The bulk of the evidence for the 3β-hydroxylation of GA_{12} aldehyde has come from studies using intact cultures of the fungus. These studies have shown that GA_{14} aldehyde is an intermediate in the formation of GA_{14} (Hedden, MacMillan, & Phinney, 1974) and that GA_{12} is not a good substrate for 3β-hydroxylation (Bearder, MacMillan, & Phinney, 1975; Evans & Hanson, 1975). GA_{12} aldehyde, therefore, can be oxidized either at C-7 (to GA_{12}) or at C-3 (to GA_{14} aldehyde) and lies at a branch point in the pathway. The oxidation of GA_{14} aldehyde to GA_{14} must proceed rapidly in fungal microsomes, although this has not been investigated.

Ropers et al. (1978) found that GA_{12} aldehyde was converted to numerous products when incubated with a cell-free extract from immature seeds of *P. sativum*. Three of these products were identified by gas chromatography-mass spectrometry (GC-MS) as GA_{12}, GA_{53}, and GA_{44}. The formation of these GAs was catalyzed by the 2,000-g supernatant but not by the 200,000-g supernatant, indicating that at least the initial reaction(s) is pelletable at 200,000 g. NADPH was found not to stimulate activity in this system, but as the system had not been dialyzed it may be assumed that sufficient cofactors were already present or were generated during the incubation. The microsomes have now been confirmed as the location of the GA_{12} aldehyde oxidases (Kamiya & Graebe, 1983). GA_{12} aldehyde is oxidized by a microsomal fraction to GA_{12} and GA_{53}, and GA_{12} is converted to GA_{53}. Thus, both C-7 and C-13 hydroxylating activities are present in the microsomes. GA_{12} appears to be a better substrate for 13-hydroxylation *in vitro* than GA_{12} aldehyde and may be the intermediate between GA_{12} aldehyde and GA_{53}. The other possible intermediate—GA_{53} aldehyde formed by 13-hydroxylation of GA_{12} aldehyde—has been detected in only small amounts as a product from incubations with GA_{12} aldehyde (MacMillan, 1978). However, as is the case for GA_{14} aldehyde in the fungal cell-free system, GA_{53} aldehyde may not accumulate. The existence of an early branch in the GA biosynthetic pathway in *P. sativum* seeds, resulting in the formation of 13-hydroxy GAs (GA_{20}) and 13-deoxy-GAs (GA_9), had previously been suggested by Sponsel and MacMillan (1977) on the basis of results from *in vivo* experiments. However, it was not possible to determine the exact location of the branch.

Microsomal preparations from *C. maxima* endosperm were shown to oxidize GA_{12} aldehyde to GA_{12} in a reaction that requires NADPH and is stimulated by Mg^{2+} (Graebe & Hedden, 1974). GA_{12} has also been tentatively identified as a product of *ent*-kaurenoic acid metabolism in the cell-free preparation from *M. macrocarpus* endosperm (West, 1973). In common with the microsomal preparations from *G. fujikuroi* and *P. sativum*, microsomes from *C. maxima* have been found to contain a GA_{12} aldehyde hydroxylating activity (Hedden et al., in preparation). The product of this reaction was shown to be 12α-hydroxy GA_{12} aldehyde, which is further converted by the high-speed supernatant fraction to endogenous 12α-hydroxylated GAs. A peculiarity of the 12α-hydroxylase is its relatively low pH optimum (6.2) compared with the other microsomal oxidases in *C. maxima* endosperm, which have pH optima between 7.0 and 7.5. Figure 3.6 shows the pH dependence of the oxidation of GA_{12} aldehyde at C-12 and C-7 by *C. maxima* microsome preparations. The relative yield of the products, 12α-hydroxy GA_{12} aldehyde and GA_{12}, is influenced by pH, low pH favoring 12α-hydroxylation and higher pH promoting oxidation to GA_{12}. This pH effect is apparently

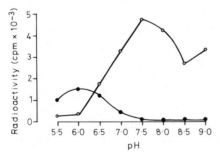

Figure 3.6 The effect of pH on the oxidation of $[^{14}C]GA_{12}$ aldehyde to 12α-hydroxy GA_{12} aldehyde (—●—) and GA_{12} (—○—) by resuspended microsomes from *C. maxima* endosperm.

peculiar to the endosperm as maximum rates of both 12α- and 12β-hydroxylation of GA_{12} aldehyde occur at pH values above 7.0 in cell-free preparations from roots of C. maxima seedlings (Froneberg, unpublished results in this laboratory). Surprisingly, GA_{12} is 13-hydroxylated to GA_{53} by microsome preparations from C. maxima endosperm (Hedden et al., in preparation), a reaction analogous to that in P. sativum. Since 13-hydroxylated GAs are found only in trace amounts in C. maxima endosperm, 13-hydroxylation can have little significance in vivo. However, it does demonstrate the high substrate-specificity of the microsomal oxidases in C. maxima. It also provides circumstantial evidence that GA_{12} is the true substrate for 13-hydroxylation in higher plants.

3.3.6 Branches from the GA pathway

It is clear from the wide distribution of ent-kaurenoid compounds as secondary plant products that the pathway leading to GAs as far as ent-7α-hydroxykaurenoic acid is not exclusive to GA biosynthesis. Two groups of metabolites—the kaurenolides and the oxidation products of ent-$6\alpha,7\alpha$-dihydroxykaurenoic acid—have been found together with GAs in seeds of higher plants, and their formation appears to be closely associated with GA biosynthesis. Both groups of metabolites are also found in cultures of the GA-producing fungi, G. fujikuroi and Sphaceloma manihoticola.

The kaurenolides are neutral compounds containing a 19-6 lactone (9–11). All kaurenolides so far characterized are 7β-hydroxylated, a fact that suggests that 7β-hydroxykaurenolide (Fig. 3.1) is the first kaurenolide intermediate. Subsequent hydroxylation of this precursor would give rise to the various dihydroxykaurenolides that have been identified. Kaurenolide biosynthesis has been most frequently studied in cultures of G. fujikuroi, from which evidence was obtained for the intermediacy of ent-7α-hydroxykaurenoic acid (Evans, Hanson, & White, 1970; Hanson, Hawker, & White, 1972). Thus, it was assumed that 7β-hydroxykaurenolide was formed directly from ent-7α-hydroxykaurenoic acid by ent-6β-hydroxylation. Microsomes from G. fujikuroi were found to produce 7β-hydroxy-kaurenolide when incubated with ent-kaurenoic acid, but the details of this conversion were not investigated (West, 1973). The complete pathway for kaurenolide biosynthesis has, however, been worked out in microsome preparations from C. maxima endosperm, and it does not agree with the view that ent-7α-hydroxykaurenoic acid is a precursor (Hedden & Graebe, 1981). In this sequence (steps 9–11

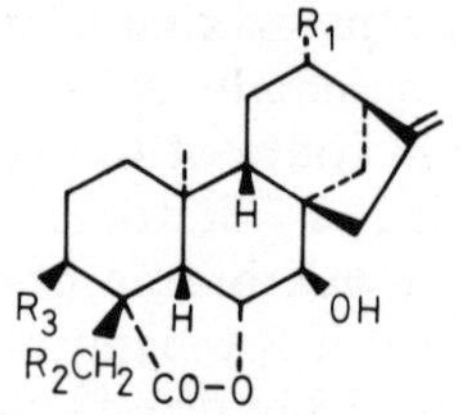

(9) R_1 = OH, R_2 = H , R_3 = H
(10) R_1 = H , R_2 = OH , R_3 = H
(11) R_1 = H , R_2 = H , R_3 = OH

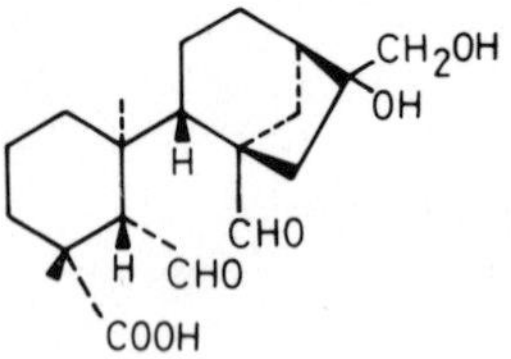

(12)

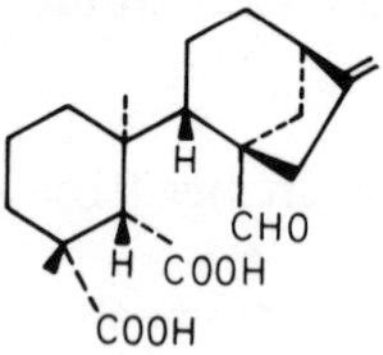

(14)

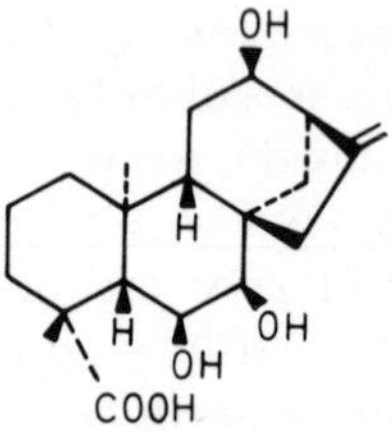

(15)

Structures (9)–(12), (14)–(15)

in Fig. 3.1) the branch point from the GA pathway lies at *ent*-kaurenoic acid, which is converted to *ent*-kaura-6,16-dienoic acid and hence to 7β-hydroxykaurenolide. The final step in the pathway in *C. maxima*, a 12α-hydroxylation, produces the endogenous product 7β,12α-dihydroxykaurenolide (*9*). All steps in the sequence are catalyzed by NADPH-requiring oxidases of the same general type that participate in the main GA pathway. Indeed, in some cases at least, the same oxidases may catalyze reactions in both pathways. Mixed-function oxygenases that hydroxylate saturated carbon atoms are known to react with double bonds to give epoxides. It might be expected, therefore, that the *ent*-7α-hydroxylase for which *ent*-kaurenoic acid is a substrate would convert *ent*-kauradienoic acid to the *ent*-6α,7α-epoxide. This latter compound was not detected as an intermediate, but it is known to be very unstable in aqueous solution, being readily hydrolyzed to 7β-hydroxykaurenolide (Beale et al., 1982). It is, therefore, difficult to determine whether the hydrolysis of the epoxide to 7β-hydroxykaurenolide is an enzymatic process or not. The final 12α-hydroxylation is also common to the GA pathway where GA_{12} aldehyde is a substrate.

There is now evidence that the pathway to 7β-hydroxykaureno-lide in *G. fujikuroi* is the same as in *C. maxima*. Beale et al. (1982) obtained only very low incorporation of radioactivity from *ent*-7α-hydroxy[$6α$-^{2}H,17-^{3}H$_2$]kaurenoic acid into kaurenolides, and no detectable incorporation of deuterium, after incubation with cultures of the mutant fungus B1-41a. In contrast, *ent*-kauradienoic acid was efficiently converted to kaurenolides. The major kaurenolide produced by *G. fujikuroi* is 7β,18-dihydroxykaurenolide (*10*) (Cross, Hanson, & Galt, 1963c), which has been shown to be formed by 18-hydroxylation of 7β-hydroxykaurenolide in fungal cultures (Cross, Galt, & Norton, 1968). Nothing is known about the nature of the 18-hydroxylase. However, the hydroxylation of GA_{12} aldehyde at C-3, which is close to C-18, occurs in the fungal microsomes (West, 1973). Furthermore, 7β,18-dihydroxykaurenolide is accompanied by smaller amounts of 3β,7β-dihydroxykaurenolide (*11*) in the fungal culture medium (Bateson & Cross, 1973; Hedden, MacMillan, & Grinstead, 1973). A single enzyme activity could hydroxylate at both the 3β- and 18-positions, the difference in hydroxylation pattern obtained with GA_{12} aldehyde (3β) and 7β-hydroxykaurenolide (3β and 18) reflecting differences in the binding of the two substrates at the active site. On this basis it would be expected that the 18-hydroxylase is also a microsomal enzyme.

A second major branch results from the formation of *ent*-6α,7α-dihydroxykaurenoic acid from *ent*-7α-hydroxykaurenoic acid (step 8 in Fig. 3.1) in association with ring contraction (see Section 3.3.4). In *C. maxima* microsome preparations, almost 50% conversion of *ent*-7α-hydroxykaurenoic acid to the dihydroxy acid occurs (Graebe & Hedden, 1974). Further oxidation of the dihydroxy acid by *C. maxima* microsomes results in the formation of an endogenous product, which, on the basis of its mass spectrum, has been tentatively assigned structure (*12*) (MacMillan, Gaskin, Hedden, and Graebe, unpublished information). An intermediate between *ent*-6α,7α-dihydroxykaurenoic acid and (*12*) has also been obtained and assigned structure (*13*) (see Fig. 3.7) from its mass spectrum and its identity with the product obtained from the oxidation of *ent*-6α,7α-dihydroxykaurenoic acid with sodium metaperiodate (Hedden and Graebe, unpublished results). The conversion of *ent*-6α,7α-dihydroxy-kaurenoic acid to the seco-ring B product (*13*) is a single-step oxida-

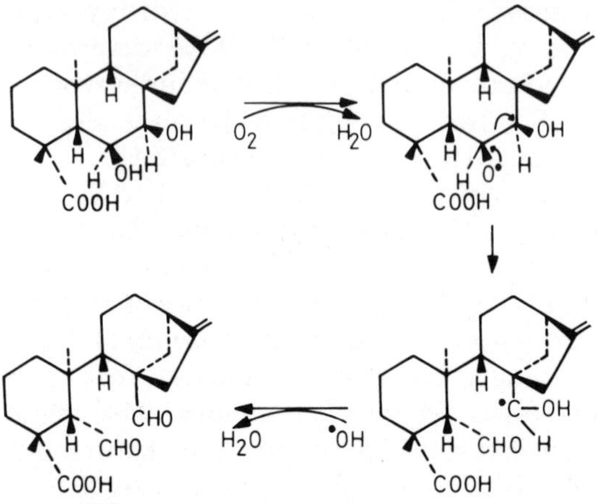

Figure 3.7   Scheme proposed for the oxidative cleavage of ring B in *ent*-6α,7α-dihydroxykaurenoic acid to give the dialdehyde (*13*).   The formation of (*13*) from the dihydroxy acid is catalyzed by microsome preparations from *C. maxima* endosperm.

tion. A possible mechanism for the ring opening based on a free-radical reaction is presented in Fig. 3.7. The initial oxidation could take place equally well at the *ent*-7α hydroxy group. Epoxidation at the 16,17-double bond followed by hydration of the epoxide would then give rise to the end product (*12*). The cell-free synthesis of (*13*) requires NADPH and is probably the result of a mixed-function oxygenase similar to those discussed throughout this section.

The branch pathway to seco-ring B compounds also exists in *G. fujikuroi. Ent*-6α,7α-dihydroxykaurenoic acid was tentatively identified as one of the products resulting from incubations of *ent*-kaurenoic acid with microsomes prepared from the fungal mycelia (West, 1973). The further conversion of the dihydroxy acid was not investigated, but in fungal cultures it is metabolized to fujenal, the anhydride of the di-acid (*14*) (Cross, Stewart, & Stoddard, 1970; Hanson et al., 1972). Fujenal and (*14*) are major terpenoid components of *G. fujikuroi* cultures (Cross, Galt, & Hanson, 1963a). The di-acid (*14*) differs from the *C. maxima* intermediate (*13*) only in that C-6 is oxidized one step further to a carboxylic acid. Interestingly, small amounts of (*14*) have been detected in *C. maxima* endosperm (Blechschmidt et al., in preparation) although nothing is known about its biosynthesis in this system.

An alternative metabolic fate for *ent*-6α,7α-dihydroxykaurenoic acid in *C. maxima* endosperm is indicated from the identification of *ent*-6α,7α,12α-trihydroxykaurenoic acid (*15*) as a major metabolite in this tissue (Blechschmidt et al., in preparation). However, the *ent*-12α-hydroxylation of *ent*-6α,7α-dihydroxykaurenoic acid has not yet been observed directly. Mature seeds of *C. maxima* and *C. pepo* contain both 12β- and 12α-hydroxylated GAs—e.g., GA_{49} and GA_{48} (Blechschmidt et al., in preparation; Gaskin et al., in preparation; Fukui et al., 1977)—a fact that demonstrates the complexity of the *Cucurbita* system. Whether or not separate oxidases catalyze hydroxylation at the 12β and 12α positions remains to be determined.

3.4 REACTIONS CATALYZED BY SOLUBLE ENZYMES

The later steps of the GA biosynthetic pathway are catalyzed by enzymes that remain in the supernatant after centrifugation at 200,000 g. The soluble nature of these enzymes might have been anticipated since their substrates are more highly oxidized and, therefore, more hydrophilic than the earlier intermediates, which are oxidized by microsomal enzymes. The available information on the soluble oxidases is less extensive than that for the microsomal enzymes, but new results, to be discussed later in this section, have shed some light on their nature. Complete pathways to polar GAs *in vitro* have been demonstrated in cell-free extracts from *C. maxima* endosperm (Graebe et al., 1974a, 1974b), immature seeds of *P. sativum* (Kamiya & Graebe, 1983), and suspensors of *Ph. coccineus* (Ceccarelli, Lorenzi, & Alpi, 1981b). In addition, a single-terminal reaction has been studied in a cell-free system from germinating seeds of *Ph. vulgaris* (Patterson & Rappaport, 1974; Patterson et al., 1975; Hoad et al., 1982).

3.4.1 The pathways in *Cucurbita maxima* endosperm

GA_{12} aldehyde is converted by the 200,000-g supernatant fraction from *C. maxima* endosperm to mainly GA_{43} (Graebe & Hedden, 1974) and a small amount of GA_4 (Graebe et al., 1974b). Since GA_{43} is an endogenous component of *C. maxima* endosperm, the observed conversions *in vitro* parallel those operating *in vivo*. When the substrate concentration is increased or suboptimal cofactor concentrations are used, intermediates between GA_{12} aldehyde and GA_{43} accumulate. These intermediates were identified by GC-MS and re-incubated with the cell-free system to establish the pathways illustrated in Fig. 3.8 (Graebe et al., 1974a; Graebe, Hedden, &

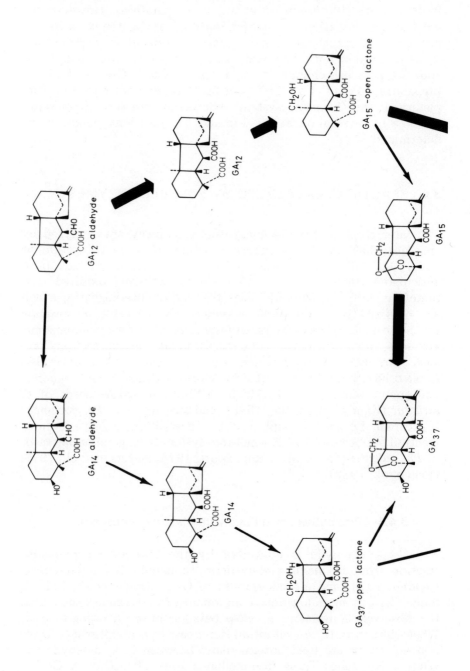

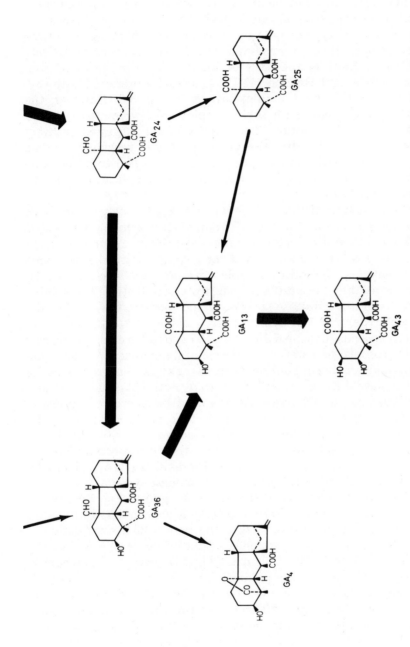

Figure 3.8 Pathways for the conversion of GA_{12} aldehyde to GA_{43} and GA_4 by the 200,000-g supernatant fraction of the *C. maxima* cell-free system. Thick arrows denote the "preferred" pathway.

MacMillan, 1974c; Graebe & Hedden, 1974). This network of pathways represents a metabolic grid in which numerous routes are possible between GA_{12} aldehyde and GA_{43}. Some of these intermediates—e.g., GA_{14} aldehyde—were obtained only at high substrate concentrations; others—for example, GA_{25}—were converted relatively poorly. Therefore, on the basis of the relative efficiencies of conversion, Graebe and Hedden (1974) suggested a preferred pathway, which is indicated by the thick arrows in Fig. 3.8. However, the results of experiments with such complex enzyme mixtures should be interpreted with caution. The validity of the suggested pathway can be checked only after the isolation of the individual enzyme activities when their relative specificities with regard to all possible substrates can be determined without the complications due to competing reactions.

The oxidation of GA_{12} aldehyde to GA_{12} is catalyzed by both soluble and microsomal enzymes. Two distinct enzymes are indicated from their different cofactor requirements; the microsomal enzyme requires only NADPH and O_2, and the soluble enzyme has the same requirements as the other soluble oxidases (see Section 3.4.4). Furthermore, they have different substrate specificities. The soluble oxidase has a low substrate specificity and oxidizes GA_{12} aldehyde, GA_{14} aldehyde, and 12α-hydroxy GA_{12} aldehyde to the respective dicarboxylic acids (Graebe & Hedden, 1974; Hedden et al., in preparation). Microsomal preparations, however, only oxidize GA_{12} aldehyde (Hedden and Graebe, unpublished results). The two enzyme activities may thus have different functions *in vivo*. Both GA_{12} aldehyde and 12α-hydroxy GA_{12} aldehyde are produced by microsome preparations, the former metabolite being further oxidized to GA_{12} by these preparations. Although GA_{12} aldehyde is also efficiently oxidized by the 200,000-g supernatant, the true substrate for the soluble C-7 oxidase *in vivo* may be 12α-hydroxy GA_{12} aldehyde, which is metabolized only by the soluble oxidase system.

The further reactions of the metabolic grid in Fig. 3.8 involve sequential oxidation of C-20 to a carboxylic acid, and hydroxylation at the 3β- and 2β-positions. Within this series, 2β-hydroxylation is specific to GA_{13} but the sequence of the other reactions (i.e., the oxidation level of C-20 at which 3β-hydroxylation can occur) is relatively unimportant. One point of interest is the oxidation of the C-20 alcohol. GAs that have C-20 at the alcohol oxidation level are found exclusively as $19 \rightarrow 20$ lactones. However, there is evidence that the C-20 alcohol group must be free for further oxidation to occur. GA_{15} and GA_{37} were identified as products after incubations of GA_{12} aldehyde with the *C. maxima* cell-free system (Graebe et al.,

1974a), but GA_{15} was converted by the same system only to GA_{37} (by 3β-hydroxylation) and GA_{37} was not metabolized (Graebe et al., 1974c). If, however, the δ-lactone of GA_{15} is opened by alkaline hydrolysis prior to incubation, conversion occurs to GA_{43}, the end product of the pathway (Hedden & Graebe, 1982). Thus, the free-alcohol forms of the δ-lactone GAs appear to be the true biosynthetic intermediates.

The 200,000-g supernatant also metabolizes 12α-hydroxy GA_{12} aldehyde, which is produced by the microsomes, to 12α-hydroxy GA_{14} (16) and 12α-hydroxy GA_{37} (17) (Hedden et al., in preparation),

(16)   (17)

Structures (16)–(17)

both of which are endogenous to *C. maxima* endosperm (Blechschmidt et al., in preparation; Gaskin et al., in preparation). Thus, a second set of pathways originating from 12α-hydroxy GA_{12} aldehyde must also be operating *in vivo*. The pH optimum for this conversion is *ca.* 6.2, which is similar to that for the microsomal 12α-hydroxylation reaction but lower than that for the conversion of GA_{12} aldehyde to GA_{43} to GA_4 (pH 7.5). This difference is the pH optima for the two sets of pathways may indicate that they are segregated within the cell.

3.4.2   The loss of carbon-20

The C_{19}-GA, GA_4, was identified as a minor product from an incubation of GA_{12} aldehyde with the *C. maxima* cell-free system (Graebe et al., 1974b). Although it is formed in only low yield (5%), its biosynthesis has provided an important opportunity to study the formation of the C_{19}-GAs. The C_{19}-GAs, which are generally recognized as the physiologically active hormones, are formed from C_{20}-GAs by the loss of C-20. This conversion has been the subject of a great deal of research and perhaps even more speculation. Serious consideration of possible mechanisms for the loss of C-20 has been hampered because the identity of the immediate C_{20} precursor was unknown. Experiments with the cultures of *G. fujikuroi* gave inconclusive results, perhaps due to differential permeability of the fungal cell membrane to the substrates applied (Bearder et al., 1975). One important result to emerge from such experiments, however, showed conclusively that both oxygen atoms in the γ-lactone of C_{19}-GAs originate from the 19-oic acid group of the C_{20}-precursor (Bearder, MacMillan, & Phinney, 1976). Thus, an intermediate with a free-hydroxy group at C-10 can be ruled out.

Recent investigations using the *C. maxima* cell-free system have shown that GA_4 is formed from GA_{36} but not from GA_{13} (see Fig. 3.8) (Böse, 1980; Graebe, Hedden, & Rademacher, 1980). The enzymatic properties for this conversion are discussed in Section 3.4.4. This result verifies an early suggestion that C-20 was lost at the aldehyde oxidation level (Hanson & White, 1969) but contradicts recent results from experiments with *G. fujikuroi* that supported loss of C-20 as CO_2 from a carboxylic acid (Dockerill & Hanson, 1978). In these experiments, $^{14}CO_2$ was recovered from an incubation of *ent*-$[^{14}C]$kaurene labeled, *inter alia*, at C-20. No ^{14}C-labeled formaldehyde or formic acid was detected. However, this is not conclusive evidence that C-20 is lost at the carboxylic acid oxidation level. Cellular oxidation of C-20 subsequent to its loss as formaldehyde or formic acid would give the same result. Attempts to trap formaldehyde or formic acid might only be successful if the products are released as such from the cell or cell compartment where metabolism had occurred. For this reason *in vitro* systems should prove more suitable than intact tissue for testing possible mechanisms for the loss of C-20. The *C. maxima* endosperm system is potentially very useful for this purpose. Although the non-12-hydroxylation pathway in this system produces predominantly the C_{20}-GA, GA_{43}, and GA_4 only in low yield, the end products of the 12-hydroxylation pathway are mainly C_{19}-GAs, for example, GA_{58} (12α-hydroxy GA_4). Therefore, the use of 12-hydroxylated precursors with the *C. maxima* system should yield valuable information on the formation of C_{19}-GAs. Other suitable systems that produce predominantly C_{19}-GAs are those from *P. sativum* seeds and *Ph. coccineus* suspensors (see Section 3.4.3).

The demonstration that C-20 is lost at the aldehyde oxidation level limits the number of possible mechanisms for this reaction. Demethylation reactions involving aldehydes are known in steroid metabolism (Arigoni et al., 1975; Pascal, Chang, & Schoepfer, 1980) but must be mechanistically different from the reaction in GA biosynthesis. In the steroid reactions the loss of the carbon atom as formic acid is assisted by the movement of a β,γ double bond to the

α,β-position. However, corresponding unsaturated intermediates are not involved in GA demethylation (Hanson & White, 1969). The direct participation of the 19-oic acid may provide the driving force in this case. Another biological reaction that may be analogous to the GA demethylation is the oxidation of camphor to the 1,2-campholide by a mixed-function oxygenase from *Pseudomonas* (Gunsalus, Conrad, & Trudgill, 1964). This reaction closely parallels the chemical Baeyer-Villiger oxidation of camphor (Meinwald & Frauenglass, 1960). However, a normal Baeyer-Villiger mechanism in GA demethylation as proposed by Hanson and White (1969) must be discounted in the light of the results obtained by Bearder et al. (1976), although a modified reaction in which the intermediate ester is cleaved by alkyl-oxygen fission as suggested by Bearder and Sponsel (1977) would be permissable. The loss of C-20 still remains the major unsolved problem in GA biosynthesis.

3.4.3 Other systems

The 2β-hydroxylation of $[^3H]GA_1$ to $[^3H]GA_8$ has been shown to occur in homogenates of germinating seeds of *Ph. vulgaris* (Patterson & Rappaport, 1974). This activity, which was confined to the cotyledons, remained in the supernatant after centrifugation at 95,000 g. The reaction appears to be catalyzed by the same type of soluble oxidase as those in *C. maxima* endosperm. The substrate, which was prepared by catalytic tritiation of GA_3 (Nadeau & Rappaport, 1974), was assumed to contain ^3H at the 1β- and 2β-positions as well as at random positions in the molecule. 2β-Hydroxylation was accompanied by the release of ^3H, which could be recovered quantitatively as ^3HOH. The reaction was found to have at least some substrate specificity; 3-*epi*-GA_1 (3α-hydroxy GA_{20}) and C/D ring-rearranged GA_1, also labeled at the 1β- and 2β-positions, were not metabolized under conditions in which GA_1 was hydroxylated (Patterson et al., 1975). The enzyme properties for the conversion were studied in some detail and will be discussed in the following section.

Recent experiments using homogenates from immature seeds of *P. sativum* have extended the pathway in this system to the endogenous C_{19}-GAs, namely GA_9, GA_{20}, GA_{29}, and GA_{51} (Kamiya & Graebe, 1983). The results are summarized in Fig. 3.9; each step has been confirmed by GC-MS identification of the products. The two parallel pathways from GA_{12} to GA_{51} and from GA_{53} to GA_{29} are catalyzed by the 200,000-g supernatant fraction. GA_{15} and

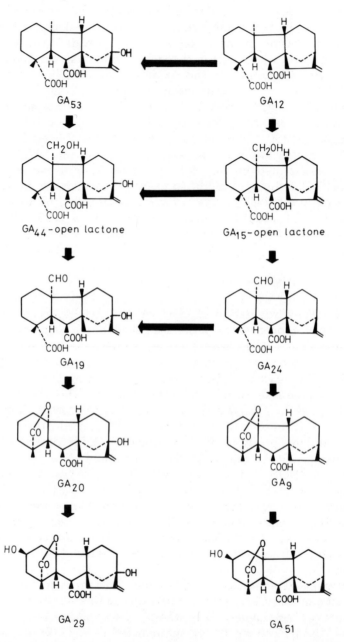

Figure 3.9 Conversions observed *in vitro* in immature seeds of *P. sativum*. Horizontal arrows represent reactions catalyzed by the 200,000-g supernatant; vertical arrows represent a 13α-hydroxylation, which is catalyzed by the 200,000-g pellet.

GA_{44} were converted only as the open lactones produced by base hydrolysis. The pathways are connected by a 13-hydroxylation step. The enzyme(s) involved is located in the microsomal fraction and exhibits limited specificity with respect to the GAs in the pathway, although GA_{15} and GA_9 undergo 13-hydroxylation only as open lactones. *In vivo* 13-hydroxylation is expected to occur early.

Ceccarelli et al. (1981b) have recently identified three C_{19}-GAs, namely GA_1, GA_5, and GA_8, as products from an incubation of *ent*-7α-hydroxykaurenoic acid with the 2,000-g supernatant from *Ph. coccineus* suspensor homogenates. No intermediates were detected, so it was not possible to draw any conclusions about the pathway. Since GA_1 has been identified as the major GA in *Ph. coccineus* suspensor (Alpi et al., 1979), the *in vitro* metabolism observed by the authors very probably represents an endogenous process. Furthermore, the suspensor is established as an important site of GA biosynthesis in immature *Ph. coccineus* seeds.

3.4.4 Properties of the enzymes

The 2β-hydroxylase in *Ph. vulgaris* cotyledons has been studied quite extensively (Patterson & Rappaport, 1974; Patterson et al., 1975). The enzyme requires O_2, Fe^{2+}, and a reducing agent, NADPH or ascorbate, for activity. The activity could be inhibited by the addition of EDTA and subsequently restored by adding Fe^{2+} or Fe^{3+}. Fe^{3+} was more effective in the presence of ascorbate, whereas Fe^{2+} was equally effective with or without the reducing agent. Ascorbate appears to act, therefore, by reducing Fe^{3+} to Fe^{2+}, which is the form of iron necessary for enzyme activity. The inhibition by EDTA indicates that Fe^{2+} is loosely bound to the enzyme, in contrast to the cytochrome P_{450} oxidases where it is covalently bound as part of a heme system. An inhibition of activity by 2-mercaptoethanol was partially overcome by Fe^{2+}, which lead Patterson et al. (1975) to suggest that sulfhydryl groups may be involved in binding the metal ion. Very high levels of CO (90–100%) were required to inhibit the enzyme activity, and the inhibition was the same in light and dark. Thus, again in contrast to the microsomal oxidases, a cytochrome-type enzyme is not indicated. A peak of activity was obtained after centrifugation of the 95,000-g supernatant through a sucrose density gradient corresponding to a sedimentation coefficient of 4.5 S. Thus, the enzyme consists of a single protein or group of proteins of similar size.

The soluble oxidases from *C. maxima* endosperm were also found to have an absolute requirement for Fe^{2+} and to be stimulated by NADPH (Graebe & Hedden, 1974). Enzyme activity is inhibited by the divalent metal ions Mn^{2+}, Co^{2+}, Ni^{2+}, and Cu^{2+} (Hedden & Graebe, 1982). Although Cu^{2+} and Co^{2+} were also found to inhibit the *Ph. vulgaris* hydroxylase, Mn^{2+} had little effect on the activity (Patterson et al., 1975). Since the Mn^{2+} inhibition in the *C. maxima* system could be reversed by the addition of Fe^{2+}, it is possible that the two metal ions compete for a binding site.

Recent results obtained using *C. maxima* endosperm have now given some indication as to the nature of the soluble oxidases (Hedden & Graebe, 1982). After gel filtration of the 200,000-g supernatant through Sephadex G-25, enzyme activity for GA_{12} aldehyde metabolism to GA_{43} was lost, but could be restored by adding back the filtrate. The necessary component of the filtrate was identified as 2-oxoglutaric acid, for which the enzyme activity had an absolute and specific requirement. Pyruvic and oxalacetic acids were ineffective. The only other necessary cofactor was Fe^{2+}, although ascorbate and, to some extent, NADPH stimulated activity when Fe^{2+} was limiting or when only Fe^{3+} was available. This supports the contention that the reducing agents act by maintaining Fe^{2+} in the reduced state. The conversion of GA_{12} aldehyde was also inhibited by a N_2 atmosphere, demonstrating the need for O_2 at least for the first step in the metabolic sequence. Table 3.2 lists the effects of cofactors on the conversion of GA_{12} aldehyde to GA_{13} and GA_{43}. The requirement for 2-oxoglutaric acid, Fe^{2+}, O_2, and a reducing agent, are properties shared by a group of dioxygenases (Abbott & Udenfriend, 1974) for which 2-oxoglutaric acid is a cosubstrate, being oxidized to succinic acid and CO_2 according to the following equation:-

$$HO_2C \cdot CH_2 \cdot CH_2 \cdot CO \cdot CO_2H + O_2 + RH$$
$$\rightarrow HO_2C \cdot CH_2 \cdot CH_2 \cdot CO_2H + CO_2 + ROH$$

The mechanism of the dioxgenase reaction and the role of Fe^{2+} is still uncertain, although the formation of an oxenoid intermediate from the reaction of Fe^{2+}, O_2, and 2-oxoglutaric acid has been proposed (Hamilton, 1974; Siegel, 1979; Visser, 1980). An alternative mechanism in which the substrate is oxidized to a peroxide, which then attacks the 2-oxoglutaric acid, has also been suggested (Lindblad et al., 1969). One of the few examples of this type of enzyme isolated from a higher plant is peptidyl proline hydroxylase, which was purified from carrot storage roots (Sadava & Chrispeels, 1971). However, this enzyme could utilize pyruvate or oxalacetate

Table 3.2. Cofactor requirements for the conversion of GA_{12} aldehyde to GA_{13} and GA_{43} by the 200,000-g supernatant from a homogenate of *C. maxima* endosperm after gel filtration through Sephadex G-25

α-Keto acid (0.5 mM)	Cation (0.5 mM)	Reductant	GA_{13}	GA_{43}	Total
	Cofactors		Yield of products (pmol)		
2-Oxoglutarate	Fe^{2+}	ascorbate[a]	306	163	469
2-Oxoglutarate	Fe^{2+}	NADPH[b]	372	89	461
2-Oxoglutarate	Fe^{2+}	—	186	61	247
2-Oxoglutarate	—	ascorbate[a]	27	28	55
2-Oxoglutarate	Fe^{3+}	—	6	29	35
2-Oxoglutarate	Fe^{3+}	NADPH[b]	5	14	19
2-Oxoglutarate	Fe^{3+}	ascorbate[a]	363	129	492
—	Fe^{2+}	ascorbate[a]	4	13	17
Pyruvate	Fe^{2+}	ascorbate[a]	3	7	10
Oxalacetate	Fe^{2+}	ascorbate[a]	7	17	24

NOTE: $[^{14}C]GA_{12}$ aldehyde (500 pmol) was incubated with the protein fraction (200 μl, half original concentration) at $30°C$ for 1 hr.
[a] 5 mM.
[b] 0.5 mM.

almost as well as 2-oxoglutarate. A 2-oxoglutarate-dependent dioxygenase has also been shown to be involved in anthocyanin biosynthesis in *Matthiola incana* flowers (Forkmann, Heller, & Grisebach, 1980). In fact, this system resembled GA biosynthesis in that both a microsomal NADPH-dependent monooxygenase and the soluble dioxygenase were involved.

Each reaction in the sequence between GA_{12} aldehyde and GA_{43} in *C. maxima* has been shown to be 2-oxoglutarate-dependent (Hedden & Graebe, 1982), as has also the formation of GA_4 from GA_{36} (Böse, unpublished results in this laboratory). Thus, the enzyme activity responsible for the loss of C-20 has similar properties to the other soluble oxidases. This has also been confirmed for *P. sativum*, in which the soluble oxidases have the same cofactor requirements as the *C. maxima* enzymes (Kamiya & Graebe, 1983). The 2β-hydroxylase from *Ph. vulgaris* cotyledons has now been extensively purified and shown also to be a 2-oxoglutarate-dependent dioxygenase (Hoad et al., 1982). Thus it can be predicted with resonable confidence that all the enzymes for the later stages of GA biosynthesis will prove to be of this type.

3.5 SUMMARY AND CONCLUSIONS

Five cell-free systems capable of synthesizing GAs from MVA have been described. Four are from higher plants (*M. macrocarpus*, *C. maxima*, *P. sativum*, and *Ph. coccineus*) and the fifth from the fungus *G. fujikuroi*. Although these examples are still few, a comparison of the systems reveals some interesting differences and similarities. The biosynthetic pathway to GA_{12} aldehyde is the same in all systems, and parts of the sequence have also been demonstrated in other systems (see Table 3.1). It is probable that GA_{12} aldehyde is synthesized by a pathway common to all organisms that produce GAs. The further metabolism of GA_{12} aldehyde is, thus, the earliest point at which differences between species can be observed. GA_{12} aldehyde is subject to two oxidation reactions by membrane-associated enzymes; oxidation at C-7 to give GA_{12}, which is a reaction common to the four systems in which the individual steps have been examined, or hydroxylation at a position that varies depending on the organisms. In *P. sativum* the hydroxy group is introduced at C-13. This hydroxylation shows relatively low substrate specificity, so that GA_{12} and more polar GAs also serve as substrates *in vitro*. However, if it is reasonable to assume that those intermediates that are synthesized in the microsomes are the most probable substrates for microsomal enzymes, then GA_{12} aldehyde or GA_{12} are the most likely substrates for 13-hydroxylation *in vivo*. The further metabolism of GA_{12} and 13-hydroxy GA_{12} aldehyde or 13-hydroxy GA_{12} (GA_{53}) by soluble oxidases to more polar GAs, including C_{19}-GAs, results in two parallel pathways, one for 13-hydroxylated intermediates and the second for non-13-hydroxylated intermediates.

Since 13-hydroxylated GAs have been identified in many species (see Bearder, 1980), early (microsomal) 13-hydroxylation is expected to be a common feature of GA biosynthesis in higher plants. Microsomal preparations from *C. maxima* endosperm hydroxylate GA_{12} aldehyde at C-12, but GA_{12} at C-13. The similarity of the 12- and 13-hydroxylases, and the close proximity of C-12 and C-13 suggest that the two enzymes are related. The 13-hydroxylation of GA_{12}, a reaction analogous to that found in *P. sativum*, can be of little importance *in vivo* since only traces of 13-hydroxylated GAs are produced. GA_{12} is rapidly metabolized by soluble oxidases, as is the other major product of the microsomes, 12α- hydroxy GA_{12} aldehyde. Another difference between *P. sativum* and *C. maxima* is the presence of a soluble 3β-hydroxylase in the latter. However, in this case *P. sativum* appears to be the exception, since 3β-hydroxylated GAs are of widespread occurrence.

The branch at GA_{12} aldehyde is also observed in the fungus *G. fujikuroi*. Microsomal oxidases convert GA_{12} aldehyde either to GA_{12} or to 3β-hydroxy GA_{12} aldehyde (GA_{14} aldehyde), the latter not accumulating but being rapidly oxidized to GA_{14}. Thus, the molecular site of hydroxylation by fungal microsomes is quite distinct from that in the higher plant species, *P. sativum* and *C. maxima*. Unfortunately, since the steps beyond GA_{12} and GA_{14} in the fungus could not be obtained in active form, it is not known whether or not the enzymes involved in this part of the pathway are of the same type as those in higher plants. However, 13-hydroxylation is known to occur late in the pathway in *G. fujikuroi*; in fact, it is the final step in the biosynthesis of GA_3. Thus, although the fungus hydroxylates at the same positions as many higher plant species, it does so in reverse order and probably uses a different type of enzyme for the respective reactions. This suggests separate origins for GA biosynthesis in the fungus and in higher plants.

The consistent pattern that the first hydroxylation after formation of the GA skeleton occurs on microsomes may be due to a requirement that the intermediates attain sufficient polarity to leave the lipid environment in which they are synthesized and become available for oxidation by soluble enzymes. Conversion of GA_{12} aldehyde to GA_{12} could have the same result.

A common feature of GA biosynthesis in *C. maxima* endosperm and *G. fujikuroi* is the formation of kaurenolides and the oxidation products of *ent*-$6\alpha,7\alpha$-dihydroxykaurenoic acid as side products. The synthesis of these products is confined to the microsomes, and in *C. maxima* it has been shown that they are not substrates for the soluble oxidases. It may turn out that the formation of these compounds is an unavoidable consequence of GA biosynthesis. There are indications that many of the reactions involved in the synthesis of the side products are catalyzed by enzymes related to those of the main pathway to the GAs. Hydroxylation, when it occurs, is at a similar position to that observed in the GA pathway. Thus $7\beta,12\alpha$-dihydroxykaurenolide is produced in *C. maxima* and $3\beta,7\beta$- and $7\beta,18$-dihydroxykaurenolides are found in *G. fujikuroi*. C-18 is sufficiently close to C-3 for a single enzyme to hydroxylate at both positions. Although no such biproducts have been found in *P. sativum*, it is possible that they are further metabolized to very polar products in this species since *ent*-6α,7α-dihydroxykaurenoic acid is rapidly converted to polar products when incubated with the cell-free system from *P. sativum* seeds (Ropers, 1978). In mature seeds of *Ph. coccineus*, a species that produces 13-hydroxylated GAs, both 7β,13-dihydroxykaurenolide and *ent*-6α,7α,13-trihydroxykaurenoic acid have been identified as conjugates (Gaskin & MacMillan, 1975).

The advances in GA metabolism research *in vitro* have been made largely using immature seeds or tissues derived from them, generally for reasons of convenience. Seeds of dicotyledonous plants have a relatively high GA content, and the immature seeds a high level of biosynthetic activity. However, the role of GAs in seed development is still poorly understood. Cell-free systems from vegetative tissue, in which GAs are known to be important for internode elongation, have been much less successful. What, then, is the relevance of studying GA metabolism *in vitro* to physiological problems? To understand how the GA content of a particular tissue might be regulated, it is important to know the individual reactions in the metabolism of the hormone and the nature of the enzymes that catalyze these reactions. Although it cannot be assumed that the GA content of a developing leaf or other tissue does not differ qualitatively as well as quantitatively from that of immature seeds, biosynthetic pathways established in seeds can serve as models from which to base experiments with other tissues.

It has been shown that GAs can vary both qualitatively and quantitatively during the development of a plant (see, for example, Frydman, Gaskin, & MacMillan, 1974). Knowledge of the metabolic relations between different GAs, in conjunction with their biological activities, can form the basis for understanding the significance of changing GA patterns. An example of the control of GA biosynthesis through the apparent regulation of a single step is suggested from the results of Metzger and Zeevaart (1980). These workers found a change in GA content of *Spinacea oleracea* after the introduction of a long-day photoperiod. A drop in the level of GA_{19} was accompanied by a rise in the level of GA_{20}. Other GAs monitored maintained constant levels. The proposed biosynthetic relationship, $GA_{19} \rightarrow GA_{20}$, is supported by the results with the *C. maxima in vitro* system in which it was shown that a C_{20}-GA with C-20 at the aldehyde oxidation level (such as GA_{19}) is indeed the direct precursor of C_{19}-GAs (Graebe et al., 1980). The possibility that a particular reaction in the GA pathway is under photoperiodic control is an important concept and warrants further investigation.

Detailed exploration of the enzymological aspects of GA metabolism is only just beginning. Knowledge of the enzymes involved could be of great benefit. For example, the sites of action of inhibitors of GA biosynthesis, some of which have agricultural applications as plant growth retardants, have been determined using cell-free systems. A better understanding of the mechanisms of action of these substances at the molecular level could lead to the design of new inhibitors with increased potency and specificity. This could have both economical and ecological benefits. But, apart from anything else, the complexity of the GA biosynthetic pathway (about 20 steps are required for the biosynthesis of a C_{19}-GA from MVA) provides the plant biochemist with a fascinating challenge.

ACKNOWLEDGMENT

The author wishes to thank Professor J. E. Graebe for his useful suggestions on the manuscript.

REFERENCES

Abbot, M. T., & Udenfriend, S. (1974). α-Ketoglutarate-coupled dioxygenases. In: Molecular Mechanisms of Oxygen Activation, pp. 167–214, Hayaishi, O., ed. Academic Press, New York.

Alpi, A., Lorenzi, R., Cionini, P. G., Bennici, A., & D'amato, F. (1979). Identification of gibberellin A_1 in the embryo suspensor of *Phaseolus coccineus*. Planta *147*, 225–228.

Arigoni, D., Battaglia, R., Aktar, M., & Smith, T. (1975). Stereospecificity of oxidation at C-19 in oestrogen biosynthesis. J. Chem. Soc., Chem. Commun., 185–186.

Bateson, J. H., & Cross, B. E. (1973). 3β,7β-Dihydroxykaurenolide. A new metabolite of *Gibberella fujikuroi*. Tetrahedron Lett., 3407–3408.

Beale, M. H., Bearder, J. R., Down, D. H., Huchison, M., MacMillan, J., & Phinney, B. O. (1982). The biosynthesis of kaurenolide-diterpenoids by *Gibberella fujikuroi*. Phytochem. *21*, 1279–1287.

Bearder, J. R. (1980). Plant hormones and other growth substances—their background, structures and occurrence. In: Hormonal Regulation of Development I. Molecular Aspects of Plant Hormones. Encyclopedia of Plant Physiology New Series, Vol. 9, pp. 9–112, MacMillan, J., ed. Springer-Verlag, Berlin, Heidelberg, New York.

Bearder, J. R., MacMillan, J., & Phinney, B. O. (1975). Fungal products. Part XIV. Metabolic pathways from *ent*-kaurenoic acid to the fungal gibberellins in mutant B1-41a of *Gibberella fujikuroi*. J. Chem. Soc., Perkin Trans. I, 721–726.

Bearder, J. R., MacMillan, J., & Phinney, B. O. (1976). Origin of the oxygen atoms in the lactone bridge of C_{19}-gibberellins. J. Chem. Soc., Chem. Commun., 834–835.

Bearder, J. R., MacMillan, J., Wels, C. M., Chaffey, M. B., & Phinney, B. O. (1974). Position of the metabolic block for gibberellin biosynthesis in mutant B1-41a of *Gibberella fujikuroi*. Phytochem. *13*, 911–917.

Bearder, J. R., & Sponsel, V. M. (1977). Selected topics in gibberellin metabolism. Biochem. Rev. *5*, 569–582.

Becker, B. (1974). Die Biosynthese von GA_{12}-aldehyd und *ent*-6α,7α-Dihydroxy-kaurensaüre aus *ent*-7α-Hydroxykaurensaüre in einem Zell-freiem System aus Cucurbita-endosperm. Diplomarbeit. Georg-August-Universität, Göttingen, F.R.G.

Blechschmidt, S., Castel, U., Gaskin, P., Graebe, J. E., Hedden, P., & MacMillan, J. (in preparation). GC-MS analysis of the plant hormones in seeds of *Cucurbita maxima*.

Böse, G. (1980). Zur Frage der Zwischenstufen bei dem Übergang von C_{20}- zu C_{19}-gibberellinen in der Gibberellinbiosynthese. Diplomarbeit. Georg-August-Universität, Göttingen, F.R.G.

Boulton, K., & Cross, B. E. (1981). Inhibitors of the biosynthesis of gibberellins. Part 1. 7-Fluoro-10β-fluoromethyl-1β, 8-dimethylgibbane-1α,4aα-carbolactone. J. Chem. Soc., Perkin Trans. I, 427–432.

Bowen, D. H., MacMillan, J., & Graebe, J. E. (1972). Determination of specific radioactivity of [^{14}C]-compounds by mass spectrometry. Phytochem. *11*, 2253–2257.

Buchenauer, H. (1977). Mode of action and selectivity of fungicides which interfere with ergosterol biosynthesis. In: Proceedings of the 1977 British Crop Protection Conference—Pests and Diseases, pp. 699–711. British Crop Protection Council, London.

Ceccarelli, N., Lorenzi, R., & Alpi, A. (1979). Kaurene and kaurenol biosynthesis in cell-free system of *Phaseolus coccineus* suspensor. Phytochem. *18*, 1657–1658.

Ceccarelli, N., Lorenzi, R., & Alpi, A. (1981a). Kaurene metabolism in cell-free extracts of *Phaseolus coccineus* suspensor. Plant Sci. Lett. *21*, 325–332.

Ceccarelli, N., Lorenzi, R., & Alpi, A. (1981b). Gibberellin biosynthesis in *Phaseolus coccineus* suspensor. Z. Pflanzenphysiol. *102*, 37–44.

Coolbaugh, R. C., & Durst, R. W. (1978). Formation of *ent*-kaurene-2-mercapto-ethanol complex *in vitro*. Pacific Div., AAAS Abstracts, 7.

Coolbaugh, R. C., & Hamilton, R. (1976). Inhibition of *ent*-kaurene oxidation and growth by α-cyclopropyl-α-(p-methoxyphenyl)-5-pyrimidine methyl alcohol. Plant Physiol. *57*, 245–248.

Coolbaugh, R. C., Heil, D. R., & West, C. A. (1982a). Comparative effects of substituted pyrimidines on growth and gibberellin biosynthesis in *Gibberella fujikuroi*. Plant Physiol. *69*, 712–716.

Coolbaugh, R. C., Hirano, S. S., & West, C. A. (1978). Studies on the specificity and site of action of α-cyclopropyl-α-[p-methoxypropyl]-5-pyrimidine methyl alcohol (ancymidol), a plant growth regulator. Plant Physiol. *62*, 571–576.

Coolbaugh, R. C., & Moore, T. C. (1971). Metabolism of kaurene in cell-free extracts of immature pea seeds. Phytochem. *10*, 2401–2412.

Coolbaugh, R. C., Moore, T. C., Barlow, S. A., & Ecklund, R. R. (1973). Biosynthesis of *ent*-kaurene in cell-free extracts of *Pisum sativum* shoot tips. Phytochem. *12*, 1613–1618.

Coolbaugh, R. C., Swanson, D. I., & West, C. A. (1982b). Comparative effects of ancymidol and its analogs on growth of peas and *ent*-kaurene oxidation in cell-free extracts of immature *Marah macrocarpus* endosperm. Plant Physiol. *69*, 707–711.

Coon, M. J., Ballou, D. P., Haugen, D. A., Krezoski, S. O., Nordblom, G. D., & White, R. E. (1977). Purification of membrane-bound oxygenases: Isolation of two electrophoretically homogeneous forms of liver microsomal cytochrome P-450. In: Microsomes and Drug Oxidations, pp. 82–94, Ullrich, V., Roots, I., Hildebrandt, A., Estabrook, R. W., & Conney, A. H., eds. Pergamon Press, Oxford.

Cross, B. E., Galt, R. H. B., & Hanson, J. R. (1963a). New metabolites of

Gibberella fujikuroi. Part V. The structures of fujenal and fujenoic acid. J. Chem. Soc., 5052-5081.

Cross, B. E., Galt, R. H. B., Hanson, J. R., Curtis, P. J., Grove, J. F., & Morrison, A. (1963b). New metabolites of *Gibberella fujikuroi.* Part II. The isolation of fourteen new metabolites. J. Chem. Soc., 2937-2943.

Cross, B. E., Galt, R. H. B., & Norton, K. (1968). The biosynthesis of the gibberellins—II. Tetrahedron *24*, 231-237.

Cross, B. E., Hanson, J. R., & Galt, R. H. B. (1963c). New metabolites of *Gibberella fujikuroi.* Part IV. The structures of 7,18-dihydroxy- and 7,16,18-trihydroxykaurenolides. J. Chem. Soc., 3783-3790.

Cross, B. E., Stewart, J. C., & Stoddard, J. L. (1970). $6\beta,7\beta$-Dihydroxykaurenoic acid: Its biological activity and possible role in the biosynthesis of gibberellic acid. Phytochem. *9*, 1065-1071.

Dennis, D. T., & West, C. A. (1967). Biosynthesis of gibberellins. III. The conversion of (—)-kaurene to (—)-kauren-19-oic acid in endosperm of *Echinocystis macrocarpa* Greene. J. Biol. Chem. *242*, 3293-3300.

Dockerill, B., Evans, R., & Hanson, J. R. (1977). Removal of C-20 in gibberellin biosynthesis. J. Chem. Soc., Chem. Commun., 919-921.

Dockerill, B., & Hanson, J. R. (1978). The fate of C-20 in C_{19}-gibberellin biosynthesis. Phytochem. *17*, 701-704.

Evans, R., & Hanson, J. R. (1972). The formation of (—)-kaurene in a cell-free system from *Gibberella fujikuroi.* J. Chem. Soc., Perkin Trans. I, 2382-2385.

Evans, R., & Hanson, J. R. (1975). Studies in terpenoid biosynthesis. Part XIII. The biosynthetic relationship of the gibberellins in *Gibberella fujikuroi.* J. Chem. Soc., Perkin Trans. I, 663-666.

Evans, R., Hanson, J. R., & White, A. F. (1970). Studies in terpenoid biosynthesis. Part VI. The stereochemistry of some stages in tetracyclic diterpene biosynthesis. J. Chem. Soc. C, 2601-2603.

Forkmann, G., Heller, W., & Grisebach, H. (1980). Anthocyanin biosynthesis in flowers of *Matthiola incana.* Flavanone 3- and flavanoid 3'-hydroxylases. Z. Naturforsch. *35c*, 691-695.

Frydman, V. M., Gaskin, P., & MacMillan, J. (1974). Qualitative and quantitative analyses of gibberellins throughout seed maturation in *Pisum sativum* cv. Progress No. 9. Planta *118*, 123-132.

Fukui, H., Nemori, R., Koshimizu, N., & Yamazaki, Y., (1977). Structures of gibberellin A_{39}, A_{48}, A_{49}, and a new kaurenolide in *Cucurbita pepo* L. Agr. Biol. Chem. *41*, 181–187.

Gaskin, P., Huchison, M., Lewis, N., MacMillan, J., & Phinney, B. O. (in preparation). Microbiological conversion of 12-oxygenated, and other derivatives, of *ent*-kaur-16-en-19-oic acid by *Gibberella fujikuroi*, mutant B1-41a.

Gaskin, P., & MacMillan, J. (1975). Polyoxygenated *ent*-kauranes and water-soluble conjugates in seed of *Phaseolus coccineus*. Phytochem. *14*, 1575–1578.

Graebe, J. E. (1968). Biosynthesis of kaurene, squalene and phytoene from mevalonate-2-^{14}C in a cell-free system from pea fruits. Phytochem. 7, 2003–2020.

Graebe, J. E. (1969). The enzymatic preparation of ^{14}C-kaurene. Planta *85*, 171–174.

Graebe, J. E. (1972). The biosynthesis of gibberellin precursors in a cell-free system from *Cucurbita pepo* L. In: Plant Growth Substances, 1970, pp. 151–157, Carr, D. J., ed Springer-Verlag, Berlin, Heidelberg, New York.

Graebe, J. E., Bowen, D. H., & MacMillan, J. (1972). The conversion of mevalonic acid into gibberellin A_{12}-aldehyde in a cell-free system from *Cucurbita pepo*. Planta *102*, 261–271.

Graebe, J. E., Dennis, D. T., Upper, C. D., & West, C. A. (1965). Biosynthesis of gibberellins. I. The biosynthesis of (−)-kaurene, (−)-kauren-19-ol and *trans*-geranylgeraniol in endosperm nucellus of *Echinocystis macrocarpa* Greene. J. Biol. Chem. *240*, 1847–1854.

Graebe, J. E., & Hedden, P. (1974). Biosynthesis of gibberellins in a cell-free system. In: Biochemistry and Chemistry of Plant Growth Regulators, pp. 1–16, Schreiber, K., Schütte, H. R., & Sembdner, G., eds. Acad. Sci. German Democratic Republic, Inst. Plant Biochem., Halle, G.D.R.

Graebe, J. E., Hedden, P., Gaskin, P., & MacMillan, J. (1974a). Biosynthesis of gibberellins A_{12}, A_{15}, A_{24}, A_{36}, and A_{37} by a cell-free system from *Cucurbita maxima*. Phytochem. *13*, 1433–1440.

Graebe, J. E., Hedden, P., Gaskin, P., & MacMillan, J. (1974b). The biosynthesis of a C_{19}-gibberellin from mevalonic acid in a cell-free system from a higher plant. Planta *120*, 307–309.

Graebe, J. E., Hedden, P., Kukulenz, M., Bearder, J. R., & MacMillan, J. (in preparation). The conversion of 7β-hydroxy-*ent*-kaurenoic acid to GA_{12}-aldehyde and 6β,7β,dihydroxy-*ent*-kaurenoic acid in GA biosynthesis.

Graebe, J. E., Hedden, P., & MacMillan, J. (1974c). Gibberellin biosynthesis: New intermediates in the *Cucurbita* system. In: Plant Growth Substances, 1973, pp. 260–266. Hirokawa, Tokyo.

Graebe, J. E., Hedden, P., & MacMillan, J. (1975). The ring contraction step in gibberellin biosynthesis. J. Chem. Soc., Chem. Commun., pp. 161–162.

Graebe, J. E., Hedden, P., & Rademacher, W. (1980). Gibberellin biosynthesis. In: Gibberellins—Chemistry, Physiology and Use, pp. 31–47, Lenton, J. R., ed. British Plant Growth Regulator Group, Wantage.

Graebe, J. E., & Ropers, H.-J. (1978). Gibberellins. In: Phytohormones and Related Compounds—A Comprehensive Treatise, Vol. 1, pp. 107–204, Lethem, D. S., Goodwin, P. B., & Higgins, J. J. V., eds. Elsevier/North-Holland Biomedical Press, Amsterdam.

Groves, J. T., McClusky, G. A., White, R. E., & Coon, M. J. (1978). Aliphatic hydroxylation by highly purified liver microsomal cytochrome P-450. Evidence for a carbon radical intermediate. Biochem. Biophys. Res. Commun. *81*, 154–160.

Gunsalus, I. C., Conrad, H. E., & Trudgill, P. W. (1964). Generation of active oxygen for mixed-function oxidation. In: Oxidases and Related Redox Systems, Vol. 1, pp. 417–447, King, T. E., Mason, H. S., & Morrison, M., eds. Wiley, New York, London, Sydney.

Hamilton, G. A. (1974). Chemical models and mechanisms for oxygenases. In: Molecular Mechanisms of Oxygen Activation, pp. 405–451, Hayaishi, O., ed. Academic Press, New York.

Hanson, J. R., Hawker, J., & White, A. F. (1972). Studies on terpenoid biosynthesis. Part IX. The sequence of oxidation on ring B in kaurene-gibberellin biosynthesis. J. Chem. Soc., Perkin Trans. I, 1892–1895.

Hanson, J. R., & White, A. F. (1969). Studies in terpenoid biosynthesis. Part IV. Biosynthesis of the kaurenolides and gibberellic acid. J. Chem. Soc. (C), 981–985.

Hanson, J. R., Willis, C. L., & Parry, K. P. (1980). The inhibition of gibberellic acid biosynthesis by *ent*-kauran-16β,17-epoxide. Phytochem. *19*, 2323–2325.

Hasson, E. P., & West, C. A. (1976a). Properties of the system for the mixed function oxidation of kaurene and kaurene derivatives in microsomes of the immature seed of *Marah macrocarpus*. Cofactor requirements. Plant Physiol. *58*, 473–478.

Hasson, E. P., & West, C. A. (1976b). Properties of the system for the mixed function oxidation of kaurene and kaurene derivatives in microsomes of the immature seed of *Marah macrocarpus*. Electron transfer components. Plant Physiol. *58*, 479–484.

Hedden, P., & Graebe, J. E. (1981). Kaurenolide biosynthesis in a cell-free system from *Cucurbita maxima* seeds. Phytochem. *20*, 1011–1015.

Hedden, P., & Graebe, J. E. (1982). The cofactor requirements for the soluble oxidases in the metabolism of the C_{20}-gibberellins. J. Plant Growth Regulation *1*, 105–116.

Hedden, P., Graebe, J. E., Beale, M. H., Gaskin, P., & MacMillan, J. (in preparation). The biosynthesis of 12α-hydroxylated gibberellins in a cell-free system from *Cucurbita maxima* endosperm.

Hedden, P., MacMillan, J., & Grinstead, M. J. (1973). Fungal products. Part VIII. New Kaurenolides from *Gibberella fujikuroi*. J. Chem. Soc., Perkin Trans. I, 1773–1778.

Hedden, P., MacMillan, J., & Phinney, B. O. (1974). Fungal products . Part XII. Gibberellin A_{14}-aldehyde, an intermediate in gibberellin biosynthesis in *Gibberella fujikuroi*. J. Chem. Soc., Perkin Trans. I, 587–592.

Hedden, P., MacMillan, J., & Phinney, B. O. (1978). The metabolism of the gibberellins. Ann. Rev. Plant Physiol. *29*, 149–192.

Hedden, P., & Phinney, B. O. (1979). Comparison of *ent*-kaurene and *ent*-isokaurene synthesis in cell-free systems from etiolated shoots of normal and dwarf-5 maize seedlings. Phytochem. *18*, 1475–1479.

Hildebrandt, A., & Estabrook, R. W. (1971). Evidence for the participation of cytochrome b_5 in hepatic microsomal mixed-function oxidation reactions. Arch. Biochem. Biophys. *143*, 66–79.

Hildebrandt, E., Graebe, J. E., Rademacher, W., & Jung, J. (1982). Mode of action of new potent plant growth retardants: Bas 106 . .W and triazole compounds. In: Proceedings of the Eleventh International Conference on Plant Growth Substances, abstracts, p. 65.

Hoad, G. V., MacMillan, J., Smith, V. A., & Taylor, D. A. (1982). Metabolic deactivation of gibberellins. In: Proceedings of the Eleventh Conference on

Plant Growth Substances, pp. 91–100, Wareing, P. F., ed. Springer-Verlag, Berlin, Heidelberg, New York.

Kamiya, Y., & Graebe, J. E. (1983). The biosynthesis of all major pea gibberellins in a cell-free system from *Pisum sativum*. Phytochem. *22*, 681–690.

Leopold, A. C. (1971). Antagonism of some gibberellin actions by a substituted pyrimidine. Plant Physiol. *48*, 537–540.

Lew, F. T., & West, C. A. (1971). (−)-Kaur-16-en-7β-ol-19-oic acid, an intermediate in gibberellin biosynthesis. Phytochem. *10*, 2065–2076.

Lindblad, B., Linstedt, G., Linstedt, S., & Tofft, M. (1969). The mechanism of α-ketoglutarate oxidation in coupled enzymatic oxygenations. J. Amer. Chem. Soc. *91*, 4604–4606.

Loomis, W. D., & Battaile, J. (1966). Plant phenolic compounds and the isolation of plant enzymes. Phytochem. *5*, 423–438.

MacMillan, J. (1978). Gibberellin metabolism. Pure Appl. Chem. *50*, 995–1004.

Meinwald, J., & Frauenglass, E. (1960). The Baeyer-Villiger oxidation of bicyclic ketones. J. Amer. Chem. Soc. *82*, 5235–5239.

Mellon, J. E., & West, C. A. (1979). Diterpene biosynthesis in maize seedlings in response to fungal infection. Plant Physiol. *64*, 406–410.

Metzger, J. D., & Zeevaart, J. A. D. (1980). Effect of photoperiod on the levels of endogenous gibberellins in spinach as measured by combined gas chromatography-selected ion current monitoring. Plant Physiol. *66*, 844–846.

Moore, T. C., Barlow, S. A., & Coolbaugh, R. C. (1972). Participation of non-catalytic "carrier" protein in the metabolism of kaurene in cell-free extracts of pea seeds. Phytochem. *11*, 3225–3233.

Murphy, G. J. P., & Briggs, D. E. (1973). *Ent*-kaurene, occurrence and metabolism in *Hordeum distichon*. Phytochem. *12*, 2509–2605.

Murphy, G. J. P., & Briggs, D. E. (1975). Metabolism of *ent*-kaurenol-[17-^{14}C], *ent*-kaurenal-[17-^{14}C] and *ent*-kaurenoic acid-[17-^{14}C] by germinating *Hordeum distichon* grains. Phytochem. *14*, 429–433.

Murphy, P. J., & West, C. A. (1969). The role of mixed function oxidases in kaurene metabolism in *Echinocystis macrocarpa* Greene endosperm. Arch. Biochem. Biophys. *133*, 395–407.

Nadeau, R., & Rappaport, L. (1974). The synthesis of [^{3}H]gibberellin A$_3$ and [^{3}H]gibberellin A$_1$ by the palladium-catalyzed actions of carrier-free tritium on gibberellin A$_3$. Phytochem. *13*, 1537–1545.

Noshiro, M., Harada, N., & Omura, T. (1980). Immunochemical study on the route of electron transfer from NADH and NADPH to cytochrome P-450 of liver microsomes. J. Biochem. *88*, 1521–1535.

Pascal, R. A., Chang, P., & Schoepfer, G. J. (1980). Possible mechanisms of demethylation of 14α-methyl sterols in cholesterol biosynthesis. J. Amer. Chem. Soc. *102*, 6599–6601.

Patterson, R. J., & Rappaport, L. (1974). The conversion of gibberellin A$_1$ to gibberellin A$_8$ by a cell-free system. Planta *119*, 183–191.

Patterson, R. J., Rappaport, L., & Breidenbach, R. W. (1975). Characterization of an enzyme from *Phaseolus vulgaris* seeds which hydroxylates GA$_1$ to GA$_8$. Phytochem. *14*, 363–368.

Phinney, B. O. (1979). Gibberellin biosynthesis in the fungus *Gibberella fujikuroi* and in higher plants. In: Plant Growth Substances. ACS Symp. Ser. 111, pp. 57–78, Mandava, N. B., ed. American Chemical Society, Washington, D.C.

Rademacher, W., & Jung, J. (1981). Comparative potency of various synthetic plant growth retardants on the elongation of rice seedlings. Z. Acker-Pflanzenbau *150*, 363–371.

Robinson, D. R., & West, C. A. (1970). Biosynthesis of cyclic diterpenes in extracts from seedlings of *Ricinus communis* L. 1. Identification of diterpene hydrocarbons formed from mevalonate. Biochem. *9*, 70–79.

Ropers, H.-J. (1978). Gibberellinbiosynthese in höheren Pflanzen. Dissertation. Georg-August-Universität, Göttingen, F.R.G.

Ropers, H.-J., Graebe, J. E., Gaskin, P., & MacMillan, J. (1978). Gibberellin biosynthesis in a cell-free system from immature seed from *Pisum sativum*. Biochem. Biophys. Res. Commun. *80*, 690–697.

Sadava, D., & Chrispeels, M. J. (1971). Hydroxyproline biosynthesis in plant cells. Peptidyl proline hydroxylase from carrot disks. Biochim. Biophys. Acta *227*, 278–287.

Scallen, T. J., Srikantaiah, M. V., Seetharam, B., Hansburg, E., & Gavey, K. L. (1974). Sterol carrier protein hypothesis. Fed. Proc. *33*, 1733–1746.

Schechter, I., & West, C. A. (1969). Biosynthesis of gibberellins. IV. Biosynthesis of cyclic diterpenes from *trans*-geranylgeranyl pyrophosphate. J. Biol. Chem. *244*, 3200–3209.

Sembdner, G., Gross, D., Liebisch, H.-W., & Schneidner, G. (1980). Biosynthesis and metabolism of plant hormones. In: Hormonal Regulation of Development I. Molecular Aspects of Plant Hormones. Encyclopedia of Plant Physiology New Series, vol. 9, pp. 281–444, MacMillan, J., ed. Springer-Verlag, Berlin, Heidelberg, New York.

Siegel, B. (1979). α-Ketoglutarate dependent dioxygenases: A mechanism for proplyl hydroxylase reaction. Bioorg. Chem. *8*, 219–226.

Sponsel, V. M., & MacMillan, J. (1977). Further studies on the metabolism of gibberellins (GAs) A_9, A_{20}, A_{29} in immature seeds of *Pisum sativum* cv. Progress No. 9. Planta *135*, 129–136.

Sponsel, V. M., & MacMillan, J. (1978). Metabolism of gibberellin A_{29} in seeds of *Pisum sativum* cv. Progress No. 9; use of [^{2}H] and [^{3}H] GAs, and the identification of a new GA catabolite. Planta *144*, 69–78.

Vermilion, J. L., Ballou, D. P., Massey, V., & Coon, M. J. (1981). Separate roles for FMN and FAD in catalysis by liver microsomal NADPH-cytochrome P-450 reductase. J. Biol. Chem. *256*, 266–277.

Vermilion, J. L., & Coon, M. J. (1978). Purified liver microsomal NADPH-cytochrome P-450 reductase. J. Biol. Chem. *253*, 2694–2704.

Visser, C. M. (1980). Role of ascorbate in biological hydroxylation. Origin of life considerations and the nature of the oxenoid species in oxygenase reactions. Bioorg. Chem. *9*, 261–271.

Wada, K. (1978). New gibberellin biosynthesis inhibitors, 1-*n*-decyl- and 1-geranylimidazole: Inhibitors of (−)-kaurene 19-oxidation. Agric. Biol. Chem. *42*, 2411–2413.

West, C. A. (1973). Biosynthesis of gibberellins. In: Biosynthesis and Its Control in Plants, pp. 473–482, Milborrow, B. V., ed. Academic Press, London, New York.

West, C. A. (1980). Hydroxylases, monooxygenases, and cytochrome P-450. In: The Biochemistry of Plants. A Comprehensive Treatise Vol. 2, Metabolism and Respiration, pp. 317–364, Davies, D. D., ed. Academic Press, New York, London, Toronto, Sydney, San Francisco.

White, R. E., & Coon, M. J. (1980). Oxygen activation by cytochrome P-450. Ann. Rev. Biochem. *49*, 315–356.

Wurtele, E. S., Hedden, P., & Phinney, B. O. (1982). *In vitro* metabolism of the gibberellin-precursors *ent*-kaurene, *ent*-kaurenol and *ent*-kaurenal in normal maize seedling shoots. J. Plant Growth Regulation *1*, 15-24.

Yafin, Y., & Schechter, I. (1975). Comparison between biosynthesis of *ent*-kaurene in germinating tomato seeds and cell suspension cultures of tomato and tobacco. Plant Physiol. *56*, 671-675.

Yoshida, Y., & Kumaoka, H. (1975). Studies on the substrate-induced spectral change of cytochrome P-450 in liver microsomes. J. Biochem. *78*, 455-468.

4

In Vivo Gibberellin Metabolism in Higher Plants

Valerie M. Sponsel

4.1 INTRODUCTION

Most higher plant tissues so far examined contain several of the 66 known gibberellins. GAs have been shown to vary qualitatively and quantitatively with time in those plants which have been extracted at periodic intervals. Thus, it is inferred that GAs are being actively metabolized in plant tissues. Many of the currently known GAs are probably precursors or catabolites of a few "primary" GAs, that is, those GAs having hormonal function per se (MacMillan, 1977; Hedden, MacMillan, & Phinney, 1978; Rappaport & Adams, 1978). Indeed, of the 55 GAs so far identified in higher plant tissues only a small proportion have significant biological activity in standard bioassays and are, thus, likely candidates for hormonal status. A study of biosynthesis and metabolism of GAs in higher plants will help to pin-point which are the primary GAs in a given species or tissue. Furthermore, investigations into the control mechanisms operative in the pathway may determine how the levels of active hormones can be finely regulated within a plant and throughout its life cycle. This in turn may shed light on the mode of action of these hormones.

Biosynthesis and metabolism have been studied extensively in the fungus *Gibberella fujikuroi*, as described in Chapters 2 and 5. A wealth of information has been obtained on the nature of the pathways operating in the fungus, both from a structural and a mechanistic point of view. Furthermore, some of the enzymes catalyzing important reactions in the pathway have been isolated and studied in partially purified form. Although *G. fujikuroi* provides a good model system, the plant physiologist must eventually turn to higher plants to study those aspects of metabolism and its regulation which are involved in the production of biologically active hormones.

Metabolic studies in higher plants can present difficulties that are not encountered in studies using *G. fujikuroi*. The amounts of GAs in higher plants, especially vegetative tissue, are low and care should be taken in metabolic studies to feed "physiological" concentrations. Also, the substrate must be administered in such a way as to avoid damage to the plant, but in a manner that still allows adequate uptake. Within the plant, translocation may pose problems for the recovery of residual substrate and metabolites. Finally, technical difficulties encountered in purifying plant extracts to enable identification of metabolites may in some cases be formidable.

Many of these problems can be circumvented by preparing cell-free homogenates from higher plant tissues, and indeed many biochemical aspects of the problem are best studied in this way. The recent excellent progress in this field is described in Chapter 3. Nevertheless, some constraints normally operating in an intact plant are lost in the preparation of cell-free systems. For instance, compartmentation is disrupted, possibly allowing the access of enzymes to substrates from which they are normally spatially separated. Work with intact plants is therefore necessary to ascertain that a given pathway does operate *in vivo*. Also, since many developmental responses are correlative phenomena—that is, those that involve an interaction of different tissues and organs within the plant—individual plant parts cannot be considered in isolation. These types of developmental phenomena are best studied, at least initially, at the whole plant level.

Several excellent reviews of GA metabolism have recently been published (e.g., Graebe & Ropers, 1978; Hedden et al., 1978;

Sembdner et al., 1980). The present chapter concentrates, initially, on some practical aspects of studying GA metabolism in intact higher plants and then describes and attempts to assess recent work in this area.

4.2 TECHNIQUES

4.2.1 Design of experiments

There have been several approaches to the study of GA metabolism in intact higher plants. The earliest approach was to apply radio-labeled GA_3 to a plant and then follow its disappearance together with the appearance of new, radio-labeled, but unidentified compounds (e.g., Nitsch & Nitsch, 1963; Rappaport et al., 1967). It soon became apparent that most plant tissues possess enzymes capable of metabolizing applied GAs and, thus, in order to obtain definitive information, more specific experiments had to be planned. Also, techniques had to be developed for the preliminary rigorous purification of substrates, as well as for the identification of metabolites.

A more rational approach has been to feed a variety of radio-labeled GAs to plants that are known to respond to the applied GAs and to follow their metabolism (e.g., Nadeau, Rappaport, & Stolp, 1972; Durley & Pharis, 1973; Davies & Rappaport, 1975a, 1975b; Silk, Jones, & Stoddart, 1977). Invariably, the plant material chosen was that employed in standard GA bioassays, namely, *Hordeum vulgare* (barley) half-seeds, *Oryza sativa* (rice) cv. Tanginbozu seedlings, *Zea mays* (maize) dwarf *d-5* seedlings and *Lactuca sativa* (lettuce) hypocotyls. The characteristics of these bioassay systems are described by Reeve and Crozier (1974) and in Volume II (Chapter 2). Growth responses or other physiological phenomena could be related to the disappearance of radio-labeled substrate and appearance of identifiable metabolites (Section 4.3.3). An extension of this approach has been to look at the effect of environmental factors, such as light, on growth responses and GA metabolism (Durley et al., 1976). This approach, however, ignores the metabolism of native GAs, whether this metabolism is altered by the applied GA, and, if it is, whether the alteration in native GA content has any effect on the physiological response being measured (e.g., Lance et al., 1976a).

Another approach has been to apply to a given plant material only those GAs known to be endogenous constituents (Yamane,

Murofushi, & Takahashi, 1975; Yamane et al., 1977; Frydman & MacMillan, 1975; Sponsel & MacMillan, 1977, 1978, 1980). This necessitates the preliminary identification of GAs in a given plant material (normally immature seeds since they contain higher levels of GAs than vegetative tissue), and the preparation of those same GAs in isotopically labeled form. Frydman and MacMillan (1975) suggested several criteria for conducting feeds to ensure that the metabolism of applied substrate reflected the metabolism of native GAs. More recently, methods have been developed using stable isotope-labeled GAs to show that metabolism of applied GA is indeed equivalent to that of the same native GA. This technique enables the pathways operative *in vivo* to be determined and quantitated. Once accurate data for turnover rates become available, the effects on the metabolism of native GAs of environmental factors, other hormones, and so forth can be quantitatively assessed. In turn, the metabolism of native GAs can be knowingly and measurably modified with GA antagonists, enzyme inhibitors, etc. and the effects on developmental responses can be studied.

4.2.2 Choice of Substrates

Historically, the choice of substrate for feeding was largely determined by the availability of radio-labeled GAs. In the last decade, however, numerous methods have been published for the preparation of radio-labeled GAs, and these methods are discussed in Chapter 7. There is now a considerable choice as to the nature of the isotope and its position in the molecule. Thus, it is no longer necessary to plan experiments around the available substrates, since it is often possible to prepare specifically labeled substrates for a given experiment. Some factors that must be taken into account when determining the type of substrate to be used are discussed below.

Tritium- and ^{14}C-labeled GAs are still routinely used as substrates for metabolic work. These radioisotopes can be readily traced throughout the work-up procedure to locate metabolites and to determine their recoveries. To a certain extent the sensitivity of radio-analysis determines the amount of substrate that must be fed. GAs of reasonably high specific activity are therefore preferable since it is advisable to feed as little mass as possible in order to maintain a "physiological" concentration of substrate.

All substrates should be of confirmed radio-nuclidic purity before use, since analysis of plant material after feeding normally entails radio-scanning of extracts to detect the presence of radioactive compounds other than the initial substrate. A combination of purification and separatory techniques in conjunction with radio-counting is employed. Ultimately, extracts should be suitable for mass-spectrometric analysis. In practice, combined gas chromatography-mass spectrometry (GC-MS) is most suitable since the initial gas chromatography separatory step allows the mass spectrometric identification of GAs in relatively impure extracts. Unfortunately there are many instances where GC-MS identification of metabolites has not been achieved.

There are several examples in the literature of particular GAs being fed to plants in which they are not known to be native, and where GC-MS identification of residual substrate and/or metabolites has been achieved (e.g., Durley, Railton, & Pharis, 1974a; Silk et al., 1977). GC-MS analysis of extracts in these cases is relatively straightforward, since the only GAs identified are the substrate and those GAs derived from it. Failure to detect the native GAs occurs either because the substrate has been fed at a higher dose than the level of endogenous GAs, or because the substrate and/or metabolites are different from the native GAs. However, since it is advisable to feed GAs that are known to be native to a particular plant, and to feed them at the same level as that of the native GAs, one is faced with a problem of having to identify the substrate and its metabolites by GC-MS in the presence of the *same* endogenous GAs (Frydman & MacMillan, 1975). A general solution to this problem, which is discussed in detail below, is the use of substrates containing sufficiently high incorporation of either a radioactive or a stable isotope, such that the isotope can be detected by MS. The applied substrate can then be distinguished from the same endogenous GA by MS. Provided that the substrate is not diluted in the plant by a vast excess of the same native GA, then the isotope is observed in the mass spectrum of the metabolites too. This proves formation of the metabolite from the added substrate and, in addition, can provide considerable quantitative information, since measurements of isotope intensities in the mass spectra can be made. Thus, for radioactive substrates, accurate specific activities of metabolites can be calculated by mass spectrometry without recourse to isolation of metabolites. This technique was demonstrated by Bowen, MacMillan, and Graebe (1972) for kaurenoids produced from $10.3 \, \mu\text{Ci} \, \mu\text{mol}^{-1}$ [^{14}C]mevalonic acid in cell-free systems from *Cucurbita maxima* (pumpkin). In practice, this technique is not suitable for tritium-labeled substrates, since for ^{3}H- and ^{14}C-labeled compounds of the same specific activity, the proportion of labeled to unlabeled molecules is 1000-fold less for the ^{3}H-labeled compound than for the ^{14}C-labeled one. Thus, one would require an impracticably high specific activity in a ^{3}H-labeled substrate in order to detect the ^{3}H isotope in its mass spectrum.

As an alternative, the use of substrates containing a stable isotope (e.g., ^{2}H, ^{13}C, or ^{18}O) is feasible since a very high incorporation of stable isotope can be achieved (Beale et al., 1980; Beale & MacMillan, 1981a). Stable isotope-labeled GAs make excellent internal standards, but used on their own they are of limited utility in metabolic studies. However, GAs that are doubly labeled with a stable isotope and a radioisotope are particularly useful in metabolic studies. The radioisotope (e.g., ^{3}H) can be used as a tracer to follow the recovery of metabolites, and the content of stable isotope (e.g., ^{2}H, ^{13}C or ^{18}O) in the metabolites can be detected and quantitated by MS. The relative proportions of labeled and unlabeled metabolites, and hence the relative rates of metabolism of externally applied and native GA, can then be calculated. Thus depending on the type of analysis to be used and the amount of information required from an experiment, a choice can be made as to whether the substrate should contain radio- and/or stable-isotope.

When preparing a substrate, with no *a priori* knowledge of its potential metabolites, it is sometimes difficult to decide on the most suitable position in the molecule for the isotope to be located. There are many examples in the literature where radioactivity is lost during the metabolic process or during subsequent work-up of extracts (e.g., Bown, Reeve, & Crozier, 1975; Lance et al. 1976a). If the position of the isotope within the substrate is determined prior to use, then loss of isotope can give definitive information on the molecular transformation occurring during metabolism and/or work-up. Patterson and Rappaport (1974) observed that ^{3}H was lost during the metabolism of $[1,2\text{-}^3\text{H}_2]\text{GA}_1$ in a cell-free system from *Phaseolus vulgaris* (French bean) seeds. Measurement of the amount of tritiated water formed gave a direct measure-of the amount of metabolism occurring. The product, $[^3\text{H}]\text{GA}_8$, arising by 2β-hydroxylation of $[^3\text{H}]\text{GA}_1$ with resulting loss of ^{3}H from the 2β-position, still contained ^{3}H in the 1-position and thus could be traced. In other instances, metabolism and/or work-up results in complete loss of isotope, rendering the metabolite unlabeled and untraceable (Sponsel & MacMillan, 1978). However, since the position and stereochemistry of isotope in the $[1\beta,3\alpha\text{-}^2\text{H}_2]$ $[1\beta,3\alpha\text{-}^3\text{H}_2]\text{GA}_{20}$ substrate used by Sponsel and MacMillan (1978) had been determined prior to use (Beale et al., 1980), considerable information could be obtained on the type of transformation occurring and on its location in the molecule.

If an isotope (e.g., ^{3}H) is lost from a molecule during metabolism, then breakage of a bond involving the ^{3}H atom must have occurred. Metabolism in this case would be subject to a kinetic isotope effect; for example, bond breakage would be expected to proceed more slowly for the C—^{3}H bond than for the corresponding C—^{1}H bond in unlabeled substrate. Thus, nonlabeled substrate would be metabolized faster than labeled substrate. Graebe (1980) has observed a kinetic isotope effect during the metabolism of *ent-*[6α-^{3}H]7α-hydroxykaurenoic acid in a cell-free system from *C. maxima*, since ^{1}H/^{3}H is lost from the *ent-*6α-position of the substrate during metabolism to *ent-*6α,7α-dihydroxykaurenoic acid and GA_{12} aldehyde. Similarly, Sponsel and MacMillan (1978) observed an apparent isotope effect during the metabolism of $[2α-^2H_1][2α-^3H_1]$-GA_{29} and thus deduced that metabolism involved oxidation at C-2.

Comment has been made in a recent review (Durley, 1983) that feeds of stable isotope-labeled GAs can give misleading results due to isotope effects and/or loss of label. It is the position of the isotope in the molecule that determines whether it is lost during metabolism and/or whether metabolism is subject to an isotope effect. The amount of information obtained by MS analysis of feeds of stable isotope-labeled substrates can indeed yield very complicated results. However, the advantage of the technique is that this wealth of information allows not only conclusive identification of metabolites to be made, but also enables accurate quantitation to be conducted without isolation of metabolites. Indeed, stable isotopes are now widely used in many types of biosynthetic studies (Tanabe, 1973). In the field of plant hormones they have also been used in metabolic studies on auxins (Bandurski, 1980).

4.2.3   Feed procedures

The means of applying isotopically labeled substrates to plant tissues are many and varied. Labeled substrates are normally applied onto the surface of, or injected into, a plant part in a small volume of organic or aqueous solvent. In practice, it is advisable to adopt a method that allows optimal substrate uptake yet minimizes damage to the plant tissue.

The translocation of substrate or metabolites throughout the plant must be taken into account. In some instances, the radioactivity is localized only in the plant part to which it was fed; for example, GAs applied to fruits tend to be retained there (Sponsel

& MacMillan, 1977). In other cases, substrate and/or metabolites may become widely distributed throughout the plant, necessitating analysis of the whole plant (Durley et al., 1976).

In metabolic studies with callus cultures, substrate may either be added to the medium (Rappaport et al., 1974) or applied directly to the callus surface (Lance et al., 1976a). The substrate should normally be filter-sterilized. On occasions, organ culture (e.g., of developing seeds) has been used as a convenient model system (Frydman & MacMillan, 1975; Yamane et al., 1977). Here there must be adequate insurance that development in culture proceeds as it does *in vivo* (Frydman & MacMillan, 1973).

4.2.4 Analysis of extracts

When using radioactive substrates, repeated radio-counting should be conducted throughout the analysis of extracts to monitor recovery of radioactive components. In theory, all the radioactivity applied to the plant should be recovered in the subsequent methanolic extract. Any deficiency at this stage would be an indication of isotope loss during metabolism. Similarly, all radioactivity in the methanolic extract should be accounted for after fractionation and purification of the extract. Unexpected losses may arise from chemical modification of a metabolite during work-up. The analysis of a particular extract should only be considered to be satisfactory if all the radioactivity applied at the beginning of the experiment can be accounted for at the end. Only then can one be certain that all metabolic products have been detected. In Sections 4.3 and 4.4 many examples are discussed in which this thorough type of analysis has not been conducted.

The amount of purification required for a given extract will depend on the number, nature, and quantity of metabolites present relative to the other components in the plant extract. Normal work-up procedure involves fractionation of the extract by partitioning, followed by a variety of purification and separation procedures. Generally, a compromise is reached between the amount of purification of metabolites which is required and the number of fractions which are generated for further analysis. In a simple situation only thin layer chromatography (TLC) may be necessary. In other cases, repeated column chromatography is required. Scintillation counting of aliquots from column fractions is essential to locate and quantify radioactive peaks. The more sophisticated the separatory procedure the more fractions will be generated, so that manual radio-analysis becomes time-consuming and tedious. Reeve and Crozier (1977) have devised two on-line systems for continuous flow monitoring, particularly for use with high performance liquid chromatography (HPLC). The first is a heterogeneous system in which the column eluant moves to a flow cell packed with solid scintillator; the sample is not mixed with scintillant and can easily be recovered. Although useful for ^{14}C-labeled samples, this system is of little utility for ^{3}H-labeled samples because of the poor counting efficiency for low-energy β-emitters. The other technique is a homogeneous system in which the eluate is directly mixed with scintillation cocktail. With the provision of a splitter, a proportion only of the sample need be counted and the remainder can be recovered. The scintillator-eluate mixture passes through a spiral flowcell and the parameters needed to give efficient counting of ^{3}H and ^{14}C without loss of resolution have been discussed in detail (Reeve & Crozier, 1977). Despite the excellent separatory potential of HPLC for GAs and the utility of this technique for radio-counting, HPLC has been little used to date for the fractionation of extracts containing radioactive GA metabolites.

GC also provides excellent separation of GAs. For instance, Durley, Sassa, and Pharis (1979) distinguished 42 of the known GAs by separation on three GC columns. There are two methods whereby GC can be conducted in conjunction with radio-counting (GC-RC). In one method (MacMillan & Wels, 1974; Wels, 1977), the effluent from the flame ionization detector (FID), containing tritiated water or [^{14}C]carbon dioxide from the combustion of ^{3}H- or ^{14}C-labeled compounds, respectively, is manually trapped into methoxyethanol (plus phenyl ethylamine for ^{14}C-labeled compounds) at 0.25 or 0.5 min intervals. Scintillation cocktail is then added, and fractions are counted by liquid scintillation counting. This method is extremely sensitive since (a) there is no split and (b) ^{3}H and ^{14}C can be collected almost quantitatively. Thus, as little as twice background radioactivity can be detected. Recently, this method has been modified for use with capillary GC (Hamnett & Pratt, 1978). The alternative method for GC-RC involves splitting the GC column effluent such that a little (e.g., 10%) goes to the FID and the remainder goes to a gasflow radio-monitor (Railton, 1976a). This technique is more rapid than the former, with FID and radioactivity traces being displayed simultaneously on a dual pen chart recorder, but it is less sensitive (Yokota, Murofishi, & Takahashi, 1980). The utility of GC-RC is based on its easy coordination with GC-MS which is ultimately required for the identification of metabolites. GC-MS is preferentially conducted using the same GC conditions as for GC-RC, thus enabling the retention times of mass peaks identified by GC-MS to be correlated with retention times of radio-active peaks on GC-RC. GC-RC and GC-MS should be conducted and coordinated on several GC columns to confirm the homogeneity of radioactive peaks and to establish their identity. When a radioactive or stable isotope is present at sufficient incorporation to be visible directly in the mass spectrum, then GC-RC and GC-MS correlation on multiple columns may no longer be necessary.

Comparative HPLC and GC-RC require access to reference GAs in order to standardize retention volumes/retention times. Often these reference GAs are not easily accessible, and this limits the potential utility of the chromatographic system for GA identification. On the other hand, reference mass spectra are available from the literature, so that GC-MS identification can be achieved even in the absence of reference compounds. Furthermore, MS surpasses HPLC and GC-RC in that it provides considerable information regarding molecular structure, and thus it has great diagnostic value in the analysis of hitherto unknown metabolites.

Reeve and Crozier (1980) and Crozier (1980) have discussed some theoretical aspects of the identification and quantitation of plant hormones in extracts. Their discussion is equally relevant to the analysis of labeled metabolites in plant extracts. The object of extract analysis is to accumulate sufficient information (measured in bits) such that there is a high probability that the identification made is an accurate one. The number of components in an extract, and the nature of the analytical procedure, will determine how much information is obtained. TLC and paper chromatography yield up to *ca.* 5 bits, HPLC can yield up to *ca.* 100 bits, capillary GLC up to *ca.* 300 bits, and MS a theoretical value of 400 bits per 100 mass units. In practice, the actual amount of information obtained is far less than the theoretical prediction, especially when extracts contain many components. Since the amount of informa-tion obtained by each procedure is additive, Crozier (1980) implies that several analytical HPLC or GC-RC runs may cumulatively yield more information than one poorly recorded mass spectrum. While making a valid point, he is nevertheless comparing the very best results from one technique with the worst result from another. It is to be hoped that the expertise applied to HPLC/GC-RC could equally be applied to GC-MS. In experienced hands and with adequately purified extracts, GC-MS is a powerful analytical tool and its use is to be recommended whenever possible. Analytical procedures that are used to analyze GAs are discussed in detail in Chapter 8.

4.2.5   Quantitation of metabolites and treatment of data

Results of metabolic feeds of radioactive substrates are most readily expressed as a percentage conversion of the applied substrate to a given metabolite. This percentage is also referred to as the percentage yield of metabolite. Classically, it was normal practice to isolate metabolites to confirm their radio-nuclidic purity and estimate their specific radioactivities. However, little of the work discussed in this chapter has been conducted on a scale that allowed isolation of metabolic products. Instead it has become accepted practice to estimate the amount of radioactivity associated with a given metabolite during chromatography, preferably HPLC or GC. Repeated chromatography with different mobile and/or stationary phases should be conducted to confirm that the radioactivity is indeed associated with the putative metabolite. Ideally, GC-MS should be performed to identify the metabolite and to confirm the purity of mass peaks observed in radio-chromatography. However, although rigorous purification of extracts is desirable, the more extensive the work-up procedure the lower will be the calculated percentage conversion, as a result of metabolite losses during purification. Thus, Sponsel and MacMillan (1977) gave two values for the percentage conversion of [^{3}H]GA$_9$ to each of three metabolites, one value based on the amount of metabolite obtained at the end of the experiment, and a higher value which took into account measurable losses associated with repeated TLC, GC, and enzyme hydrolysis. In general, the figures in the literature are not adjusted for losses and will thus vary greatly depending on the extent of purification of extracts prior to metabolite estimation.

When expressing the results of a radioactive feed, it is essential to specify not only the percentage conversion to each of the observed metabolites, but also the percentage of unmetabolized substrate remaining at the end of the feed. In addition, any loss of radioactivity that could not be accounted for should be discussed. There are some papers in the literature where conversion of substrate to metabolite(s) of *ca.* 5% was observed, but it is impossible to ascertain whether the residual radioactivity was associated with unmetabolized substrate or was lost from the system (e.g., Durley et al., 1974a, 1974b). In such cases, very different interpretations of the results are possible.

It is instructive in some instances to be able to determine whether the metabolism of an applied GA is equivalent to the metabolism of the same native GA. To do so entails calculating the relative dilutions of the added substrate (before and after metabolism) and of the metabolite with native GA. This can readily be achieved by MS, without recourse to isolation, as long as substrates containing sufficiently high incorporation of a radioactive or stable isotope are used. In theory, ^{14}C-labeled substrates of high specific activity are most readily employed, since the ^{14}C isotope is used both as a radioactive tracer, and as a "heavy" isotope for quantitation by mass spectrometry. However, the use of ^{14}C-labeled substrates in metabolic studies with intact plants is not well documented in the literature. Sponsel and MacMillan (1978) have outlined the procedure for using doubly labeled substrates to determine the absolute amounts of exogenous and endogenous substrate and metabolite(s) in plant extracts. In the $[1\beta,3\alpha\text{-}^2H_2][1\beta,3\alpha\text{-}^3H_2]GA_{20}$ substrate used by them, the incorporation of radioisotope (i.e., ^3H) was sufficient to enable recovery and percentage conversion to be monitored by scintillation counting, while the stable isotope (i.e., ^2H) at high incorporation was detected by MS. Thus, the percentage of deuterated and nondeuterated GA_{20} and deuterated and nondeuterated metabolite (i.e., GA_{29}) could be calculated from the relative intensities of ions in the molecular ion clusters of mass spectra of recovered derivatized GA_{20} and GA_{29}. Knowing, from ^3H monitoring, the amount of unmetabolized radioactive substrate remaining and the amount of radioactive metabolite formed, and knowing from MS the relative amounts of deuterated and non-deuterated GA present, it was then possible to calculate the absolute amounts of native substrate (GA_{20}) and metabolite (GA_{29}) present. Thus, it was confirmed that transformation of applied $[1\beta,3\alpha\text{-}^2H_2][1\beta,3\alpha\text{-}^3H_2]GA_{20}$ to $[1\beta,3\alpha\text{-}^2H_2][1\beta,3\alpha\text{-}^3H_2]GA_{29}$ and of native GA_{20} to native GA_{29} proceeded simultaneously.

When substrates labeled only with stable isotope are used for feeds, the relative amounts of labeled and unlabeled substrate and metabolite can still be calculated by MS. However, the percentage conversion of substrate to metabolite is not measurable, and so absolute amounts of unlabeled substrate and metabolite cannot be calculated. Sponsel and MacMillan (1980) encountered this when feeding $[17\text{-}^{13}C_1]GA_{29}$, and they overcame the problem by adding putative metabolite (i.e., GA_{29}-catabolite) which was labeled with ^{18}O to resulting extracts as an internal standard. Thus, the percentages of $[^{12}C][^{16}O]GA_{29}$-catabolite (i.e., native), of $[^{13}C_1][^{16}O]GA_{29}$-catabolite (i.e., formed from applied $[^{13}C_1][^{16}O]GA_{29}$ substrate), and of $[^{12}C][^{18}O_1]GA_{29}$-catabolite (i.e., internal standard which was added to the extract) could be calculated from the relative intensities of ions in the molecular ion clusters of mass spectra of recovered, derivatized GA_{29}-catabolite. This is possible since the spectra of the two labeled species, $[^{13}C_1][^{16}O]$- and $[^{12}C][^{18}O_1]GA_{29}$-catabolite, are shifted one and two units up the mass scale respectively from that of the unlabeled species, $[^{12}C][^{16}O]GA_{29}$-catabolite. From the known amount of internal standard added, the absolute amount of $[^{13}C_1]$labeled and endogenous GA_{29}-catabolite can thus be calculated. However, by this method there is still no means of calculating the absolute amount of residual, unmetabolized substrate.

Results expressed as a percentage conversion, either of applied or native GA, give no indication of the dynamics of a system, since they give a measure only of the metabolite that has accumulated. If time-course feeds are conducted (by extracting plants at intervals of time after substrate application), then the disappearance of substrate and build-up of metabolite can be followed. If several metabolites are obtained, then the relative proportions at different times may give an indication of the pathway (see Durley et al., 1974a, 1974b). However, putative intermediates should always be re-fed to confirm the proposed pathway.

Once the direct precursor of a substance has been ascertained it is, in theory at least, possible to calculate the "turnover" of that substance. Turnover, as defined by Reiner (1953a), is that part of the incorporation of new molecules into a substance which is balanced by the removal of molecules of that substance which were previously present. In other words, it is the "flux" of molecules "through" a substance. Thus, the rate of turnover of a substance is equal to the number of molecules of that substance which are newly formed per unit time, or the number of molecules of that substance already present which are transformed into a metabolite per unit time, whichever is the smaller. It follows that in the steady state the two values are equal.

Theoretical and practical aspects of the measurement of turnover rates in intact organisms are discussed by Reiner (1953a, 1953b). Experimental application of the theory has been limited by the many practical complexities. For instance, if the substance of interest has more than one precursor because it is part of a metabolic network or grid, then not only must the structure of the metabolic network be fully understood, but the specific activity and concentration of each component of the network must be ascertained. Fortunately this no longer requires quantitative isolation of components, if isotope data can be calculated by MS. Secondly, calculation of turnover rates should always be based on time-course data since no valid conclusions can be drawn from data pertaining to a single time point. Thirdly, there is the problem of the heterogeneity of metabolic pools within organisms, since it is implicit in Reiner's formulation that the organism or tissue under study is effectively homogenous. In view of the amount of experimental work involved, it is perhaps not surprising that no definitive measurement has been made of GA turnover in higher plants. Bandurski (1980), however, has attempted to measure the turnover of IAA in *Z. mays* kernels.

4.3 METABOLISM OF GIBBERELLINS

4.3.1 Overview

The biosynthetic pathway from mevalonic acid to the first-formed gibberellin, GA_{12} aldehyde, has been demonstrated in the fungus *G. fujikuroi* and in cell-free systems derived from a number of angiosperms (see Chapters 2, 3, and 5). The biosynthetic pathway has not yet been confirmed in *intact* higher plants, but this is undoubtedly due to technical limitations. It has been assumed, without serious contention, that the pathway demonstrated *in vitro* also operates *in vivo*.

The further conversion of GA_{12} aldehyde to GAs in *G. fujikuroi* is discussed in detail in Chapter 5. Since metabolic studies in *G. fujikuroi* provide the background for all higher plant studies, a very brief outline of GA_{12} aldehyde metabolism in *G. fujikuroi* is given here. There is a dichotomy at GA_{12} aldehyde, with one route leading directly to non-3-hydroxylated GAs (non-3-hydroxylation pathway); the other, beginning with the 3-hydroxylation of GA_{12} aldehyde to GA_{14} aldehyde, leads to 3-hydroxylated GAs (early 3-hydroxylation pathway). In both pathways the aldehydes are oxidized to the C-7 acids, GA_{12} and GA_{14}. The next steps in both pathways are the oxidation of C-20, giving the lactonized alcohol GA_{15}, the aldehyde GA_{24}, and the acid GA_{25} in the non-3-hydroxylated pathway. In the 3-hydroxylated pathway the counterparts are GA_{37}, GA_{36}, and GA_{13}. At a crucial step in the pathway, the angular C-20 is lost, giving rise to GAs that contain only 19 carbon atoms (C_{19}-GAs). In *G. fujikuroi* the oxidative state of C-20 at the time of its loss is not known, and the immediate precursor of C_{19}-GAs has yet to be recognized. The first-formed C_{19}-GA in the non-3-hydroxylation pathway is GA_9, and the comparable GA in the early 3-hydroxylation pathway is GA_4. Detailed discussion of this part of the pathway is provided in Chapter 5. Recent progress in the understanding of C_{20}- to C_{19}-GA conversion has been made with cell-free systems from higher plants. In the *C. maxima* system the C-20 aldehyde GA_{36}

is an efficient precursor of GA_4 (Graebe, Hedden, & Rademacher, 1980). Also, in the *Pisum sativum* cell-free system the C-20 aldehydes, GA_{24} and GA_{19}, are the immediate precursors of GA_9 and GA_{20}, respectively (Kamiya & Graebe, 1983; Chapter 3, this volume).

In *G. fujikuroi* it is only at a late stage in the biosynthetic pathway—i.e., after the loss of C-20—that further hydroxyl groups or other functionalities are introduced into the GA molecule. In either the non-3-hydroxylation or the early 3-hydroxylation pathway 1α- or β-, 2α- and/or 13-hydroxylation may occur and/or dehydro-genation may take place giving an array of structurally divergent C_{19}-GAs (15 known to date). The only structural modification of C_{20}-GAs to occur in *G. fujikuroi*, apart from oxidation of C-20, is hydration of the exocyclic methylene of GA_{13} and GA_{14} giving the 16α-alcohols, GA_{41} and GA_{42}, respectively. However, these GAs may be formed non-enzymatically since hydration can occur spontaneously at low pH.

There is as yet no explanation why C_{20}-GAs do not become further structurally modified in the fungus. Enzymes early in the pathway *can* metabolize exogenously supplied hydroxylated sub-strates. Thus 2-, 12-, 13-, and 15-hydroxylated kaurenoids fed to *G. fujikuroi* will undergo ring contraction, C-20 oxidation, and even eventual loss of C-20 (Bearder et al., 1975a, 1975b; Lunnon, MacMillan, & Phinney, 1977; Gaskin et al., in preparation). In fact, enzymes catalyzing these reactions are quoted as having remarkably low substrate selectivity (Hedden et al., 1978). One must therefore infer that enzymes involved in 1-, 2-, and 13-hydroxylations in *G. fujikuroi* can only accept a C_{19}-GA as substrate.

Metabolic studies with higher plants have lagged behind those with the fungus. In *intact* higher plants there has been a dearth of well-planned and well-executed studies on the formation of C_{19}-GAs. Indeed, virtually all the information on probable biosynthetic pathways from GA_{12} aldehyde to C_{19}-GAs in intact plants is specu-lative and comes from a consideration of the type of naturally occurring C_{20}-GAs in a given species. In contrast to *G. fujikuroi*, in angiosperms there is an array of structurally divergent C_{20}-GAs (Table 4.1), with 2β-, 3-, 12α-, and/or 13-hydroxylations occurring before the loss of C-20. The existence of these GAs suggests the existence of several parallel pathways from GA_{12} aldehyde to C_{19}-GAs in higher plants. Indeed this suggestion is corroborated by metabolic studies with cell-free systems from *C. maxima* and from *P. sativum* (Chapter 3). In *C. maxima* there are non-3-hydroxylation and early 3-hydroxylation pathways that have some features in common with the fungal pathways (Graebe et al., 1974a, 1974b). In addition,

Table 4.1. Gibberellins from fungal and some higher plant sources with respect to the oxidative state of C-20, and the position of hydroxylation

Hydroxyl groups	Oxidative state of C-20	GA	Sources[a]
None	CH_3	GA_{12}	*Gibberella, Pyrus, Cucurbita*
	CH_2OH	GA_{15}	*Gibberella, Pyrus, Triticum*
	CHO	GA_{24}	*Gibberella, Marah, Secale, Triticum*
	COOH	GA_{25}	*Gibberella, Marah, Pyrus, Cucurbita*
	OH	GA_9	*Gibberella, Pyrus, Triticum, etc.*
C-3	CH_3	GA_{14}	*Gibberella*
	CH_2OH	GA_{37}	*Gibberella, Phaseolus*
	CHO	GA_{36}	*Gibberella*
	COOH	GA_{13}	*Gibberella, Enhydra, Gossipium, Cucurbita*
	OH	GA_4	*Gibberella, Phaseolus, Pyrus, Marah, Cucurbita, etc.*
C-13	CH_3	GA_{53}	*Vicia, Spinacia, Agrostemma,[b] Zea[c]*
	CH_2OH	GA_{44}	*Pisum, Phaseolus, Vicia, Zea, Spinacia, Agrostemma,[b] Pharbitis,[d] etc.*
	CHO	GA_{19}	*Pisum,[e] Phaseolus, Vicia, Spinacia, Agrostemma,[b] Zea,[c] Pharbitis,[d] etc.*
	COOH	GA_{17}	*Pisum, Phaseolus, Vicia, Zea, Spinacia, Agrostemma,[b] Pharbitis,[d] etc.*
	OH	GA_{20}	*Gibberella,[f] Pisum, Phaseolus, Vicia, Spinacia, Agrostemma,[b] Zea,[c] Pharbitis, etc.*
C-3, C-13	CH_3	GA_{18}	*Lupinus, Wisteria*
	CH_2OH	GA_{38}	*Phaseolus*
	CHO	GA_{23}	*Lupinus, Wisteria*
	COOH	GA_{28}	*Lupinus, Phaseolus*
	OH	GA_1	*Gibberella, Phaseolus, Triticum, Secale, etc.*

Table 4.1. *(Continued)*

Hydroxyl groups	Oxidative state of C-20	GA	Sources[a]
C-2(β)	CH_3	—	—
	CH_2OH	—	—
	CHO	—	—
	COOH	GA_{46}	*Marah*
	OH	GA_{51}	*Pisum*
C-2(β), C-3	CH_3	—	—
	CH_2OH	GA_{27}	*Pharbitis, Calonyction*
	CHO	—	—
	COOH	GA_{43}	*Cucurbita, Marah*
	OH	GA_{34}	*Calonyction, Phaseolus*
C-12(α)	CH_3	12α-OH GA_{12}	*Cucurbita*[g]
	CH_2OH	—	—
	CHO	—	—
	COOH	12α-OH GA_{25}	*Cucurbita*[g]
	OH	—	—
C-3, C-12(α)	CH_3	12α-OH GA_{14}	*Cucurbita*[g]
	CH_2OH	12α-OH GA_{37}	*Cucurbita*[g]
	CHO	—	—
	COOH	GA_{39}	*Cucurbita*[g]
	OH	GA_{58}	*Cucurbita*[h]
C-2(β), C-3 C-12(α)	CH_3	—	—
	CH_2OH	—	—
	CHO	—	—
	COOH	GA_{52}	*Lagenaria*
	OH	GA_{49}	*Cucurbita*

[a] Unless otherwise indicated literature references can be found in Bearder (1980, Table 1.7).

[b] Jones & Zeevaart (1980b).

[c] Hedden et al. (1981).

[d] Jones et al. (1980).

[e] Ingram & Browning (1979).

[f] McInnes et al. (1977).

[g] Bleschmidt et al. (in preparation)

[h] Graebe et al. (1980).

there appears to be a pathway in *C. maxima* in which 12α-hydroxylation of GA_{12} aldehyde occurs (Hedden et al., in preparation). In *P. sativum* there are two parallel pathways to C_{19}-GAs. In the major pathway (Ropers et al., 1978; Kamiya & Graebe, 1983) early 13-hydroxylation gives the intermediates GA_{53}, GA_{44}-open lactone, and GA_{19}, with the first-formed C_{19}-GA being GA_{20}. The minor pathway, analogous to the non-3-hydroxylation pathway in *G. fujikuroi* gives rise to GA_9 (Kamiya & Graebe, 1983). There is no evidence for 3-hydroxylation of GAs in cell-free systems derived from *Pisum*.

Most of the metabolic work with intact plants has been concerned with the conversion of one C_{19}-GA into another. Sometimes only a single metabolic step has been demonstrated in a particular species. Since there is evidence that related genera contain structurally similar C_{19}- and C_{20}-GAs, work on related species or genera is grouped together in the following discussion into subsections (4.3.2 to 4.3.4). More general and physiological aspects of the work are considered in the final sections (4.4 and 4.5).

4.3.2   Legumes

Metabolic studies in seeds of legumes merit consideration in detail for several reasons. First, immature seeds contain relatively high levels of GAs, which allowed their identification and quantitation to be conducted prior to metabolic work. These native GAs have been fed at specific stages of development in order to correlate metabolism with seed maturation. Second, various useful techniques have been illustrated—for example, stable isotope labeling—to determine rates of metabolism of exogenous and endogenous GAs. Third, in each species several C_{19}-GAs have been fed, allowing a comprehensive picture of C_{19}-GA conversions to be built up. Thus, many fundamental metabolic reactions—for example, 2β-hydroxylation—have been demonstrated, and a 2β-hydroxylating enzyme in high-speed supernatant fractions of *Ph. vulgaris* seed extracts has been partially characterized. The phenomenon of GA conjugation (glycosylation) in maturing *Phaseolus* seeds is described below, and a catabolic pathway in *Pisum* is also discussed. The physiological significance of GA conjugation and GA catabolism in relation to seed germination and seedling growth is also discussed. The information accumulated from this work has allowed the reappraisal of several physiological phenomena in legume seedlings (Section 4.4).

One difference between the feeds to legume seeds discussed here and those feeds discussed in following subsections is that in legume seeds the applied GAs have no visible effect on growth. Baldev, Lang, and Agatep (1965) showed that GA levels in immature pea seeds can be substantially reduced without adverse effects on seed development and germination. Thus, although these metabolic studies in legume seeds are the most complete to date, and provide examples of many aspects of metabolic work, a role for GAs in developing seeds has yet to be established.

4.3.2.1 *Phaseolus* species

The number and diversity of GAs in immature and mature seeds of *Ph. coccineus* and *Ph. vulgaris* have made these species ideal subjects for GA metabolic studies. The free GAs identified in both species of *Phaseolus* are shown in Table 4.2, along with the GAs identified in other legumes. Structures of the free GAs in *Phaseolus* are shown in the metabolic scheme in Fig. 4.1, and structures of the GA conjugates known to occur in *Ph. vulgaris* are shown in Fig. 4.2. Durley, MacMillan, and Pryce (1971) using GC, followed the distribution of the C_{19}-GAs in different sized seeds of both species, and Hiraga et al. (1974a, 1974c) isolated the free GAs and GA conjugates from immature and mature seeds and seedlings of *Ph. vulgaris*. The data indicated a fairly widespread distribution of free GAs throughout seed development, but a decline in free GAs and accumulation of conjugated GAs in mature seeds.

Definitive information on GA biosynthesis from GA_{12} aldehyde in *Phaseolus* species is lacking. However, from the types of C_{20}-GAs known to be present one can infer the existence of three parallel pathways from GA_{12} aldehyde to C_{19}-GAs (Fig. 4.1). The presence of GA_{37} indicates a pathway where early 3-hydroxylation takes place, and the possible presence of other 3-hydroxylated C_{20}-GAs such as GA_{14}, GA_{36}, and GA_{13} (Table 4.1) can be predicted. This pathway, analogous to that in the fungus, would give rise to the C_{19}-GA, GA_4. In a second potential pathway in *Phaseolus*, early 13-hydroxylation may take place as evidenced by the presence of GA_{19} and GA_{17} in *Ph. coccineus*, and GA_{44} and GA_{17} in *Ph. vulgaris*. An early 13-hydroxylation pathway would give rise to the C_{19}-GA, GA_{20}. The potential for early 13-hydroxylation is a departure from fungal metabolism, and 13-hydroxylated C_{20}-GAs have only been isolated from higher plants (Table 4.1). A third possible pathway is suggested by the presence of the 3,13-dihydroxylated C_{20}-GAs, GA_{28} and GA_{38}. The first-formed C_{19}-GA in an early 3,13-dihydroxylation pathway would be GA_1. This third potential pathway has

Table 4.2. Comparison of gibberellins from *Leguminosae* with respect to 3- and 13-hydroxylation[a]

	Without 13- and/or 3-OH		3-OH only		13-OH only		3- and 13-OH[b]	
	C_{19}-GAs	C_{20}-GAs	C_{19}-GAs	C_{20}-GAs	C_{19}-GAs	C_{20}-GAs	C_{19}-GAs	C_{20}-GAs
Phaseolus coccineus	—	—	GA_4, GA_{34}	GA_{37}[c]	GA_{20}	GA_{17}, GA_{19}	GA_1, GA_3, GA_5, GA_6, GA_8	GA_{28}, GA_{38}
Phaseolus vulgaris	—	—	GA_4	GA_{37}	GA_{20}, GA_{29}	GA_{17}, GA_{44}	GA_1, GA_5, GA_6, GA_8	GA_{38}
Vigna unguiculata	—	—	GA_4	—	GA_{20}, GA_{29}	GA_{17}, GA_{19}	GA_1, GA_5, GA_6, GA_8	—
Pisum sativum	GA_9, GA_{51}	—	—	—	GA_{20}, GA_{29}	GA_{17}, GA_{19},[d] GA_{44}	GA_1[e]	—
Vicia faba	—	—	—	—	GA_{20}, GA_{29}	GA_{17}, GA_{19}, GA_{44}, GA_{53}	GA_5	—
Wisteria floribunda	—	—	—	—	—	—	—	GA_{18}, GA_{23}
Lupinus luteus	—	—	—	—	—	GA_{19}	—	GA_{18}, GA_{23}, GA_{28}

[a] Unless otherwise indicated, literature references can be found in Bearder (1980, Table 1.7).
[b] Or other oxidative function at C-3.
[c] Sponsel (unpublished).
[d] Ingram & Browning (1979).
[e] Ingram, Reid, Potts, & Murfet (unpublished).

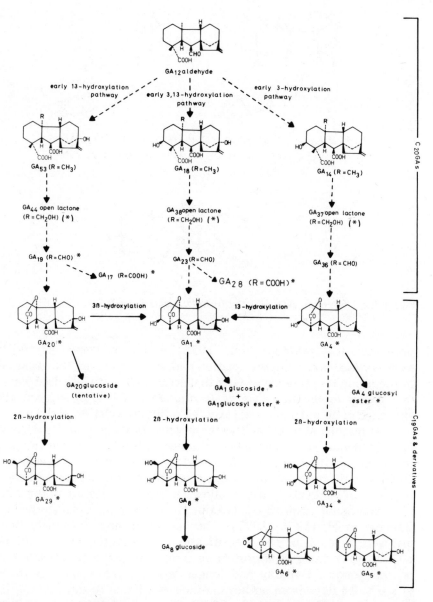

Figure 4.1 GA metabolic pathways in *Ph. vulgaris* seeds. Pathways for C_{20}-GA metabolism are based on those established in *P. sativum* cell-free systems by Kamiya and Graebe (1983). ⟶ Metabolic conversion established by feeding studies; ------→ speculative metabolic conversion; * GA known to be native to seeds of *Phaseolus* spp.; (*) the lactones (GA_{44}, GA_{38}, and GA_{37}) are present in plant extracts.

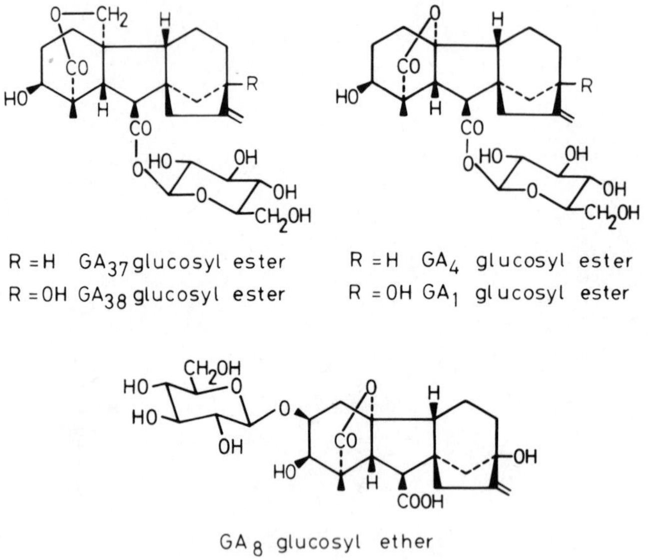

R = H   GA$_{37}$ glucosyl ester
R = OH GA$_{38}$ glucosyl ester

R = H   GA$_4$ glucosyl ester
R = OH GA$_1$ glucosyl ester

GA$_8$ glucosyl ether

Figure 4.2   GA conjugates isolated from *Ph. vulgaris* seeds.

been regarded (Hedden et al., 1978) as a convergence of the 3-hydroxylated and 13-hydroxylation pathways. However, the exclusive or predominant presence of GAs that are both 3- *and* 13-hydroxylated in *Lupinus luteus* and *Wisteria floribunda* (Table 4.2) suggests the existence of an independent, early 3,13-dihydroxylation pathway.

Numerous studies on the metabolic conversion of C$_{19}$-GAs in both *Phaseolus* species have been published. Studies using developing seeds will be discussed first, followed by those on mature seeds and seedlings. Although this does not necessarily follow the chronological sequence of publication, it does allow a logical organization of the data.

The metabolism of C$_{19}$-GAs in *Ph. vulgaris* has been studied by Yamane et al. (1975, 1977). This group, having previously been involved in much of the identification work (Hiraga et al., 1974a, 1974c), took the precaution of feeding only those GAs known to be endogenous. Two-day and longer term *in vivo* feeds were made to seeds, 18 days from anthesis, which continued to develop on the parent plant (Yamane et al., 1975). Unfortunately the amount of radioactivity recovered in methanolic extracts of seeds was exceptionally low in most instances, and losses were not accounted for. Possibly substrate and/or metabolites were translocated out of treated fruits. Indeed, somewhat improved recovery of radioactivity was reported for *in vitro* feeds (Yamane et al., 1977). In these experiments, seeds harvested 10 days after anthesis were cultured for 2 to 10 days on agar containing radioactive substrates. In both feeds some loss of radioactivity from the substrate by metabolism (yielding volatile radioactive products and unlabeled GA metabolites) cannot be excluded and was not discussed by the authors. In the following discussion of the results of Yamane et al. (1975, 1977) the yields of products have been calculated, as a percentage of radioactivity fed, from data given in Tables 1 and 2 of their papers.

$[1,2-^3H_2]GA_4$ was completely metabolized in all feeds. $[^3H]GA_1$, identified by GC-RC, was the major metabolite in short-term *in vitro* feeds, with 32.1% yield. In longer-term feeds a decline in the amount of $[^3H]GA_1$ present (7.9%) indicated its further conversion to $[^3H]GA_8$ (22.7%), $[^3H]GA_1$ glucoside (2.8%) and $[^3H]GA_8$ glucoside (5.5%) (Fig. 4.1). Unidentified polar compounds accounted for 20% of recovered radioactivity. *In vivo* yields of all metabolites were low (1–2%), with 90% of applied radioactivity not being recovered in methanolic extracts.

In vivo feeds of $[2,3-^2H_2]GA_{20}$ again gave $[^3H]GA_1$ as the major metabolite (13.8%), identified by isolation and co-crystallization to constant specific radioactivity. A later repeat of this work showed that $[^3H]GA_{29}$ was also formed as a minor product. Metabolites from the further conversion of $[^3H]GA_1$—for example, $[^3H]GA_8$, $[^3H]GA_1$ glucoside, and $[^3H]GA_8$ glucoside—were observed in low yield. Some $[^3H]GA_{20}$ glucoside (tentative identification) was detected in long-term feeds. Radioactivity losses of 60–75% were evident.

In the 2-day *in vitro* feed of $[^3H]GA_{20}$, $[^3H]GA_1$ and $[^3H]GA_{29}$ were formed in approximately equal proportion, although no details of their separation and estimation are given. $[^3H]GA_8$ was detected in long-term feeds (3%), but no GA conjugates were formed *in vitro*.

Thus $[^3H]GA_4$ can be 13-hydroxylated and $[^3H]GA_{20}$ can be 3-hydroxylated, both to give $[^3H]GA_1$ (Fig. 4.1), but there is no indication from the feeds which might be the predominant route to native GA_1. Stable isotope studies would be pertinent to compare rates of metabolism of exogenous and endogenous GA_4 and GA_{20} to GA_1. An additional route to GA_1 directly from a 3,13-dihydroxylated C_{20}-GA precursor is also feasible (Fig. 4.1) but has not been experimentally studied.

$[1,2-^3H_2]GA_1$ applied to seeds developing *in vivo* was converted in very low yield (*ca.* 1%) to $[^3H]GA_8$ and its glucoside, and also to $[^3H]GA_1$ glucoside and glucosyl ester. In the long-term feed,

3% $[^3H]GA_8$ glucoside accumulated although 90% of the applied radioactivity was not recovered. During *in vitro* feeds too, nearly 80% radioactivity was lost after 10 days. Nevertheless 23.8% conversion to $[^3H]GA_8$ was recorded in the 3-day feed.

$[1-^3H_1]GA_5$ was metabolized to $[^3H]GA_8$ (2–5%) and $[^3H]GA_8$ glucoside (2.3%) *in vivo*. $[^3H]GA_8$ was identified by co-crystallization. In long-term *in vivo* feeds, small quantities of the glucosyl ether and ester conjugates of $[^3H]GA_5$ were tentatively identified. *In vitro*, however, neither $[^3H]GA_8$ nor $[^3H]GA_8$ glucoside were obtained as metabolites of $[^3H]GA_5$. Instead, there was poor recovery of applied radioactivity and there were several unidentified metabolites.

$[1-^3H_1]GA_8$ was metabolized *in vivo* to a high proportion (74%) of unidentifiable products. $[^3H]GA_8$ glucoside was a minor product. In long-term *in vivo* feeds only 3% of applied radioactivity was recovered. No *in vitro* feeds of $[^3H]GA_8$ were performed.

The foregoing feeds have been discussed in detail because the results were not as straightforward as has been suggested in previous reviews (e.g., Hedden et al., 1978; Graebe & Ropers, 1978). Indeed, many points deserve further attention and may give new insight into GA metabolism in *Phaseolus*. First, the cause of radioactivity loss should be established, since nonlabeled metabolites may be escaping detection. Second, several major products remained unidentified; these may be artifacts from feeding nonphysiological doses (up to 100 μg/seed) but might equally well indicate metabolic conversions hitherto unknown in *Phaseolus*.

Despite these qualifications the feeds to *Ph. vulgaris* described here are the most extensive metabolic studies carried out in a single (intact) higher plant, and the comprehensive picture of C_{19}-GA metabolism obtained is shown schematically in Fig. 4.1. The results are summarized very briefly below.

$[^3H]GA_4$ and $[^3H]GA_{20}$ were metabolized to $[^3H]GA_1$ at both stages of development at which they were fed. $[^3H]GA_1$ and $[^3H]GA_{20}$ were 2β-hydroxylated to give $[^3H]GA_8$ and $[^3H]GA_{29}$, respectively. 2β-Hydroxylaction of $[^3H]GA_4$ was not observed, but GA_{34} is endogenous to *Ph. coccineus* (Sponsel, Gaskin, & MacMillan, 1979). Thus, the three pathways can converge at $[^3H]GA_1$, and indeed GA_1 and GA_8 are quantitatively the major C_{19}-GAs in maturing *Ph. vulgaris* seeds. 2β-Hydroxylated GAs have very low to zero biological activity (Reeve & Crozier, 1974) and are formed during the later stages of seed maturation. A 2β-hydroxylating enzyme obtained in high-speed-supernatant fractions of *Ph. vulgaris* seed extracts requires NADPH and Fe^{3+}, and has the characteristics of a mono-oxygenase (Patterson & Rappaport, 1974).

The biosynthetic origin of GA_5 in *Phaseolus* has not yet been determined, since neither of the potential precursors, $[^3H]GA_{20}$ or $[^3H]GA_1$, were converted to $[^3H]GA_5$ in feeds to maturing *Ph. vulgaris* seeds. Applied $[^3H]GA_5$ was converted to $[^3H]GA_8$ only in older seeds, and there was no indication of the β-epoxide, GA_6, being an intermediate. $[^3H]GA_6$ had previously been shown to be converted to $[^3H]GA_8$ in maturing seeds of *Ph. coccineus* (Sembdner et al., 1968). The structures of GA_5 and GA_6 are shown in Fig. 4.1, but they are not included in the metabolic scheme.

Evidence, albeit tenuous in some cases, was presented for the formation of $[^3H]GA_1$, $[^3H]GA_5$, $[^3H]GA_8$, and $[^3H]GA_{20}$ glucosides (glucosyl ethers) and for $[^3H]GA_1$, $[^3H]GA_4$, and $[^3H]GA_5$ glucosyl esters. Not all these conjugates are native to *Ph. vulgaris* (see Fig. 4.2), and some may arise by the action of nonspecific glucosylating enzymes on excessive levels of applied $[^3H]GAs$. $[^3H]GAs$ were *not* found to be conjugated in short-term feeds to young (10-day-old) seeds. Thus, it appears that these glucosylating enzymes are only present or are only activated at a later stage in development. This correlates well with the occurrence of native GA conjugates, since these are isolated predominantly from older seeds.

GA conjugates. GA_8 glucoside (Schreiber, Weiland, & Sembdner, 1967, 1970) and unspecified conjugates of GA_{17}, GA_{20}, and GA_{28} (Gaskin & MacMillan, 1975) are present in mature seeds of *Ph. coccineus* (runner bean). Sembdner et al. (1968) observed the metabolism of applied $[^3H]GA_3$ to $[^3H]GA_3$ and $[^3H]GA_8$ glucosides during the final stages of seed maturation. Müller et al. (1974) have obtained a glucosyl transferase enzyme from maturing seeds of *Ph. coccineus* which catalyzes glucosylation of GA_3. GA conjugates isolated from *Ph. vulgaris* are shown in Fig. 4.2.

Some general comments on GA conjugates are required here, although the subject is treated in detail in Chapter 6. The most common naturally occurring GA conjugates are those in which the GA is covalently bound to glucose either via a hydroxyl group (ether linkage) or carboxyl group (ester linkage). The first naturally occurring glucosyl ether (glucoside) to be characterized was GA_8 glucoside (Schreiber et al., 1967, 1970), and the first glucosyl esters were isolated and characterized by Hiraga et al. (1972). The structures are shown in Fig. 4.2. Neither type of GA conjugate is known to occur in *G. fujikuroi*, either naturally or as a metabolite of applied GAs. This suggests a fundamental difference between higher plant and fungal GA metabolism.

The biological activity of GA glucosyl ethers and esters has been tested often. After a comprehensive series of bioassays, Yokota et al. (1971a), Hiraga, Yamane, and Takahashi, (1974b), and Sembdner et al. (1976) concluded that all conjugates had low to zero activity per se. However, GA conjugates may exhibit some biological activity in bioassay as a result of their hydrolysis, if the aglycones formed on hydrolysis are biologically active GAs.

All the naturally occurring GA glucosyl esters are conjugates of biologically active GAs, namely GA_1, GA_4, GA_9, GA_{37}, and GA_{38}. Thus, whilst the conjugates per se have little biological activity, the aglycones exhibit high activity. Alternatively, four of the six naturally occurring GA glucosyl ethers are conjugates of 2β-hydroxylated GAs, namely GA_8, GA_{26}, GA_{27}, and GA_{29}. In these instances, the GA conjugates and the aglycones have similarly low activity.

During bioassay, plant enzymes and those of contaminating microorganisms are able to hydrolyze GA glucosyl esters more readily than the corresponding GA glucosyl ethers. However, hydrolysis of glucosyl esters is neither universal nor complete. In rice seedlings some GA glucosyl esters have activity equivalent to that of their aglycones, and indeed complete hydrolysis of GA_3 glucosyl ester applied to rice seedlings apparently occurs in 3-4 days. GA conjugates usually show less activity when applied to dwarf maize and dwarf pea seedlings where hydrolysis is less extensive (Liebisch, 1974).

The conjugation of a biologically active GA, and its subsequent hydrolysis, constitutes, at least in theory, a reversible mechanism for GA deactivation/reactivation. A depot or storage role for GA conjugates during periods of low physiological activity has therefore been envisaged. For instance, Sembdner et al. (1968) proposed that GA conjugates, synthesized in maturing seeds of *Ph. coccineus* and stored in mature seeds, were subsequently hydrolyzed in germinating seeds to yield active GAs for utilization in seedling growth. With hindsight, it is unlikely that GA_8 glucoside—the major GA conjugate in *Ph. coccineus* seeds—has a storage function, since the aglycone, GA_8, has no higher biological activity than the conjugate. Other GA conjugates known to be present in *Ph. vulgaris*, such as the glucosyl esters of GA_1, GA_4, GA_{37}, and GA_{38}, are more likely candidates for the storage or depot compounds envisaged by Sembdner et al., since the aglycones have considerably higher biological activity than the conjugates themselves. However, although the evidence for the accumulation of GA conjugates in mature seeds of *Phaseolus* species is incontrovertible, definitive evidence for conjugate hydrolysis during germination and early seedling growth is lacking. There are many reports in the literature of [^{3}H]GA conjugates, formed during seed maturation, being hydrolyzed during germination (e.g, Barendse, Kende, & Lang, 1968; Sembdner et al., 1968), but these claims are based on relative proportions of radioactivity in ethyl acetate and aqueous or butanol fractions. While the results give an indication that transformation of a polar compound into a less polar one is occurring, they cannot be construed as conclusive evidence for the conversion of GA conjugates to free GAs. Yamane et al. (1975) gave evidence for the production of GA_1 glucosyl ester during seed maturation of *Ph. vulgaris* and reported that during germination the radioactivity associated with this compound decreased from 0.5% of applied radioactivity to 0.3%. Although this result is encouraging it is unfortunate that more than 90% of applied radioactivity was unaccounted for, and so quantitative measurements of identified metabolites based on a fraction of 1% of the applied radioactivity are not convincing.

Seeds of *Ph. coccineus* will germinate even if imbibed in AMO-1618 (2'-isopropyl-4'-[trimethylammonium chloride]5'-methyl-phenylpiperidine carboxylate) (Sembdner et al., 1968). AMO-1618 is a potent inhibitor of GA biosynthesis in many systems. Similarly, *N,N,N*-trimethyl-1-methyl-(2',6',6',-trimethylcyclohex-2'-en-1'-yl)-prop-2-enyl ammonium iodide (Haruta et al., 1974; Hedden et al., 1977) does not inhibit germination of either *Ph. coccineus* or *Ph. vulgaris* (Sponsel, 1980a). Although results with growth retardants must be viewed with caution (see Lang, 1970), these results do suggest that germination in *Phaseolus* species is independent of *de novo* biosynthesis. However, more definitive experiments are required to determine whether germination can proceed independently of *de novo* GA synthesis, and/or whether hydrolysis of existing conjugated GAs is essential.

Several workers have followed the metabolism of applied GAs in germinating seeds and seedlings of *Phaseolus* species. Surprisingly, conjugate formation continues well on into germination and seedling growth—a feature which is difficult to reconcile with a role for glucosyl conjugates as a source of free GAs during germination. Thus, Nadeau and Rappaport (1972) applied [1,2-^{3}H$_2$]GA$_1$ to germinating seeds of *Ph. vulgaris* and observed 28% conversion to [^{3}H]GA$_8$ glucoside in 30 hr. [^{3}H]GA$_8$ itself scarcely accumulated. Reeve et al. (1975) applied [1,2-^{3}H]GA$_1$ to 6-day-old seedlings of *Ph. coccineus* and observed very low (0.2%) conversion to [^{3}H]GA$_8$. They did not examine polar fractions which would contain the glucoside.

4.3.2.2 *Pisum sativum*

The GAs identified in immature seeds of *Pisum sativum* (garden pea) cv. Progress No. 9 are GA_9, GA_{17}, GA_{20}, GA_{29}, GA_{44}, GA_{51}, and a catabolite of GA_{29} (see Frydman, Gaskin, & MacMillan, 1974; Sponsel, 1980b). GA_{19} has been identified in the photoperiodically sensitive G2 line of pea (Ingram & Browning, 1979). *Vicia faba* (broad bean), of the same tribe as pea, contains GA_{53} in addition to the above GAs (Sponsel et al., 1979). The presence of the 13-hydroxylated C_{20}-GAs, GA_{53}, GA_{44}, GA_{19}, and GA_{17}, suggests the existence of an early 13-hydroxylation pathway (Table 4.1, Fig. 4.3). This suggestion has been partially substantiated, since GA_{12} aldehyde is metabolized, inter alia, to GA_{53} and GA_{44} in cell-free systems from pea (Ropers et al., 1978). Other results further substantiate the intermediacy of GA_{44}-open lactone and GA_{19} in the pathway to GA_{20} (Kamiya & Graebe, 1983; Chapter 3, this volume). An early 13-hydroxylation pathway is also thought to occur in *Phaseolus*, and it may well be of common occurrence in higher plants. The presence of the nonhydroxylated C_{19}-GAs, GA_9 and GA_{51}, suggests the existence of a nonhydroxylation pathway comparable to that in the fungus. No nonhydroxylated C_{20}-GAs have so far been identified in pea (Table 4.1). With the possible exception of GA_{38}, whose presence in pea was reported (Frydman, Gaskin, & MacMillan, 1974) and later withdrawn (Sponsel et al., 1979), no 3-hydroxylated GAs have been identified in immature seeds of pea to date.

Extensive metabolic studies have been conducted with *P. sativum*. In parallel with the previous discussion on *Phaseolus* metabolism, studies using developing pea seeds will be described first, followed by work on germinating seeds and seedlings.

The successive accumulation of GA_9, GA_{20}, and GA_{29} in developing seeds of Progress No. 9 (Fig. 4.4) (Frydman et al., 1974; Sponsel & MacMillan, 1978) indicated the metabolic sequence $GA_9 \rightarrow GA_{20} \rightarrow GA_{29}$. This conversion was also postulated to occur in pea seedlings (Railton, Durley, & Pharis, 1974a) and is discussed later. Frydman and MacMillan (1975) and Sponsel and MacMillan (1977, 1978) applied $[^3H]GA_9$, $[^3H]GA_{20}$, and $[^3H]GA_{29}$ initially to seeds cultured *in vitro* and, in later studies, to seeds that continued to develop *in vivo*. $[17-^3H_2]GA_9$ was initially fed to 20-day-old seeds when endogenous GA_9 levels were maximal, and treated seeds were extracted after 2 days when the level of the putative metabolite, GA_{20}, was known to be maximal (Fig. 4.4). At first $[^3H]GA_9$ was fed at approximately the same dose as the endogenous level, but this dose had to be increased 10- to 50-fold to enable

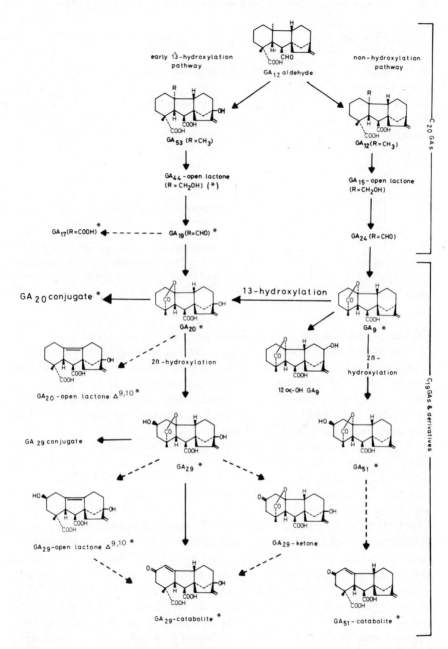

Figure 4.3 GA metabolic pathways in *P. sativum* seeds. Pathways for C_{20}-GA metabolism have been observed in cell-free systems only (Kamiya & Graebe, 1983). ⟶ Metabolic conversion established by feeding studies; ┈┈┈▶ speculative metabolic conversion; ∗ GA known to be native to seeds of *P. sativum*; (∗) the lactone (GA_{44}) is present in plant extracts.

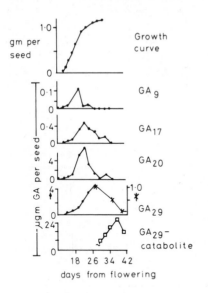

Figure 4.4 Quantitation of GAs throughout seed maturation in *P. sativum* cv. Progress No. 9. —●— Data from Frydman et al. (1974); —X— data from Sponsel and Macmillan (1978); —□— data from Sponsel and MacMillan (1980).

GC-MS identification of metabolites. For *in vitro* feeds the uptake of radioactivity was 60-80%. The residual radioactivity in the medium accounted for the remaining activity fed, indicating that none was lost (cf. *Phaseolus* feeds). This residual radioactivity in the medium was assumed to be [³H]GA₉, although later work showed the medium to contain not only unmetabolized substrate but also radioactive metabolites leached from the seeds.

The results of feeding [³H]GA₉ to developing pea seeds were complex (Fig. 4.3). [³H]GA₉ was metabolized to either [³H]GA₂₀ *or* 2β-OH[³H]GA₉ (identified by GC-RC and GC-MS) in yields of up to 8 and 26% of radioactivity taken up, respectively. 12α-OH[³H]GA₉* also accumulated in all feeds while a conjugate of 12α-OH[³H]GA₉, which formed in yields of up to 60%, was the major metabolite. In subsequent work, 2β-OH GA₉ was found to be native to developing pea seeds and was named GA₅₁ (Sponsel & MacMillan, 1977). On the other hand, 12α-OH GA₉ is not known to be native to pea and is assumed to be an artifact (see later). 2β-Hydroxylation of [³H]GA₉ to give [³H]GA₅₁ is restricted to seeds weighing *ca.* 0.67 g or more. Since endogenous GA₉ and GA₅₁ are present in seeds of this size, Sponsel and MacMillan (1977) concluded that native GA₉ is probably metabolized to GA₅₁. On the other hand, 13-hydroxylation of

---

*12α-OH[³H]GA₉ was referred to as H₂-[³H]GA₃₁ in the original publications. For structural determination see Railton, Durley, and Pharis (1974b).

$[^3H]GA_9$ to give $[^3H]GA_{20}$ appears to be restricted to small seeds in which neither GA_9 nor GA_{20} are detectable. The reason for the apparent absence of these GAs may be that their turnover is so rapid in very small seeds that neither can accumulate. However, Sponsel and MacMillan (1977) suggested instead that the small conversion (<8%) of $[^3H]GA_9$ to $[^3H]GA_{20}$ observed in feeding experiments to small seeds was catalyzed by a 13-hydroxylating enzyme present during the early stages of seed development whose natural substrate was *not* native GA_9. As an alternative, they suggested that native GA_{20} was probably formed directly from a C_{20}-GA precursor which had been 13-hydroxylated at an early stage of development. Thus, GA_9 and GA_{20} would be the first-formed C_{19}-GAs on two parallel pathways (see Fig. 4.3). This suggestion has recently been substantiated with cell-free preparations derived from *P. sativum* cv. Kleinrheinlander seeds (Kamiya & Graebe, 1983). Thus, $[^{14}C]GA_9$ is formed from $[^{14}C]GA_{12}$ aldehyde via non-hydroxylated intermediates with the C-20 aldehyde, GA_{24}, being the immediate precursor. On the other hand, $[^{14}C]GA_{20}$ is formed from the same substrate via the 13-hydroxylated intermediates GA_{53}, GA_{44}-open lactone, and GA_{19}.

The formation of the seemingly unnatural 12α-OH$[^3H]GA_9$ in all feeds of $[^3H]GA_9$ to developing seed of Progress No. 9 (Frydman & MacMillan, 1975) is not readily explained. The authors postulated that 12α-hydroxylation of $[^3H]GA_9$ may arise due to unnatural orientation of $[^3H]GA_9$ with the 2β-hydroxylating enzyme, since the 2β- and 12α-protons occupy similar positions in space if the GA molecule is rotated $180°$ on the axis of the C-7 bond. The general symmetry of the GA molecule and the stereochemical equivalence of the 2β- and 12α-positions is self-evident (see Stoddart & Venis, 1980), but the possibility that the same enzyme can hydroxylate in both positions has not yet been critically examined. There is, however, some recent evidence against this possibility. In cell-free enzyme systems from pea, $[^3H]GA_9$ is *not* metabolized to 12α-OH$[^3H]GA_9$ under incubation conditions that allow the metabolism of $[^3H]GA_9$ to $[^3H]GA_{51}$. By analogy with the *C. maxima* cell-free system, in which 2β- and 12α-hydroxylating enzymes have recently been shown to have different pH optima (Hedden et al., in preparation), 2β- and 12α-hydroxylation in pea seeds may also be brought about by two separate enzymes. Whatever the biochemical basis for the formation of 12α-OH$[^3H]GA_9$ in feeds to pea, there is still no evidence for 12α-OH GA_9 being native to pea and, thus, serious reservations should be placed on the significance of the results of all $[^3H]GA_9$ feeds.

$[17\text{-}^3H_2]GA_{20}$ fed to 22- to 24-day-old seeds of Progress No. 9 cultured *in vitro* was metabolized by 2β-hydroxylation to $[^3H]GA_{29}$ (identified by GC-RC and GC-MS) in 3 days (43% yield) (Frydman & MacMillan, 1975); 30% of the $[^3H]GA_{20}$ taken up was unmetabolized, but in longer-term feeds $[^3H]GA_{20}$ was almost completely metabolized. $[^3H]GA_{20}$ was efficiently metabolized to $[^3H]GA_{29}$ in seeds varying in age from 12–28 days from anthesis. Thus 2β-hydroxylation of $[^3H]GA_{20}$ can occur throughout a fairly wide phase of seed development.

$[^3H]GA_{29}$, either applied directly or generated *in situ* from applied $[^3H]GA_{20}$, was scarcely metabolized in 4–10-day *in vivo* or *in vitro* feeds (Frydman & MacMillan, 1975; Sponsel & MacMillan, 1977). Thus, at the end of a 10-day *in vitro* feed of $[^3H]GA_{20}$, 73% of the radioactivity taken up was accounted for as $[^3H]GA_{29}$, 9% as a $[^3H]GA_{29}$ conjugate, and 1% as a $[^3H]GA_{20}$ conjugate. Sponsel and MacMillan (1978) showed that the levels of native GA_{29} declined more slowly than was previously reported (Frydman et al., 1974), and with hindsight the above feeds should have been conducted for longer.

Durley, Sassa, and Pharis (personal communication to J. MacMillan, 1976) fed $[2,3\text{-}^3H]GA_{20}$ to developing fruits of cv. Meteor and showed by GC-RC on three GC columns and GC-MS on two GC columns that it was converted to $[^3H]GA_{29}$ and to four other unidentified compounds, for one of which they obtained a mass spectrum. This compound was known to be endogenous to seeds of Progress No. 9 (Sponsel & MacMillan, 1976), but its structure was unknown. Thus, the work of Durley et al. was not published immediately and was not further extended by them. To investigate the metabolism of GA_{29} more fully, Sponsel and MacMillan (1978) fed GA_{20} and GA_{29} which were labeled with both stable isotopes and radioisotopes, thus allowing a comparison of the rates of metabolism of exogenous and endogenous GA_{20} and GA_{29} (see Section 4.2.5).

In an *in vivo* time-course feed of $[2\alpha\text{-}^2H_1][2\alpha\text{-}^3H_1]GA_{29}$, seeds were extracted at 2–3-day intervals for 13 days (Sponsel & MacMillan, 1978). Even after 13 days, nearly 90% of the applied radioactivity was recovered from the treated seeds. The sole radioactive component present in all extracts was identified by GC-RC and GC-MS as $[^2H]\text{-}[^3H]GA_{29}$. Ratios of nondeuterated to deuterated GA_{29} were computed from the molecular ion clusters in mass spectra. Throughout the feed endogenous GA_{29} declined relative to $[^2H]GA_{29}$; that is, endogenous GA_{29} was being metabolized faster than exogenous deuterated substrate. Two possible reasons for the non-equivalence of exogenous and endogenous GA_{29} were postulated: (1) the inability of the applied substrate to reach the site within the seed for metabolism to occur or, (2) the presence of a kinetic isotope effect due to metabolism of applied GA_{29} involving cleavage of the $C-{}^2H$ and $C-{}^3H$ bonds. Further results indicated that the second possibility was more likely (see Fig. 4.5).

In a similar feed of $[1\beta,3\alpha\text{-}^3H_2][1\beta,3\alpha\text{-}^3H_2]GA_{20}$ (Sponsel & MacMillan, 1978), $[^2H][^3H]GA_{29}$ was formed in good yield but there was no further build-up of other radioactive metabolites. Instead, some radioactivity appeared to be lost during metabolism and during work up, thereby preventing the detection of further metabolites. Isotope ratios stabilized by day 5 of the feed and remained constant, which indicated that endogenous and exogenous GA_{20} and GA_{29} were being metabolized in an equivalent manner.

A logical conclusion from both these feeds was that native GA_{29} was further metabolized by oxidation at C-2. Large-scale extractions of germinating pea seeds were conducted in order to isolate the putative metabolite of native GA_{29}. A compound was isolated from 3-day-old seeds of Progress No. 9 which had a similar mass spectrum to that obtained by Durley, Sassa, and Pharis

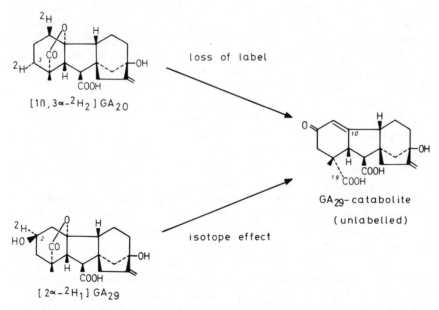

Figure 4.5   The metabolic conversion of isotopically labeled GA_{20} and GA_{29} to unlabeled GA_{29}-catabolite.

(unpublished) for a putative metabolite of $[^3H]GA_{20}$. Partial structural determination was conducted (Sponsel & MacMillan, 1978), and the structure of a GA-catabolite (see Fig. 4.5) was confirmed by synthesis (Kirkwood, 1979; Gaskin, Kirkwood, & MacMillan, 1981). The isotope data obtained from feeding doubly labeled GA_{20} and GA_{29} are quite consistent with this α,β-unsaturated ketone being the hitherto untraced metabolite of GA_{29} in maturing pea seeds (Fig. 4.5) (Sponsel & MacMillan, 1978). The compound is referred to as a GA-catabolite since it no longer possesses the γ-lactone characteristic of other C_{19}-GAs.

After the structure of the GA_{29}-catabolite had been determined (Sponsel & MacMillan, 1978; Kirkwood, 1979), the results of feeds conducted in 1976 by Durley, Sassa, and Pharis demonstrating the conversion of $[^3H]GA_{20}$ to the now-known GA_{29}-catabolite were published (Durley et al., 1979). The $[^3H]GA_{20}$-substrate used by Durley et al. was labeled at C-2 and C-3, and ^3H would therefore be lost during metabolism to the GA_{29}-catabolite, and during work-up. The actual conversion of $[^3H]GA_{20}$ to the GA_{29}-catabolite is undoubtedly higher than that reported, since the kinetic data presented by Durley et al. (1979) are based on the unknown amount of ^3H randomly incorporated into the substrate during its preparation (Murofushi, Durley, & Pharis, 1977).

$[17\text{-}^{13}C_1]GA_{29}$ has recently been prepared (Kirkwood, 1979; Gaskin et al., 1981) to confirm that the GA-catabolite is indeed a metabolite of GA_{29}, and to provide accurate quantitative data. $[17\text{-}^{13}C_1]GA_{29}$ was fed to 28-day-old seeds developing *in vivo*, for up to nine days (Sponsel & MacMillan, 1980). Since radio-labeled GA_{29} was not available for feeding along with $[^{13}C]GA_{29}$, $[19\text{-}^{18}O_1]$-GA_{29}-catabolite was added to *extracts* as an internal standard. This enabled the quantitation of native and $[^{13}C]GA_{29}$-catabolite to be conducted by MS. The accumulation of endogenous and exogenous GA_{29}-catabolite in resulting extracts was parallel over the first 7 days of the experiment, suggesting that metabolism of the applied $[^{13}C]GA_{29}$ reflected that of native GA_{29}. At least 50% of the applied $[^{13}C_1]GA_{29}$ was converted to $[^{13}C]GA_{29}$-catabolite in 7 days. There was evidence toward the end of the feed that the GA_{29}-catabolite itself was being further metabolized.

Isomerization of the GA_{29}-catabolite occurs during derivatization for GC, resulting in the formation of several artifacts that complicate the analysis of extracts (Sponsel & MacMillan, 1980; Gaskin et al., 1981). These artifacts could unknowingly be mistaken for additional metabolites if extracts were analyzed solely by GC-RC. Variability in plant material is another factor leading to inconsistencies in analysis of the GA_{29}-catabolite. There have been marked differences in accumulation of the GA_{29}-catabolite, with a corresponding variation in the levels of GA_{29}, in different batches of Progress No. 9 seeds (Sponsel & MacMillan, 1977, 1978, 1980).

The feeds to maturing pea seeds so far described are summarized briefly below, and the main results are shown in the metabolic scheme in Fig. 4.3. [^{3}H]GA_9 is metabolized to three monohydroxylated GAs, namely [^{3}H]GA_{20}, [^{3}H]GA_{51}, and 12α-OH[^{3}H]GA_9. [^{3}H]GA_{20} is formed only in feeds to small seeds before endogenous GA_9 and GA_{20} are detectable, and it is concluded that native GA_9 is an unlikely precursor of native GA_{20}. GA_9 and GA_{20} are probably the first-formed C_{19}-GAs on two parallel pathways—a nonhydroxylation and an early 13-hydroxylation pathway, respectively. The natural metabolite of GA_9 is thought to be the 2β-hydroxylated GA_{51}, since [^{3}H]GA_9 is metabolized to [^{3}H]GA_{51} at a developmental stage when both native GA_9 and GA_{51} are present. These results suggest a specificity of GA metabolism correlated with seed size/age. Thus, 13-hydroxylation appears to be restricted to small/young seeds, while 2β-hydroxylation occurs in larger/older seeds. The third metabolite of [^{3}H]GA_9, 12α-OH[^{3}H]GA_9, becomes readily conjugated and is the major metabolite in all [^{3}H]GA_9 seeds. It appears to be an artifact since it is not known to be naturally occurring in pea. The formation of 12α-OH[^{3}H]GA_9 causes problems in the interpretation of the results of all these [^{3}H]GA_9 feeds. Conclusions based on these results should be questioned until the biochemical basis for 12α-OH[^{3}H]GA_9 formation is fully understood.

[^{3}H]GA_{20} is efficiently metabolized by 2β-hydroxylation to give [^{3}H]GA_{29}, throughout a wide phase of seed development. [^{3}H]GA_{29} is metabolized in good yield toward the later stages of seed development to a hitherto unknown type of GA-catabolite. This GA_{29}-catabolite is the major GA derivative in mature and germinating seeds of Progress No. 9. GA conjugates are also formed, but in a very low yield.

Recent work on the localization of GAs in maturing pea seeds has shown that GA_{20} and GA_{29} are predominantly localized in the cotyledons, while the GA_{29}-catabolite is almost exclusively located in the testa (Sponsel, in preparation). Work is in progress to determine where the conversion of GA_{29} to the catabolite occurs, and what are the intermediates in this conversion. Preliminary results from feeds to isolated seed parts suggest that cotyledons cultured *in vitro* have only a limited capacity to metabolize [17-^{13}C][$15,17$-^{3}H]GA_{29}. Conversely, [^{13}C][^{3}H]GA_{29} is efficiently metabolized to the catabolite in excised testas, suggesting that *in vivo* native GA_{29} may be translocated to the testas before catabolism occurs. Two possible alternative intermediates in the conversion of GA_{29} to the catabolite [GA_{29}-open lactone $\Delta^{9,10}$ and GA_{29}-ketone (see Fig. 4.3)] have been discussed briefly (Sponsel, 1980b). The open lactone $\Delta^{9,10}$ compound is known to be endogenous, but the ketone has not so far been identified in pea. However, due to problems of derivatization the ketone may have escaped detection using normal procedures. Another complicating factor is the spontaneous conversion of synthetic GA_{29}-ketone to GA_{29}-catabolite (Gaskin et al., 1981). Sponsel and MacMillan (1978) and Sponsel (1980b) presented further evidence that GA_{29}-ketone is *not* present in extracts, and from that they inferred that the GA_{29}-catabolite is the true metabolite of GA_{29}.

What is the physiological significance of the GA_{29}-catabolite in peas? The catabolite has low to zero biological activity in lettuce hypocotyl, pea stem, and barley half-seed assays. Also, it is unable to promote the growth of cultured embryonic axes excised from imbibing pea seeds. Although an indirect effect on axis growth via an effect of cotyledonary metabolism cannot be excluded, it would appear unlikely in view of its low activity in the barley assay. Thus, Sponsel and MacMillan (1978) and Sponsel (1980b) have concluded that 2β-hydroxylation followed by conversion to the GA_{29}-catabolite is a mechanism for the deactivation of GA_{20}. The 13-deoxy analogue of the GA_{29}-catabolite is also present in pea and is a potential catabolite of GA_{51} (Sponsel & MacMillan, 1980; Gaskin et al., 1981). Thus, a fate for GA_9 parallel to that for GA_{20} can be envisaged (see Fig. 4.3). The GA_{29}-catabolite is also known to be a major GA component of *Vicia faba* seeds, and the 3-hydroxy analogue (a potential catabolite of GA_1) is present in trace amounts in *Ph. coccineus* and *Vigna unguiculata* (cowpea) (Sponsel et al., 1979). Thus, catabolism appears to occur in a variety of legumes and may be a general means for GA deactivation.

Conjugation of GAs is known to occur in pea, but to a limited extent only (Frydman & MacMillan, 1975). Isolation of native GA conjugates on the scale employed with *Phaseolus* (Hiraga et al., 1972, 1974c) has not been conducted for *Pisum*. Instead, fractions of extracts expected to contain GA conjugates have been hydrolyzed with crude enzyme preparations. The aglycones released by hydrolysis have been identified by GC-MS (Sponsel & MacMillan, 1977), but this procedure gives no indication as to the nature of the conjugate itself.

Sponsel et al. (1979) and Sponsel (1980a) suggested that *Pisum* and the closely related genus *Vicia* may deactivate GAs predominantly by catabolism, whereas in *Phaseolus* the major deactivation mechanism seems to be that of conjugation. If these speculations are confirmed, physiological studies on the role of GAs in germination of *Pisum* and *Phaseolus* would be in order. Conjugation is a potentially reversible means of GA deactivation, while catabolism appears to be irreversible, suggesting the possibility of a different role for GAs in seed germination in the two genera.

Early data on the GA content of pea seedlings (cv. Progress No. 9) were obtained by Kende and Lang (1964) who reported two biologically active fractions with chromatographic and biological properties consistent with those of GA_1 and GA_5. Neither of these have subsequently been identified in developing seed of Progress No. 9, but the structurally and hence chromatographically similar GA_{29} and GA_{20} are present. Quantitation of GA_{20}, GA_{29}, and the GA_{29}-catabolite throughout the final stages of seed maturation and on into germination (Sponsel, in preparation) suggest that these GAs are carried over from the maturation phase. Several other GA derivatives are known to be present in germinating seeds and seedlings, namely GA_{20}-open lactone, GA_{29}-open lactone, and 9-hydroxy GA_{29}-catabolite (Kirkwood, 1979). GA_{29}-open lactone $\Delta^{9,10}$ is a potential precursor of the GA_{29}-catabolite, and the 9-hydroxyl derivative is a potential further metabolite of the latter. Several GAs and derivatives can also be released, normally in trace amounts, from butanol-soluble fractions of pea seedling extracts by enzyme hydrolysis. These include GA_{17}, GA_{20}, GA_{29}, GA_{51}, GA_{29}-catabolite, GA_{20}- and GA_{29}-open lactone $\Delta^{9,10}$ derivatives, and 9-hydroxy GA_{29}-catabolite (Kirkwood, 1979). Also present, as presumed conjugates, are several polyhydroxykauranoic acids and dihydrophaseic acid (Sponsel & MacMillan, 1977). To date, there is no conclusive evidence for the presence of free or conjugated GA_9 in seedlings of Progress No. 9 and several other cultivars of pea (Sponsel, unpublished; Hedden, unpublished; Ingram, unpublished).

Extensive metabolic studies have been conducted with germinating seeds and seedlings of pea. Durley et al. (1974a, 1974b) fed the 3-hydroxylated C_{20}-GA, GA_{14}, to 5-day-old etiolated seedlings of cv. Meteor. Several radioactive metabolites were detected and were identified on the basis of GC-RC on three GC columns. In a 20-hr feed $[17-^3H_2]GA_{14}$ was metabolized to $[^3H]GA_{18}$ (4.8% of the radioactivity in the tissue), $[^3H]GA_{38}$ (1.43%), $[^3H]GA_{23}$ (0.36%), $[^3H]GA_{28}$ (0.01%), and $[^3H]GA_8$ (0.02%) (for structures, see Table 4.1 and Fig. 4.1). In a 40-hr feed yields were $[^3H]GA_{18}$ (1.74%), $[^3H]GA_{38}$ (0.8%), $[^3H]GA_{23}$ (0.96%), $[^3H]GA_{28}$ (0.14%), $[^3H]GA_1$ (0.57%), and $[^3H]GA_8$ (0.50%). GC-MS identification of $[^3H]GA_{18}$ and $[^3H]GA_1$ was reported by Durley et al. (1974a).

No radioactive products other than those identified were detected, from which it must be concluded that more than 90% of the $[^3H]$-GA_{14} taken up was unmetabolized. The authors placed significance on the *relative* proportions of the metabolites, and from this quantitative data, from biological data, and with *a priori* considerations of chemical structure the authors proposed the pathway $[^3H]GA_{14} \rightarrow [^3H]GA_{18} \rightarrow [^3H]GA_{38} \rightarrow [^3H]GA_{23} \rightarrow [^3H]GA_1 \rightarrow [^3H]GA_8$. Only $[^3H]GA_1$ was re-fed (see below), and further feeding studies with other putative intermediates are required to confirm the proposed pathway. Nevertheless, the results do indicate immediate 13-hydroxylation of the applied substrate, followed by sequential oxidation at C-20. The authors suggested that the C-20 aldehyde, $[^3H]GA_{23}$, was the immediate precursor of $[^3H]GA_1$, with the C-20 acid $[^3H]GA_{28}$ being off the main route to the C_{19}-GAs. The proposed pathway may well be the major GA pathway in *Phaseolus*, *Lupinus*, and *Wisteria* and other legumes in which 3,13-dihydroxylated GAs predominate. However, none of the 3-hydroxylated metabolites obtained by Durley et al. (1974a, 1974b) have been shown to be native to Meteor seedlings. Kamiya and Graebe (1983) have recent evidence that the pathway $[^3H]GA_{12} \rightarrow [^3H]GA_{53} \rightarrow [^3H]GA_{44}$-open lactone $\rightarrow$ $[^3H]GA_{19} \rightarrow [^3H]GA_{20}$ operates in cell-free systems from developing pea seed.

In an extension of the above work, Durley et al. (1974a) fed $[^3H]GA_1$ and obtained several products, one of which was $[^3H]GA_8$ (1.5% yield in 40 hr). Interestingly, there was no evidence for the presence of $[^3H]GA_8$ glucoside which is the major product from $[^3H]GA_1$ feeds to *Phaseolus* seedlings (Nadeau & Rappaport, 1972).

Several other feeds of C_{19}-GAs to pea seedlings have been conducted, but many were performed *before* the identification of endogenous GAs had been achieved. Kende (1967), and Musgrave and Kende (1970), having TLC and bioassay evidence for the presence of GA_1 and GA_5 in seedlings of Progress No. 9, fed both these GAs in tritiated form. No metabolites could be identified but quantiative aspects of the metabolism were compared in dark- and light-grown seedlings; this work is discussed in detail in Section 4.4.1. Durley, Railton, and Pharis (1973) fed $[1,2\text{-}^3H_2]GA_5$ to dark-grown seedlings of cv. Meteor and reported 0.8% conversion to $[^3H]GA_3$. It is difficult to reconcile any of these results with the metabolism of native GAs in pea seedlings.

Railton et al. (1974a, 1974b) obtained bioassay and chromatographic evidence for the presence of GA_9 in seedlings of cv. Meteor, and at the same time GA_9 was shown to be endogenous to developing seeds of cv. Progress No. 9 (Frydman et al., 1974). When Railton et al. (1974a, 1974b) applied $[17\text{-}^3H_2]GA_9$ to 5-day-old dark-grown Meteor seedlings, a considerable amount of substrate remained unmetabolized after 20 hr, but several metabolites were observed in low yield. Three of these were identified by GC-RC and GC-MS of a large-scale feed as $[^3H]GA_{20}$, $[^3H]GA_{10}$, and the hitherto unknown $12\alpha\text{-OH}[^3H]GA_9$. Railton et al. (1974b) provided evidence for the structure of this metabolite which, as described earlier, was later found by Frydman and MacMillan (1975) in feeds of $[^3H]GA_9$ to developing seeds. The data (Railton et al., 1974b) demonstrate a conversion of 3.6% of applied $[^3H]GA_9$ to $[^3H]GA_{20}$, 5.6% conversion to $12\alpha\text{-OH}[^3H]GA_9$, and 0.4% conversion to $[^3H]GA_{10}$ after 20 hr. There was no evidence for the formation of $[^3H]GA_{51}$ (cf. $[^3H]GA_9$ feeds to developing seeds). From a time-course feed the authors concluded that $[^3H]GA_{20}$ and $12\alpha\text{-OH}[^3H]GA_9$ are formed concurrently from applied $[^3H]GA_9$. Railton (1976a) suggested that since $12\alpha\text{-OH}$ GA_9 did not appear to be native to pea, it was possibly formed as a result of the 13-hydroxylating enzyme failing to distinguish between the 12- and 13-positions on the $[^3H]GA_9$ substrate. $[^3H]GA_{10}$ is also a possible artefact since it can arise non-enzymatically from $[^3H]GA_9$. Nevertheless, despite the low conversion of $[^3H]GA_9$ to $[^3H]GA_{20}$ and the formation of several other metabolites, Railton et al. (1974a, 1974b; Railton, 1976a) concluded that GA_9 was a genuine precursor of GA_{20} in pea seedlings. Railton (1974a, 1974b) studied the effects of light further and applied benzyl adenine on the metabolism of $[^3H]GA_9$ to $[^3H]GA_{20}$ in pea seedlings, even though the yield of $[^3H]GA_{20}$ relative to the yields of "extraneous" metabolites continued to be very low. This work is discussed in detail in Sections 4.4.1 and 4.4.4.

The apparently conflicting results from feeds of $[^3H]GA_9$ to developing seeds and seedlings of pea arise not from substantially different results in the two systems, but from different interpretation of the data obtained. Thus, Frydman and MacMillan (1975) and Sponsel and MacMillan (1977) suggested that neither their own data nor those of Railton et al. (1974b) support the original assumption that GA_9 is the precursor of native GA_{20}. Conversely, Railton et al. (1974b) and Railton (1976a) explain the low yields of $[^3H]GA_{20}$ from $[^3H]GA_9$ in terms of rapid turnover of $[^3H]GA_{20}$, and conclude that $GA_9 \rightarrow GA_{20}$ is a genuine metabolic step in pea seedlings. Recent work by Kamiya and Graebe (1983) has shown that $[^{14}C]GA_{12}$ aldehyde and $[^{14}C]GA_{12}$ are readily 13-hydroxylated in cell-free systems derived from pea seeds, but $[^{14}C]GA_9$ is only 13-hydroxylated if it is first converted to the open lactone form by alkaline treatment. From these results, Kamiya and Graebe (1983) concluded that the metabolism of GA_9 to GA_{20} is not a natural step in pea seeds.

Railton et al. (1974c) extended their metabolic work with pea seedlings to study the metabolism of $[2,3\text{-}^3H_2]GA_{20}$ in germinating seeds and seedlings. It was converted to $[^3H]GA_{29}$ in good yield in germinating seeds (34% of radioactivity fed), and in 3.8% yield in seedlings. $[^3H]GA_{29}$, isolated after a feed of $[^3H]GA_{20}$ to seeds, was re-fed to seeds and seedlings (Railton, 1976b). Limited conversion to metabolites more polar than $[^3H]GA_{29}$ was observed (3% yield). Little radioactivity was detected in butanol fractions, suggesting that little conjugate formation took place. Sponsel and MacMillan (unpublished) have fed $[^{13}C]GA_{29}$ to germinating seeds and observed little metabolism, despite the high conversion of $[^{13}C]GA_{29}$ to $[^{13}C]GA_{29}$-catabolite in developing seeds. Feeds of $[^3H]GA_9$ and $[^3H]GA_{20}$ to chloroplast preparations from pea seedlings are discussed in Section 4.5.

Railton (1980a) has attempted to correlated the 2β-hydroxylation of $[2,3\text{-}^3H_2]GA_{20}$ with growth of Meteor seedlings. $[^3H]GA_{20}$ (0.5 μg applied per 4-day-old seedlings) induced elevated growth relative to the controls for 11 days. $[^3H]GA_{20}$ was metabolized most rapidly over the first 24 hr, by which time 70% of applied $[^3H]GA_{20}$ had disappeared. Thereafter, the level of unmetabolized $[^3H]GA_{20}$ gradually declined, until it reached zero at day 15. $[^3H]GA_{29}$, identified by GC-RC, accumulated slowly and reached a maximal level equivalent to 9% of the applied radioactivity after 10 days. The $[^3H]GA_{29}$ level also declined to zero at day 15, indicating its further metabolism, but neither the GA_{29} catabolite nor any other potential metabolite of $[^3H]GA_{29}$ was detected. The losses of radioactivity observed in the experiment were great (65–100% of applied radioactivity) and are unlikely to be due entirely to the metabolism of $[^3H]GA_{29}$ to unlabeled products. It is thus difficult to assess the significance of the small accumulation of $[^3H]GA_{29}$ observed, without a complete understanding of the metabolic fate of $[^3H]GA_{20}$ in this experiment. Nevertheless, Railton (1980a) concluded from his results that growth and 2β-hydroxylation were positively correlated since the accumulation of $[^3H]GA_{29}$ was coincident with the decline in stem growth. It is perhaps worth noting that GA_{29} is inactive, *not* inhibitory, and that the disappearance of biologically active substrate rather than the appearance of inactive metabolites might have more relevance to growth.

The results of feeds to germinating seeds and seedlings can be summarized in that enzymes responsible for C-20 oxidation and for 2β-, 12α-, and 13-hydroxylation have been demonstrated, but to what extent the reported feeds reflect the metabolism of native GAs is unknown. Many of the GAs so far identified in germinating seeds and seedlings appear to have been formed in, and carried over from, developing seeds. Novel compounds detected in seedlings are predominantly derivatives of inactive GAs. Since the catabolic pathway appears to be irreversible and little conjugate formation occurs, the only source of biologically active GAs in germinating seeds and seedlings would be via *de novo* synthesis. Barendse et al. (1968) studied the germination and seedling growth of Progress No. 9 seeds imbibed for 24 hr in either water or AMO-1618. Although AMO-1618 did not inhibit germination, the growth of seedlings was suppressed. Sponsel (1980a, and unpublished data) obtained similar results using another growth retardant (Haruta et al., 1974; Hedden et al., 1977). The specificity of these inhibitors for GA biosynthesis is not absolute, but the results do indicate that seed germination in pea may proceed independently of *de novo* GA biosynthesis, whereas seedling growth is dependent on a newly synthesized supply of GAs. The feed of Durley et al. (1974a, 1974b) of $[^3H]GA_{14}$ to 5-day-old seedlings demonstrated that the enzymes required for sequential oxidation and eventual loss of C-20 are present. Thus, one might expect to detect newly formed GAs in seedlings of this age, but their presence has yet to be recorded.

Of relevance here is a brief mention of some recent work on the identity and metabolism of GAs in tall and dwarf genotypes of pea. Tall peas, expressing the gene *Le* exhibit polar GA-like biological activity which is absent from dwarf plants which are homozygous for the recessive gene *le* (Potts, Reid, & Murfet, 1982). Further work on the identity of this polar biological activity involving GC-MS has revealed the presence of GA_1 in an extract from 20-day-old light-grown pea shoots (Ingram, Reid, Potts, and Murfet, unpublished data). Furthermore, $[17-^{13}C][15,17-^3H]GA_{20}$ can be converted to $[^{13}C][^3H]GA_1$ in addition to $[^{13}C][^3H]GA_{29}$ and $[^{13}C][^3H]GA_{29}$-catabolite, in expanding tissue of tall pea shoots (Ingram, Reid, Murfet, and MacMillan, unpublished data). Since GA_1 has high biological activity in the pea stem assay, the metabolic conversion of GA_{20} to GA_1 in tall peas may be of profound physiological significance and warrants further study.

Other metabolic work concerned with the physiological bases of light-induced dwarfism in pea, and the retardation of apical senescence in pea, are discussed in Sections 4.4.1 and 4.4.2. Effects of abscisic acid and the cytokinin, benzyl adenine, on GA metabolism in pea seedlings are reviewed in Section 4.4.4.

4.3.3 Cereals

In the metabolic studies conducted with cereals, the metabolism of an applied GA has been correlated with a growth, or other physiological, response—e.g., the elongation of stems of dwarf lines of *Oryza sativa* and *Zea mays*, or the formation of α-amylase in half-seeds of *Hordeum vulgare*. The metabolic studies described in this subsection were conducted before the native GAs of these mono-cotyledonous plants were known. Nevertheless, the metabolism of applied GAs and the measured physiological responses are probably not influenced by endogenous GAs since the particular systems chosen (embryoless seeds or dwarf genotypes) are apparently GA-deficient.

The most extensively studied of all systems has been the barley half-seed or isolated aleurone system. The aleurone layer is a "target tissue" for GAs where the direct response to an applied GA (i.e., the production and release of α-amylase) can be measured. In a series of experiments Musgrave, Kays, and Kende (1972), Nadeau et al. (1972), and Stolp, Nadeau, and Rappaport (1973, 1977) have tried to elucidate the importance of GA metabolism in the barley system.

Musgrave et al. (1972) compared the uptake, metabolism, and activity of $[^3H]GA_1$, $[^3H]GA_5$, and $[^3H]GA_5$ methyl ester in isolated aleurone layers after 20 hr. Uptake was least for $[^3H]GA_1$ (4% versus 18% and 22% for $[^3H]GA_5$ and its ester, respectively), but the metabolism of $[^3H]GA_1$ was also less than for either $[^3H]GA_5$ or its ester. Thus, the amount of unmetabolized $[^3H]GA_1$ remaining in the tissue after 20 hr was greater (0.7% of applied radioactivity) than the amount of residual $[^3H]GA_5$ (0.2%) and of residual $[^3H]GA_5$ methyl ester (also 0.2%). The higher levels of unmetabolized $[^3H]GA_1$ may account for the greater biological activity of GA_1 versus GA_5 and its ester. However, it is difficult to rationalize the relative activities of GA_5 and its ester on this basis, since GA_5 has moderate activity and its methyl ester is inactive. Identification of metabolites was not attempted.

Nadeau et al. (1972) studied the activity and metabolism of $[^3H]GA_1$ in isolated aleurone layers in the presence and absence of abscisic acid (ABA). ABA inhibits GA_1-induced α-amylase synthesis, but it enhanced $[^3H]GA_1$ uptake (9% for $[^3H]GA_1$ and 24% for $[^3H]GA_1$ with ABA). Uptake was proportional to the amount of $[^3H]GA$ metabolism, and, in turn, metabolism was found to be proportional to the log of the ABA concentration. $[^3H]GA_1$ was metabolized to $[^3H]GA_8$, $[^3H]GA_8$ glucoside (both identified by comparative GC-RC on two GC columns), possibly $[^3H]GA_1$ glucoside, and a fourth product originally named [^{3}H]GA-X and later referred to as ampho-GA$_1$ (Nadeau & Rappaport, 1974). This metabolite has amphoteric properties and is thought to be a GA-peptide conjugate. The amount of unmetabolized [^{3}H]GA$_1$ was approximately the same in all treatments, so that ABA does not appear to be modifying the activity of GA$_1$ by reducing the amount of unchanged [^{3}H]GA$_1$ remaining in the tissue. Neither does ABA appear to promote the formation of metabolites that are *inhibitory* to α-amylase production, since application of GA$_8$ glucoside with GA$_1$ does not modify the activity of GA$_1$.

Stolp et al. (1973) compared the effect of ABA on the uptake and metabolism of [^{3}H]GA$_1$ and its biologically inactive 3-epimer (Fig. 4.6) in barley half-seeds. ABA promoted the uptake of [^{3}H]GA$_1$ and its epimer, and promoted the metabolism of [^{3}H]GA$_1$. The epimer was unmetabolized, both in the presence and absence of ABA, leading the authors to conclude that ABA-enhanced uptake of GAs is not a consequence of enhanced GA metabolism.

The uptake and metabolism of several other biologically inactive derivatives of [^{3}H]GA$_1$ (Fig. 4.6) in isolated aleurone layers in the presence and absence of ABA was examined by Stolp et al. (1977). ABA enhanced the uptake of [^{3}H]GA$_1$ and 3-*epi*-[^{3}H]GA$_1$, inhibited the uptake of rings-C/D-rearranged [^{3}H]GA$_1$, and had no effect on

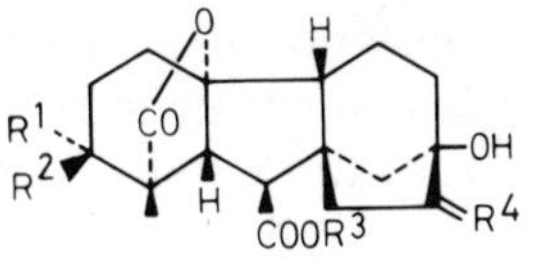

R^1	R^2	R^3	R^4	
OH	H	H	CH$_2$	3-epi-GA$_1$
H	OH	H	CH$_2$	GA$_1$
H	OH	CH$_3$	CH$_2$	GA$_1$ methyl ester
H	OH	H	CH$_3$, H	16,17-dihydro-GA$_1$

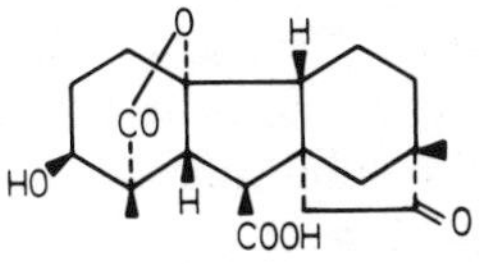

Figure 4.6 GA$_1$ and derivatives fed to *Hordeum vulgare* half-seeds and isolated aleurone layers.

rings-C/D-rearranged GA$_1$

the uptake of $[^3H]GA_1$ methyl ester and $H_2[^3H]GA_1$. The latter three derivatives were metabolized to unidentified products, in the presence and absence of ABA, and $3\text{-}epi\text{-}[^3H]GA_1$ remained unmetabolized.

The results of these experiments (Musgrave et al., 1972; Nadeau et al., 1972; Stolp et al., 1973, 1977) are complex and difficult to rationalize. There must of course be several reasons why GAs are active or inactive in a particular test system. There is no evidence that GA_5 is active per se in barley, and indeed it may be converted to another more active GA before a response is elicited. GA_5 methyl ester may be inactive because esterification prevents metabolism to the active GA. The activities of GA_5 and its methyl ester would, thus, not be directly related to the amount of unmetabolized GA remaining but to the amount, if any, of active metabolite formed. On the other hand, in barley $[^3H]GA_1$ is metabolized only to products that are themselves inactive, suggesting that GA_1 is active per se and that the amount of unmetabolized GA_1 remaining in the tissue is important. The metabolic fate of the inactive GA_1 derivatives tested by Stolp et al. (1973, 1977) is probably irrelevant if their inactivity is inherent, i.e., if it is due to a failure to fulfill the structural requirements of an active molecule. Thus, on the basis of the available evidence, it is very difficult to make any generalizations about GA metabolism vis-à-vis biological activity in the barley aleurone system.

The effect of GAs on stem growth has been studied using a series of mutants of *Zea mays* and *Oryza sativa*. The *d-5* mutant of *Z. mays* is a single-gene recessive mutant which has a dwarf growth habit. Application of GAs restores growth to that of normal (wild type) plants (Phinney, 1956). Extraction has shown *d-5* plants to contain lower amounts of extractable GA-like substances than normals (Phinney, 1961). Although evidence obtained by Katsumi et al. (1964) suggested that this was due to reduced biosynthesis of GAs in the mutant, Davies and Rappaport (1975a, 1975b) proposed that enhanced GA deactivation (for example, by enhanced 2β-hydroxylation and conjugation) might be responsible for reducing the levels of biologically active GAs in the mutant. Thus, Davies and Rappaport (1975a, 1975b) compared the metabolism of $[^3H]GA_1$ and $[^3H]GA_3$ in normal and *d-5* plants. Neither GA was known to be endogenous to *Zea* but both can restore normal growth in the *d-5* mutant. $[^3H]GA_1$ was metabolized to the inactive $[^3H]GA_8$, $[^3H]GA_8$ glucoside (identified by comparative GC-RC with standards on two GC columns), and to ampho-$[^3H]GA_1$, the putative GA_1-peptide conjugate also known as a metabolite of $[^3H]GA_1$ in barley. Metabolites of $[^3H]GA_3$ were not identified. Because of gross morphological differences between tall and dwarf plants, metabolism was studied and compared in isolated plant parts. The amount of metabolism of $[^3H]GA_1$ varied from organ to organ, but there were no consistent differences in metabolism between the same organ or tissue derived from tall versus dwarf plants. It was concluded that reduced growth in the d-5 mutant was not due to enhanced GA deactivation.

More recent work on the molecular basis of dwarfism in Zea has confirmed that GA biosynthesis is indeed altered in the d-5 mutant (Hedden & Phinney, 1979). In cell-free systems derived from etiolated shoots of normal plants, the major diterpene formed from mevalonic acid is ent-kaur-16-ene (ent-kaurene). An isomer, ent-kaur-15-ene (ent-isokaurene), is formed in minor proportion. In cell-free systems from the d-5 mutant, the synthesis of ent-kaurene is much reduced, whereas the synthesis of ent-isokaurene is increased to such an extent that it is the major diterpene product. In G. fujikuroi, ent-isokaurene is not a precursor of GAs (Hedden et al., 1977), and thus GA biosynthesis appears to be blocked by the d-5 mutation in Zea. The mutant d-5 gene has been shown to alter the B activity of ent-kaurene synthetase so that the cyclization of copalylpyrophosphate (CPP) to the normal product, ent-kaurene, is prevented. The altered enzyme catalyzes the conversion of CPP to ent-isokaurene instead (Hedden & Phinney, 1979; see also Chapter 2, this volume). Since d-5 plants respond to a variety of C_{20}- and C_{19}-GAs (Katsumi et al., 1964; Phinney, unpublished), this dwarf mutant appears to have an unimpaired capacity for GA metabolism.

The major GAs of normal Z. mays plants have recently been identified as GA_{17}, GA_{19}, GA_{20}, GA_{44}, and GA_{53} (Hedden et al., 1982) (Tables 4.1 and 4.3). In addition, GA_1, GA_8, and GA_{29} have been tentatively identified by gas chromatography-selected ion current monitoring (GC-SICM) (Hedden et al., 1982). Thus Zea appears to have an early 13-hydroxylation pathway, comparable to that in legumes, in which the first-formed C_{19}-GA is GA_{20}.

Further metabolic work should enable the biochemical bases of several other single-gene mutations in Zea to be established, in turn confirming the presumptive pathway in wild-type plants. Bioassay of GA_{12} aldehyde, GA_{53} aldehyde, GA_{53}, GA_{20}, and GA_1 has indicated that GA metabolism is blocked in the d-1, d-2, and d-3 dwarf mutants (Phinney, 1982). In d-3 plants GA_{12} aldehyde is inactive whereas GA_{53} aldehyde has some activity, suggesting that the d-3 mutation affects early 13-hydroxylation. In d-2 plants GA_{12} and GA_{53} aldehydes are inactive, whereas the acid, GA_{53}, is active. This indicates that the d-2 mutation affects C-7 oxidation.

Table 4.3. Gibberellins from *Graminae*

Species	Plant part	Identity of GAs	Reference
Triticum aestivum	developing grains	GA_{15}, GA_{17}, GA_{19}, GA_{20}, GA_{24}, GA_{44}, GA_{54}, GA_{55}, GA_{60}, GA_{61}	Gaskin et al. (1980, and unpublished)
	germinating grains	GA_1, GA_3	Gaskin et al. (unpublished)
	chloroplasts	GA_4, GA_7	Browning & Saunders (1977)
Zea mays	developing seeds	GA_{17}, GA_{19}, GA_{20}, GA_{44}, GA_{53}	Hedden et al. (1981)
Secale cereale	immature fruits	GA_8, GA_{16}, GA_{24}	Dathe et al. (1978)
Avena sativa	inflorescences	GA_3	Kaufman et al. (1976)
Oryza sativa	roots shoots	GA_{19}	Kurogochi et al. (1978)
Hordeum vulgare	developing grains	GA_1, GA_4, GA_8, GA_{12}, GA_{17}, GA_{20}, GA_{25}, GA_{34}, GA_{48}, 18-OH GA_4, 12β-OH GA_9	Gaskin et al. (1982, 1983)
	germinating grains	GA_1, GA_{17}, GA_{19}, GA_{20}, GA_{34}, GA_{48}	Gaskin et al. (1982, 1983)

Of all the GAs tested on *d-1* plants only GA_1 is active, suggesting that the *d-1* mutation is unable to convert GA_{20} to GA_1, that is, it cannot carry out a 3-hydroxylation step. Definitive evidence on the GA pathway and the position of metabolic blocks in the dwarf mutants can only come from extensive metabolic work. However, two features of GA metabolism in *Zea* are emerging from the bioassay data: first, there appears to be a single pathway from GA_{12} aldehyde to GA_1 and, second, GA_1 appears to be the only native GA which is active per se.

Durley and Pharis (1973) investigated the metabolism of $[1,2\text{-}^3H_2]GA_4$ in Tan-ginbozu rice over 24 hr, by which time enhanced growth of the treated plants was evident. By comparative GC-RC with authentic standards on three GC columns, they identified three metabolites: namely, $[^3H]GA_1$ (0.85% of applied radioactivity), $[^3H]GA_2$ (0.32%), and $[^3H]GA_{34}$ (0.43%). A large peak of unmetabolized $[^3H]GA_4$ was observed together with three other unidentified radioactive peaks. The amount of radioactivity in the fractionated extract was *ca.* 30% of applied radioactivity, and unmetabolized substrate and observed metabolites after GC-RC accounted for only *ca.* 12% of the applied substrate. Some radioactivity would be lost from the 2β-position in the conversion of $[^3H]GA_4$ to $[^3H]GA_{34}$, but in view of the small conversion to $[^3H]GA_{24}$ this would be negligible. The authors gave no explanation for the substantial loss of radioactivity, but it is possible that metabolic transformations of $[^3H]GA_4$ other than those described may be occurring. Nevertheless, the data do demonstrate the ability of dwarf rice seedlings to metabolize $[^3H]GA_4$ by 13-hydroxylation to $[^3H]GA_1$, by 2β-hydroxylation to $[^3H]GA_{34}$, and by hydration to $[^3H]GA_2$, although whether the latter conversion is enzymatic is not certain. The low ($<1.0\%$) conversion to $[^3H]GA_1$ over 24 hr suggests that $[^3H]GA_4$ is active per se in rice, although GA_1 is also active in this assay. The 2β-hydroxylated GA_{34} has very low activity (Reeve & Crozier, 1974).

Applied $[1,2\text{-}^3H_2]GA_1$ is also 2β-hydroxylated in rice, with 1.2% conversion to $[^3H]GA_8$ observed in 24 hr (Railton, Durley, & Pharis, 1973). Recently, Railton (1980b) has attempted to correlate metabolism of $[^3H]GA_1$ in rice seedlings with growth, in a comparable study to that described in Section 4.3.2.2 on $[^3H]GA_{20}$ metabolism and the growth of pea seedlings (Railton, 1980a). In rice, $[^3H]GA_1$ promoted leaf sheath elongation for 48 hr, while the level of $[^3H]GA_1$ in the tissue declined very rapidly with only 23% of the applied $[^3H]GA_1$ remaining after 24 hr. $[^3H]GA_8$, which accounted for 2% of the applied radioactivity after 24 hr, was identified by comparative GC-RC on two GC columns. The disappearance of $[^3H]GA_1$ far exceeded the observable metabolism to $[^3H]GA_8$, even taking into account the loss of radioactivity from the 2β-position of $[^3H]GA_1$ during metabolism. Although Railton (1980b) suggested that $[^3H]GA_8$ may itself be metabolized, no metabolites of applied $[1-^3H]GA_8$ were identified. Railton (1980b) concluded from the data that growth and 2β-hydroxylation were *not* correlated, although the incomplete identification and quantitation of metabolites must make any conclusion concerning metabolism open to question. In fact, growth may be more directly correlated with the levels of biologically active substrate rather than with the levels of metabolite, since 2β-hydroxylated GAs are inactive, *not* inhibitory.

Of interest in this context is the biological activity of some 2-substituted GA_4 and GA_9 derivatives prepared by Beale and MacMillan (1981b). If biologically active GAs are deactivated in plant tissue by 2β-hydroxylation, then GA derivatives in which potential 2β-hydroxylation is blocked by the presence of an appropriate substituent might be expected to have more, or longer-lasting, activity. Of the many GA_4 and GA_9 derivatives tested, only 2β-methyl GA_4 and 2,2-dimethyl GA_4 had higher activity than the unsubstituted GA. 2,2-Dimethyl GA_4 had *ca.* 100 times the activity of GA_4 in Tan-ginbozu dwarf rice, oat first leaf, *d-5* maize (Hoad et al., 1981), and α-amylase bioassays (Hoad, unpublished). The effect of substitution at C-2 is exemplified by the response of rice seedlings. Both GA_4 and dimethyl GA_4 promoted elongation of the second leaf sheath, but only dimethyl GA_4 promoted growth of the third leaf sheath. This effect is consistent with the basic premise that GA_4 is deactivated in the plant tissue whereas the substituted GA_4 is not. However, other possible explanations for the enhanced activity of dimethyl GA_4 have also been put forward (Hoad et al., 1981).

The enhanced activity of dimethyl GA_4 is only manifested in assays using monocotyledonous plants. The biochemical basis for this selective activity is, at present, unknown. Although several metabolic conversions have been observed in monocotyledons, most are conversions that are also commonly observed in dicotyledonous plants and, therefore, would not appear to impart any specificity or selectivity for GA activity in monocotyledons. The only metabolic reaction purported to be specific for monocotyledons is the formation of the putative GA_1-peptide conjugate, which according to Rappaport et al. (1974) is observed in barley, maize, and wheat but not in any of the dicotyledonous plants tested (e.g., bean, olive, tobacco). Since much of this metabolic work was conducted, the endogenous GAs of several *Gramineae* have been identified (Kaufman et al., 1976; Dathe, Schneider, & Sembdner, 1978; Kurogochi et al., 1978; Gaskin et al., 1980, 1982; Hedden et al., 1981) (Table 4.3). Many of these GAs are frequently identified in dicotyledonous plants too, but some—for example, 1-hydroxylated GAs such as GA_{54}, GA_{55}, GA_{60}, and GA_{61}—have not been identified in dicotyledonous plants. Only further work, both in terms of establishing the identity and biosynthesis of native GAs in monocotyledonous plants and studying the metabolism of GAs in relation to selective growth responses in monocotyledons, will indicate whether there are inherent differences in GA physiology in the two groups of plants.

4.3.4 Conifers

GAs have been implicated in the control of growth and development of gymnosperms, as well as of angiosperms. Thus, in conifers, applied GAs can, inter alia, promote seed and pollen grain development, vegetative growth, and the induction and/or development of ovulate and/or staminate strobili (see Pharis & Kuo, 1977; see also Volume II, Chapter 7). Moreover qualitative and quantitative changes in the levels of endogenous GA-like substances have been correlated with some of these developmental phenomena. Endogenous GAs have been identified in three members of the *Pinaceae*, and GA metabolism has been studied in relation to the known developmental responses to applied GAs.

Kamienska et al. (1974, 1976a) identified GA_4 and GA_7 by GC-MS in dormant pollen grains of *Pinus attenuata*. On germination the levels of GA_4 and GA_7 decreased, with a concomitant increase in a more polar GA, identified by GC-MS as GA_3. Kamienska et al. (1976b) went on to study the metabolism of $[1,2\text{-}^3H_2]GA_4$ in germinating pollen grains. Most of the $[^3H]GA_4$ (fed at *ca.* 100 times the level of native GA_4) was unmetabolized, although two metabolites, $[^3H]GA_1$ and $[^3H]GA_{34}$, were identified by comparative GC-RC on three GC columns, with 0.3 and 2.4% conversion from absorbed $[^3H]GA_4$, respectively. Some radioactivity would be lost during metabolism to $[^3H]GA_{34}$ so that the above figure is a slight underestimation of the yield of $[^3H]GA_{34}$. $[^3H]GA_2$ was formed by hydration with *ca.* 0.5% yield, and a further radioactive compound was tentatively identified as 16-*epi*-GA_2. In addition there were four unidentified metabolites. No radioactive GA_7 or GA_3 was observed despite the previous observations that these GAs are endogenous to *Pinus* pollen (Kamienska et al., 1974, 1976a). $[1,2\text{-}^3H_2]$-

GA_4, by analogy with $[1,2\text{-}^3H_2]GA_1$, is labeled predominantly in the 1β- and 2β-positions (Patterson & Rappaport, 1974). Dehydration of GA_4 to GA_7 results in loss of the 1α- and 2α-protons (Evans, Hanson, & White, 1970) and would therefore not lead to loss of label in this experiment. Thus, in the absence of any $[^3H]GA_7$ in these feeds, Kamienska et al. (1976b) concluded that GA_4 is not a precursor of GA_7 in *Pinus* pollen. In addition, it was concluded that since GA_4 is endogenous to *Pinus* pollen the observed metabolites, GA_1 and GA_{34}, must also be endogenous. It is far from certain, however, that the metabolism of applied $[^3H]GA_4$ is identical to the metabolism of native GA_4 and so the status of GA_1 and GA_{34} awaits further experimentation. The effect of neither GA_1 nor GA_{34} has been tested on pollen grain germination. The latter is unlikely to be active since it has low activity in other bioassay systems (Reeve & Crozier, 1974).

GAs have been implicated in the control of vegetative growth of conifers. The 15,16-double-bond isomer of GA_9 and GA_9 glucosyl ester are endogenous to vegetative shoots of mature trees of *Picea sitchensis* (Sitka spruce) (Lorenzi et al., 1975, 1976, 1977). The isomer has less activity than GA_9 in lettuce hypocotyl and dwarf rice bioassays (Lorenzi et al., 1977), but its activity does not appear to have been assessed in any assay based on *Picea* or other member of the *Pinaceae*. Nevertheless, high levels of $\Delta^{15,16}GA_9$ are associated with phases of active vegetative growth in *Picea sitchensis*. There is a reciprocal relationship between the levels of $\Delta^{15,16}GA_9$ and GA_9 glucosyl ester, but the metabolic relationship of the two is unclear. There is no evidence to date for the presence of GA_9 itself in Sitka spruce, and the metabolism of $\Delta^{15,16}GA_9$ has not been studied.

In *Picea abies* (Norway spruce) the levels of a relatively polar GA_1- or GA_3-like compound increase dramatically during the annual period of maximum elongation growth of 8-year-old ramets (i.e., grafted scions from mature trees which have not yet resumed flowering after grafting). Simultaneously, the levels of less-polar GA-like substances decrease, again suggesting a metabolic relationship (Dunberg, 1976). Perhaps rather prematurely, since the identity of none of these GA-like substances in *Picea abies* has been established, Dunberg (1981) and Dunberg et al. (1983) studied the metabolism of applied $[17\text{-}^3H_2]GA_9$, $[1,2\text{-}^3H_2]GA_4$, and $[1,2\text{-}^3H_2]GA_1$. $[17\text{-}^3H_2]GA_9$ was applied to 9-year-old ramets toward the end of their annual growth flush. None of the metabolites present after 2 and 7 days were identified, but the major metabolite (12%) had chromatographic properties similar, but *not* identical, to GA_4. $[^3H]GA_4$, applied at the beginning of the growth flush, was extensively metabolized with up to 70% of the applied radioactivity being unaccounted for after 7 days. Fourteen different metabolites were observed in a series of 15 extracts, although 6 of these metabolites were minor and were present in less than half the extracts. Two of the major metabolites were identified by comparative GC-RC on three GC columns as $[^3H]GA_2$ and $[^3H]GA_{34}$. The latter was formed in at least 10% yield, loss of radioactivity during its formation leading to an underestimation of its yield. None of the other metabolites were identified, and extensive GC-RC showed them to be different from any of the 35 standards available to the authors. The major peak of biological activity present in extracts of rapidly growing shoots was *not* identical to any of the observed metabolites from applied $[^3H]GA_4$. Dunberg (1981) also fed $[1,2\text{-}^3H_2]GA_1$ to 5-week-old seedlings of *Picea abies*. Approximately 15% conversion to $[^3H]GA_8$ (tentative identification based on elution volume from silica columns) was observed.

Wample, Durley, and Pharis (1975) have studied the metabolism of $[1,2\text{-}^3H_2]GA_4$ in the shoots of 2-year-old seedlings of *Pseudotsuga menziesii* (Douglas fir) at three different stages of development—namely bud break, elongation, and bud set. $[^3H]GA_4$ was injected beneath the bark under the new whorl, and needles and stems from the new whorl were harvested 1, 3, and 6 days later. Translocation of radioactivity from the point of application into the new tissue varied with development. $[^3H]GA_2$ and $[^3H]GA_{34}$ were identified by comparative GC-RC on three GC columns, but up to six unidentified metabolites were also observed. The amount of unmetabolized $[^3H]GA_4$ was highest during elongation, which indicated to the authors that GA_4 was probably involved in the control of elongation growth. Wample et al. (1975) also tried to correlate the amount of radioactive metabolites with a requirement or nonrequirement for active GA at a particular developmental stage. However, since this required basic assumptions on the relative biological activities of unidentified metabolites the conclusions are, at best, tenuous.

The metabolic studies described above (Dunberg et al., 1983; Dunberg, 1981; Wample et al., 1975) have two limitations. First, the $[^3H]GAs$ fed were not known to be endogenous; second, many unidentified metabolites were obtained. The first problem can be resolved by undertaking large-scale extracts to establish the identity of the native GAs. The second problem—a large number of minor metabolites—seems to be a particular feature of $[^3H]GA$ feeds to conifers. It is possible that some may be artifacts, formed non-enzymatically during work-up. Thus, Kamienska et al. (1976a) observed that when $[^3H]GA_4$ was added to *Pinus* pollen extracts as an internal standard, up to 17% of the radioactivity was separable from GA_4 on counter-current chromatography. $[^3H]GA_4$ alone chromatographed as a discrete peak. The possibility therefore exists that some components of conifer extracts, and particularly of vegetative extracts, may be capable of modifying GAs during work-up. Extraction "blanks" should always be conducted with every feeding experiment; that is, radioactive substrate should either be added to plants that are then extracted immediately, or added directly to a plant extract. Any radioactive components, other than the added substrate, that are present in the extracts after work-up and analysis can be assumed to be artifacts. Indeed, when Durley et al. (1976) fed $[^3H]GA_4$ to germinating lettuce seeds, two of the three radioactive peaks, originally thought to be metabolites, were shown to be artifacts of the work-up procedure.

Strobilus development can be promoted in conifers by various cultural practices (e.g., girdling) and by hormone treatment. GA_3 application can induce strobilus formation in numerous species within the *Cupressaceae* and *Taxodiaceae* (see Table 4.4 and Volume II, Chapter 7, for examples), but it has little activity on members of the *Pinaeceae*. Within the *Pinaceae*, less-polar gibberellins have striking effects on strobilus development. Thus, a GA_4/GA_7 mixture has been found to promote the development of ovulate strobili in thirteen *Pinaceae* species from five different genera (see Volume II, Chapter 7; Pharis, Ross, & McMullan, 1980; Wheeler, Wample, & Pharis, 1980 for references, and Table 4.4 for examples). In five of these species $GA_{4/7}$ also promotes the development of staminate strobili. The propensity to produce female versus male strobili appears to be controlled by hormone concentration, time of application, and environmental factors (Pharis & Kuo, 1977; Wheeler et al., 1980) and has not been investigated fully for all species. GA_9 can act synergistically with GA_4/GA_7 in some species (Table 4.4). Within several *Pinaceae* species (e.g., *Pseudotsuga menziesii*), cultural treatments that promote strobilus development can also lead to increased levels of nonpolar GA-like substances (Pharis, Ebell, and Morf, unpublished, quoted in Pharis & Kuo, 1977). Moreover, the metabolism of applied $[^3H]$GAs may also be modified. Plastic-covered grafts of *Picea* species have both a greater propensity to strobilate (Tompsett & Fletcher, 1979; Brøndbo, 1969) and a reduced ability to metabolize applied $[^3H]GA_4$ (Dunberg et al., 1983). The stability of $[^3H]GA_4$ in covered grafts may reflect the stability of native GA_4 or similar GA in this tissue, an accumulation of which may in turn be responsible for enhanced strobilus production.

Some of the data discussed in this subsection are summarized in Table 4.4. Physiological and biochemical aspects of GAs and conifers are reviewed in detail in Volume II, Chapter 7.

4.4 EFFECT OF ENVIRONMENTAL FACTORS ON GIBBERELLIN METABOLISM

4.4.1 Light-induced inhibition of growth

4.4.1.1 Stem growth of *Pisum*, *Pharbitis*, and *Phaseolus* seedlings

White or red light can inhibit the stem growth of seedlings of a variety of species, and this effect is often GA-reversible. In *P. sativum*, growth of the tall cultivar, Alaska, is only slightly retarded in the light, but in other cultivars—for example, Progress No. 9—stem growth is so greatly reduced in the light that the plants assume a dwarf growth habit (Lockhart, 1956). The term "physiological dwarf" has been applied to Progress No. 9 and similar cultivars (Jones, 1973) since dwarfism is only manifested in the light. Applied GAs can reverse the light-induced inhibition of stem elongation in tall and physiologically dwarf cultivars (Lockhart, 1956), and this phenomenon forms the basis of several bioassays for GAs (Brian & Hemming, 1955; Köhler & Lang, 1963; Phillips & Jones, 1964).

Since applied GAs can overcome the effect of light, Lockhart (1956) suggested that light may be acting by inhibiting the biosynthesis or promoting the deactivation of native GAs, or by reducing the sensitivity of the tissue to its endogenous GA complement. Kende and Lang (1964), examining these possibilities, first compared the endogenous GAs of dark- and light-grown seedlings of Progress No. 9. Extracts were bioassayed using four types of test plants, namely d-5 dwarf maize, Alaska peas grown in red light and dwarfed with AMO-1618, Progress No. 9 peas grown in darkness and dwarfed with AMO-1618, and Progress No. 9 peas grown in red light. The d-5 maize assay showed that extracts from both dark- and light-grown seedlings of Progress No. 9 contained approximately equal amounts of two GA-like fractions designated fraction 1 and fraction 2. Fraction 1 had chromatographic properties similar to those of GA_5, and fraction 2 was similar to GA_1. Fraction 1, however, had no activity when tested on Progress peas grown in red light, although it was active on AMO-dwarfed etiolated Progress or Alaska seedlings. These biological properties were also consistent with the proposed identities of

Table 4.4. Identification, metabolism, and effect of gibberellins in conifers

Family and species	Endogenous GAs	Plant part	Metabolism of applied GAs	Effect of applied GAs	
				GA	Response
Pinaceae					
Picea sitchensis	$\Delta^{15,16}$ GA$_9$[a], GA$_9$ glucosyl ester[b]	vegetative shoots	—	GA$_3$/GA$_4$/GA$_7$, GA$_4$/GA$_7$/GA$_9$	strobilus development[c]
Picea abies	GA$_1$-like[d]	vegetative shoots	[^{3}H]-GA$_9$ → unidentified products[e]	GA$_4$/GA$_7$	strobilus development[g]
			[^{3}H]GA$_4$ → [^{3}H]GA$_2$, [^{3}H]GA$_{34}$ and unidentified products[e]; [^{3}H]GA$_1$ → [^{3}H]GA$_8$[f]	GA$_4$/GA$_7$/GA$_9$	
Pseudotsuga menziesii	GA$_3$[h]	vegetative shoots	[^{3}H]GA$_4$ → [^{3}H]GA$_2$, [^{3}H]GA$_{34}$ and unidentified products[i]	GA$_{4/7}$, GA$_5$, GA$_9$	strobilus development[j]
Pinus attenuata	GA$_4$, GA$_7$[k]	dormant pollen	[^{3}H]GA$_4$ → [^{3}H]GA$_1$, [^{3}H]GA$_2$, [^{3}H]GA$_{34}$ and unidentified products[l]	GA$_3$	pollen tube growth[m]
	GA$_3$[k]	germinating pollen			

Taxon / Species				strobilus induction
Cupressaceae				
Cupressus arizonica	GA₃[n]	shoots	GA₃	strobilus induction[o]
Juniperus scopulorum	GA₃[p]	shoots	GA₃	strobilus induction[q]
Taxodiaceae				
Cryptomeria japonica	?	—	GA₃	strobilus induction[r]

[a] Lorenzi et al. (1977).
[b] Lorenzi et al. (1976).
[c] Tompsett & Fletcher (1979).
[d] Dunberg (1976).
[e] Dunberg et al. (1983). Product identification based on GC-RC.
[f] Dunberg (1981). Product identification based on column chromatography.
[g] Dunberg (1980).
[h] Durley & Pharis (unpublished) quoted in Pharis & Kuo (1977).
[i] Wample et al. (1975). Product identification based on GC-RC.
[j] Ross & Pharis (1976).
[k] Kamienska et al. (1976b).
[l] Kamienska et al. (1976a). Product identification based on GC-RC.
[m] Ching & Ching (1959).
[n] Durley, Glenn, Morf, Pharis (unpublished) quoted in Pharis & Kuo (1977).
[o] Owens & Pharis (1967) and many others cited in Pharis & Kuo (1977).
[p] Durley, McGraw, Pharis (unpublished) quoted in Pharis & Kuo (1977).
[q] McGraw & Pharis (unpublished) quoted in Pharis & Kuo (1977).
[r] Hashizume (1959) and many others cited in Pharis & Kuo (1977).

fractions 1 and 2, since GA_1 is 10 times more active than GA_5 in red-light-grown Progress No. 9 seedlings, whereas GA_1 and GA_5 are equally active in both etiolated Alaska and Progress seedlings. In view of the similar GA levels in dark- and light-grown seedlings, Kende and Lang (1964) concluded that GA biosynthesis was not inhibited by light. However, they did suggest that light either interfered with the responsiveness of the tissue to the native GA designated fraction 1, or inhibited the conversion of the relatively inactive fraction 1 to the more biologically active fraction 2.

Musgrave and Kende (1970) examined the latter possibility by comparing the metabolism of $[1,2\text{-}^3H_2]GA_5$ (thought to be equivalent to fraction 1) in dark- and light-grown shoots of Progress No. 9. In the dark, up to 8% conversion of $[^3H]GA_5$ to an unidentified radioactive metabolite chromatographically similar to fraction 2 was observed, with a corresponding conversion in the light of *ca.* 1%. Despite this apparent difference, it was concluded that the amount of metabolite formed in either dark or light was insufficient to account for the response to applied GA_5. Furthermore, they provided additional bioassay evidence that GA_5 was active per se, thus giving little credence to the idea that fraction 1 must be converted to fraction 2 for growth promotion to occur. Kende (1967) observed no qualitative or quantitative differences in the metabolism of $[1,2\text{-}^3H_2]GA_1$ in dark- and light-grown seedlings of Progress No. 9.

Subsequent work has shown that GA_5 and GA_1 are not detectable in immature seeds of Progress No. 9 and that the major GAs are the chromatographically similar GA_{20} and GA_{29} (Frydman & MacMillan, 1973). This raised the possibility that the GA-like substances in Progress No. 9 seedlings tentatively identified by Kende and Lang (1964) as GA_5 and GA_1 were GA_{20} and GA_{29}, respectively. Indeed, more recently GA_{20} and GA_{29} have been identified in seedlings of Progress No. 9 (see Section 4.3.2.2), with no GA_5 or GA_1 being detectable. GA_{29} has low to zero biological activity, so the bioassay data presented by Kende and Lang (1964) are not easily reconciled with the proposed identity of fraction 2 as GA_{29}. GA_{20} and GA_5 have similar activities in most assays, but the comparative activity of GA_{20} in red-light- and dark-grown Progress seedlings is not known. Nevertheless if GA_{20} and GA_{29} are indeed fractions 1 and 2 of Kende and Lang, the proposal that fraction 1 is a less biologically active precursor of fraction 2 clearly does not hold.

Railton et al. (1974b) showed that $[^3H]GA_9$, applied to dark-grown seedlings of dwarf pea cv. Meteor, was converted, inter alia, to $[^3H]GA_{20}$, and that $[^3H]GA_{20}$ was in turn converted to $[^3H]GA_{29}$ (see Section 4.3.2.2). Railton (1974a) extended these metabolic studies to compare the metabolism of $[^3H]GA_9$ and $[^3H]GA_{20}$ in dark- and light-grown seedlings of Meteor. In light, the conversion of $[^3H]GA_9$ to the radioactive metabolite(s) chromatographing at Rf 0.7 was reduced, and GC-RC on a single GC column showed the two components of this TLC zone, previously identified as $[^3H]GA_{20}$ and 12α-OH$[^3H]GA_9$, to be reduced in level to a similar extent. It was estimated that the conversion of $[^3H]GA_9$ to $[^3H]GA_{20}$ was reduced by 50% in the light. However, insufficient data are given in the paper to allow the reader to calculate the actual conversion of $[^3H]GA_9$ to $[^3H]GA_{20}$ in dark or light. In the previous publication (Railton et al., 1974b), the yield of $[^3H]GA_{20}$ in dark-grown seedlings after 24 hr was 3.6% of applied $[^3H]GA_9$. A reduction of 50% in a conversion as small as that is of questionable significance. Moreover, no estimate of reproducibility between feeds was given.

Railton (1974a) also compared the metabolism of $[^3H]GA_{20}$ in dark- and light-grown Meteor seedlings. Formation of the sole metabolite, which had TLC properties akin to those of the previously identified $[^3H]GA_{29}$, was reduced by 65% in the light. Again, a figure for actual conversion can only be obtained from a previous publication (Railton et al., 1974c), namely 3.8% for the conversion of $[^3H]GA_{20}$ to $[^3H]GA_{29}$ in the dark. Similar reservations therefore apply to the significance of the reported reduction to $[^3H]GA_{20}$ metabolism as those given above for $[^3H]GA_9$ metabolism. From the data obtained, Railton (1974a) concluded that light reduced the "turnover" of GA_{20} since it inhibited both the formation of $[^3H]GA_{20}$ from $[^3H]GA_9$, and the metabolism of $[^3H]GA_{20}$ to $[^3H]GA_{29}$. Railton further suggested that this reduction in turnover of GA_{20} might be responsible for light-induced growth inhibition. More data are required before these suggestions can be critically assessed (see Section 4.2.5).

Barendse and Lang (1972) have compared the endogenous GA-like substances of light- and dark-grown seedlings of tall and physiologically dwarf cultivars (cv. Violet and cv. Kidachi, respectively) of *Pharbitis nil* (Japanese morning glory). Light reduces the stem growth of both cultivars, especially that of cv. Kidachi, and the effect is GA_3-reversible. Ogawa (1962) had previously shown that light-grown seedlings of cv. Violet contained higher levels of GA-like activity, as estimated by bioassay, than their dark-grown counterparts. Barendse and Lang (1972) reported that the levels of free GA-like substances, as estimated by the *d-5* corn assay, were similar in both cultivars under both growing conditions. In contrast, however, the level of conjugated GAs, as estimated by bioassay of acid-hydrolysates, varied between cultivar and between treatment.

Thus, the levels were higher in cv. Violet versus those in cv. Kidachi, and were higher in light-grown seedlings of both cultivars relative to their dark-grown counterparts.

Many free and conjugated GAs have been identified in developing seeds of *Ph. nil*, namely GA_3, GA_5, GA_8, GA_{17}, GA_{19}, GA_{20}, GA_{26}, GA_{27}, GA_{29}, and GA_{44} (Murofushi et al., 1968; Yokota et al., 1971b; Jones, Metzeger, & Zeevart, 1980), and glucosyl ether conjugates of GA_3, GA_8, GA_{26}, GA_{27}, and GA_{29} (Yokota et al., 1971c). In view of the number and diversity of GAs and GA conjugates in *Pharbitis* seeds, it is difficult to rationalize the bioassay data for seedling extracts obtained by Barendse and Lang (1972). Of the GA conjugates known to be present in *Pharbitis* seeds, only GA_3 glucosyl ether would yield a biologically active aglycone on hydrolysis. However, the hydrolysis procedure employed by Barendse and Lang (1972) would lead to the formation of some acid rearrangement products of GA_3, having reduced biological activity relative to that of GA_3. Thus, the quantitative data for conjugated GAs presented by Barendse and Lang (1972) are inadequate.

Although the metabolism of $[1,2\text{-}^3H_2]GA_1$ was shown to be similar in light- and dark-grown seedlings of both cultivars (Barendse & Lang, 1972), Barendse (1974) did observe quantitative differences in the metabolism of $[17\text{-}^{14}C_1]GA_3$ in dark- and light-grown seedlings of cv. Violet. $[^{14}C]GA_3$ was metabolized to a single product (Barendse, 1974) that had chromatographic and biological properties consistent with it being $[^{14}C]GA_3$ glucosyl ether (Barendse & de Klerk, 1975). Metabolism was greater in light-grown seedlings, with 84% of radioactivity being accounted for as the putative conjugate after 8 hr, versus 67% in dark-grown seedlings. This enhanced metabolism to the putative conjugate correlated with the higher levels of conjugated GAs, as determined by bioassay, in light-grown seedlings (Barendse & Lang, 1972). Barendse (1974) therefore concluded that enhanced GA-deactivation in light-grown tissue may explain the inhibition of stem growth by light. However, the higher levels of conjugated GAs, again determined by bioassay, in cv Violet (tall cultivar) versus those in cv. Kidachi (dwarf) are not readily explained by this hypothesis. Applied $[2,3\text{-}^3H_2]GA_{20}$ is also metabolized in seedlings of cv. Violet to two polar products, one of which has TLC properties similar to those of GA_{29} (Barendse, de Klerk, & v. Mierlo, 1977).

Light also inhibits the growth of *Phaseolus coccineus* cv. Prizewinner seedlings and this light-induced inhibition can be partially reversed by applied GAs (Bown et al., 1975). The GAs endogenous to light-grown seedlings are GA_1, GA_4, GA_5, and GA_{20} (Bowen et al., 1973). Only GA_4 has been conclusively identified in dark-grown seedlings (Crozier et al., 1971), although Bown et al. (1975) estimated by bioassay that, unlike the situation in *Pisum* and *Pharbitis*, dark-grown seedlings of *Ph. coccineus* contain at least three times the GA content of light-grown seedlings.

Bown et al. (1975) studied the metabolism of $[1,2-^3H]GA_4$ in dark- and light-grown seedlings of *Ph. coccineus*. $[^3H]GA_4$ was metabolized considerably faster in the light, with only 16% of applied $[^3H]GA_4$ remaining after 48 hr versus 50% remaining in the dark. A reduced capacity for metabolism of applied substrate in dark-grown seedlings was not due to greater dilution of the substrate by the larger pool of endogenous GAs in dark-grown tissue. Nor were there any differences in penetration into the two types of tissue. Radioactivity was lost during metabolism such that only 40% of applied radioactivity was recovered in ethyl acetate fractions after 48 hr metabolism in the light. Furthermore, radioactivity in the aqueous phase of extracts was found to be associated with volatile components. By comparison, 80% of applied radioactivity was recovered in ethyl acetate fractions of dark-grown tissue after 48 hr, and no activity was evident in the aqueous phase. Two metabolites accumulated in dark- and light-grown tissue. One, at Rf 0.5-0.6, built up to a maximum level at 2 hr in the light and thereafter declined with further accumulation of a polar metabolite at Rf 0.1-0.25. In the dark, the less polar metabolite accumulated more slowly, reaching its maximal level after 8 hr. Neither metabolite was identified, nor could any explanation be given for the loss of radioactivity. Thus, although metabolism was clearly shown to be greater in the light than in the dark, no conclusions could be reached as to the physiological significance of this. One can perhaps infer that if the metabolites of $[^3H]GA_4$ are inactive, the greater metabolism of $[^3H]GA_4$ in the light may offer a possible explanation for the reduced levels of biologically active GA-like substances in light-grown seedlings.

At first sight, the results summarized in Table 4.5 for *P. sativum* (Railton, 1974a), *Pharbitis nil* (Barendse, 1974), and *Ph. coccineus* (Bown et al., 1975) appear to be contradictory, with light inhibiting metabolism in *Pisum* seedlings and enhancing it in *Pharbitis* and *Phaseolus*. Light may, however, be reducing the availability of biologically active GA in all systems, by inhibiting anabolic metabolism in *Pisum*, and by enhancing catabolic metabolism in *Phaseolus* and conjugation in *Pharbitis*. Alternatively, the mechanism of light-

Table 4.5. Effects of light on gibberellin metabolism

Species and plant part	Effect of light	Endogenous GAs	Metabolism of applied GAs	
			Substrate and products	Effect of light
Pisum sativum seedlings	inhibition of stem growth	GA_{20}, GA_{29},[a] etc.	$[^3H]GA_5 \rightarrow [^3H]GA_1$-like	reduces metabolism[b]
			$[^3H]GA_1 \rightarrow$ unidentified products	metabolism similar in light and dark[c]
			$[^3H]GA_9 \rightarrow [^3H]GA_{20}$ and 12α-OH $[^3H]GA_9$	reduces metabolism[d]
			$[^3H]GA_{20} \rightarrow [^3H]GA_{29}$	reduces metabolism[d]
Pharbitis nil seedlings	inhibition of stem growth	GA_3, GA_3-glucoside,[e] etc.	$[^3H]GA_1 \rightarrow$ unidentified products	metabolism similar in light and dark[f]
			$[^{14}C]GA_3 \rightarrow [^{14}C]GA_3$ glucoside	enhances metabolism[g]
Phaseolus coccineus seedlings	inhibition of stem growth	GA_4,[h] etc.	$[^3H]GA_4 \rightarrow$ unidentified products	enhances metabolism[i]

Lactuca sativa seeds	promotion of germination	?	$[^3\text{H}]\text{GA}_4$	no metabolism in light or dark[j]
seedlings	inhibition of hypocotyl growth	?	$[^3\text{H}]\text{GA}_4 \rightarrow [^3\text{H}]\text{GA}_1$	reduces metabolism in hypocotyls only[j]
Nicotiana tabacum callus	inhibition of growth	?	$[^3\text{H}]\text{GA}_{20} \rightarrow [^3\text{H}]\text{GA}_1 \rightarrow$?	reduces overall metabolism, allowing $[^3\text{H}]\text{GA}_1$ to accumulate[k]

[a] Sponsel (unpublished).
[b] Musgrave & Kende (1970).
[c] Kende (1967).
[d] Railton (1974a). Product identification based on TLC, GC-RC and Railton et al. (1974b, 1974c).
[e] Jones et al. (1980), Yokota et al. (1971c).
[f] Barendse & Lang (1972).
[g] Barendse (1974), Barendse & de Klerk (1975). Product identification based on TLC and bioassay.
[h] Crozier et al. (1971), Bowen et al. (1973).
[i] Bown et al. (1975).
[j] Durley et al. (1976). Product identification based on GC-RC.
[k] Lance et al. (1976a). Product identificaton by GC-RC and GC-MS.

inhibited stem growth may be different in the three species, as indicated by differences in the endogenous GA status and in the responsiveness to applied GAs of the three species. Rigorous identification of metabolites and precise analysis of rates of metabolism are required before any definitive conclusions can be reached. While being sufficient to meet these requirements, the level of substrate feeding must nevertheless be kept low, for if the dose of [^{3}H]GA fed is such that the light-induced growth inhibition is reversed, one might not expect to demonstrate differences in GA metabolism in dark- and light-grown tissue. This aspect of the work has not been discussed in any of the papers considered here, although clearly in the work of Bown et al. (1975) the level of [^{3}H]GA$_4$ fed was insufficient to overcome the light-effect.

4.4.1.2 Lettuce seedlings

The germination of *Lactuca sativa* cv. Grand Rapids seeds can be promoted by light, and applied GAs can substitute for this light requirement. Subsequent seedling growth in several cultivars is, however, inhibited by light, and applied GAs can overcome the inhibitory effect of light. Durley et al. (1976) have studied the metabolism of applied [1,2-^{3}H$_2$]GA$_4$, in relation both to germination and to seedling growth, in dark- and light-grown lettuce cv. Grand Rapids. The level of [^{3}H]GA$_4$ applied was such that it nullified the effects of light, so that germination rate was the same and hypocotyl extension were similar in both treatments.

[^{3}H]GA$_4$ was rapidly taken up by the seeds, but it was scarcely metabolized after 24 hr, by which time both light- and dark-grown seeds had germinated. It was, therefore, concluded that GA$_4$ per se must be active. ABA inhibits GA-induced germination in the dark, but there was no effect of ABA on [^{3}H]GA$_4$ metabolism during the germination period.

After a further 3 days, extensive metabolism of [^{3}H]GA$_4$ had occurred, with the single major metabolite being identified by comparative GC-RC as [^{3}H]GA$_1$. The metabolic conversion to [^{3}H]GA$_1$ appeared to be similar (*ca.* 40%) in dark- and light-grown seedlings. However, when the distribution of substrate and metabolite throughout the seedling parts was compared after 5 days, Durley et al. (1976) showed that the conversion of [^{3}H]GA$_4$ to [^{3}H]GA$_1$ was 60% in dark-grown hypocotyls and 43% in light-grown. No such differences in metabolism in roots or cotyledons was observed, although the roots were a major site of $[^3H]GA_4$ metabolism. As in previous studies of this type (e.g., Railton, 1974a) there is no documentation of the reproducibility of these effects of light on GA metabolism. However, the reported difference correlates well with the observation that only the hypocotyls respond to applied GA_4 and that the response is different in light- and dark-grown seedlings (hypocotyl lengths $= 11.2$ mm in dark, 7.0 mm in light). GA_4 and GA_1 are reported to have similar activities on hypocotyl extension, but in studies with radio-labeled GAs the uptake of $[^3H]GA_4$ was three times that of $[^3H]GA_1$, suggesting that GA_1 is considerably more active than GA_4. Thus, a 30% conversion of applied GA_4 to GA_1 would account for the observed growth response. ABA (applied to germinated seeds) markedly reduced the conversion of $[^3H]GA_4$ to $[^3H]GA_1$, and also reduced hypocotyl extension, further suggesting that GA_1 is responsible for the effect of applied GA_4.

These results indicate a small though seemingly specific effect of light on $[^3H]GA_4$ metabolism in hypocotyls of light- and dark-grown lettuce seedlings. If, however, less substrate had been fed, such that germination did not proceed in the dark and/or inhibition of hypocotyl growth in the light had been more pronounced, then differences in $[^3H]GA$ metabolism under the two growing conditions may have been more apparent. Moreover, they may also have given a better indication of the differences in metabolism of native GAs in light- and dark-grown seedlings, although to date no definitive work on the identity of GAs in light- and dark-grown lettuce seedlings has been conducted.

Silk et al. (1977) and Nash, Jones, & Stoddart (1978) have also correlated $[^3H]GA$ metabolism with lettuce hypocotyl growth, and although the work is not primarily concerned with light effects it is discussed here. The growth response of hypocotyl sections can be more clearly defined than that of intact hypocotyls (Silk & Jones, 1975). GA_3, at concentrations as low as $10^{-10} M$, causes elevated growth of sections for at least 24 hr and the effect is solely attributable to cell elongation. Hypocotyl sections also respond to GA_1, and Silk et al. (1977) have studied the uptake and metabolism of $[1,2-^3H_2]GA_1$ in relation to section growth.

In the continued presence of $[^3H]GA_1$, sections accumulated radioactivity for 24 hr. After extraction radioactivity was approximately equally distributed between ethyl acetate and aqueous phases. The former contained predominantly unmetabolized $[^3H]GA_1$ (GC-MS identification) with some (*ca.* 7%) $[^3H]GA_8$ (tentative identification). The activity in the aqueous phase was later shown to be associated with two polar components tentatively identified after high-voltage electrophoresis as [^{3}H]GA$_1$ glucosyl ester and [^{3}H]GA$_1$ glucosyl ether (Stoddart & Jones, 1977).

When [^{3}H]GA$_1$ was applied as a 2 hr pulse, sections still retained 59% of the radioactivity taken up during the pulse after 21 hr; 31% of this activity was in the acidic, ethyl acetate fraction and was presumably unmetabolized [^{3}H]GA$_1$ (data not given).

Silk et al. (1977) discussed the means whereby the response to a short pulse of GA$_1$ could be sustained over many hours. They concluded that hypocotyl sections have a continuous requirement for GA$_1$. In view of the relatively slow metabolism of GA$_1$ in the tissue, hypocotyl sections are apparently able to accumulate enough GA during the pulse to maintain sufficient unmetabolized, biologically active GA$_1$ in the tissue during the subsequent 21 hr. This is borne out by the observation that as the duration of a pulse decreased, so the concentration of GA required to elicit a specific response increased.

Hypocotyl sections also respond to GA$_9$ at concentrations of 10^{-9} M and above. [2,3-^{3}H$_2$]GA$_9$ was rapidly taken up by sections and was efficiently metabolized such that after 24 hr only 6% of the tissue radioactivity was still associated with [^{3}H]GA$_9$ (Nash et al., 1978). The major metabolite had TLC and HPLC properties akin to those of GA$_{20}$. The addition of GA$_9$ and GA$_4$/GA$_7$ at concentrations up to 50 μM decreased both the uptake of [^{3}H]GA$_9$ and its metabolism to the putative [^{3}H]GA$_{20}$ from 14% of applied radioactivity to 0.7%. This precluded the possibility of adding cold carrier GA$_9$ to increase the yield of metabolite for GC-MS identification. GA$_1$ and GA$_3$ had no consistent effects on [^{3}H]GA$_9$ uptake and metabolism, indicating to the authors that metabolism did involve 13-hydroxylation.

The enzyme inhibitor 2,2'-dipyridyl inhibited both the uptake and metabolism of [^{3}H]GA$_9$ and also the growth response. The effect of dipyridyl on [^{3}H]GA$_9$ metabolism could be overcome by simultaneous application of FeSO$_4$, but whether FeSO$_4$ could also overcome the growth inhibition is not known. Applied GA$_{20}$ could not overcome the inhibitory effect of dipyridyl on growth (Stoddart, personal communication), although GA$_{20}$ is biologically active from 10^{-8} M.

Since hypocotyl sections continue to elongate after virtually all the [^{3}H]GA$_9$ has been metabolized, the growth response to applied GA$_9$ does not apparently require the continued presence of GA$_9$. GA$_9$ may either trigger a series of self-sustaining events, or growth may be correlated with the formation and continued presence of a biologically active metabolite. The formation and accumulation of putative $[^3H]GA_{20}$ gives credence to the second possibility, since GA_{20} itself is known to be active. The inability of GA_{20} to overcome the growth inhibition caused by 2,2′-dipyridyl is not in serious conflict with this suggestion, for dipyridyl is a chelating agent and, thus, would be expected to inhibit a multitude of metallo-enzymes in addition to those involved in GA metabolism. Nash et al. (1978) nevertheless dismissed GA_{20} as a potential growth-sustaining factor and postulated instead the existence of another transiently formed metabolite. The authors had no evidence for the occurrence of this hypothetical metabolite, and indeed the suggestion would appear to be quite illogical.

Whatever the biochemical basis for light-induced inhibition of lettuce hypocotyl growth, the work of Durley et al. (1976), Silk et al. (1977), and Nash et al. (1978) demonstrates that light-grown hypocotyls still retain their capacity to metabolize applied GAs. Thus, $[^3H]GA_4$ is efficiently metabolized to $[^3H]GA_1$, and indeed GA_4 may not be active per se. GA_9 is active, but perhaps only by virtue of its metabolism to a compound tentatively identified as GA_{20}. GA_1 appears to be active per se, and $[^3H]GA_1$ persists in the tissue for a considerable time. The binding of $[^3H]GA_1$ to cell-wall components in lettuce hypocotyls may be related to its activity (Stoddart, 1979a, 1979b) and is discussed in detail in Volume II, Chapter 1.

4.4.1.3 Callus cultures

The effect of exogenously applied GA_3 on cultured tissue varies from growth promotion to growth inhibition although a good many tissues show no response at all (Nickell & Tulecke, 1959). The lack of responsiveness to GAs may well be due to cultural conditions.

Nicotiana tabacum (tobacco) callus does not require GAs for growth, but under certain cultural conditions GA_3 can promote growth and affect organogenesis (Murashige, 1964; Helgeson & Upper, 1970). The amount of growth induced by GA_3 depends partially on the cytokinin content of the medium (Lance et al., 1976b). In dark-grown cultures, $10^{-7} M$ kinetin is optimal for growth; at this cytokinin concentration $10^{-6} M$ GA_3 further enhances growth. Light inhibits the growth of tobacco callus, and cytokinins have little growth-promoting activity in the light. However, at a combination of $10^{-8} M$ kinetin and $10^{-6} M$ GA_3 slight growth-promotion (ca. 10% of that observed in dark-grown callus) is observed. Dark-grown callus contains higher GA levels (as estimated by bioassay) than light-grown, although the positive correlation between growth rate and GA content does not hold completely.

Lance et al. (1976a) studied the metabolism of $[2,3\text{-}^3H_2]$-GA_{20} applied to dark- and light-grown tobacco callus, to determine whether light was inhibiting growth by altering the metabolism or utilization of applied GA. GA_{20} is not known to be endogenous to tobacco, but it does promote growth of dark-grown callus. After 6 days, 55% of the applied radioactivity was recovered from dark-grown cultures (callus plus agar) versus 70% recovery from light-grown cultures. Only in dark-grown cultures did the agar contain significant amounts of radioactivity, indicative of very polar and perhaps volatile metabolites.

The distribution of radioactivity and biological activity was studied after TLC using a solvent system of ethyl acetate-chloroform-formic acid (50:50:1). There was some residual substrate at Rf 0.6–0.7 with biological activity associated with it. The major radioactive and biologically active metabolite at Rf 0.3–0.4 accumulated in light-grown callus only. Most significant was the biological activity in low Rf zones in dark-grown callus extracts which was *not* associated with radioactivity. In a large-scale feed of $[^3H]GA_{20}$ to light-grown callus, $[^3H]GA_1$ was identified by GC-RC and GC-MS as the major metabolite. No $[^3H]GA_{29}$ was present, despite the initial observation by GC-RC that some might be formed in dark-grown callus.

Lance et al. (1976a) postulated that metabolism of $[^3H]GA_{20}$ was faster in dark-grown callus than in light, as evidenced by the greater loss of activity and the presence of putative polar metabolites in dark cultures. However, the sole identifiable metabolite, namely $[^3H]GA_1$, accumulated in *light-grown* callus only. Applied GA_1 is more active than GA_{20} in promoting growth of tobacco callus (only tested on dark-grown), so that it is surprising that a growth response to applied $[^3H]GA_{20}$ was not observed in light-grown callus. Lance et al. (1976a) suggested that the further metabolism of $[^3H]GA_1$ may be blocked in light-grown callus, thus offering a possible explanation for the inactivity of $[^3H]GA_1$ in the light. The presence of polar, biologically active though nonradioactive compounds in dark-grown callus suggested to the authors that further metabolism of $[^3H]GA_1$ occurred with loss of label. While this is a distinct possibility, none of the five other potential metabolites of $[^3H]GA_{20}$ suggested by Lance et al. (i.e., GA_{29}, GA_{29} glucoside, 3-*epi*-GA_1, GA_8, and GA_8 glucoside) have substantial biological activity. GA_5, suggested by them as another potential metabolite, has chromatographic properties that are inconsistent with those observed for the putative metabolite(s).

An alternative suggestion was that these polar, biologically active GAs were not metabolites of [^{3}H]GA_{20}, but were produced by dark-grown callus in response to the applied GA (Lance et al., 1976a). These polar GAs are not present in untreated dark-grown callus, or in light-grown callus. They may, indeed, have some significance in the growth effect produced by [^{3}H]GA_{20} in dark-grown callus, but without further experimentation this explanation is entirely speculative. An indirect effect of applied GAs on tobacco callus growth via a modification of native GA metabolism is consistent with the observation that applied GA will only enhance callus growth at optimal cytokinin concentration. At this cytokinin concentration callus is fast growing and contains its highest levels of extractable GA-like substances (Lance et al., 1976b). Fry and Street (1980) have also defined conditions under which cell suspension cultures best respond to applied GA_3. Two auxin-independent cultures—one of *Spinacia oleracea* (spinach), the other of *Rosa* spp ("Paul's Scarlet" rose)—responded to GA_3 with enhanced cell expansion. The response of *Rosa* cells was cytokinin-dependent. In addition, cultural conditions that promoted growth also enhanced the response to GA_3. Conversely, GA_3 became inhibitory under conditions that permitted only slow growth.

Rappaport et al. (1974) have studied the metabolism of [^{3}H]-GA_1 in tobacco and *Olea europea* (olive) callus, grown in diffuse light. [^{3}H]GA_8, identified by comparative GC-RC, was formed first, with the build-up of an unknown metabolite of [^{3}H]GA_8 occurring later. In neither case was metabolism correlated with growth, and insufficient details of the work are available for it to be critically assessed.

Attempts to correlate the metabolism of a GA, applied under different environmental conditions, with an observed growth response in a system that contains native GAs can be extremely complicated (see Section 4.2.1). The type of investigation conducted by Lance et al. (1976a) would be better performed (a) by initially identifying the endogenous GAs of tobacco callus, (b) by feeding appropriately labeled GAs such that isotope is not lost during the experiment, and (c) by comparing metabolism of both applied and native GAs under different environmental conditions.

4.4.2 Photoperiodic control of apical senescence

The genetic control of flowering in *P. sativum* has recently been reviewed (Murfet, 1977a, 1977b). Photoperiodic sensitivity is coded for by the dominant alleles of the gene *Sn* (Murfet, 1971a), since in short days *Sn* governs the production of a graft-transmissible flower inhibitor in cotyledons and leaves (Murfet, 1971b). Plants of the genotype *SnHr* are obligate long-day plants (*Hr* maintains the high response to photoperiod by preventing the ageing of *Sn*; Murfet, 1973). Photoperiod-insensitive I-type plants are recessive for the gene *Sn*, and the well-known cultivars Progress No. 9 and Meteor are of the genotype *snhr* (I_3-type).

The expression of the genes *SnHr* can be modified by other genes—namely *E*, the major gene for early flowering, and *Lf*, a series of at least four alleles conferring increasing lateness to flower irrespective of photoperiod (Murfet, 1971a, 1975). Plants of the genotype *LfESnHr* (G-type) are late flowering and have an obligate long-day requirement (Marx, 1968). Alternatively, plants of the genotype *lfESnHr* (G2-type) are early flowering in short or long days, although some flower abortion is evident especially in short days. In addition, G2 plants show extended vegetative growth in short days despite the presence of flowers and fruits (Marx, 1968; Proebsting, Davies, & Marx, 1976; Davies, Proebsting, & Gianfanga, 1977). In short days only, G2 plants produce a graft-transmissible substance that delays senescence (Proebsting, Davies, & Marx, 1977). Furthermore, applied GA_3 can cause a similar retardation of apical senescence of G2 plants held in long days (Proebsting, Davies, & Marx, 1978). This observation has prompted a comparative study of GA metabolism in G2-, I-, and G-type plants (Proebsting et al., 1978; Proebsting & Heftman, 1980).

GA_9, applied to shoot apices of G2 plants in long days, was without activity, whereas GA_{20} had a moderate senescence-delaying effect (Proebsting et al., 1978). $[2,3-^3H_2]GA_9$, applied to fully expanded leaflets of G2 plants in short days, was extensively metabolized to up to seven unidentified metabolites. The least-polar metabolite coeluted with GA_{20} on silica gel column chromatography, while the major metabolite, designated GA_E, had a similar polarity to GA_1. Insufficient data are presented in the paper for percentage conversions to be calculated. A reduced capacity for $[^3H]GA_9$ metabolism was evident in G2 plants that had received more than one long day, and this effect was reversible on the return of plants to short days. $[^3H]GA_9$ was extensively metabolized in photoperiodic-insensitive I_2-type plants, but the metabolite designated GA_E was not formed in these plants.

These results prompted Proebsting et al. (1978) to investigate the endogenous GA-like substances in G2- and I-type plants. In all extracts the major biologically active GA (in lettuce hypocotyl assays) co-eluted with GA_{20}, but G2 plants maintained in short days also contained a zone of biological activity at the retention volume of GA_E. It was concluded that a biologically active polar gibberellin, $[^3H]GA_E$, is formed from $[^3H]GA_9$ in G2 plants under short days and that GA_E promoted the continued apical growth displayed by G2 plants in short days. Under long days, applied GA_9 would not be metabolized to GA_E and hence would be inactive. GA_E was suggested to be GA_1, on the basis of chromatographic properties. Proebsting et al. (1978) drew an analogy between the senescence-retarding factor in G2 plants and the flower inhibitory factor also produced in short days by photoperiodically-sensitive lines (Murfet, 1971b).

Proebsting and Heftmann (1980) studied the metabolism of $[^3H]GA_9$ in G-type plants under long and short days. G-type plants (*LfESnHr*) are late-flowering obligate long-day plants which have an extended vegetative period in short days. $[^3H]GA_9$ was extensively metabolized in short days with the apparent production of the metabolite previously designated GA_E by Proebsting et al. (1978). One long-day treatment prevented the production of $[^3H]GA_E$ from $[^3H]GA_9$ and promoted flowering, again suggesting to the authors that GA_E might be the floral inhibitor.

A critical assessment of whether GA_E is the putative senescence-retarding factor in G2 plants and the floral inhibitor in this and other photoperiodically-sensitive plants is beyond the scope of this chapter. The proposal that GA_E is GA_1 (Proebsting et al., 1978) has not yet been critically examined. GA_1 is not known to be present in immature seeds or seedlings of Progress No. 9 although the isomeric GA, GA_{29}, is present (Sections 4.3.2.2 and 4.4.1.1). GA_{29} has negligible activity in the lettuce assay and is unlikely to be GA_E. In young seeds of Progress No. 9 and in germinating seeds and seedlings of Meteor, $[^3H]GA_9$ can be metabolized, inter alia to $[^3H]GA_{20}$, and in turn $[^3H]GA_{20}$ can be metabolized to $[^3H]GA_{29}$ (Section 4.3.2.2). There is no evidence for the formation of $[^3H]GA_1$. Both cultivars are, however, of the I_3-type and from the results of Proebsting et al. (1978) might not be expected to contain GA_E or, indeed, have the capacity to produce it. These authors therefore suggested that only plants of the genotype *SnHr* can produce GA_E, while those lines recessive at either locus would instead produce the isomeric GA_{29}. This suggestion requires a critical examination of the endogenous GAs in lines of the $SnHr$ genotype, and definitive metabolic studies with conclusive identification of metabolites.

Ingram and Browning (1979) examined extracts of developing seeds of G2 ($lfESnHr$) plants, grown under long and short days. The major GAs, like those in Progress No. 9 ($snhr$) were identified by GC-MS as GA_{20} and GA_{29}. GA_1 was not detected although GA_{19}, which had not previously been identified in pea (see Section 4.3.2.2), was found to occur in extracts of seeds from both photoperiods. GA_{19} has similar chromatographic properties to GA_1 and GA_{29} on columns of silica gel, and Ingram and Browning suggested that GA_{19} might be GA_E. GA_{19} however has only 1–5% of the activity of GA_1 in the lettuce hypocotyl assay (Reeve & Crozier, 1974), and so unprecedentedly high levels of GA_{19} would be required to elicit the biological activity reported for GA_E by Proebsting et al. (1978). Moreover [^{3}H]GA_9 was reported to be a precursor of GA_E (Proebsting et al., 1978), which is again not consistent with GA_{19} being GA_E.

Ingram (1980) has further examined the endogenous GAs in developing seeds of nine lines of pea by SICM. Seven of these lines were recessive for one or both of the genes Sn and Hr. G2 ($lfESnHr$) and line 63 ($lfeSnHr$) were also examined. GA_{20}, GA_{29}, and GA_{29}-catabolite were the major GAs in all lines. GA_{19} and GA_{44} were present in eight of the lines, GA_5 in five lines, GA_{17} in three lines, and GA_9 in one. There were no consistent qualitative differences with genotype. When GAs were quantitated throughout seed development in a few selected lines, G2 seeds contained more GA_{19} and less GA_{17} than those lines that had recessive alleles for one or both of the genes Sn and Hr. There was, however, no evidence for a correlation with photoperiod. In preliminary investigations with seedling tissue, Ingram (1980) has again not observed qualitative differences in GA content with genotype.

4.4.3 Photoperiodic control of bolting and flowering

GAs have been studied in relation to the photoperiodic control of bolting and flowering in several long-day plants and in the long-short–day plant *Bryophyllum daigremontianum* by Zeevaart and associates (e.g., Jones & Zeevaart, 1980a, 1980b; Metzger & Zeevaart, 1980a, 1980b; Durley, Pharis, & Zeevaart, 1975).

In the long-day plant *Spinacia oleracea* (spinach), there were marked qualitative differences in the content of GA-like substances, as estimated by bioassay, in plants that had been exposed to either long or short days. Also GA turnover, as measured by the rate of decline in endogenous GA-like substances after the application of AMO-1618, appeared to be greater in long days, leading Zeevaart (1971) to suggest that GAs control the increased stem elongation in long days which is associated with eventual flower production. More recently, GA_{17}, GA_{19}, GA_{20}, GA_{29}, GA_{44}, and GA_{53} have been identified by GC-MS in spinach (Metzger & Zeevaart, 1980a) and quantitation of five of these GAs by GC-SICM has been conducted in tissue exposed to increasing numbers of long days (Metzger & Zeevaart, 1980b). The absolute amount only of GA_{20} could be calculated; amounts of the other GAs relative to the maximal level of each individual GA was estimated. After 4 long days the level of GA_{19} had decreased by 80%, while that of GA_{20} had increased from $0.8 \mu g \ g^{-1}$ dry weight in short days to $5.5 \mu g \ g^{-1}$ dry weight after 14 long days. The level of GA_{29} also increased, whereas GA_{44} and GA_{17} remained fairly constant throughout. Metzger and Zeevaart (1980b), on the assumption that GA_{19} is the immediate precursor of GA_{20}, suggested that the metabolic conversion of GA_{19} to GA_{20} was under photoperiodic control. Enhanced conversion of GA_{19} under long days would thus explain the elevated levels of GA_{20}. $[^{2}H]GA_{53}$ has recently been fed to spinach plants grown under short and long days, and treated with AMO-1618 to reduce the levels of endogenous GAs (Gianfagna, Zeevaart, & Lusk, 1983a). The substrate was obtained by the microbiological conversion of $[^{2}H]$steviol (Gianfagna, Zeevaart, & Lusk, 1983b). The labeling of steviol acetate norketone by a Wittig reaction gives scrambling of the deuterium with up to 4 atoms ^{2}H per molecule (Gianfagna et al., 1983b). Under short days GA_{44} and GA_{19} were identified by selected ion current monitoring of four prominent ions in their mass spectra. In addition the relative intensities of ions in the molecular ion cluster of GA_{44} and in the base peak cluster of GA_{19} were comparable to the molecular ion cluster of $[^{2}H]GA_{53}$, indicating that both GA_{44} and GA_{19} contained ^{2}H. After 2 long days $[^{2}H]GA_{20}$, but *not* $[^{2}H]GA_{44}$ or $[^{2}H]GA_{19}$, was identified by selected ion current monitoring and comparison of its molecular ion cluster. Unfortunately, the amount of metabolism could not be quantified, nor was any indication of reproducibility given. Nevertheless, the results provide additional evidence that GA_{19} is the immediate precursor of GA_{20} in spinach, and that this metabolic step is under photoperiodic control.

The photoperiodic response of *Silene armeria* has also been correlated with GAs (Cleland & Zeevaart, 1970). In more recent studies with the closely related *Agrostemma githago* (corn cockle), Jones and Zeevaart (1980a) have shown that application of GA_{20} to plants maintained under short days can mimic the effect of inductive long days on stem elongation. GA_1 and its 3-epimer, together with GA_{17}, GA_{19}, GA_{20}, GA_{44}, and GA_{53}, were identified in extracts of shoots and leaves of *Agrostemma* by GC-MS (Jones & Zeevart, 1980b). Levels of GAs as estimated by GC-SICM were followed for 16 days after the transfer of plants from short to long days. GA_{17}, GA_{19}, GA_{20}, and GA_{44} underwent a transient 5-fold increase, peaking after exposure to 8 long days, when plants were just entering a phase of rapid stem elongation. GA_1 and *epi*-GA_1, present in a ratio of 1:2-3, both increased approximately 10-fold, peaking after exposure to 12 long days. GA_{53} levels, which were fairly constant for the first 12 days, then rose concomitantly with a second increase in GA_{17}, GA_{19}, and GA_{44} levels. Thus, in *Agrostemma* complex changes in GA levels, which cannot easily be rationalized, occur during inductive long-day treatment. GA_1 is the most active of the native GAs in promoting stem elongation in short days, while its epimer has little activity. GA_{20} has substantial activity. GA_{17}, the only C_{20}-GA tested, is inactive.

$[2,3\text{-}^3H_2]GA_{20}$ has been fed to *Agrostemma* plants after exposure to 8 long days, and $[1,2\text{-}^3H_2]GA_1$ and 3-*epi*-$[1,2\text{-}^3H_2]GA_1$ were fed after 12 long days (Jones & Zeevart, 1982). All GAs were also fed to plants maintained under constant short days. $[^3H]GA_{20}$ was more rapidly metabolized under long days, with only 12% of the applied $[^3H]GA_{20}$ remaining after 2 long days, versus 50% after 2 short days. A single major metabolite, obtained by HPLC of acidic ethyl acetate fractions, was further resolved by analytical TLC into two components tentatively identified on the basis of chromatographic properties as *epi*-$[^3H]GA_1$ and $[^3H]GA_1$. They were formed in a ratio of 50:1. More *epi*-$[^3H]GA_1$ accumulated in long days relative to short, although percentage conversions could not be calculated from data presented. A polar metabolite, observed under both photoperiods, was not identified.

The metabolism of *epi*-$[^3H]GA_1$, studied in long days only, was slow with 56% of the applied substrate remaining unmetabolized after 4 days. There was no evidence of epimerization to $[^3H]GA_1$. The only minor metabolite observed was not identified. In contrast, $[^3H]GA_1$ was extensively metabolized in long days (16% remaining unmetabolized after 2 long days, with 72% remaining after 2 short days). Metabolites were not identified, although one was similar (but not identical) to $[^3H]GA_8$.

The epimer of GA_1 has not previously been shown to be naturally occurring, but Jones and Zeevaart (1980b) presented evidence that it was not formed in *Agrostemma* extracts by epimerization of GA_1 during work-up. From metabolic studies, it would appear to be a genuine metabolite of GA_{20} in this plant. The precursor of GA_1 in *Agrostemma* is not known. $[^3H]GA_{20}$ and $[^3H]GA_1$, which are both biologically active GAs, are metabolized more rapidly in long days than in short, while the inactive *epi*-$[^3H]GA_1$ is slowly metabolized in long days. *Epi*-$[^3H]GA_1$ is scarcely metabolized in barley half seeds (Stolp et al., 1973) or in cell-free systems derived from *Ph. vulgaris* seeds (Patterson, Rappaport, & Breidenbach, 1975). Similarly, biologically inactive 2β-hydroxylated GAs are only slowly metabolized. No 2β-hydroxylated GAs have been identified in *Agrostemma*, and 3α-hydroxylation may constitute an equivalent deactivating process in this species.

GA_{20} is also native to the long-short–day plant *Bryophyllum daigremontianum* (Gaskin, MacMillan, & Zeevaart, 1973) and levels, as estimated by GC, were correlated with increased stem elongation and subsequent flowering induced by transfer of plants from long to short days (Zeevaart, 1973). Applied GA_{20} is, however, less active than several seemingly nonnative GAs (e.g., GA_1, GA_3, GA_5, and GA_7) in inducing flowering in *Bryophyllum* plants maintained under constant short days (Gaskin, MacMillan, & Zeevaart, 1973). These authors, therefore, concluded that GA_{20} may not be active per se and may be metabolized in inductive conditions to a more polar biologically active GA. $[2,3\text{-}^3H_2]GA_{20}$, injected into mature leaves of *Bryophyllum* plants under long and short days, was metabolized slowly with 12% of the substrate remaining unmetabolized after 51 short days (Durley et al., 1975). Of the applied substrate, 46% was metabolized while 42% was unaccounted for. Three metabolites were identified by comparative GC-RC in extracts of treated leaves and growing shoots, namely *epi*-$[^3H]GA_1$, $[^3H]GA_{29}$, and rings-C/D-rearranged $[^3H]GA_{20}$. At least some of the latter metabolite was thought to be originally present as a highly labile conjugate. Metabolism of $[^3H]GA_{20}$ was initially more rapid in long days than short. Because the stereochemistry of ^3H-labeling in the substrate had not been determined (Murofushi, Durley, & Pharis, 1974, 1977), metabolic conversions to *epi*-$[^3H]GA_1$ and $[^3H]GA_{29}$ could not be calculated accurately. Durley et al. (1975) estimated that *ca.* 10% of applied $[^3H]GA_{20}$ was converted to $[^3H]GA_{29}$ in 20 long days, with a similar amount converted in 51 short days. Likewise, *ca.* 4% conversion of $[^3H]GA_{20}$ to *epi*-$[^3H]GA_1$ was observed in 20 long days and in 51 short days. Most of the *epi*-$[^3H]GA_1$ was present in the treated leaves. Rings-C/D-rearranged $[^3H]GA_{20}$ was a minor metabolite, and may indeed have been an artifact.

Despite the initial suggestion that native GA_{20} may be metabolized in *Bryphyllum* to a more highly active GA, all the observed metabolites of $[^3H]GA_{20}$ have low to zero activity in standard bioassays. However, none of these metabolites are known to be native to *Bryophyllum* and the significance of this metabolism in flower induction in *Bryophyllum* is, therefore, difficult to assess.

While elevated levels of biologically active GAs have been correlated with stem elongation following the transfer of plants of *Spinacia*, *Agrostemma*, and *Bryophyllum* from noninductive to inductive conditions, metabolic studies have predominantly demonstrated deactivation of the applied $[^3H]GAs$ (Durley et al., 1975, Jones & Zeevaart, 1982). Perhaps more relevant would be a comparison of the *formation* of active GAs in inductive and noninductive photoperiods. Although metabolic studies of this type would be technically difficult (for instance, suitably labeled substrates are not readily available), they would nevertheless be timely, since evidence does suggest that enhanced synthesis of GA_{20} occurs when the three species discussed here are transferred from noninductive to inductive photoperiods.

4.4.4 Effects of plant water balance on GA metabolism, and related effects of other hormones

Reid and Crozier (1971) presented some preliminary evidence that water-logged plants of *Lycopersicon esculentum* (tomato), which exhibited reduced growth and other flooding injury, contained reduced levels of GA-like substances. Railton and Reid (1973) reported that spraying leaves of water-logged tomato plants with the cytokinin N^6-benzyl adenine could relieve some of the flooding injuries. The same investigators later examined the GA-like activity in extracts of benzyl adenine-treated flooded and nonflooded tomato plants (Reid & Railton, 1974). Flooding reduced the total GA content, as estimated by lettuce hypocotyl, tomato hypocotyl, and barley half-seed assays. Quantitative estimates varied with the different assays, but flooding at least halved the GA-like activity in extracts, and foliar application of benzyl adenine restored the activity at least to that in extracts of nonflooded plants. The three different assays, however, recorded different qualitative effects of flooding on GAs, undoubtedly due to the differing sensitivities of the three assay plants to tomato gibberellins. To date the native GAs of tomato are unidentified, so that accurate quantitation has not been possible nor have metabolic studies been conducted using tomato.

Reid and Railton (1974) did suggest that cytokinins were able to maintain normal or elevated levels of shoot GAs in tomato under adverse physiological conditions, but the authors were not able to establish whether flooding had a direct effect on GA levels, or whether there was an initial effect on cytokinin production (possibly in the roots) leading to a depletion of shoot GAs. Railton (1974b) used pea seedlings, cv. Meteor, to investigate further the effects of benzyl adenine on GA metabolism. $[^3H]GA_9$, applied to seedlings that had been sprayed for three consecutive days with benzyl adenine, was reported to be more rapidly metabolized than in untreated seedlings. Thus, from TLC profiles, the amount of radioactivity associated with unmetabolized $[^3H]GA_9$ appeared to be less in extracts of cytokinin treated plants, but no quantitative data were presented nor was any indication of the reproducibility given. Metabolites were detected by TLC only, but they were assumed to be the same as those identified in previous studies as $[^3H]GA_{20}$, $12\alpha\text{-OH}[^3H]GA_9$, and $[^3H]GA_{10}$ (Railton et al., 1974b). The metabolism of applied $[^3H]GA_{20}$ also appeared to be increased in cytokinin-treated plants, although again only TLC data were presented. Although it is not possible to draw firm conclusions from this work, it was inferred that turnover of $[^3H]GA_{20}$ in pea shoots was enhanced by cytokinin application. While enhanced turnover of GA_{20} would increase the amount of GA_{20} available for growth, it would presumably not result in increased accumulation of biologically active GA, as was observed in cytokinin-treated tomato plants.

Crozier and Reid (1971, 1972) reported the effects of root excision on GA-like activity in *Ph. coccineus* shoots. The accumulation of a relatively nonpolar GA (assumed now to have been GA_5 or GA_{20}; see Bowen et al., 1973) in shoots of de-rooted seedlings was thought to be due to its restricted translocation to the roots where, in intact seedlings, its conversion to a more polar GA (probably GA_1) was suggested to occur. Railton (1979) determined that $[^3H]GA_{20}$ applied to apical buds of *Ph. coccineus* was scarcely translocated over 24 hr, leading him to question the idea put forward by Crozier and Reid (1972). Root excision enhanced $[^3H]GA_{20}$ metabolism, as estimated by the amount of unmetabolized substrate remaining after 24 hr. However, cytokinins also enhanced $[^3H]GA_{20}$ metabolism and, moreover, applied cytokinins could not restore the shoot inhibition caused by root excision. Root excision is therefore not depleting the shoot of root-produced cytokinins, with accompanying effects on the levels of shoot GAs.

Wilting causes elevated levels of abscisic acid in seedlings—for example, pea (Simpson & Saunders, 1972)—and a concomitant reduction in shoot growth, although whether there is a corresponding reduction in GA levels is not known. Taylor and Railton (1977) studied the metabolism of $[^3H]GA_{20}$ in pea seedlings that had previously been sprayed for 2 days with ABA, or had been wilted to 10% reduction in fresh weight. The amount of metabolism of $[^3H]GA_{20}$ was expressed as a ratio of the radioactivity associated on TLC with $[^3H]GA_{20}$ to that associated with its putative metabolite $[^3H]GA_{29}$. ABA and wilting increased the ratio of $[^3H]GA_{20}$ to $[^3H]GA_{29}$-like metabolite, leading the authors to conclude that approximately 50% reduction in metabolism had occurred. Severe wilting to 30–37.5% reduction in fresh weight was purported to decrease the metabolism of $[^3H]GA_{20}$ by 80%. Results, expressed solely as a ratio of substrate to metabolite, do not however take into account any differences in uptake of the substrate in differently treated plants. ABA is known to enhance the uptake of $[^3H]GA_1$ and some related compounds by barley half seeds (Stolp et al., 1973, 1977). Taylor and Railton (1977) gave some data on the amount of $[^3H]GA_{20}$ washed-off treated seedlings, which were purported to show that moderate wilting did not affect uptake of $[^3H]GA_{20}$ and that ABA application only enhanced uptake slightly. However, no figures were given for the amount of substrate applied, or taken up, nor were any raw data for metabolism presented. Thus, the reader cannot assess the independent effects of ABA on $[^3H]GA_{20}$ uptake and on metabolism. Nevertheless, the data do indicate that ABA may reduce 2β-hydroxylation of $[^3H]GA_{20}$. While it is impossible to gain an overall view of effects on GA metabolism from the study of a single metabolic conversion, the results obtained by Taylor and Railton (1977) indicate that biologically active GA_{20} would accumulate in wilted and ABA-treated plants. These results are therefore not consistent with the reduction in stem growth observed in these treatments.

To date no cohesive picture for the interrelationship of root- and shoot-produced GAs has emerged from these studies on the effects of flooding, wilting, and root surgery on the levels of endogenous GA-like substances, and on the metabolism of applied GAs. Neither has the involvement of other hormones been clarified. Precise and extensive metabolic studies are required to elucidate the sites of GA biosynthesis and metabolism in intact, unstressed plants. The effects of physiological stress and the involvement of other hormones could then perhaps be more readily defined.

4.5 COMPARTMENTATION

Compartmentation of GA metabolism can be considered at many levels: organ, tissue, cellular, and subcellular. Gross quantitative differences in GA content are evident in different plant organs, and quantitative differences in the metabolism of applied GAs have also been observed (e.g., Davies & Rappaport, 1975a; Durley et al., 1976; Nash & Crozier, 1975). At the tissue level there is evidence that within pea seeds GA_{20} and GA_{29} are present in cotyledons while GA_{29}-catabolite is located in the testa. This reflects a difference in the metabolic ability of the cotyledons and testa (Sponsel, in preparation). At the subcellular level, specific stages of GA biosynthesis or metabolism may be located within subcellular organelles. Early stages of the biosynthesis of terpenes, including that of GAs, have been demonstrated within chloroplasts. Furthermore, GA-like substances are known to occur in sonicated chloroplasts, and chloroplast preparations are capable of metabolizing applied GAs. Other evidence suggests that later stages of GA metabolism may be located in the vacuoles. At a biochemical level, cell-free work also indicates that different enzymes in the GA biosynthetic pathway may be located in different cellular fractions, as evidenced by differential centrifugation. Furthermore, specific cofactor requirements and pH optima of particular enzymes are indicative of specific locations within the cell. This latter aspect is discussed in detail in Volume II, Chapter 1, while the ensuing discussion centers on evidence for the potential compartmentation of GA metabolism in chloroplasts and vacuoles.

Stoddart (1968) demonstrated GA-like biological activity in sonicated chloroplast preparations from spinach and barley. Several other reports of biological activity in chloroplasts from a variety of species followed, but the identity of GA-like substances in chloroplasts proved difficult to establish since the low recovery of intact plastids from plant material precluded GC-MS identification of the GAs. It was only when Browning and Saunders (1977) developed a technique for detergent (Triton X-100) extraction of chloroplasts that GA_4 and GA_9 were conclusively identified in *Triticum aestivum* (wheat) plastids. GAs are supposedly extracted *ca.* 10^3 times more efficiently by Triton than by aqueous methanol. However Saunders (unpublished information, quoted by Yokota et al., 1980) has disclosed that the reported efficacy of Triton for extracting GAs from wheat is not reproducible in other systems. Indeed, Railton and Rechav (1979) demonstrated little difference in the levels of GA-like substances extracted from pea chloroplasts by methanol and Triton, and MacMillan (1978) reported that the levels of GA_9, GA_{20}, and GA_{29} (estimated by internal standardization) were the same in Triton and methanol extracts of immature seeds of pea. No conclusive identity of chloroplast GAs other than that of GA_4 and GA_9 in Triton-extracted *Triticum* chloroplasts has been reported.

Railton and Reid (1974a) reexamined the observations by Jones and Lang (1968) that of the two GA-like substances thought to be present in pea seedlings only the GA_1-like factor would diffuse into agar. Railton and Reid (1974a) showed that both the GA_1- and GA_5-like factors were extractable from chloroplasts of pea seedlings cv. Alaska. However, they presented evidence that either the chloroplast membrane was preferentially permeable to the GA_1-like factor or that this compound was also present in the cytoplasm. Thus, they suggested that exclusive localization of the GA_5-like factor in chloroplasts might explain its nondiffusibility from pea shoots. Such potential compartmentation of different GAs is an obvious controlling factor for metabolism.

In view of the observation that GA_{20} and GA_{29} were the major GAs in Progress No. 9 seeds (Frydman & MacMillan, 1973) and that the pathway $GA_9 \rightarrow GA_{20} \rightarrow GA_{29}$ was postulated to occur in pea seedlings (Railton et al., 1974) (Section 4.3.2), Railton and Reid (1974b) fed $[^3H]GA_9$ and $[^3H]GA_{20}$ to isolated pea chloroplasts (cv. Meteor). They observed 4% conversion of $[^3H]GA_9$ to a very polar unidentified compound and 2.7% conversion to a compound with TLC properties comparable to those of GA_{20} in 40 min. No identification of metabolites was initially undertaken (Railton & Reid, 1974b), but reexamination of TLC fractions of the original feed was later undertaken (Railton, 1977a, 1977b). $[^3H]GA_{10}$ and 16,17-dihydro-16,17-dihydroxy$[^3H]GA_9$ were identified by GC-MS. $[^3H]GA_{10}$ may be non-enzymatically formed, and in turn may be hydroxylated to give the diol. Neither of the major metabolites of $[^3H]GA_9$ observed in feeds to intact seedlings, namely $[^3H]GA_{20}$ and 12α-OH$[^3H]GA_9$, were identified in chloroplast preparations. When $[^3H]GA_{20}$ was similarly fed to isolated chloroplasts, 13% conversion to a metabolite with TLC properties comparable to those of GA_{29} were observed (Railton & Reid, 1974b), but no further information on the identity of this metabolite has been published.

Railton (1977c) later examined the permeability of isolated chloroplasts of cv. Meteor to $[^3H]GA_{20}$ and was unable to demonstrate appreciable uptake of $[^3H]GA_{20}$ into intact chloroplasts over 15 min. Over 99% of the radioactivity associated with the initial chloroplast pellet following incubation with $[^3H]GA_{20}$ was found to be removed by repeated washing, indicating that the chloroplast membranes were relatively impermeable to $[^3H]GA_{20}$. Thus, Railton (1977c) concluded that, contrary to previous assumptions, the metabolism of $[^3H]$GAs applied to intact leaves (e.g., Railton et al., 1974a, 1974b, 1974c) was unlikely to have occurred within the chloroplasts.

The permeability of the plastid envelope is known to be selective, but this selectivity does alter during plastid development. Thus, while Rogers, Shah, and Goodwin (1966) showed that the envelope of photosynthetically competent chloroplasts was impermeable to mevalonic acid (MVA), Wellburn and Hampp (1976) demonstrated that etioplast envelopes were permeable to MVA. The ability of plastids to take up MVA is lost after 2–4 hr illumination. Alternatively, Cooke, Saunders, and Kendrick (1975) and Evans and Smith (1976) reported that short-term illumination of etioplasts with red light increased the permeability of etioplast envelopes to GAs, causing an efflux of GA-like activity. This alteration of membrane permeability and other possible effects of illumination on GA biosynthesis and/or metabolism have been suggested to be under phytochrome control (see Volume II, Chapter 1).

Although these preliminary observations on GA localization and metabolism in plastids (Railton & Reid, 1974; Railton, 1977a, 1977b) and associated work implicating a fundamental control of GA levels in plastids by light (Cooke et al., 1975; Evans & Smith, 1976) were most encouraging, the hoped-for advances in recent years have not been forthcoming. Technical difficulties (for instance, problems of scaling-up plastid preparations to obtain definitive qualitative and quantitative information on native GAs and of conducting metabolic studies on a scale that would allow conclusive identification of all metabolites) have undoubtedly limited progress in the field. Although some evidence (Railton, 1977c) indicates that metabolism of $[^3H]$GAs applied to intact leaves is most probably extrachloroplastidic, one might logically assume that effects of different light regimes on GA levels may still be mediated by light-perceived events within the plastid.

Ohlrogge et al. (1980) have studied the uptake and accumulation of $[^3H]GA_1$ by means of barley and *Vigna unguiculata* (cowpea) seedlings. Protoplasts were isolated from leaves that had previously been incubated with the radioactive GA and were treated in a variety of ways. When protoplasts of barley leaves were mechanically lysed, 80% of the chloroplasts were recovered intact but only 5% of the recovered radioactivity was associated with the chloroplast pellet. When protoplasts were osmotically lysed and the contents were fractionated by density-gradient centrifugation, a considerable, though variable, amount of radioactivity was associated with the vacuoles. Approximately 50% of the radioactivity taken up by the leaves was lost during protoplast preparation, and yields of vacuoles were in the range 5–15%. It is impossible to calculate from the data presented what proportion of [^{3}H]GA$_1$ applied to the leaves was localized in the vacuoles, and further quantitative work is required. Osmotic breakage of the organelles caused radioactivity to be readily released, suggesting to the authors that it was present in the vacuolar sap, rather than being associated with the tonoplast.

In a separate publication Rappaport and Adams (1978) reported the incubation of isolated vacuoles with [^{3}H]GA$_1$. Metabolism was dependent on the pH of the medium, with approximately 22.9% of the radioactivity supplied being present as [^{3}H]GA$_8$ (identified by comparative GC-RC on two GC columns) after 5 hr at pH 5.7. No [^{3}H]GA$_8$ glucoside was detected, although in parallel feeds to isolated protoplasts, 8.9% of applied radioactivity was attributable to [^{3}H]GA$_8$ glucoside after 6 hr. In addition, several unidentified metabolites were formed from [^{3}H]GA$_1$ by protoplasts but not by vacuoles. Rappaport and Adams (1978) thus, suggested that spatial separation of 2β-hydroxylation and glucosylation may occur within the cell, with 2β-hydroxylation taking place within the vacuole and glucosylation occurring elsewhere. Hartmann, Fonteneau, and Benveniste (1977) demonstrated that alkylation of the triterpene, cycloartenol, is mediated by a microsomal enzyme, whereas glucosylation is mediated by a plasmalemma-associated glucosyl-transferase. While the solubility of the 2β-hydroxylating enzyme isolated from germinating *Phaseolus* seeds (Patterson & Rappaport, 1974; Patterson et al., 1975) is not inconsistent with its possible compartmentation within the vacuole (i.e., in the vacuolar sap, *not* associated with the tonoplast), the neutral pH optimum for this enzyme (see Chapter 3) is not consistent with a vacuolar location. Moreover, since very efficient 2β-hydroxylation of GAs applied to intact plants or plant parts is readily observed (e.g., Nadeau & Rappaport, 1972; Durley & Pharis, 1973; Frydman & MacMillan, 1975), it would appear unlikely that 2β-hydroxylation is exclusively localized in plant vacuoles.

4.6 CONCLUSION

A considerable amount of metabolic work has been conducted with intact plants in recent years, although not all of it has been of the highest standards in terms of basic experimental design, execution and analysis. Clearly, some of it merits repeating or reworking. Most of the work described has centered on the metabolism of applied C_{19}-GAs. Some experiments have demonstrated only single metabolic conversions; extended metabolic sequences have been established in others. The metabolism of applied C_{19}-GAs has been shown to be different in different developmental systems and it can be altered by environmental factors. In general, much of the work has demonstrated deactivation of the applied GAs, either by 2β-hydroxylation or conjugation. Neither of these metabolic processes take place in *G. fujikuroi*, in which GAs have no known function. Thus, it has been inferred that these deactivation processes may be specifically concerned with regulating the levels of biologically active GAs in higher plant tissues. However, evidence is accumulating to suggest that the *production*, rather than the destruction of biologically active GAs has more relevance to growth and development.

Several different pathways to C_{19}-GAs probably operate in higher plants. The proposed pathways, while being fundamentally similar, differ from each other in the degree and position of hydroxylation of the C_{20}-GAs. An early 13-hydroxylation pathway leading to the C_{19}-GA, GA_{20}, appears to be of especially widespread occurrence in higher plants. The production of biologically active C_{19}-GAs, particularly GA_{20}, has been implicated in the control of seedling growth, in the bolting and flower induction of photoperiodic plants, and in the retardation of apical senescence.

The past decade has seen a proliferation of work on C_{19}-GA metabolism in intact higher plants. It is now timely to redirect efforts. A realistic goal for the next decade would be to endeavor to understand the control of C_{19}-GA formation during higher plant development.

ACKNOWLEDGMENTS

Thanks are due to Professor J. MacMillan FRS, Professor B. O. Phinney, and Dr. M. G. Jones for reading the manuscript, and to Miss M. E. Panes for typing it. Financial support from the Agricultural Research Council is gratefully acknowledged.

REFERENCES

Baldev, B., Lang, A., & Agatep, A. O. (1965). Gibberellin production in pea seeds developing in excised pods: Effect of growth retardant AMO-1618. Science *147*, 155-157.

Bandurski, R. S. (1980). Homeostatic control of concentrations of indole-3-acetic acid. In: Plant Growth Substances 1979. Proceedings of the 10th International Conference on Plant Growth Substances, pp. 37-49, Skoog, F., ed. Springer-Verlag, Berlin, Heidelberg, New York.

Barendse, G. W. M. (1974). Accumulation and metabolism of radioactive gibberellic acid in seedlings of *Pharbitis nil* Chois. In: Plant Growth Substances 1973. Proceedings of the 8th International Conference on Plant Growth Substances, pp. 332-341. Hirokawa, Tokyo.

Barendse, G. W. M., Kende, H., & Lang, A. (1968). Fate of radioactive gibberellin A_1 in maturing and germination seeds of peas and Japanese morning glory. Plant Physiol. *43*, 815-822.

Barendse, G. W. M., & de Klerk, G. J. M. (1975). The metabolism of applied gibberellic acid in *Pharbitis nil* Chois. Tentative identification of its sole metabolite as gibberellic acid glucoside and some of its properties. Planta *126*, 25-35.

Barendse, G. W. M., de Klerk, G. J. M., & v. Mierlo, J. (1977). Metabolism of radioactive gibberellins in *Pharbitis nil* Chois. In: Plant Growth Regulators, pp. 161-164, Kudrev, T., Ivanova, I., & Karanov, E., eds. Academy of Sciences, Sofia.

Barendse, G. W. M., & Lang, A. (1972). Comparison of endogenous gibberellins and of the fate of applied radioactive gibberellin A_1 in a normal and a dwarf strain of Japanese morning glory. Plant Physiol. *49*, 836-841.

Beale, M. H., Gaskin, P., Kirkwood, P. S., & MacMillan, J. (1980). Partial synthesis of gibberellin A_9 and [3α- and 3β-$^2H_1$]gibberellin A_9; gibberellin A_5 and [$1\beta,3$-$^2H_2$ and -$^3H_2$]gibberellin A_5; and gibberellin A_{20} and [$1\beta,3\alpha$-$^2H_2$ and -$^3H_2$]gibberellin A_{20}. J. Chem. Soc., Perkin Trans. I, 885-891.

Beale, M. H., & MacMillan, J. (1981a). Partial syntheses of [2α-^2H] and [2α-^3H]-

gibberellin A_{29} and [2α-^{2}H,-15,17-^{3}H$_4$]gibberellin A_{51} from gibberellin A_3. J. Chem. Soc., Perkin Trans. I, 394–400.

Beale, M. H., & MacMillan, J. (1981b). Preparation of 2- and 3-substituted gibberellins A_9 and A_4 for bioassay. Phytochem. *20*, 693–701.

Bearder, J. R. (1980). Plant hormones and other growth substances—their background, structures and occurrence. In: Hormonal Regulation of Development I. Molecular Aspects of Plant Hormones. Encyclopedia of Plant Physiology New Series, vol. 9, pp. 9–112, MacMillan, J., ed. Springer-Verlag, Berlin, Heidelberg, New York.

Bearder, J. R., Dennis, F. G., MacMillan, J., Martin, G. C., & Phinney, B. O. (1975a). A new gibberellin (A_{45}) from seed of *Pyrus communis* L. Tetrahedron Lett., 669–670.

Bearder, J. R., MacMillan, J., Wels, C. M., & Phinney, B. O. (1975b). The metabolism of steviol 13-hydroxylated *ent*-gibberellanes and *ent*-kauranes. Phytochem. *14*, 1741-1748.

Blechschmidt, S., Castel, U., Gaskin, P., Hedden, P., Graebe, J. E., & MacMillan, J. (in preparation). GC/MS analysis of the plant hormones in seeds of *Cucurbita maxima*.

Bowen, D. H., Crozier, A., MacMillan, J., & Reid, D. M. (1973). Characterization of gibberellins from light-grown *Phaseolus coccineus* seedlings by combined GC-MS. Phytochem. *12*, 2935-2941.

Bowen, D. H., MacMillan, J., & Graebe, J. E. (1972). Determination of specific radioactivity of [^{14}C]-compounds by mass spectroscopy. Phytochem. *11*, 2253-2257.

Bown, A. W., Reeve, D. R., & Crozier, A. (1975). The effect of light on the gibberellin metabolism and growth of *Phaseolus coccineus* seedlings. Planta *126*, 83-91.

Brian, P. W., & Hemming, H. G. (1955). The effect of gibberellic acid on shoot growth of pea seedlings. Physiol. Plant. *8*, 669-681.

Brøndbo, P. (1969). Induction of flowering by high temperature treatment in grafts of Norway Spruce (*Picea abies*/L./Karst.). Medd. Nor. Skogforsoksves *98*, 299-311.

Browning, G., & Saunders, P. F. (1977). Membrane localised gibberellins A_9 and A_4 in wheat chloroplasts. Nature (London) *265*, 375-377.

Ching, K. K., & Ching, T. M. (1959). Extracting Douglas-fir pollen and effects of gibberellic acid on its germination. Forest Sci. *5*, 74-80.

Cleland, C. F., & Zeevaart, J. A. D. (1970). Gibberellins in relation to flowering and stem elongation in the long day plant *Silene armeria.* Plant Physiol. *46,* 392–400.

Cooke, R. J., Saunders, P. F., & Kendrick, R. E. (1975). Red light induced production of gibberellin-like substances in homogenates of etiolated wheat leaves and in suspensions of intact etioplasts. Planta *124,* 319–328.

Crozier, A. (1980). Analysis of gibberellins: verification of accuracy. In: Gibberellins—Chemistry, Physiology and Use. British Plant Growth Regulator Group, Monograph 5, pp. 17–24, Lenton, J. R., ed. British Plant Growth Regulator Group, Wantage.

Crozier, A., Bowen, D. H., MacMillan, J., Reid, D. M., & Most, B. H. (1971). Characterization of gibberellins from dark-grown *Phaseolus coccineus* seedlings by gas liquid chromatography-mass spectrometry. Planta *97,* 142–154.

Crozier, A., & Reid, D. M. (1971). Do roots synthesize gibberellins? Can. J. Bot. *49,* 967–975.

Crozier, A., & Reid, D. M. (1972). Gibberellin metabolism in the roots of *Phaseolus coccineus* seedlings. In: Plant Growth Substances 1970. Proceedings of the 7th International Conference on Plant Growth Substances, pp. 414–419, Carr, D. J., ed. Springer-Verlag, Berlin, Heidelberg, New York.

Dathe, W., Schneider, G., & Sembdner, G. (1978). Endogenous gibberellins and inhibitors in caryopses of rye. Phytochem. *17,* 963–966.

Davies, P. J., Proebsting, W. M., & Gianfagna, T. J. (1977). Hormonal relationships in whole plant senescence. In: Plant Growth Regulation. Proceedings of the 9th International Conference on Plant Growth Substances, pp. 273–280, Pilet, P. E., ed. Springer-Verlag, Berlin, Heidelberg, New York.

Davies, L. J., & Rappaport, L. (1975a). Metabolism of tritiated gibberellins in *d-5* dwarf maize. 1. In excised tissues and intact dwarf and normal plants. Plant Physiol. *55,* 620–625.

Davies, L. J., & Rappaport, L. (1975b). Metabolism of tritiated gibberellins in *d-5* dwarf maize. II. [^{3}H]gibberellin A_1, [^{3}H]gibberellin A_3, and related compounds. Plant Physiol. *56,* 60–66.

Dunberg, A. (1976). Changes in gibberellin-like substances and indole-3-acetic acid in *Picea abies* during the period of shoot elongation. Physiol. Plant. *38,* 186–190.

Dunberg, A. (1980). Stimulation of flowering in *Picea abies* by gibberellins. Silvae Genet. *29*, 51-53.

Dunberg, A. (1981). Metabolism of tritiated gibberellin A_1 in seedlings of Norway spruce, *Picea abies*. Physiol. Plant. *51*, 349-352.

Dunberg, A., Malmberg, G., Sassa, T., & Pharis, R. P. (1983). Metabolism of GA_4 and GA_9 in *Picea abies*. Plant Physiol. *71*, 257-262.

Durley, R. C. (1983). Biosynthesis of gibberellins in higher plants. In: Aspects of Physiology and Biochemistry of Plant Hormones, Purohit, S. S., ed. Kalyani Publishers, Ludhiana (India).

Durley, R. C., Bewley, J. D., Railton, I. D., & Pharis, R. P. (1976). Effects of light, abscisic acid, and [6]N-benzyladenine on the metabolism of [^{3}H] gibberellin A_4 in seeds and seedlings of lettuce, cv. Grand Rapids. Plant Physiol. *57*, 699-703.

Durley, R. C., MacMillan, J., & Pryce, R. P. (1971). Investigation of gibberellins and other growth substances in the seed of *Phaseolus multiflorus* and of *Phaseolus vulgaris* by gas chromatography-mass spectrometry. Phytochem. *10*, 1891-1908.

Durley, R. C., & Pharis, R. P. (1973). Interconversion of gibberellin A_4 to gibberellin A_1 and A_{34} by dwarf rice, cultivar Tan-ginbozu. Planta *109*, 357-361.

Durley, R. C., Pharis, R. P., & Zeevaart, J. A. D. (1975). Metabolism of [^{3}H] gibberellin A_{20} by plants of *Bryophyllum daigremontianum* under long- and short-day conditions. Planta *126*, 139-149.

Durley, R. C., Railton, I. D., & Pharis, R. P. (1973). Interconversion of gibberellin A_5 to gibberellin A_3 in seedlings of dwarf *Pisum sativum*. Phytochem. *12*, 1609-1612.

Durley, R. C., Railton, I. D., & Pharis, R. P. (1974a). The metabolism of gibberellin A_1 and gibberellin A_{14} in seedlings of dwarf *Pisum sativum*. In: Plant Growth Substances 1973. Proceedings of the 8th International Conference on Plant Growth Substances, pp. 285-293. Hirokawa, Tokyo.

Durley, R. C., Railton, I. D., & Pharis, R. P. (1974b). Conversion of gibberellin A_{14} to other gibberellins in seedlings of dwarf *Pisum sativum*. Phytochem. *13*, 547-551.

Durley, R. C., Sassa, T., & Pharis, R. P. (1979). Metabolism of tritiated gibberellin A_{20} in immature seeds of dwarf pea, cv. Meteor. Plant Physiol. *64*, 214-219.

Evans, R., Hanson, J. R., & White, A. F. (1970). Studies in terpenoid biosynthesis. Part VI. The stereochemistry of some stages in tetracyclic diterpene biosynthesis. J. Chem. Soc. C, 2601–2603.

Evans, A., & Smith, H. (1976). Localization of phytochrome in etioplasts and its regulation *in vitro* of gibberellin levels. Proc. Nat. Acad. Sci. (U.S.A.) *73*, 138–142.

Fry, S. C., & Street, H. E. (1980). Gibberellin-sensitive suspension cultures. Plant Physiol. *65*, 472–477.

Frydman, V. M., Gaskin, P., & MacMillan, J. (1974). Qualitative and quantitative analyses of gibberellins throughout seed maturation in *Pisum sativum* cv. Progress No. 9. Planta *118*, 123–132.

Frydman, V. M., & MacMillan, J. (1973). Identification of gibberellins A_{20} and A_{29} in seed of *Pisum sativum* cv. Progress No. 9 by combined gas chromatography-mass spectrometry. Planta *115*, 11–15.

Frydman, V. M., & MacMillan, J. (1975). The metabolism of gibberellins A_9, A_{20} and A_{29} in immature seeds of *Pisum sativum* cv. Progress No. 9. Planta *125*, 181–195.

Gaskin, P., Gilmour, S. J., Lenton, J. R., MacMillan, J., & Sponsel, V. M. (1982). Endogenous gibberellins and related compounds in developing grain and germinating seedlings of barley. Abstract for the 11th International Conference on Plant Growth Substances, p. 50.

Gaskin, P., Gilmour, S. J., Lenton, J. R., MacMillan, J., & Sponsel, V. M. (1983). Endogenous gibberellins and kaurenoids identified from developing and germinating seedlings of barley. J. Plant Growth Reg., in press.

Gaskin, P., Hutchinson, M., Lewis, N., MacMillan, J., Phinney, B. O. (in preparation). Microbiological conversion of 11- and 12-oxygenated derivatives of *ent*-kaur-16-en-19-oic acid by *Gibberella fujikuroi*, mutant B1-41a.

Gaskin, P., Kirkwood, P. S., Lenton, J. R., MacMillan, J., & Radley, M. E. (1980). Identification of gibberellins in developing wheat grain. Agr. Biol. Chem. *44*, 1589–1593.

Gaskin, P., Kirkwood, P. S., & MacMillan, J. (1981). Partial synthesis of *ent*-13-hydroxy-2-oxo-20-norgibberella-1(10),16-diene-7,19-dioic acid, a catabolite of gibberellin A_{29}, and of related compounds. J. Chem. Soc., Perkin Trans. I., 1083–1091.

Gaskin, P., & MacMillan, J. (1975). Polyoxygenated *ent*-kauranes and water-soluble conjugates in seed of *Phaseolus coccineus.* Phytochem. *14*, 1575–1578.

Gaskin, P., MacMillan, J., & Zeevaart, J. A. D. (1973). Identification of gibberellin A_{20}, abscisic acid, and phaseic acid from flowering *Bryophyllum daigremontianum* by combined gas chromatography-mass spectrometry. Planta *111*, 347–352.

Gianfagna, T., Zeevart, J. A. D., & Lusk, W. J. (1983a). Effect of photoperiod on the metabolism of deuterium-labeled gibberellin A_{53} in spinach. Plant Physiol. *72*, 86–89.

Gianfagna, T., Zeevart, J. A. D., & Lusk, W. J. (1983b). Synthesis of deuterium-labeled gibberellins from steviol using the fungus *Gibberella fujikuroi.* Phytochem. *22*, 427–430.

Graebe, J. E. (1980). GA-biosynthesis: The development and application of cell-free systems for biosynthetic studies. In: Plant Growth Substances 1979. Proceedings of the 10th International Conference on Plant Growth Substances, pp. 180–187, Skoog, F., ed. Springer-Verlag, Berlin, Heidelberg, New York.

Graebe, J. E., Hedden, P., Gaskin, P., & MacMillan, J. (1974a). Biosynthesis of gibberellins A_{12}, A_{15}, A_{24}, A_{36} and A_{37} by a cell-free system from *Cucurbita maxima.* Phytochem. *13*, 1433–1440.

Graebe, J. E., Hedden, P., Gaskin, P., & MacMillan, J. (1974b). The biosynthesis of a C_{19}-gibberellin from mevalonic acid in a cell-free system from a higher plant. Planta, *120*, 307–309.

Graebe, J. E., Hedden, P., & Rademacher, W. (1980). Gibberellin biosynthesis. In: Gibberellins—Chemistry, Physiology and Use. British Plant Growth Regulator Group, Monograph 5, pp. 31–47, Lenton, J. R., ed. British Plant Growth Regulator Group, Wantage.

Graebe, J. E., & Ropers, H. J. (1978). Gibberellins. In: Phytohormones and related compounds: a comprehensive treatise. Vol. 1. The Biochemistry of Phytohormones and Related Compounds, pp. 107–204, Letham, D. S., Goodwin, P. B., & Higgins, T. J. V., eds. Elsevier, Amsterdam, Oxford, New York.

Hamnett, A. F., & Pratt, G. E. (1978). Use of automated capillary column radio gas chromatography in the identification of insect juvenile hormones. J. Chromatogr. *158*, 387–399.

Hartmann, M. A., Fonteneau, P., & Benveniste, P. (1977). Subcellular localization of sterol synthesizing enzymes in maize coleoptiles. Plant. Sci. Lett. *8*, 45–51.

Haruta, H., Yagi, H., Iwata, T., & Tamura, S. (1974). Syntheses and plant growth retardant activities of quaternary ammonium compounds derived from α-ionone and isophorone. Agr. Biol. Chem. *38*, 417–422.

Hashizume, H. (1959). The effect of gibberellin upon flower formation in *Cryptomeria japonica*. J. Jpn. For. Sci. *41*, 375–381.

Hedden, P., Graebe, J. E., Beale, M. H., Gaskin, P., & Macmillan, J. (in preparation). The biosynthesis of 12α-hydroxylated gibberellins in a cell-free system from *Cucurbita maxima* endosperm.

Hedden, P., MacMillan, J., & Phinney, B. O. (1978). The metabolism of the gibberellins. Ann. Rev. Plant Physiol. *29*, 149–192.

Hedden, P., & Phinney, B. O. (1979). Comparison of *ent*-kaurene and *ent*-isokaurene synthesis in cell-free systems from etiolated shoots of normal and *dwarf-5* maize seedlings. Phytochem. *18*, 1475–1479.

Hedden, P., Phinney, B. O., Heupel, R., Fujii, D., Cohen, H., Gaskin, P., MacMillan, J., & Graebe, J. E. (1982). Hormones of young tassels of *Zea mays* L. Phytochem. *21*, 391–394.

Hedden, P., Phinney, B. O., MacMillan, J., & Sponsel, V. M. (1977). Metabolism of kaurenoids by *Gibberella fujikuroi* in the presence of the plant growth retardant, *N,N,N*-trimethyl-1-methyl-(2′,6′,6′-trimethylcyclohex-2′-en-1′-yl)prop-2-enylammonium iodide. Phytochem. *16*, 1913–1917.

Helgeson, J. P., & Upper, C. D. (1970). Modification of logarithmic growth rates of tobacco callus tissue by gibberellic acid. Plant Physiol. *46*, 113–117.

Hiraga, K., Kawabe, S., Yokota, T., Murofushi, N., & Takahashi, N. (1974a). Isolation and characterization of plant growth substances in immature seeds and etiolated seedlings of *Phaseolus vulgaris*. Agr. Biol. Chem. *38*, 2521–2527.

Hiraga, K., Yamane, H., & Takahashi, N. (1974b). Biological activity of some synthetic gibberellin glucosyl esters. Phytochem. *13*, 2371–2376.

Hiraga, K., Yokota, T., Murofushi, N., & Takahashi, N. (1972). Isolation and characterization of a free gibberellin and glucosyl esters of gibberellins in mature seeds of *Phaseolus vulgaris*. Agr. Biol. Chem. *36*, 345—347.

Hiraga, K., Yokota, T., Murofushi. N., & Takahashi, N. (1974c). Isolation and characterization of gibberellins in mature seeds of *Phaseolus vulgaris*. Agr. Biol. Chem. *38*, 2511-2520.

Hoad, G. V., Phinney, B. O., Sponsel, V. M., & MacMillan, J. (1981). The biological activity of sixteen gibberellin A_4 and gibberellin A_9 derivatives using seven bioassays. Phytochem. *20*, 703-713.

Ingram, T. J. (1980). Gibberellins and reproductive development in peas. Ph.D. Thesis, University of East Anglia.

Ingram, T. J., & Browning, G. (1979). Influence of photoperiod on seed develop-ment in the genetic line of peas G2 and its relation to changes in endo-genous gibberellins measured by combined gas chromatography-mass spectrometry. Planta *146*, 423-432.

Jones, M. G., Metzger, J. D., & Zeevaart, J. A. D. (1980). Fractionation of gibberellins in plant extracts by reverse phase high performance liquid chromatography. Plant Physiol. *65*, 218-221.

Jones, M. G., & Zeevaart, J. A. D. (1980a). Gibberellins and the photoperiodic control of stem elongation in the long-day plant *Agrostemma githago* L. Planta *149*, 269-273.

Jones, M. G., & Zeevaart, J. A. D. (1980b). The effect of photoperiod on the levels of seven endogenous gibberellins in the long-day plant *Agrostemma githago* L. Planta *149*, 274-279.

Jones, M. G., & Zeevaart, J. A. D. (1982). Effect of photoperiod on metabolism of tritiated gibberellins applied to plants of *Agrostemma githago* L. Plant Physiol. *69*, 660-662.

Jones, R. L. (1973). Gibberellins: Their physiological role. Ann. Rev. Plant Physiol. *24*, 571-598.

Jones, R. L., & Lang, A. (1968). Extractable and diffusible gibberellins from light- and dark-grown pea seedlings. Plant Physiol. *43*, 629-634.

Kamienska, A., Durley, R. C., & Pharis, R. P. (1976a). Isolation of gibberellins A_3, A_4 and A_7 from *Pinus attenuata* pollen. Phytochem. *15*, 421-424.

Kamienska, A., Durley, R. C., & Pharis, R. P. (1976b). Endogenous gibberellins of pine pollen. III. Conversion of 1,2-[^{3}H]GA_4 to gibberellins A_1 and A_{34} in germinating pollen of *Pinus attenuata* Lemm. Plant Physiol. *58*, 68-70.

Kamienska, A., Pharis, R. P., Wample, R. L., Kuo, C. C., & Durley, R. C. (1974). Gibberellins in conifers. In: Plant Growth Substances 1973. Proceedings of the 8th International Conference on Plant Growth Substances, pp. 305-313. Hirokawa, Tokyo.

Kamiya, Y., & Graebe, J. E. (1983). The biosynthesis of all major pea gibberellins in a cell-free system from *Pisum sativum*. *Phytochem.* 22, 681–690.

Katsumi, M., Phinney, B. O., Jefferies, P. R., & Henrick, C. A. (1964). Growth response of the d-5 and an-1 mutants of maize to some kaurene derivatives. Science *144*, 849-850.

Kaufman, P. B., Ghosheh, N. S., Nakosteen, L., Pharis, R. P., Durley, R. C., & Morf, W. (1976). Analysis of native gibberellins in the internode, nodes, leaves, and inflorescence of developing *Avena* plants. Plant Physiol. *58*, 131-134.

Kende, H. (1967). Preparation of radioactive gibberellin A_1 and its metabolism in dwarf peas. Plant Physiol. *42*, 1612-1618.

Kende, H., & Lang, A. (1964). Gibberellins and light inhibition of stem growth in peas. Plant Physiol. *39*, 435-440.

Kirkwood, P. S. (1979). The partial synthesis of gibberellins and related compounds in *Pisum sativum*. Ph.D. Thesis, University of Bristol.

Köhler, D., & Lang, A. (1963). Evidence for substances in higher plants interfering with response of dwarf peas to gibberellin. Plant Physiol. *38*, 555-560.

Kurogochi, S., Murofushi, N., Ota, Y., & Takahashi, N. (1978). Gibberellins and inhibitors in the rice plant. Agr. Biol. Chem. *42*, 207-208.

Lance, B., Durley, R. C., Reid, D. M., Thorpe, T. A., & Pharis, R. P. (1976a). Metabolism of [^{3}H]gibberellin A_{20} in light- and dark-grown tobacco callus cultures. Plant Physiol. *58*, 387-392.

Lance, B., Reid, D. M., & Thorpe, T. A. (1976b). Endogenous gibberellins and growth of tobacco callus cultures. Physiol. Plant. *36*, 287-292.

Lang, A. (1970). Gibberellins: structure and metabolism. Ann. Rev. Plant Physiol. *21*, 537-570.

Liebisch, H. W. (1974). Uptake, translocation, and metabolism of labelled GA_3 glucosyl ester. In: Biochemistry and Chemistry of Plant Growth Regulators, pp. 109-113, Schreiber, K., Schütte, H. R., & Sembdner, G., eds. Institute of Plant Biochemistry, Halle.

Lockhart, J. A. (1956). Reversal of the light inhibition of pea stem growth by gibberellins. Proc. Nat. Acad. Sci. (U.S.A.) *42*, 841–848.

Lorenzi, R., Horgan, R., & Heald, J. K. (1975). Gibberellins in *Picea sitchensis* Carriers: seasonal variation and partial characterization. Planta *126*, 75–82.

Lorenzi, R., Horgan, R., & Heald, J. K. (1976). Gibberellin A$_9$ glucosyl ester in needles of *Picea sitchensis*. Phytochem. *15*, 789–790.

Lorenzi, R., Saunders, P. F., Heald, J. K., & Horgan, R. (1977). A novel gibberellin from needles of *Picea sitchensis*. Plant Sci. Lett. *8*, 179–182.

Lunnon, M. W., MacMillan, J., & Phinney, B. O. (1977). Fungal products. Part XX. Transformations of 2- and 3-hydroxylated kaurenoids by *Gibberella fujikuroi*. J. Chem. Soc., Perkin Trans. I, 2308–2316.

MacMillan, J. (1977). Some aspects of gibberellin metabolism in higher plants. In: Plant Growth Regulation. Proceedings of the 9th International Conference on Plant Growth Substances, pp. 129–138, Pilet, P. E., ed. Springer-Verlag, Berlin, Heidelberg, New York.

MacMillan, J. (1978). Gibberellin metabolism. Pure Appl. Chem. *50*, 995–1004.

MacMillan, J., & Wels, C. M. (1974). Detailed analysis of metabolites from mevalonic lactone in *Gibberella fujikuroi*. Phytochem. *13*, 1413–1417.

Marx, G. A. (1968). Influence of genotype and environment on senescence in peas, *Pisum sativum* L. Bioscience *18*, 505–506.

McInnes, A. G., Smith, D. G., Durley, R. C., Pharis, R. P., Arsenault, G. P., MacMillan, J., Gaskin, P., & Vining, L. C. (1977). Biosynthesis of gibberellins in *Gibberella fujikuroi*. Gibberellin A$_{47}$. Can. J. Biochem. *55*, 728–735.

Metzger, J. D., & Zeevaart, J. A. D. (1980a). Identification of six endogenous gibberellins in spinach shoots. Plant Physiol. *65*, 623–626.

Metzger, J. D., & Zeevaart, J. A. D. (1980b). Effect of photoperiod on the levels of endogenous gibberellins in spinach as measured by combined gas chromatography-selected ion current monitoring. Plant Physiol. *66*, 844–846.

Müller, P., Knöfel, H.-D., & Sembdner, G. (1974). Studies on the enzymatical synthesis of gibberellin-O-glucosides. In: Biochemistry and Chemistry of Plant Growth Regulators, pp. 115–119, Schreiber, K., Schütte, H. R., & Sembdner, G., eds. Institute of Plant Biochemistry, Halle.

Murashige, T. (1964). Analysis of the inhibition of organ formation in tobacco tissue culture by gibberellin. Physiol. Plant. *17*, 636–643.

Murfet, I. C. (1971a). Flowering in *Pisum*. A three-gene system. Heredity *27*, 93–110.

Murfet, I. C. (1971b). Flowering in *Pisum*: Reciprocal grafts between known genotypes. Aust. J. Biol. Sci. *24*, 1089–1101.

Murfet, I. C. (1973). Flowering in *Pisum*. *Hr*, a gene for high response to photoperiod. Heredity *31*, 157–164.

Murfet, I. C. (1975). Flowering in *Pisum*: Multiple alleles at the *lf* locus. Heredity *35*, 85–98.

Murfet, I. C. (1977a). Environmental interaction and the genetics of flowering. Ann. Rev. Plant Physiol. *28*, 253–278.

Murfet, I. C. (1977b). The physiological genetics of flowering. In: The Physiology of the Garden Pea, pp. 385–430, Sutcliffe, J. F., & Pate, J. S., eds. Academic Press, London, New York, San Francisco.

Murofushi, N., Durley, R. C., & Pharis, R. P. (1974). Preparation of radioactive gibberellins A_{20}, A_5 and A_8. Agr. Biol. Chem. *38*, 475–476.

Murofushi, N., Durley, R. C., & Pharis, R. P. (1977). Preparation of radioactive gibberellins A_{20}, A_5 and A_8. Agr. Biol. Chem. *41*, 1075–1079.

Murofushi, N., Takahashi, N., Yokota, T., & Tamura, S. (1968). Gibberellins in immature seeds of *Pharbitis nil*. Part I. Isolation and structure of a novel gibberellin, gibberellin A_{20}. Agr. Biol. Chem. *32*, 1239–1245.

Musgrave, A., Kays, S. E., & Kende, H. (1972). Uptake and metabolism of radioactive gibberellins by barley aleurone layers. Planta *102*, 1–10.

Musgrave, A., & Kende, H. (1970). Radioactive gibberellin A_5 and its metabolism in dwarf peas. Plant Physiol. *45*, 56–61.

Nadeau, R., & Rappaport, L. (1972). Metabolism of gibberellin A_1 in germinating bean seeds. Phytochem. *11*, 1611–1616.

Nadeau, R., & Rappaport, L. (1974). An amphoteric conjugate of [^{3}H] gibberellin A_1 from barley aleurone layers. Plant Physiol. *54*, 809–812.

Nadeau, R., Rappaport, L., & Stolp, C. F. (1972). Uptake and metabolism of ^{3}H-gibberellin A_1 by barley aleurone layers: response to abscisic acid. Planta *107*, 315–324.

Nash, L. J., & Crozier, A. (1975). Translocation and metabolism of [³H] gibberellins by light-grown *Phaseolus coccineus* seedlings. Planta *127*, 221-231.

Nash, L. J., Jones, R. L., & Stoddart, J. L. (1978). Gibberellin metabolism in excised lettuce hypocotyls: Response to GA_9 and the conversion of [³H]GA_9. Planta *140*, 143-150.

Nickell, L. G., & Tulecke, W. (1959). Responses of plant tissue cultures to gibberellin. Bot. Gaz. *120*, 245-250.

Nitsch, C., & Nitsch, J. (1963). Etude du mode d'action de l'acide gibbérellique au moyen de gibbérelline A_3 marquée. 1. Action sur les substances de croissance du haricot nain. Bull. Soc. Bot. France *110*, 7-17.

Ogawa, Y. (1962). Quantitative difference of gibberellin-like substances in normal and dwarf varieties of *Pharbitis nil* Chois. Bot. Mag. *78*, 474-480.

Ohlrogge, J. B., Garcia-Martinez, J. L., Adams, D., & Rappaport, L. (1980). Uptake and subcellular compartmentation of gibberellin A_1 applied to leaves of barley and cowpea. Plant Physiol. *66*, 422-427.

Owens, J. N., & Pharis, R. P. (1967). Initiation and ontogeny of the microsporangiate cone in *Cupressus arizonica* in response to gibberellin. Amer. J. Bot. *54*, 1260-1272.

Patterson, R. J., & Rappaport, L. (1974). The conversion of gibberellin A_1 to gibberellin A_8 by a cell-free enzyme system. Planta *119*, 183-191.

Patterson, R., Rappaport, L., & Breidenbach, R. W. (1975). Characterization of an enzyme from *Phaseolus vulgaris* seeds which hydroxylates GA_1 to GA_8. Phytochem. *14*, 363-368.

Pharis, R. P., & Kuo, C. G. (1977). Physiology of gibberellins in conifers. Can. J. For. Res. 7, 299-325.

Pharis, R. P., Ross, S. D., & McMullan, E. (1980). Promotion of flowering in the Pinaceae by gibberellins. III. Seedlings of Douglas fir. Physiol. Plant. *50*, 119-126.

Phillips, I. D. J., & Jones, R. L. (1964). Gibberellin-like activity in bleeding-sap of root systems of *Helianthus annuus* detected by a new dwarf pea epicotyl assay and other methods. Planta *63*, 269-278.

Phinney, B. O. (1956). The growth response of single gene dwarf mutants of *Zea mays* to gibberellic acid. Proc. Nat. Acad. Sci. (U.S.A.) *42*, 185-189.

Phinney, B. O. (1961). Dwarfing genes in *Zea mays* and their relation to the gibberellins. In: Plant Growth Regulation, pp. 489–501, Klein, R. M., ed. Iowa State University Press, Ames.

Phinney, B. O. (1982). Gibberellin (GA) metabolism in relation to dwarfism in *Zea mays* L. Abstract for the 11th International Conference on Plant Growth Substances, p. 3.

Pitel, D. W., Vining, L. C., & Arsenault, G. P. (1971). Biosynthesis of gibberellins in *Gibberella fujikuroi*. The sequence after gibberellin A_4. Can. J. Biochem. *49*, 194–200.

Potts, W. C., Reid, J. B., & Murfet, J. C. (1982). Internode length in *Pisum* 1. The effect of the *Le/le* gene difference on endogenous gibberellin-like substances. Physiol. Plantarum *55*, 323–328.

Proebsting, W. M., Davies, P. J., & Marx, G. A. (1976). Photoperiodic control of apical senescence in a genetic line of peas. Plant Physiol. *58*, 800–802.

Proebsting, W. M., Davies, P. J., & Marx, G. A. (1977). Evidence for a graft-transmissible substance which delays apical senescence in *Pisum sativum* L. Planta *135*, 93–94.

Proebsting, W. M., Davies, P. J., & Marx, G. A. (1978). Photoperiod-induced changes in gibberellin metabolism in relation to apical growth and senescence in genetic lines of peas (*Pisum sativum* L.). Planta *141*, 231–238.

Proebsting, W. M., & Heftmann, E. (1980). The relationship of $[^3H]GA_9$ metabolism to photoperiod-induced flowering in *Pisum sativum* L. Z. Pflanzenphysiol. *98*, 305–309.

Railton, I. D. (1974a). Studies on gibberellins in shoots of light-grown peas. II. The metabolism of tritiated gibberellin A_9 and gibberellin A_{20} by light- and dark-grown shoots of dwarf *Pisum sativum* var. Meteor. Plant Sci. Lett. *3*, 207–212.

Railton, I. D. (1974b). Effects of N^6-benzyladenine on the rate of turnover of $[^3H]GA_{20}$ by shoots of dwarf *Pisum sativum*. Planta *120*, 197–200.

Railton, I. D. (1976a). Aspects of gibberellin biosynthesis in higher plants. S. Afr. J. Sci. *72*, 371–377.

Railton, I. D. (1976b). The preparation of 2,3 $[^3H]$-GA_{29} and its metabolism by etiolated seedlings and germinating seeds of dwarf *Pisum sativum* (Meteor). J. S. Afr. Bot. *42*, 147–156.

Railton, I. D. (1977a). 16,17-Dihydro 16,17-dihydroxy gibberellin A_9: a metabolite of [^{3}H]gibberellin A_9 in chloroplast sonicates from *Pisum sativum* var. "Alaska." Z. Pflanzenphysiol. *81*, 323-329.

Railton, I. D. (1977b). Gibberellin metabolism in chloroplasts of *Pisum sativum* L. var. Alaska. S. Afr. J. Sci. *73*, 22-23.

Railton, I. D. (1977c). Transport of 2,3[^{3}H]-gibberellin A_{20} into chloroplasts of *Pisum sativum* L. var. Meteor. S. Afr. J. Sci. *73*, 149-150.

Railton, I. D. (1979). The influence of the root system on the metabolism of 2,3[^{3}H]-gibberellin A_{20} by shoots of *Phaseolus coccineus* L. var. Prizewinner. Z. Pflanzenphysiol. *91*, 283-290.

Railton, I. D. (1980a). Regulation of gibberellin A_{20}-induced growth by 2,β-hydroxylation in seedlings of dwarf pea. S. Afr. J. Sci. *76*, 33-35.

Railton, I. D. (1980b). A study of the role of 2,β-hydroxylation in the control of C-19 gibberellin levels during seedling growth. Z. Pflanzenphysiol. *96*, 103-114.

Railton, I. D., Durley, R. C., & Pharis, R. P. (1973). Interconversion of gibberellin A_1 to gibberellin A_8 in seedlings of dwarf *Oryza sativa*. Phytochem. *12*, 2351-2352.

Railton, I. D., Durley, R. C., & Pharis, R. P. (1974a). Studies on gibberellin biosynthesis in etiolated shoots of dwarf pea, cv. Meteor. In: Plant Growth Substances 1973. Proceedings of the 8th International Conference on Plant Growth Substances, pp. 294-304. Hirokawa, Tokyo.

Railton, I. D., Durley, R. C., & Pharis, R. P. (1974b). Metabolism of tritiated gibberellin A_9 by shoots of dark-grown dwarf pea, cv. Meteor. Plant Physiol. *54*, 6-12.

Railton, I. D., Murofushi, N., Durley, R. C., & Pharis, R. P. (1974c). Interconversion of gibberellin A_{20} to gibberellin A_{29} by etiolated seedlings and germinating seeds of dwarf *Pisum sativum*. Phytochem. *13*, 793-796.

Railton, I. D., & Rechav, M. (1979). Efficiency of extraction of gibberellin-like substances from chloroplasts of *Pisum sativum* L. Plant Sci. Lett. *14*, 75-78.

Railton, I. D., & Reid, D. M. (1973). Effects of benzyladenine on the growth of waterlogged tomato plants. Planta *111*, 261-266.

Railton, I. D., & Reid, D. M. (1974a). Studies on gibberellins in shoots of light-grown peas I. A re-evaluation of the data. Plant Sci. Lett. *2*, 157-163.

Railton, I. D., & Reid, D. M. (1974b). Studies on gibberellins in shoots of light-grown peas III. Interconversion of [^{3}H]GA$_9$ and [^{3}H]GA$_{20}$ to other gibberellins by an *in vitro* system derived from chloroplasts of *Pisum sativum*. Plant Sci. Lett. *3*, 303–308.

Rappaport, L., & Adams, D. (1978). Gibberellins: Synthesis, compartmentation and physiological process. Phil. Trans. Roy. Soc. London, Ser. B *284*, 521–539.

Rappaport, L., Davies, L., Lavee, S., Nadeau, R., Patterson, R., & Stolp, C. F. (1974). Significance of metabolism of [^{3}H]GA$_1$ for plant regulation. In: Plant Growth Substances 1973. Proceedings of the 8th International Conference on Plant Growth Substances, pp. 314–324. Hirokawa, Tokyo.

Rappaport, L., Hsu, A., Thompson, R., & Yang, S. F. (1967). Fate of ^{14}C-gibberellin A$_3$ in plant tissues. Ann. N.Y. Acad. Sci. *144*, 211–218.

Reeve, D. R., & Crozier, A. (1974). An assessment of gibberellin structure-activity relationships. J. Exp. Bot. *25*, 431–445.

Reeve, D. R., & Crozier, A. (1977). Radioactivity monitor for high-performance liquid chromatography. J. Chromatogr. *137*, 271–282.

Reeve, D. R., Crozier, A. (1980). Quantitative analysis of plant hormones. In: Hormonal Regulation of Development I. Molecular Aspects of Plant Hormones. Encyclopedia of Plant Physiology New Series, vol. 9, pp. 203–280, MacMillan, J., ed. Springer-Verlag, Berlin, Heidelberg, New York.

Reeve, D. R., Crozier, A., Durley, R. C., Reid, D. M., & Pharis, R. P. (1975). Metabolism of ^{3}H-gibberellin A$_1$ and ^{3}H-gibberellin A$_4$ by *Phaseolus coccineus* seedlings. Plant Physiol. *55*, 42–44.

Reid, D. M., & Crozier, A. (1971). Effects of waterlogging on the gibberellin content and growth of tomato plants. J. Exp. Bot. *22*, 39–48.

Reid, D. M., & Railton, I. D. (1974). The influence of benzyladenine on the growth and gibberellin content of shoots of waterlogged tomato plants. Plant Sci. Lett. *2*, 151–156.

Reiner, J. M. (1953a). The study of metabolic turnover rates by means of isotopic tracers. I. Fundamental relations. Arch. Biochem. Biophys. *46*, 53–79.

Reiner, J. M. (1953b). The study of metabolic turnover rates by means of isotopic tracers. II. Turnover in a simple reaction system. Arch. Biochem. Biophys. *46*, 80–99.

Rogers, L. J., Shah, S. P. J., & Goodwin, T. W. (1966). Intracellular localization of mevalonate-activating enzymes in plant cells. Biochem. J. *99*, 381-388.

Ropers, H.-J., Graebe, J. E., Gaskin, P., & MacMillan, J. (1978). Gibberellin biosynthesis in a cell-free system from immature seeds of *Pisum sativum*. Biochem. Biophys. Res. Commun. *80*, 690-697.

Ross, S. D., & Pharis, R P. (1976). Promotion of flowering in the Pinaceae by gibberellins I. Sexually mature, non-flowering grafts of Douglas-fir. Physiol. Plant. *36*, 182-186.

Schreiber, K., Weiland, J., & Sembdner, G. (1967). Isolierung und Struktur eines Gibberellinglucosids. Tetrahedron Lett. 4285—4288.

Schreiber, K., Weiland, J., & Sembdner, G. (1970). Isolierung von Gibberellin—A_8—$O(3)$-β-D-glucopyranosid aus Früchten von *Phaseolus coccineus*. Phytochem. *9*, 189-198.

Sembdner, G., Borgmann, E., Schneider, G., Liebisch, H.-W., Miersch, O., Adam, G., Lischewski, M., & Schreiber, K. (1976). Biological activity of some conjugated gibberellins. Planta *132*, 249-257.

Sembdner, G., Gross, D., Liebisch, H.-W., & Schneider, G. (1980). Biosynthesis and metabolism of plant hormones. In: Hormonal Regulation of Development I. Molecular Aspects of Plant Hormones. Encyclopedia of Plant Physiology New Series, vol. 9, pp. 281-444, MacMillan, J., ed. Springer-Verlag, Berlin, Heidelberg, New York.

Sembdner, G., Weiland, J., Aurich, O., & Schreiber, K. (1968). Isolation, structure and metabolism of a gibberellin glucoside. In: Plant Growth Regulators. S.C.I. Monograph 31, pp. 70-86. Society of Chemical Industry, London.

Silk, W. K., & Jones, R. L. (1975). Gibberellin response in lettuce hypocotyl sections. Plant Physiol. *56*, 267-272.

Silk, W. K., Jones, R. L., & Stoddart, J. L. (1977). Growth and gibberellin A_1 metabolism in excised lettuce hypocotyls. Plant Physiol *59*, 211-216.

Simpson, G. M., & Saunders, P. F. (1972). Abscisic acid associated with wilting in dwarf and tall *Pisum sativum*. Planta *102*, 272-276.

Sponsel, V. M. (1980a). Gibberellin metabolism in legume seeds. In: Gibberellins —Chemistry, Physiology and Use. British Plant Growth Regulator Group, Monograph 5, pp. 49-62, Lenton, J. R., ed. British Plant Growth Regulator Group, Wantage.

Sponsel, V. M. (1980b). Metabolism of gibberellins in immature seeds of *Pisum sativum*. In: Plant Growth Substances 1979. Proceedings of the 10th International Conference on Plant Growth Substances, pp. 170–179, Skoog, F., ed. Springer-Verlag, Berlin, Heidelberg, New York.

Sponsel, V. M. (in preparation). The localization, metabolism and biological activity of gibberellins in maturing and germinating seeds of *Pisum sativum* cv. Progress No. 9.

Sponsel, V. M., Gaskin, P., & MacMillan, J. (1979). The identification of gibberellins in immature seeds of *Vicia faba*, and some chemotaxonomic considerations. Planta *146*, 101–105.

Sponsel, V. M., & MacMillan, J. (1976). The metabolism of gibberellins in immature seeds of *Pisum sativum* cv. Progress No. 9. In: The 9th International Conference on Plant Growth Substances. Collected abstracts of the paper demonstrations, pp. 366–368, Pilet, P.-E., ed. Lausanne.

Sponsel, V. M., & MacMillan, J. (1977). Further studies on the metabolism of gibberellins (GAs) A_9, A_{20} and A_{29} in immature seeds of *Pisum sativum* cv. Progress No. 9. Planta *135*, 129–136.

Sponsel, V. M., & MacMillan, J. (1978). Metabolism of gibberellin A_{29} in seeds of *Pisum sativum* cv. Progress No. 9; use of [^{2}H] and [^{3}H] GAs, and the identification of a new GA catabolite. Planta *144*, 69–78.

Sponsel, V. M., & MacMillan, J. (1980). Metabolism of [$^{13}C_1$]gibberellin A_{29} to [$^{13}C_1$]gibberellin-catabolite in maturing seeds of *Pisum sativum* cv. Progress No. 9. Planta *150*, 46–52.

Stoddart, J. L. (1968). The association of gibberellin-like activity with the chloroplast fraction of leaf homogenates. Planta *81*, 106–112.

Stoddart, J. L. (1979a). Interaction of [^{3}H]gibberellin A_1 with a subcellular fraction from lettuce (*Lactuca sativa* L.) hypocotyls. I. Kinetics of labelling. Planta *146*, 353–361.

Stoddart, J. L. (1979b). Interaction of [^{3}H]gibberellin A_1 with a subcellular fraction from lettuce (*Lactuca sativa* L.) hypocotyls. II. Stability and properties of the association. Planta *146*, 363–368.

Stoddart, J. L., & Jones, R. L. (1977). Gibberellin metabolism in excised lettuce hypocotyls: evidence for the formation of gibberellin A_1 glucosyl conjugates. Planta *136*, 261–269.

Stoddart, J. L., & Venis, M. A. (1980). Molecular and sub-cellular aspects of hormone action. In: Hormonal Regulation of Development I. Molecular

Aspects of Plant Hormones. Encyclopedia of Plant Physiology New Series, vol. 9, pp. 445-510, MacMillan, J., ed. Springer-Verlag, Berlin, Heidelberg, New York.

Stolp, C. F., Nadeau, R., & Rappaport, L. (1973). Effect of abscisic acid on uptake and metabolism of [^{3}H]gibberellin A_1 and [^{3}H]pseudogibberellin A_1 by barley half-seeds. Plant Physiol. *52*, 546-548.

Stolp, C. F., Nadeau, R., & Rappaport, L. (1977). Abscisic acid and the accumulation, biological activity and metabolism of four derivatives of [^{3}H]-gibberellin A_1 in barley aleurone layers. Plant Cell Physiol. *18*, 721-728.

Tanabe, M. (1973). Stable isotopes in biosynthetic studies. In: Biosynthesis, vol. 2, pp. 241-299, Geissmann, T. A., ed. Chemical Society, London.

Taylor, C. M., & Railton, I. D. (1977). The influence of wilting and abscisic acid application on gibberellin interconversion in etiolated seedlings of dwarf *Pisum sativum* var. Meteor. Plant Sci. Lett. *9*, 317-322.

Tompsett, P. B., & Fletcher, A. M. (1979). Promotion of flowering on mature *Picea sitchensis* by gibberellin and environmental treatments. The influence of timing and hormonal concentration. Physiol. Plant. *45*, 112-116.

Wample, R. L., Durley, R. C., & Pharis, R. P. (1975). Metabolism of gibberellin A_4 by vegetative shoots of Douglas fir at three stages of ontogeny. Physiol. Plant. *35*, 273-278.

Wellburn, A. R., & Hampp, R. (1976). Uptake of mevalonate and acetate during plastid development. Biochem. J. *158*, 231-233.

Wels, C. M. (1977). High-sensitivity method of radio gas chromatography for ^{3}H- and ^{14}C-labelled compounds. J. Chromatogr. *142*, 459-467.

Wheeler, N. C., Wample, R. L., & Pharis, R. P. (1980). Promotion of flowering in the Pinaceae by gibberellins. IV. Seedlings and sexually mature grafts of lodgepole pine. Physiol. Plant. *50*, 340-346.

Yamane, H., Murofushi, N., Osada, H., & Takahashi, N. (1977). Metabolism of gibberellins in early immature bean seeds. Phytochem. *16*, 831-835.

Yamane, H., Murofushi, N., & Takahashi, N. (1975). Metabolism of gibberellins in maturing and germinating bean seeds. Phytochem. *14*, 1195-1200.

Yokota, T., Murofushi, N., & Takahashi, N. (1980). Extraction, purification and identification. In: Hormonal Regulation of Development I. Molecular Aspects of Plant Hormones. Encyclopedia of Plant Physiology New Series, vol. 9, pp. 113-202, MacMillan, J., ed. Springer-Verlag, Berlin, Heidelberg, New York.

Yokota, T., Murofushi, N., Takahashi, N., & Katsumi, M. (1971a). Biological activities of gibberellins and their glucosides in *Pharbitis nil.* Phytochem. *10*, 2943-2949.

Yokota, T., Murofushi, N., Takahashi, N., & Tamura, S. (1971b). Gibberellins in immature seeds of *Pharbitis nil.* Part II. Isolation and structures of novel gibberellins, gibberellins A_{26} and A_{27}. Agr. Biol. Chem. *35*, 573-582.

Yokota, T., Murofushi, N., Takahashi, N., & Tamura, S. (1971c). Gibberellins in immature seeds of *Pharbitis nil.* Part III. Isolation and structures of gibberellin glycosides. Agr. Biol. Chem. *35*, 583-595.

Zeevaart, J. A. D. (1971). Effects of photoperiod on growth rate and endogenous gibberellins in the long-day rosette plant spinach. Plant Physiol. *47*, 821-827.

Zeevaart, J. A. D. (1973). Gibberellin A_{20} content of *Bryophyllum daigremontianum* under different photoperiodic conditions as determined by gas-liquid chromatography. Planta *114*, 285-288.

5

In Vivo Diterpenoid Biosynthesis in *Gibberella Fujikuroi:* The Pathway after *Ent*-kaurene

John R. Bearder

5.1 INTRODUCTION

This chapter is in two main parts. The first concerns the biosynthesis and metabolism of established diterpenoid metabolites of the fungus *Giberella fujikuroi*, while the second deals with the metabolism of other compounds that are structural analogues of known metabolites. The limited knowledge of the diterpenoid biosynthesis in the fungus *Sphaceloma manihoticola* is discussed in a final part. Both main parts are subdivided into sections according to the experimental technique used: (a) radiotracer methods, (b) mutant strains of the fungus, and (c) "wild-type" fungi in conjunction with plant growth retardants.

In all sections a roughly chronological order is followed, although this has been relaxed where expedient.

A full list of gibberellins (GAs) has been compiled by Bearder (1980), and the structures of the diterpenoid metabolites of *G. fujikuroi* which have been implicated in biosynthetic studies are illustrated in Fig. 5.1. Inspection of Fig. 5.1 shows that the metabolites of *G. fujikuroi* can be grouped into three categories based on their carbon skeletons and biosynthetic origins. One group contains structures retaining the *ent*-kaurane skeleton (*13*),* the second group contains the C_{20}-GAs in which ring B of the *ent*-kaurane skeleton has undergone ring contraction to give the *ent*-gibberellane (*14*) skeleton, and the third category includes the C_{19}-GAs that have lost C-20 and contain only 19 of the original 20 diterpenoid carbon atoms. These C_{19}-GAs are characterized by the $(19{\rightarrow}10)$-γ-lactone function and generally exhibit potent biological activity (see Volume II, Chapter 2).

Previously, reviews covering aspects of diterpenoid biosynthesis in *G. fujikuroi* have been written by Cross (1968), Lang (1970), MacMillan (1971), Hanson (1971), McCorkindale (1976), Bearder and Sponsel (1977), Graebe and Ropers (1978), and Hedden, MacMillan, and Phinney (1978). Readers who are unfamiliar with the GA biosynthetic pathway should consult Figs. 5.29 and 5.16 as well as Chapters 3 and 4 for an overall view.

Unless otherwise stated the wild-type strain used in the biosynthetic studies is *Gibberella fujikuroi* (Saw.) Wr. (*Fusarium moniliforme* Sheldon), Acc 917 (CM1 58,290).

In the diagrams accompanying this chapter, biosynthetic steps are indicated by single-headed arrows, and chemical reactions by double-headed arrows.

5.2   GIBBERELLIN AND KAURENOID BIOSYNTHESIS IN *GIBBERELLA FUJIKUROI*

5.2.1   Radiotracer studies with *Gibberella fujikuroi*

The study of the biosynthesis of natural products has, in general, been based on the results of radiotracer methodology (Brown, 1972; Brown & Wetter, 1972). It is only relatively recently that stable isotope methods, particularly ^{13}C-nuclear magnetic resonance (CMR), have played a major role. The study of the GA pathway is no excep-

---

* The nomenclature is based on that published by Rowe (1968).

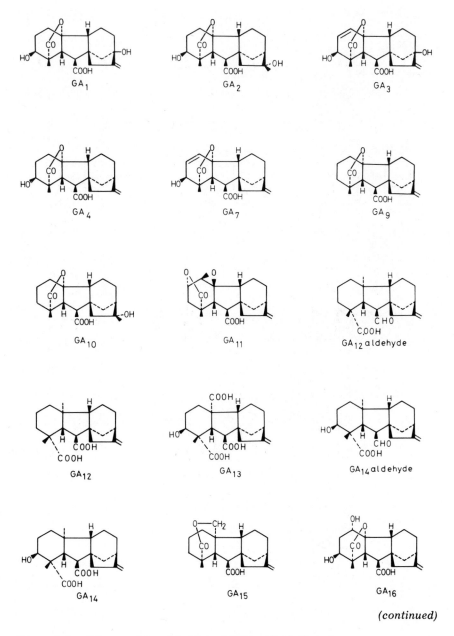

Figure 5.1   Diterpenoid metabolites of *Gibberella fujikuroi*.

(continued)

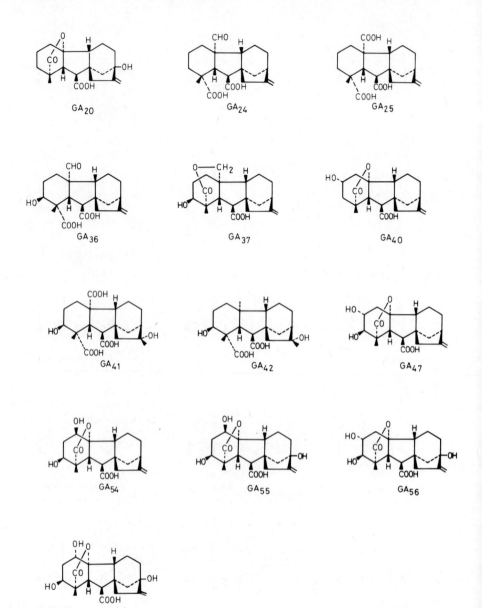

Figure 5.1 *(Continued)*

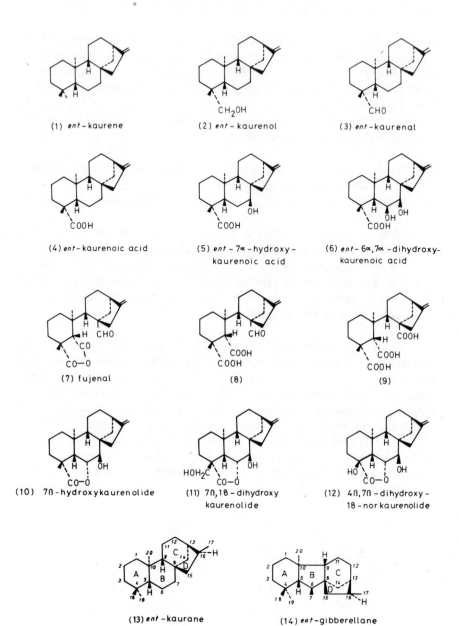

(1) *ent* - kaurene

(2) *ent* - kaurenol

(3) *ent* - kaurenal

(4) *ent* - kaurenoic acid

(5) *ent* - 7α - hydroxy - kaurenoic acid

(6) *ent* - 6α,7α - dihydroxy - kaurenoic acid

(7) fujenal

(8)

(9)

(10) 7β - hydroxykaurenolide

(11) 7β,18 - dihydroxy kaurenolide

(12) 4β,7β - dihydroxy - 18 - nor kaurenolide

(13) *ent* - kaurane

(14) *ent* - gibberellane

Figure 5.1 *Continued)*

tion, and the first insights into the biogenesis of GAs were provided by feeding experiments with $CH_3^{14}CO_2Na$ and $[2-^{14}C]$mevalonic acid (Birch et al., 1959, 1960). Inspection of the structure of gibberellic acid (GA_3) led to the formulation of a biosynthetic scheme that involved a variant of the established route to the tricyclic diterpenes, involving oxidative elaborations for which biochemical or chemical analogues were known:

1. Oxidative loss of a methyl group originally at C-10, analogous to the 14-demethylation of lanosterol.
2. Formation of ring D via a Wagner-Meerwein rearrangement (Fig. 5.2).
3. Ring contraction of ring B with extrusion of a carbon atom as a carboxylic acid via a 6,7-dioxygenated species. The mechanism for such a reaction was formally related to that of either a benzil-benzilic acid rearrangement or a Favorski-type rearrangement. Also, a monoesterified vicinal diequatorial diol could lead to a cyclopentanecarboxaldehyde.

These gross structural changes were supported by the deduced labeling pattern of GA_3 derived from radio-labeled acetate and mevalonate. It is interesting to note that of the mechanisms postulated for these three important steps, only the suggested origin of ring D is at present accepted as correct (Evans, Hanson, & Mulheirn, 1973; Honda, Shishibori, & Suga, 1980; Coates & Cavender, 1980).

$[1-^{14}C]$Acetic acid was incorporated (0.2%) into GA_3 in cultures of the fungus. The labeling pattern was supported by limited chemical degradation, the salient reactions of which are illustrated in Fig. 5.3 (Birch et al., 1959, 1960). These reactions show the absence

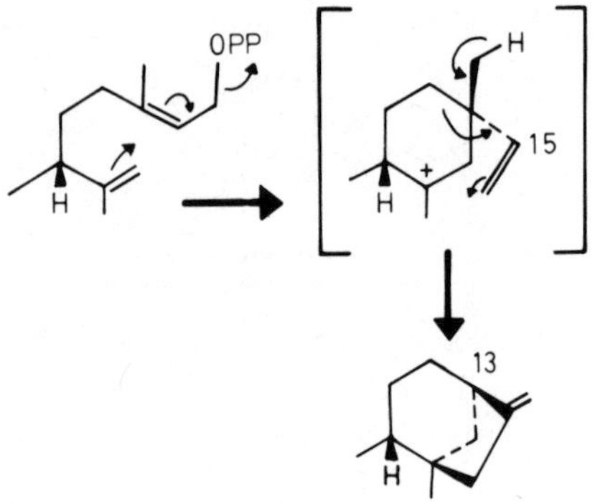

Figure 5.2 The biosynthesis of the C/D ring system in *ent*-kaurene.

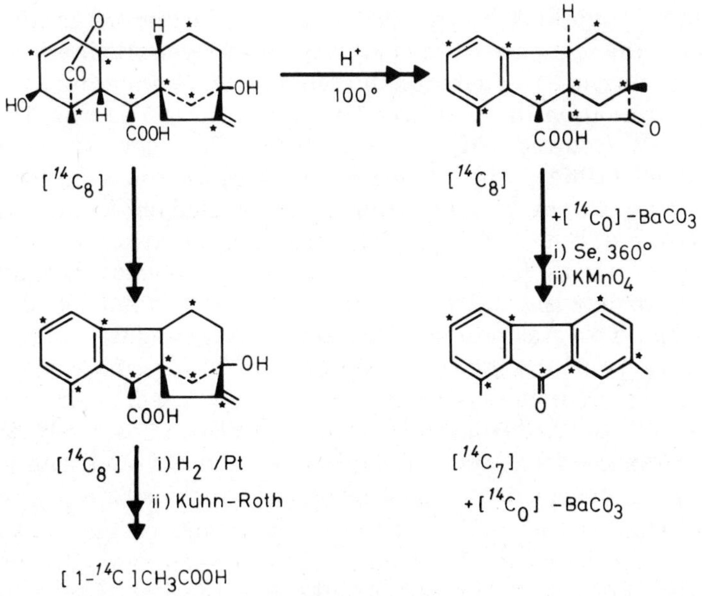

Figure 5.3 The location of label in GA_3 derived from $[1\text{-}^{14}C]\text{-}CH_3CO_2H$ in *G. fujikuroi.* (Birch et al., 1959, 1960.)

of radio-label at positions 18, 19, 7, 17, and 20, and confirm the presence of ^{14}C at C-4.

The results of the $[2\text{-}^{14}C]$mevalonate feeding experiments (Fig. 5.4) were also consistent with a diterpenoid biogenesis (i.e., the incorporation of four molecules of mevalonate). These degradations demonstrated the presence of radio-label at C-18 and C-7 and the lack of label at C-4.

GA_3 derived from $[4\text{-}^{14}C]$mevalonic acid was shown by degradation to possess ^{14}C at C-13 (Birch et al., 1960). Thus this carbon was presumably derived from "C-15" of a hypothetical tricyclic precursor (see Fig. 5.2).

The postulated biogenetic origin of ring B of the GAs was supported by the isolation (see Fig. 5.1) of a number of other metabolites co-occurring with the GAs in cultures of *G. fujikuroi* (Cross et al., 1963; Cavell & MacMillan, 1967). Many of these possessed the *ent*-kaurane carbon skeleton (having a 6-membered B ring); for example, *ent*-kaurene (*1*), 7-hydroxykaurenolide (*10*) and 7,18-dihydroxykaurenolide (*11*). Fujenal (*7*), which is obviously related biosynthetically with *ent*-kaurene, was also obtained.

Birch's hypotheses formed the basis of further tracer studies to determine the sequence of the pathway and to identify intermediates. The co-occurrence of kaurenoid metabolites in *G. fujikuroi* suggested that the formation of the [3.2.1]-bicycloheptane C/D ring structure of the GAs occurred early in the pathway. This view was tested by Cross et al. (1964), who incubated *G. fujikuroi* in the presence of *ent*-[17-^{14}C]kaurene (*1*). A portion of the labeled *ent*-kaurene (8.8%) was recovered, but 5.7% of the label was incorporated into GA$_3$. The incorporation was shown to be specific by chemical degradation involving ozonolysis and trapping the liberated formaldehyde as its dimethone. This was adopted as a standard procedure for verifying the specificity of the incorporation of [17-^{14}C]-precursors into GAs. 7-Hydroxykaurenolide (*10*) (0.05%) and 7,18-dihydroxykaurenolide (*11*) (0.44%) isolated from the culture were also radioactively labeled. Thus *ent*-kaurene (*1*) was established on the GA and kaurenolide pathways as presumably the first tetracyclic intermediate.

Bioassay results indicated the likely nature of the next metabolites in the pathway. *Ent*-kaurene (*1*), *ent*-kaurenol (*2*), and *ent*-kaurenoic acid (*4*) all showed GA-like biological activity in highly

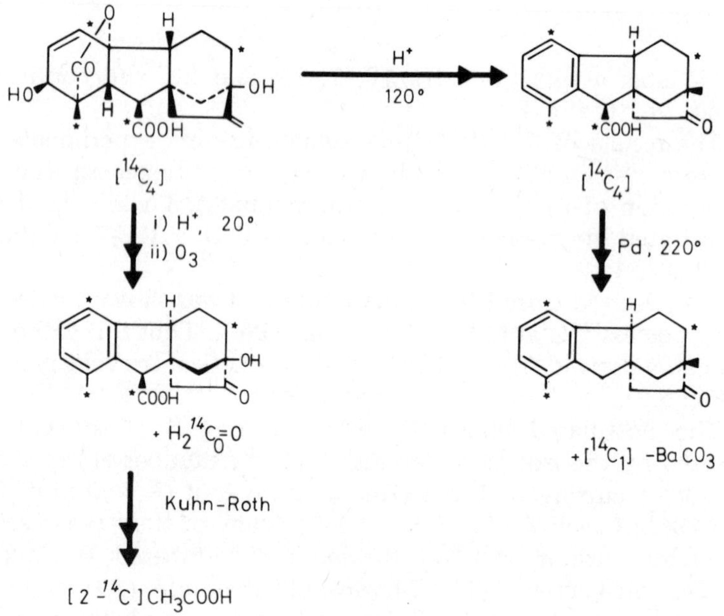

Figure 5.4 The location of label in GA$_3$ derived from [2-^{14}C]mevalonate in *G. fujikuroi*. (Birch et al., 1959.)

specific dwarf maize assays (Katsumi et al., 1964). The inference made was that these compounds were being converted into GAs in the seedlings.

A radiochemical tracer study by Galt (1965) was consistent with the idea that *ent*-kaurenol (*2*) ought to be a precursor of the GAs, but the definitive work was published by Graebe et al. (1965). *Ent*-[$^{14}C_4$]kaurene and *ent*-[$^{14}C_4$]kaurenol were prepared by incubating endosperm nucellus of *Marah macrocarpus** (wild cucumber) with [2-^{14}C]mevalonic acid. Each was then separately fed to resuspension cultures of *G. fujikuroi*. The *ent*-kaurene and *ent*-kaurenol were converted into GA_3 in 1.03% and 1.55% yield, respectively. Geissman et al. (1966) demonstrated that *ent*-[17-^{14}C]kaurenoic acid (*4*) was effectively incorporated into GA_3 (17%) after 15 days by *G. fujikuroi* (Lilly strain M-119). The time course of the metabolism of the substrate was examined by radio-thin layer chromatography (radio-TLC). After 4 hr the highest activity was associated with the $GA_4 + GA_7$ spot, but after 2 days the activity at this Rf had diminished and the $GA_1 + GA_3$ spot had the major activity. Evidence was presented which indicated that "fraction B" (later identified as GA_{14} by Jones, West, & Phinney, 1968) was on the GA pathway. Therefore, it was suggested that GA_3 biosynthesis occurred by a sequence involving progressive oxidation. This time-course study was elaborated by the same authors (Verbiscar et al., 1967), who found that *ent*-[17-^{14}C]-kaurenol (*2*) was incorporated into $GA_4 + GA_7$ and $GA_1 + GA_3$ (judging by radio-TLC criteria). Furthermore, use of the TLC solvent systems that separated GA_4 from GA_7 and GA_1 from GA_3 allowed an estimation of the changes with time of the ratios of GA_4 to GA_7 and GA_1 to GA_3, which further supported the scheme involving sequential oxidation. Radio-labeled $GA_4 + GA_7$ was isolated from a 2-day *ent*-[17-^{14}C]kaurenol feed and reincubated with the fungus for 7 days. Analysis of the derived metabolites by radio-TLC demonstrated the presence of GA_3 and GA_1. These results were consistent with work by Schmidt (1961, 1962) who claimed that addition of GA_4 to fermentation of *G. fujikuroi* increased the amount of GA_3.

The expected intermediacy of *ent*-kaurenal (*3*) was supported by work by Dennis and West (1967) who obtained *ent*-[^{14}C]kaurenal biosynthetically from the wild cucumber cell-free system discussed above and fed it to washed suspensions of the mycelium of *G. fujikuroi*. After 114 hr, an incorporation into GA_3 of 1.36% was

---

*Formerly *Echinocystis macrocarpa*.

recorded. This value is similar to that obtained with ent-[$^{14}C_4$]-kaurene and ent-[$^{14}C_4$]kaurenol under similar conditions (Graebe et al., 1965).

Research then focused on the nature of the substrate for the ring B contraction reaction. It seemed likely that this step occurred before the hydroxylation of rings A and C, as the nonhydroxylated GA, GA_{12}, had been isolated from fungal cultures (Cross & Norton, 1965). By elaborating Birch's speculations on this step, a number of candidate precursors with oxygenated B-rings could be identified. Despite having the unfavorable axial configuration, 7β-hydroxy-kaurenolide (10) might undergo lactone cleavage initiating ring-contraction and also furnishing the required C-6 carboxyl group (Cross, Galt, & Norton, 1968b). [17-^{14}C]7β-hydroxykaurenolide (10) was prepared and fed to the fungus. An incorporation into GAs of only 0.1% was registered, the bulk of the substrate (43%) being hydroxylated to 7β,18-dihydroxykaurenolide (11). In case the utilization of the substrate for GA biosynthesis was hampered by the presence of the lactone ring, the labeled triol (15) was prepared by LiAlH$_4$ reduction of [17-^{14}C]7β-hydroxykaurenolide (10). However,

Structures (15)–(21)

this compound too was not incorporated into GA_3 or GA_{13} by *G. fujikuroi*, nor was the readily prepared [17-^{14}C]7α-hydroxy-kaurenolide (*16*). Another idea was that the kaurenolides were not the precursors of GAs but that both were derived from a common intermediate. It was thought that the ideal candidate to fulfil this role might be the epoxy acid (*17*), as either heterolysis of the 6-O bond would lead to ring contraction and the GAs or internal nucleophilic attack at C-6 would give 7β-hydroxykaurenolide (*10*). This labeled epoxy acid (*17*) was not readily available, and so *ent*-[17-^{14}C]kaur-6,16-dien-19-ol (*18*) was employed. This compound, which was synthesized from [17-^{14}C]7β-hydroxykaurenolide, might act as a precursor to the required epoxy acid since it was known that in steroids, where axial hydroxylation occurs at a particular position, the corresponding unsaturated substrate is epoxidized (Bloom & Shull, 1955, Talalay, 1957). On incubation the *ent*-[17-^{14}C]kaur-6,16-dien-19-ol (*18*) underwent metabolism to GA_3 in a scant 0.48% yield, and to GA_{13} (0.19%) and fujenal (*7*) (1.2%). The specificity of the transformation was confirmed by chemical degradation and quantitative recovery of the labeled 17-[14]C. Under similar conditions, *ent*-kaurenol (*2*) was incorporated into GA_3 in 4.9%. It was therefore concluded that *ent*-kaur-6,16-dien-19-ol (*18*), unlike *ent*-kaurenol, was not on the main pathway, and no support was forthcoming for the existence of an intermediate epoxide.

By this time a number of C_{20}-GAs—including GA_{12}, GA_{13}, and GA_{14}—had been isolated from fermentations of *G. fujikuroi*, and these were soon put to the test as potential intermediates between the known 19-oxidized *ent*-kaurene precursors and the C_{19}-GAs. Cross, Norton, and Stewart (1968c) prepared ^{14}C-labeled samples of these GAs and some chemically reduced derivatives of them. These were then fed to *G. fujikuroi*, and their incorporations into GA_3 and GA_{13} were measured (see Table 5.1). Other GAs were tentatively identified as metabolites by radio-TLC.

These results were interpreted in terms of the scheme illustrated in Fig. 5.5, and the salient points are as follows: (a) GA_{12} aldehyde is an intermediate on the main pathway, but GA_{12} is not; (b) GA_{12} aldehyde is preferentially oxidized at C-3 to give GA_{14} aldehyde, rather than C-7 to give GA_{12}; (c) the metabolism of GA_{14} aldehyde to C_{19}-GAs may not necessarily go via GA_{14}; (d) GA_{13} is not on the pathway to C_{19}-GAs. The observation that 3-hydroxylation occurs before oxidation of C-20 (and C-7), explained the previous finding that the non-3-hydroxylated C_{19}-GA GA_9 did not act as a precursor to GA_3 (Cross et al., 1964, 1968b). The presumptive intermediate GA_{14}

Table 5.1. Incorporations of [17-^{14}C] labeled GAs and derivatives into GA_3 and GA_{13} by *G. fujikuroi*

[17-^{14}C] Substrate	Incorporation (%) into GA_3	GA_{13}	Fermentation time (hr)
GA_{12}	0.7	0.08	106.6
Diol (*19*)	7.5	0.24	120
GA_{12} aldehyde	15.4	0.9	69
GA_{14}	4.7	0.9	68.9
Triol (*20*)	0.23	—	70.9
GA_{13}	0	—	72.1
Tetraol (*21*)	0	0	119.4

After Cross et al. (1968c).

aldehyde was later synthesized by Hedden et al. (1974) and shown to be a metabolite of GA_{12} aldehyde and an efficient precursor to GA_3. The subject of the role of C_{20}-GAs as intermediates in the main pathway will be discussed later, particularly with reference to studies with mutants of *G. fujikuroi* (Section 5.2.2).

We will now turn our attention to a series of elegant experiments by Hanson and co-workers using stereospecifically tritiated mevalonates which defined the stereochemistry of many of the steps in GA biosynthesis. By analogy with the fundamental work by Popjak and Cornforth (1966) on the stereochemistry of steroid biosynthesis, the mode of folding of geranylgeranyl pyrophosphate to give the A and B rings of kaurenoids and GAs could be assumed. As this process should not disturb the configuration of prochiral centers, the loss or retention of tritium derived from specifically labeled mevalonates could be used to monitor the stereochemistry of biological oxidation reactions at these centers. The experimental procedure is as follows. Stereospecifically tritiated [^{14}C]mevalonates are fed, the metabolites and recovered, and their [^{3}H:^{14}C] ratios are then determined. This allows calculation of the number of tritium atoms retained in the product. The position of these labels may be discovered by chemical degradation, or inferred if unambiguous.

The number of tritium atoms and their location in four metabolites derived from [$4R$-^{3}H,2-^{14}C]mevalonates in fermentations of *G. fujikuroi* is given in Fig. 5.6 (Hanson & White, 1969).

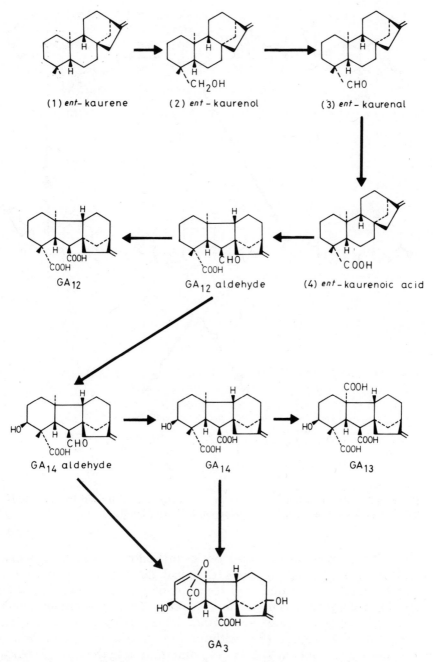

Figure 5.5 A biosynthesis scheme based on feeding studies (Table 5.1) by Cross et al. (1968c).

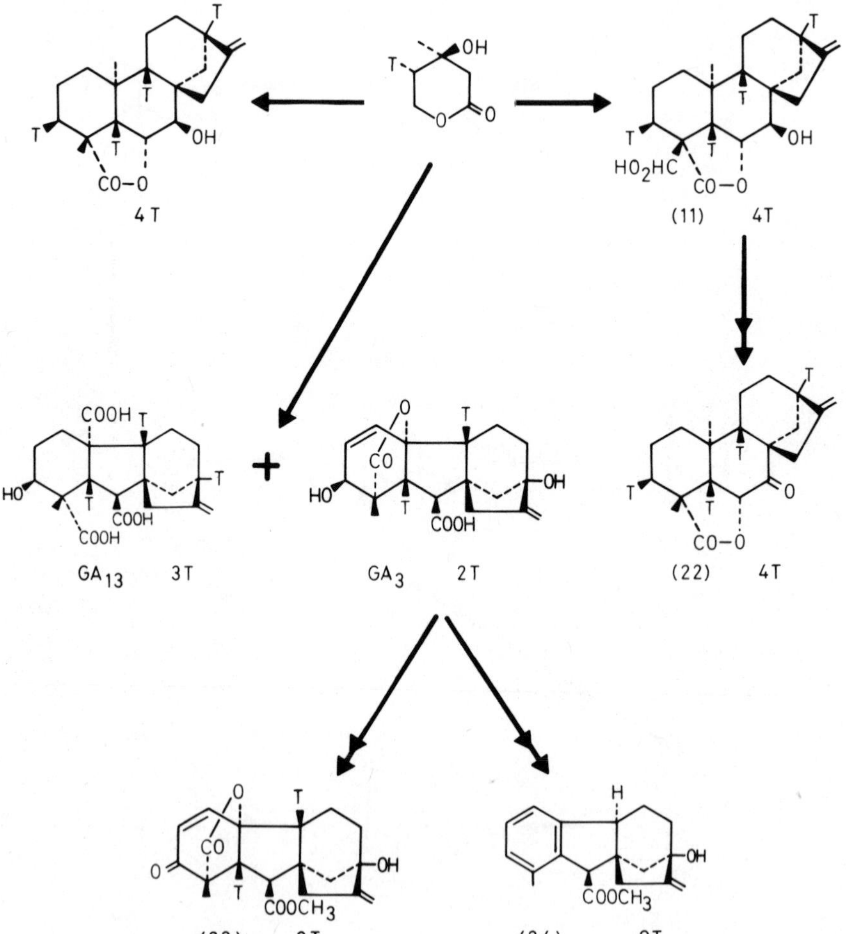

Figure 5.6 The location of tritium in four metabolites derived from $[4R\text{-}^3\text{H,-}2\text{-}^{14}\text{C}]$mevalonate in *G. fujikuroi*. (Hanson & White, 1969.)

These assignments, based on an orthodox cyclization mechanism, were supported by some chemical degradation. Tritium was shown not to be located at C-18, as oxidation and decarboxylation of 7β,18-dihydroxykaurenolide (*11*) gave 7-oxo-18-norkaurenolide (*22*) which still contained four tritium labels. Oxidation of the labeled GA$_3$ gave GA$_3$-ketone (*23*) without loss of tritium, demonstrating the absence of label at C-3 in GA$_3$ and indicating that the biological hydroxylation at C-3 in the precursor had occurred with retention of configuration. This conclusion has recently been given a firmer foundation by the work of Dawson, Jefferies, and Knox (1975) who confirmed the widely accepted hypothesis that the cyclization of geranylgeranylpyrophosphate is initiated by protonation at C-3 from the α-face of the molecule. Hanson and White (1969) demonstrated the presence of tritium at positions 5 and 9 in GA_3 by mineral acid treatment of GA_3 methyl ester to give methyl allogibberate (24) which incurred the loss of both tritium labels. The hydrogens at C-5 and C-9 are the only ones that can be lost in this reaction. The biosynthetic significance of this result will be referred to presently.

In the same publication, Hanson and White (1969) reported the results of feeding $[2\text{-}^3H_2,2\text{-}^{14}C]$mevalonate to G. fujikuroi. Six compounds were isolated, and the $^3H{:}^{14}C$ ratios measured were consistent with the patterns illustrated in Fig. 5.7. Degradation of the kaurenolides supported these assignments. On oxidation with chromic acid, 7β-hydroxykaurenolide lost one tritium label. Under the same conditions, but followed by heating, $7\beta,18$-dihydroxykaurenolide gave 7-oxo-18-norkaurenolide (22) which had lost three tritium labels. From a biosynthetic point of view the most interesting result was the finding that GA_{13} and GA_4 contained the same number of tritium atoms. Thus both labeled atoms at the C-1 position were retained on formation of the C_{19}-GAs. Remembering that tritium label from $(4R)\text{-}[4\text{-}^3H,2\text{-}^{14}C]$mevalonate was retained at C-5 and C-9 in C_{19}-GAs, it follows that no precursor to the C_{19}-GAs may have a (1,10)-, (5,10)-, or (9,10)-double bond. Thus the loss of the C-20 atom cannot occur in an analogous fashion to the 14-demethylation of lanosterol as suggested by Birch et al. (1959). It was therefore suggested that the formation of the C_{19}-GA γ-lactone involved a Baeyer-Villiger-type oxidation of a 10-carbonyl function followed by solvolysis of the remaining ester (see Fig. 5.23, pathway b), rather than by decarboxylation of an unsaturated acid.

A subsequent publication from Hanson's laboratory described biosynthetic studies with other doubly labeled mevalonates (Evans, Hanson, & White, 1970). The interpretation of the results of feeding $(2R)$- and $(2S)\text{-}[2\text{-}^3H,2\text{-}^{14}C]$mevalonate to G. fujikuroi are summarized in Figs. 5.8 and 5.9, respectively.

In the case of the $(2R)\text{-}[2\text{-}^3H,2\text{-}^{14}C]$mevalonate feed, the labels at C-18 and C-7 in the kaurenolides were established by chemical degradation. 7β-Hydroxykaurenolide (10) was oxidized to the corresponding ketone (25) with loss of one tritium label, and $7\beta,18$-dihydroxykaurenolide was oxidized and decarboxylated to give the 7-oxo-18-norkaurenolide (22) which had lost 1.5 tritium atoms. GA_3 methyl ester was converted without loss of label into methyl gibberate (26). These results provided two useful facts. First, hydroxy-

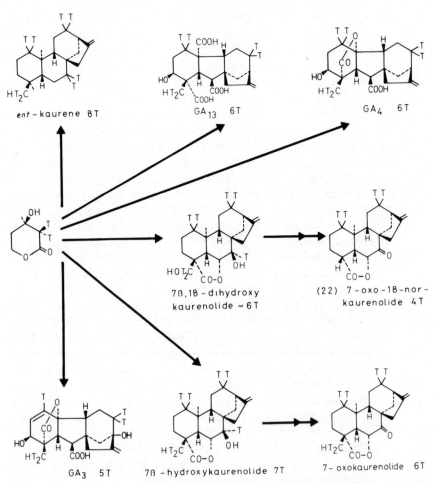

Figure 5.7 The location of tritium in six metabolites derived from [2-³H₂,-2-¹⁴C]mevalonate in *G. fujikuroi*. (Hanson & White, 1969.)

lation at C-7 to give the kaurenolides occurred with retention of configuration. Second, in the dehydrogenation giving the 1,2-double bond in GA₃ the 1α-proton of the precursor is lost.

To complement these results, an analogous experiment with (5R)-[5-³H,2-¹⁴C]mevalonate was conducted. Figure 5.10 summarizes the relevant data. Retention of the label at the 6β-position in 7β-hydroxykaurenolide was confirmed as follows. Oxidation to 7-oxo-kaurenolide (25) followed by base treatment gave a product that had lost one tritium atom. Alternatively, formation of the 7-bromo-

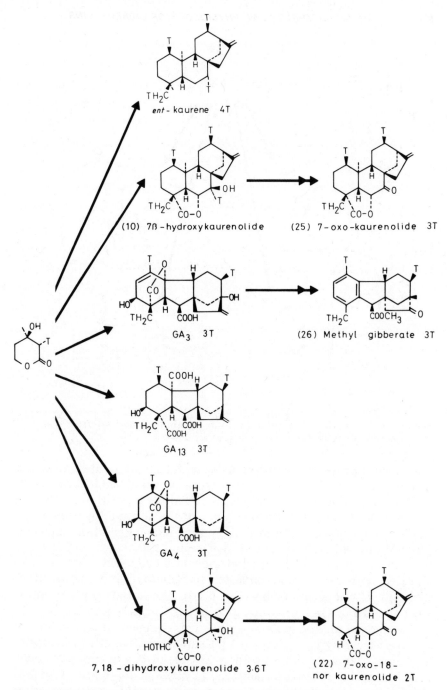

Figure 5.8 The location of tritium in six metabolites derived from [2R-³H,-2-¹⁴C]mevalonate in *G. fujikuroi*. (Evans et al., 1970.)

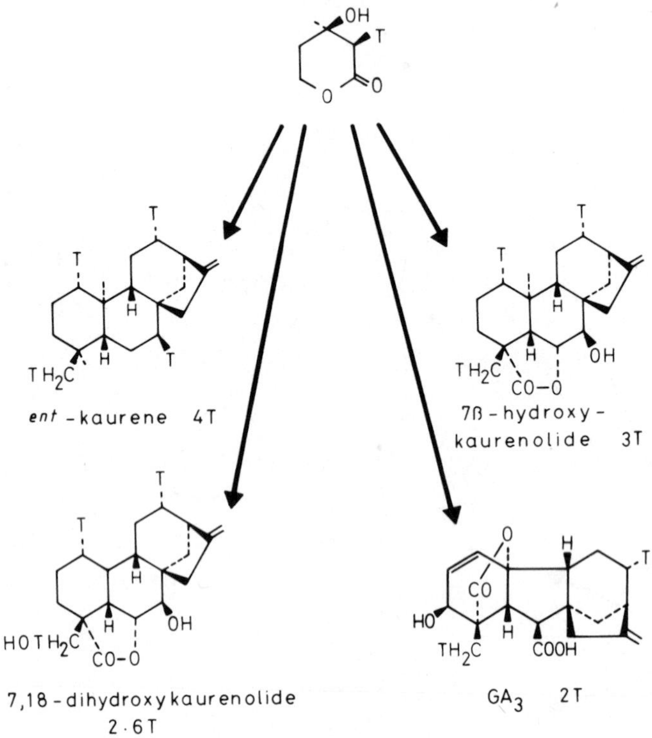

Figure 5.9 Location of tritium in four metabolites derived from [2S-³H,-2-¹⁴C]mevalonate in *G. fujikuroi*. (Evans et al., 1970.)

benzensulphonate (27) and treatment with lithium iodide in refluxing collidine gave *ent*-6-oxokaurenoic acid (28), which had lost one label. In the case of GA_{13}, label was confirmed at the 2-position by the loss of one label following base treatment of the derived 3-ketone. Label was located at position 14 in GA_3 by the following reaction sequence. [³H₂]Methyl gibberellate gave [³H₂]methyl allogibberate (24) which was converted into [³H₂]methyl gibberate (26), which was in turn oxidized with selenium dioxide to [³H₁]methyl gibberdionate (28) with loss of one tritium label (Fig. 5.10). The stereochemistry of the label at C-14 in GA_3 was shown to be α by mass spectrometry (MS) of the derived methyl gibberate (26). This stereochemical assignment has been recently corroborated by Coates and Cavender (1980). The absence of label at the 2-position in GA_3 in this experiment taken with the results of the (2R)- and (2S)-[2-³H,2-¹⁴C]mevalonic acid feed defined the stereochemistry of the dehydrogenation reaction leading to the 1,2-olefin in GA_3 as *cis*. This occurs by loss of the 1α- and 2α-hydrogens. The retention of 6β-tritium in 7β-hydroxykaurenolide

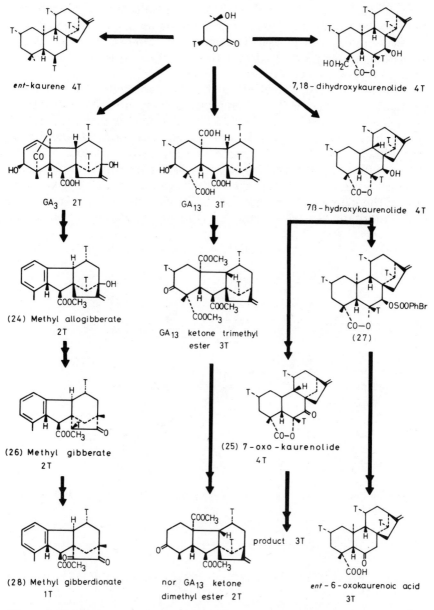

ent-kaurene 4T

7,18-dihydroxykaurenolide 4T

GA₃ 2T

GA₁₃ 3T

7β-hydroxykaurenolide 4T

(24) Methyl allogibberate 2T

GA₁₃ ketone trimethyl ester 3T

(27)

(26) Methyl gibberate 2T

(25) 7-oxo-kaurenolide 4T

product 3T

(28) Methyl gibberdionate 1T

nor GA₁₃ ketone dimethyl ester 2T

ent-6-oxokaurenoic acid 3T

Figure 5.10 Location of tritium in five metabolites derived from [5R-³H,-2-¹⁴C]mevalonate in *G. fujikuroi.* (Evans et al., 1970.)

269

showed that the 6α-functionalization resulting in 7β-hydroxy-kaurenolide occurs with retention of configuration as expected of a mixed function oxygenase. The apparent absence of label at the 6-position in GA_3 indicated that the 6β-hydrogen is not retained at position 6 after ring contraction.

Further light was thrown on the ring contraction step by feeding studies with $[1\text{-}^3H_2,2\text{-}^{14}C]$geranylpyrophosphate. The results from this experiment are given in Fig. 5.11. The label was located at C-6 in GA_3 by conversion into $[^3H_1]$methyl allogibberate (24) and thence into $[^3H_0]$methyl 6-epi-allogibberate (29) with 93% loss of label. Taken with the results of the immediately previous experiments this data suggested that the ent-6α-hydrogen is lost in B ring contraction of the kaurenoid precursor while the ent-6β-hydrogen is retained in the resultant GA. This result was later substantiated by analysis of

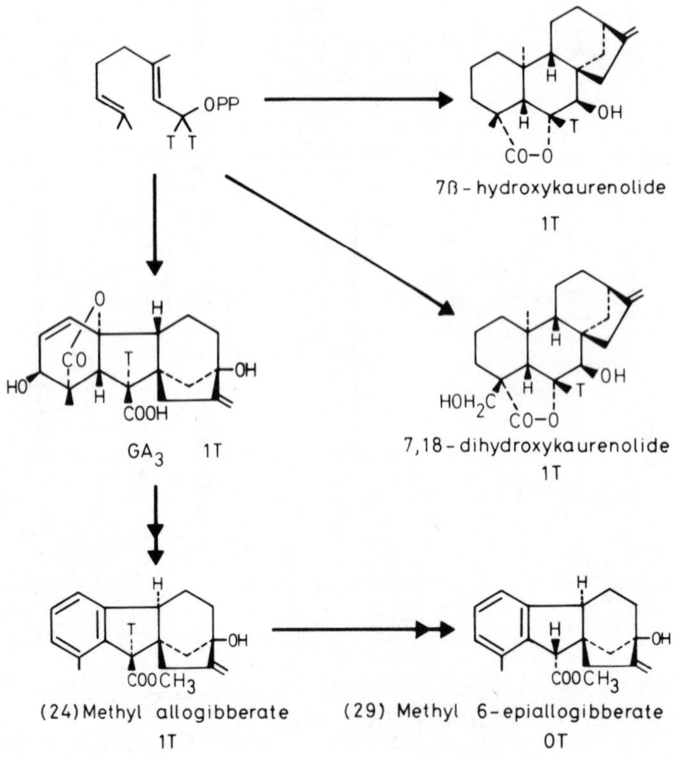

Figure 5.11 Location of tritium in three metabolites derived from $[1\text{-}^3H_2,\text{-}2\text{-}^{14}C]$geranylpyrophosphate in G. fujikuroi. (Evans et al., 1970.)

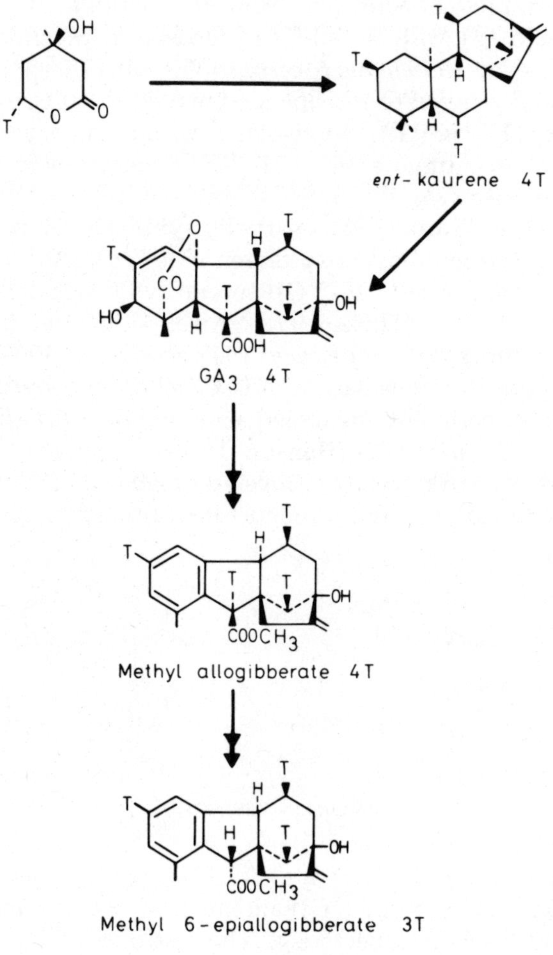

Figure 5.12 Location of tritium in GA_3 formed from *ent*-kaurene derived from [5S-^{3}H,2-^{14}C]mevalonate in *G. fujikuroi*. (Evans et al., 1973.)

the incorporation into GA_3 of *ent*-kaurene which had been enzymically derived from (5S)-[5-^{3}H$_1$]mevalonate (Evans et al., 1973). The results are summarized in Fig. 5.12. The label derived from the kaurenoid *ent*-6β-position was retained in GA_3. Its location was confirmed by loss of label on epimerization at the 6-position.

We will now return to work investigating the nature of the intermediates in the GA pathway. *Ent*-7α-hydroxykaurenoic acid (5) was identified as the next intermediate after *ent*-kaurenoic acid by two independent groups. Lew and West (1971) isolated *ent*-[17-^{14}C]-

7α-hydroxykaurenoic acid (5) from incubations of ent-[17-^{14}C]-kaurenoic acid (4) with a cell-free preparation of endosperm from immature seed of *Marah macrocarpus*. The ent-[17-^{14}C]7α-hydroxy-kaurenoic acid was fed to resuspended mycelium of *G. fujikuroi* and, after 2 days, [17-^{14}C]GA$_3$ was isolated with an incorporation of 4%. In a parallel experiment, ent-[17-^{14}C]kaurenoic acid gave a 12% incorporation into GA$_3$.

Hanson and White (1969) partially synthesized ent-7α-hydroxy-kaurenoic acid from 7β-hydroxykaurenolide (10), and this was shown to be a metabolite of ent-[17-^{14}C]kaurene by isotope dilution analysis (Hanson, Hawker, & White, 1972). After 4 days the incorporation into ent-7α-hydroxykaurenoic acid (5) was only 0.006%, but it was shown to be specific by isolation of C-17 as formaldehyde dimethone. In the same experiment unlabeled kaurenolide (30) (not a known metabolite of *G. fujikuroi*) (Hanson, 1966) was added to the incubation mixture on work-up and recovered unlabeled. Thus, kaurenolide (30) was excluded as a likely precursor of the hydroxykaurenolides.

(30) kaurenolide   (31)

Structures (30)–(31)

7β-Hydroxykaurenolide (10) and 7β,18-dihydroxykaurenolide (11) isolated from this ent-[17-^{14}C]kaurene feed showed incorporations of 0.8% and 5.45%, respectively. The position of ent-7α-hydroxy-kaurenoic acid (5) as an intermediate in the pathway to GAs was substantiated by feeding the [17-^{14}C]-compound to *G. fujikuroi*. After 4 hr the ^{14}C-label was incorporated into GA$_{12}$ aldehyde (0.21%) and GA$_{12}$ (0.29%). After a further 20 hr, these figures changed to 0.17% and 0.37%. In another experiment the potassium salt of ent-[17-^{14}C]7α-hydroxykaurenoic acid was fed and incorporations were recorded into GA$_3$ (32.3%), 7β-hydroxykaurenolide (10) (0.03%), and 7β,18-dihydroxykaurenolide (11) (0.44%). From recent results of Hedden and Graebe (1981) and Beale et al. (1982) some doubt must be cast on the incorporation of ent-7α-hydroxykaurenoic acid into the kaurenolides (see p. 313 et seq.). The much lower incorporation (4%) of ent-7α-hydroxykaurenoic acid into GA$_3$ found by Lew and West (1971) can be explained by their shorter incubation time (2 days) and their use of suspensions of mycelia in buffer which might depress *de novo* synthesis of early intermediates to "chase" the label to GA_3. The loss of the *ent*-6α-hydrogen after ring contraction, previously established in the studies with multiply labeled mevalonates, was supported by results from feeding *ent*-[17-^{14}C,6α-^{3}H]7α-hydroxykaurenoic acid. The substrate which was partially synthesized from 7β-hydroxykaurenolide (*10*) showed a 4.5% incorporation into GA_3 after 20 hr, but the GA_3 contained no tritium (Hanson & Hawker, 1972a).

Although it had been suggested that *ent*-7α-hydroxykaurenoic acid (*5*) might be the substrate for ring contraction, reaction being initiated by abstraction of the *ent*-6α-hydrogen (Evans et al., 1970) (Fig. 5.13), it was also possible that an intermediate was involved. One particularly attractive scheme involved the intervention of an *ent*-6α-(equatorial)-hydroxylation step to give *ent*-6α,7α-dihydroxykaurenoic acid (*6*) which could undergo ring contraction on conversion of the newly introduced 6β-hydroxyl into a better leaving group (Fig. 5.14).

The presumptive intermediate diol (*6*) was synthesized (Cross, Stewart, & Stoddart, 1970) in a straightforward manner from 7β-hydroxykaurenolide (*10*). The [17-^{14}C]diol (*6*) was also prepared and fed to a fermentation of *G. fujikuroi*. After 4 days GA_3 and 7β-hydroxykaurenolide were isolated but were devoid of radioactivity. However, fujenal (*7*) showed an incorporation of 9.7%. The labeled triol (*31*) derived by lithium aluminium hydride reduction of the methyl ester of the diol (*6*) was also fed and likewise was not incorporated into GA_3, while its incorporation into fujenal (*7*) was 0.36%. These results were disappointing since the diol acid (*6*) had shown more GA-like biological activity than either *ent*-7α-hydroxykaurenoic acid (*5*) or GA_{12} aldehyde in three GA bioassay systems (Cross et al., 1970). It was thought possible that the *ent*-6α,7α-dihydroxykaurenoic acid was being oxidatively cleaved before it was able to reach the ring contraction enzyme system, which may require

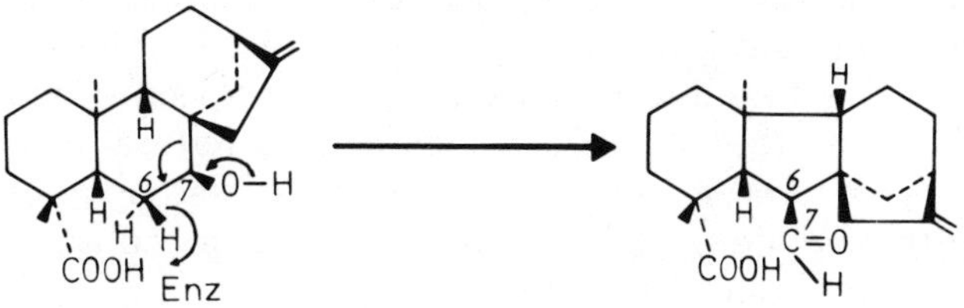

Figure 5.13 A mechanism for the ring contraction reaction initiated by abstraction of the *ent*-6α-hydrogen. (Evans et al., 1970.)

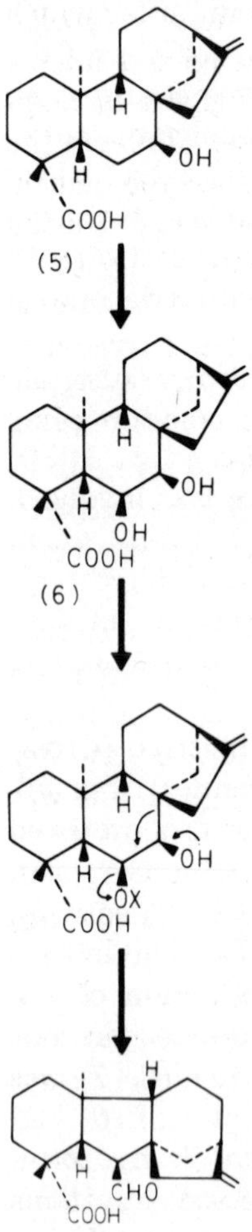

Figure 5.14 A ring contraction mechanism involving an ent-6α-oxygen function as leaving group. (Cross et al., 1970.)

prior information of a pyrophosphate. This being so, the acid (6) should be present in the culture filtrate. An isotope dilution experiment was performed by feeding ent-[17-^{14}C]kaurene to G. fujikuroi and after 24 hr diluting with unlabeled acid (6). The acid was reisolated and was found to be active, showing a specific incorporation of 0.006%. Thus the acid (6) is apparently a natural metabolite of G. fujikuroi. However, the role of the acid was not defined as a possible intermediate in the GA pathway or as a probable precursor to the metabolic shunt fujenal (7).

These data were echoed by results published by Hanson's group (Hanson et al., 1972) who also synthesized the [17-^{14}C]-labeled diol

(6) in an analogous fashion to Cross et al. (1970). When the labeled diol was fed to the fungus, GA_3 was recovered unlabeled and fujenal (7) showed a small incorporation of 0.01%. This low figure is, however, entirely compatible with Cross' results and merely reflects the different isolation procedures used. Hanson et al. studied the neutral fraction and thus the incorporation into authentic fujenal (7) leaving the diacid (8) in the acidic fraction. The result obtained by Cross et al. reflects a high incorporation into the diacid (8), as the fujenal in their work was isolated from the acidic extract of the fermentation. In fact, the bulk of fujenal isolated from cultures of G. fujikuroi is probably not a natural metabolite but an artifact of the diacid (8). It is notable that fujenal is seldom isolated from neutral extracts of the fungus but appears in silica column fractions of acidic extracts. It has been shown (Bearder, 1973) that the diacid (8) is converted into fujenal during silica gel chromatography.

The most detailed investigations yet into the ring contraction step in the fungus was the feed of $[1\text{-}^3H_2,1\text{-}^{14}C]$geranyl pyrophosphate to discover the immediate fate of the 6-hydrogens on ring contraction (Hanson et al., 1972). Six days after the feed was made, metabolites were isolated by dilution analysis and, inter alia, GA_{12} aldehyde was isolated showing a low but specific incorporation of 0.005% (Fig. 5.15). The $^3H:^{14}C$ ratio corresponded to the retention of both hydrogens from the 6-position of the kaurenoid precursor. Oxidation of the carbazone methyl ester of the GA_{12} aldehyde gave GA_{12} 19-monomethyl ester (32) with the loss of one tritium atom. These results were taken to indicate that ring contraction occurred with a shift of the ent-6α-hydrogen to the 7-position, becoming the formyl hydrogen of the GA_{12} aldehyde.

This mechanism is different from the one that has now been established unambiguously in higher plants (see Graebe et al., 1980; see Chapter 3, this volume). Bearing this in mind and considering the difficulties in accurately measuring very low levels of radioactivity (e.g., in this instance 3.95 disint. mg^{-1} min^{-1} ^3H), it seems likely that this interpretation is incorrect. However, more definitive work has yet to be done.

The latter part of the GA pathway, from GA_4 onwards (touched on by Verbiscar et al., 1967) has been studied in detail by Vining and coworkers (Pitel, Vining, & Arsenault, 1971b; McInnes et al., 1973, 1977), resulting in the pathway illustrated in Fig. 5.16. $[17\text{-}^{14}C]$-labeled gibberellins GA_4, GA_7, GA_3, and GA_1 were fed to replacement cultures of G. fujikuroi grown on synthetic media. The metabolites were separated by TLC and Sephadex partition chromatography and identified by MS. When GA_1 was fed for either 14 or 205 hr, the sub-

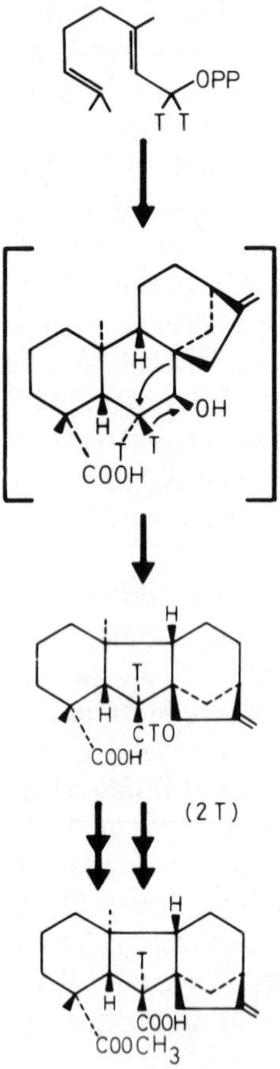

Figure 5.15 The mechanism interpretation of incorporation of $[1\text{-}^3H_2]$geranylpyrophosphate into GA_{12} aldehyde. (Hanson et al., 1972.)

strate was recovered unchanged and no metabolites were detected. It was also established that GA_1 did not inhibit the formation of GA_3. GA_1 was therefore shown to be a terminal metabolite and not on the main pathway to GA_3 as was suggested by Geissman et al. (1966). Feeding studies with GA_3 showed that it also was a terminal product. Small amounts of iso-GA_3 (*33*) and allogibberic acid (*34*) were formed but were thought to be artifacts formed non-enzymatically in the long fermentations.

The results from the feeds of GA_4 and GA_7 are shown in Tables 5.2 and 5.3, respectively. These time courses confirmed the proposals of Verbiscar et al. (1967) illustrating the conversion of GA_4 into GA_7, GA_1, and GA_3 (Fig. 5.16). GA_7 was converted very efficiently

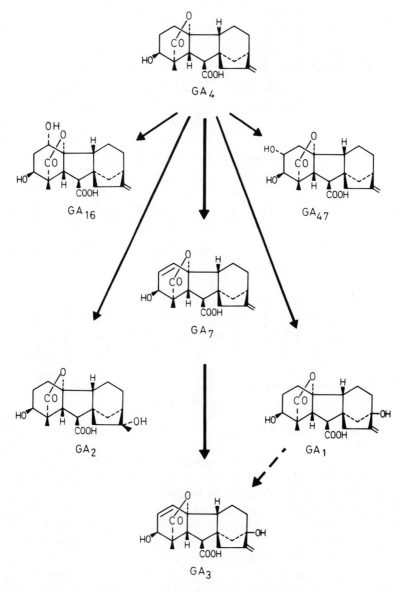

Figure 5.16 The metabolism of GA$_4$ in *G. fujikuroi*. (McInnes et al., 1977.)

Structures (33)–(38)

into GA_3 and only traces of other metabolites. No incorporation was found into the "ring A saturated" compounds, GA_1 and GA_4. GA_4 was shown to act as a precursor to GA_1 and to GA_7, but most effectively to GA_3. GA_4 was not metabolized as quickly as GA_7, and thus the low incorporation into GA_7 is understandable.

Some of these tracer experiments were also conducted in fermentations grown on a natural medium (medium B) containing corn steep liquor. In contrast to the synthetic medium, the fungus accumulated GA_1 in substantial quantities. When $[1,2-^3H]GA_1$ was fed to the fungus growing in this medium it was incorporated into GA_3 in 0.6% in 130 hr; with longer fermentation times (210 and 285 hr) no change was observed, and it was shown that GA_3 accumulated for only 50 hr after the beginning of the feed. In contrast to the results with GA_4 on the synthetic medium, feeds of this substrate on medium B (see Table 5.4) gave appreciable quantities of GA_1 as well as GA_3. In addition, a new labeled metabolite, "compound K," was formed in 0.9%, which was not detected in the work with the

Table 5.2. Incorporation of $[^{14}C]\,GA_4$ into various GAs with time

	Radioactivity (%)			
	14 hr	39 hr	96 hr	205 hr
Total extract	95.3	96.8	94.2	92.2
Acidic extract	93.6	90.5	88.6	91.2
GA_1			—	4.8
GA_3	0	1..0	2.5	70
Iso-GA_3 (*33*)			1.3	11
GA_4		87	82	1.0
GA_7	99	2.6	2.1	1.5
Iso-GA_7 (*35*)		—	0.03	0.7
Allogibberic acid (*34*)	—	—	0.1	2.3

After Pitel et al. (1971b).

synthetic medium. It was suggested that this compound might be an isomer of GA_1 formed by the nonspecific activity of the enzyme biosynthesizing GA_1. Later work from the same laboratories (McInnes et al., 1973) identified this metabolite as GA_{16}. It is unlikely that this GA is formed by the 13-hydroxylating enzyme, as GA_{16} is observed

Table 5.3. Incorporation of $[^{14}C]\,GA_7$ into various GAs with time

	Radioactivity (%)			
	14 hr	39 hr	65 hr	205 hr
Total extract	93.6	97.8	98.7	90.1
Acidic extract	91.7	97.2	98.8	89.5
GA_1		—	—	—
GA_3	28	32	73	74
Iso-GA_3 (*33*)		2.7	8.3	10
GA_4			—	—
GA_7			16	1.7
Iso-GA_7 (*35*)	63	62	0.9	2.0
Allogibberic acid (*34*)			—	2.1

After Pitel et al. (1971b).

Table 5.4. Incorporation of $[^{14}C]GA_4$ into various GAs in *G. fujikuroi* grown on medium B

	Incorporation after 70 hr (%)
Acidic extract	84
Allogibberic acid (*34*)	3.9
GA_4	—
GA_7	—
GA_1	24
GA_3	48
Iso-GA_3 (*33*)	2.2
Compound K (GA_{16})	0.9

After Pitel et al. (1971b).

in cultures of *G. fujikuroi* (mutant R9) which does not possess a 13-hydroxylating system (Bearder, MacMillan, & Phinney, 1973e; Bearder, 1973). Because McInnes et al. did not detect GA_{16} in feeds of GA_7, it is unlikely that GA_{16} has this origin as was suggested by Katsumi and Phinney (1969). Following up this work, the metabolism of radio-labeled GA_4 in *G. fujikuroi* fermentations in complex media was studied in more detail (McInnes et al., 1977). In addition to the above metabolites, radio-labeled GA_{47}, 3-acetyl-GA_1 (*36*), 3-acetyl-GA_3 (*37*), and GA_2 were also identified. As in the case of GA_{16}, GA_{47} was apparently formed not by hydration of GA_7 but by direct hydroxylation of GA_4. The formation of GA_{16} and GA_7 from GA_4—involving, as it does, removal of the 1α- and 2α-hydrogens, respectively—might also be achieved by interception of enzyme-bound intermediates of the 1,2-dehydrogenation of GA_4. McInnes et al. examined the effect of culture age on the metabolism of $[^{14}C]GA_4$ in complex media fermentations; the more significant results are summarized in Table 5.5. It can be seen that as the culture ages, the 5-day feeds of $[^{14}C]GA_4$ produce less GA_3, but more GA_1 and its isomers GA_{16} and GA_{47}. The increase in GA_2 may be related to the greater quantities of GA_4 remaining in feeds to the older cultures. These results suggested that the dehydrogenase enzyme is rather unstable. At least one enzyme early in the pathway was also found to be sensitive to culture age, as no labeled GAs were formed when $[2-^{14}C]$mevalonate was fed to 5-day-old fermentations.

After the work of Cross et al. (1968c) suggesting the intermediacy of GA_{12} aldehyde, the pathway immediately following the aldehyde was investigated in more detail by Evans and Hanson (1975). Feeds were made of $[17\text{-}^{14}C]GA_{12}$ alcohol (*38*), $[17\text{-}^{14}C]GA_{12}$ aldehyde, and $[17\text{-}^{14}C]GA_{12}$ [GA_{12} alcohol had previously been shown to be an efficient precursor of GA_3 (Bearder et al., 1973e; Hanson & Hawker, 1973)]. As a basis for interpreting the relative proportions of metabolites from these feeds, an incubation with $[2\text{-}^{14}C]$mevalonate was performed to show the natural pattern of GA production. Percentage incorporation of mevalonate into GA_3 after 5 days fermentation is shown in Table 5.6. GAs were separated by a combination of Sephadex LH-20 partition chromatography and TLC, and identified by radio-TLC, gas chromatography (GC), or MS. The results of the feeds of the three GA_{12} derivatives are shown in Tables 5.6, 5.7, 5.8, and 5.9. The feeds were made for 6 or 24 hr, but in the case of the aldehyde the fermentation was harvested after only 3 hr because the aldehyde was found to be unstable under the culture conditions, undergoing auto-oxidation to, inter alia, GA_{12}.

Of the three substrates, the results from the acid feed stand out. Although the incorporation of $[17\text{-}^{14}C]GA_{12}$ into GA_3 was not recorded, it is known from the results of Cross et al. (1968c) that this is relatively low compared to the incorporation of the aldehyde. GA_{12} therefore acts mainly as a precursor to the non-3-hydroxylated GAs, and in agreement with Cross et al. (1968c) it does not lie on the main pathway to GA_3. The $[17\text{-}^{14}C]GA_{12}$ alcohol (*38*) feed was the most similar to the $[2\text{-}^{14}C]$mevalonate feed in the distribution of radioactivity in the metabolites. The relatively high proportion of non-3-

Table 5.5. Effect of culture age on metabolism of $[^{14}C]GA_4$ in *G. fujikuroi* grown on medium B

Day of feed	Unchanged GA_4 (%)	Radioactivity (%)				
		GA_{16}	GA_{47}	GA_2	GA_1	GA_3
0	2.6	2.1	1.4	1.8	21	70
3	6.1	1.8	1.5	1.2	33	55
7	20	4.4	2.4	1.9	39	26

After McInnes et al. (1977).

Table 5.6. Incorporation of $[2\text{-}^{14}C]$ mevalonate into various metabolites in *G. fujikuroi*

	Incorporation after 5 days (%)
Ent-kaurene (*1*)	9.9
Ent-7α-hydroxykaurenoic acid (*5*)	2.6
GA_9	1.1
GA_{37}	4.3
GA_{14}	1.3
GA_4/GA_7	—
GA_{12}	5.2
GA_3	18.1

After Evans & Hanson (1975).

Table 5.7. Incorporation of $[17\text{-}^{14}C]\,GA_{12}$ alcohol (*38*) into various metabolites in *G. fujikuroi*

	Incorporation (%)	
	6 hr	24 hr
GA_9	1.3	1.2
GA_{15}	0.5	0.41
GA_{24}	1.94	2.03
GA_4	8.9	0.97
GA_7	6.7	0.35
GA_{37}	13.5	7.2
GA_{25}	3.1	2.02
GA_{14}	12.6	9.3
"Unknowns"	5.0	6.6
GA_{13}	6.7	7.8
GA_3	21.6	39.8

After Evans & Hanson (1975).

Table 5.8. Incorporation of $[17\text{-}^{14}C]\,GA_{12}$ aldehyde into various metabolites in *G. fujikuroi*

	Incorporation (%)
	3 hr
GA_9	6.8
GA_{15}	4.5
GA_{24}	9.9
GA_4	5.7
GA_7	2.6
GA_{37}	1.4
GA_{25}	4.4
GA_{14}	11.5
"Unknowns"	8.4
GA_{13}	1.73
GA_3	3.95

After Evans & Hanson (1975).

hydroxylated metabolites in the $[17\text{-}^{14}C]\,GA_{12}$ aldehyde feed versus the $[17\text{-}^{14}C]\,GA_{12}$ alcohol was ascribed to non-enzymic auto-oxidation of the aldehyde to GA_{12}. As the amount of non-3-hydroxylated GAs is normally quite small, it was evident that 3-hydroxylation of GA_{12} aldehyde occurs much faster than oxidation at C-7. Cross' conclusions were also echoed in the feed of $[17\text{-}^{14}C]\,GA_{14}$. Feeds harvested after 24 hr showed only substrate; radio-TLC revealed no other metabolites.

Table 5.9. Incorporation of $[17\text{-}^{14}C]\,GA_{12}$ into various metabolites in *G. fujikuroi*

	Incorporation (%)	
	6 hr	24 hr
GA_9	23.3	20.9
GA_{15}	9.4	10.1
GA_{24}	25.4	25.2
GA_{25}	17.7	10.5

After Evans & Hanson (1975).

Thus, it was concluded that the intermediates immediately following GA_{12} aldehyde in the main pathway either remained at the 7-aldehyde level or were held in a bound form. Further comment on the C_{20}-GA pathway will be made in Section 5.2.2 when considering similar studies to those of Evans and Hanson (1975), made by Bearder, MacMillan, and Phinney (1975b). The efficient metabolism of GA_{12} provided a useful preparation of the rare gibberellins GA_{24} and GA_{25} in 17% and 8% yield, respectively. However, this sort of micro-biological preparation is now much more readily achieved using the mutant fungus B1-41a or wild-type strains inhibited with growth retardants (see Sections 5.2.2, 5.2.3, and Chapter 7). In a similar way, $[17-^{14}C]GA_{12}$ alcohol (38) has been used for the biosynthesis of $[17-^{14}C]GA_3$ of high specific activity (Hanson & Hawker, 1973). Here again the 7-fold dilution of activity by endogenous GA_3 experienced in this work could be avoided by other fermentation techniques. Evans and Hanson (1975) also made separate feeds of $[17-^{14}C]GA_4$ and $[17-^{14}C]GA_7$ which were incorporated into GA_3 in yields of 30.3% and 43.5%, respectively—in agreement with the previously described work of Vining and co-workers (Pitel et al., 1971b).

The mechanism of the loss of C-20 in the conversion of C_{20}- into C_{19}-GAs, and the structure of the substrate for this process have been a longstanding topic of interest. It has already been mentioned that Birch's initial thoughts concerning the possible similarity of the reaction to steroid demethylation (Birch et al., 1959) had been dis-proved by Hanson's work with doubly labeled mevalonates. Further radiotracer work on this part of the pathway was also carried out by Hanson and associates, as will now be summarized. Hanson and White (1969) had suggested the possibility of a Baeyer-Villiger-type oxida-tion* of a C-10 carbonyl function as a suitable biosynthetic route to the C_{19}-GAs (Fig. 5.23, pathway b). As GA_{13} was not incorporated into GA_3, Hanson and Hawker (1972b) fed another candidate precursor, GA_{13} anhydride (39). $[17-^{14}C]GA_{13}$ was prepared by microbiological transformation of $[17-^{14}C]GA_{12}$ alcohol (38) in G. fujikuroi. The $[17-^{14}C]GA_{13}$ was isolated and extensively purified by Sephadex G25 partition chromatography and preparative TLC. Controlled pyrolysis of the acid furnished $[17-^{14}C]GA_{13}$ anhydride (39), which was incubated with a culture of G. fujikuroi for 18 hr. Although hydrolysis of the anhydride occurred over a period of hours

---

*Section 5.2.2 describes how it was subsequently shown that the Baeyer-Villiger process is not involved in the conversion of C_{20}- into C_{19}-GAs.

(39) GA_{13} anhydride   (41) GA_{13} aldehyde

Structures (*39*), (*41*)

to give GA_{13} as the major radioactive metabolite, specific incorporations into GA_4 and GA_7 (0.07%) and GA_3 (0.14%) were recorded. Under identical experimental conditions, $[17\text{-}^{14}C]GA_{13}$ was not incorporated into the C_{19}-GAs. Full experimental details of this work have not appeared, and so it is impossible to judge its full significance. It should be remembered that Cross et al. (1968b) found that *ent*-$[17\text{-}^{14}C]$kaur-6,16-dien-19-ol (*18*) was specifically incorporated into GA_3 in 0.48%, yet this compound was not considered to be a normal precursor to GA_3. A feed of GA_{13} anhydride (*39*) to *G. fujikuroi* mutant B1-41a (Bearder et al., 1975b; Section 5.2.2) could not corroborate Hanson and Hawker's result, but the experimental technique adopted might not have detected conversions of the size observed by Hanson and Hawker.

Support for Cross' theory that C_{20}-GAs in the main pathway have C-7 at the aldehyde oxidation level was provided by further work by Hanson and co-workers. Dockerill, Evans, and Hanson (1977) claimed to have isolated $[7\text{-}^3H, 17\text{-}^{14}C]GA_{13}$ aldehyde from an incubation of a 10,000-g cell-free system (supplemented by pyridine nucleotides) with $[7\text{-}^3H,17\text{-}^{14}C]GA_{12}$ aldehyde (Fig. 5.17). The identification rests on the following indirect evidence. The metabolite—which contained both tritium and ^{14}C—was readily auto-oxidized with loss of the tritium label to give GA_{13}, identified by co-chromatography and MS. A less-polar metabolite of $[7\text{-}^3H,17\text{-}^{14}C]$-$GA_{12}$ aldehyde was identified as GA_{14} aldehyde by mass spectrometry of the derived methyl ester and by oxidation and methylation to 3-oxo-GA_{14} dimethyl ester (*40*) (Fig. 5.17). Both of these metabolites were re-fed to *intact* cultures of *G. fujikuroi*. The presumptive GA_{14} aldehyde was incorporated into GA_3 in 7% yield; the "GA_{13} aldehyde" was incorporated into GA_3 (12.9%) and GA_4 and GA_7 (2.9%). Again, in the absence of detailed data such as levels of radioactivity used, the significance of these results is as yet indeterminate. However, recent work in which GA_{13} aldehyde (*41*) has been partially synthesized suggests that the substance isolated by Hanson and co-workers was not GA_{13} aldehyde (Bearder et al., 1982). Hanson's group also

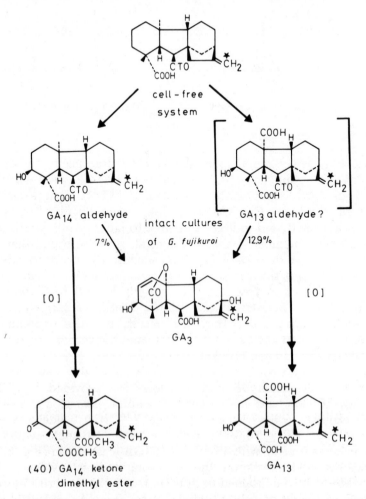

Figure 5.17 Summary of work done by Dockerill et al. (1977) postulating the intermediacy of GA_{13} aldehyde in the biosynthesis of C_{19}-GAs.

conducted an experiment that was carried out to determine the oxidation state of C-20 after it has been lost (Dockerill et al., 1977; Dockerill & Hanson, 1978). [2^{-14}C]Sodium acetate was fed to *G. fujikuroi* to give *ent*-kaurene labeled, inter alia, at C-20 (Fig. 5.18). Optimal incorporation (0.97%) was found to occur in a 12-hr incubation on day 4 of the fermentation. This labeled *ent*-kaurene was then fed to a fermentation of *G. fujikuroi* to which had been added AMO-1618 (66) to minimize dilution of label. Under these

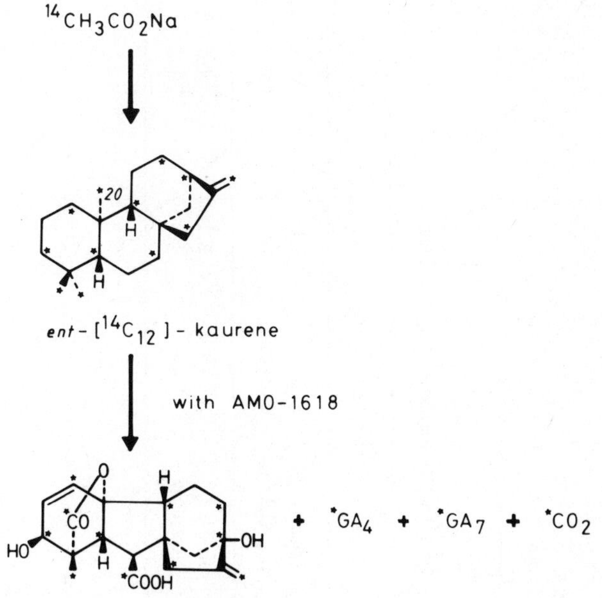

Figure 5.18 The fate of C-20 in C_{19}-GA biosynthesis. (Dockerill & Hanson, 1978.)

conditions the incorporation of *ent*-kaurene into GA_3 was raised from the normal 5% (Cross et al., 1964) to 31.6%. After optimization of the culture conditions, the *ent*-kaurene derived from sodium [2-^{14}C]acetate was incubated in sealed tubes with *G. fujikuroi* on day 4 for 24 hr. The fermentations were worked up by the injection of sulphuric acid, and the labeled CO_2 was swept with nitrogen into ethanolamine for scintillation counting. The residual culture medium was then extracted and the principal C_{19}-GAs GA_4 + GA_7 and GA_3 were isolated and counted. The results are recorded in Table 5.10. The possible origin of the $^{14}CO_2$ was limited to the C-20 of C_{20}-GAs. A blank reincubation of the [^{14}C]GA_3 evolved no radioactive CO_2, and so the conversion of GA_3 to allogibberic acid (*34*) was not the origin of the $^{14}CO_2$. The *ent*-[^{14}C]kaurene incubations were also examined for the presence of [^{14}C]formaldehyde or [^{14}C]formic acid, with no success. It seems likely, therefore, that the C-20 carbon is initially liberated as CO_2. However, as no control was performed the CO_2 detected might still have been derived from microbial oxidation of formaldehyde or formic acid. To account for this result a mechanism involving the decomposition of a peracid (Fig. 5.19) was postulated.

Table 5.10. Incubation of *Ent*-[^{14}C]kaurene with *G. fujikuroi* in the presence of AMO-1618

Experimental	*Ent*-[^{14}C]kaurene	GA$_3$	GA$_{4/7}$	Total C$_{19}$-GAs	CO$_2$ expected	CO$_2$ found	Recovery (%)
					dpm × 10^4		
1	151.3	16.96	13.19	30.15	2.740	2.182	79.6
2	151.3	15.18	12.90	28.48	2.589	1.972	76.2

After Dockerill & Hanson (1978).

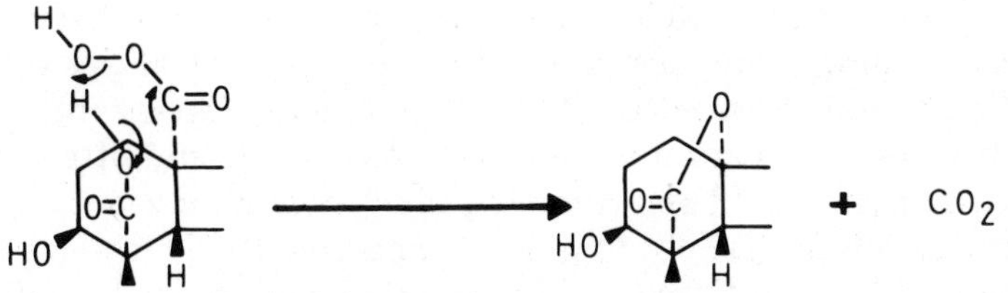

Figure 5.19 A mechanism for formation of C_{19}-GAs—decomposition of a peracid. (Dockerill & Hanson, 1978.)

5.2.2 Studies with mutant strains of *Gibberella fujikuroi*

The classic work of Beadle and Tatum (1941) with auxotrophic mutants of the fungus *Neurospora sitophila* led to the promulgation of the "one gene, one enzyme" theory. Since then, fungal mutants and bacterial mutants—especially of *E. coli*—have been used to plot the pathways of the biosynthesis of many primary (essential) microbial products. The techniques for mutant selection that were developed, such as replica plating and the use of penicillin to kill wild-type colonies, made the isolation of auxotrophic mutants a routine practice. The analysis of the shikimate pathway with auxotrophic mutants of *E. coli* is an excellent example of this type of work (Davis, 1955; Sprinson, 1960; Haslam, 1974).

As secondary metabolites are by definition not essential for growth, the simple techniques for selection of auxotrophic mutants cannot be used and each presumptive mutant colony must be individually examined to determine its phenotype. In certain circumstances (for instance, when the end product of the pathway is a pigment) presumptive mutants can be selected by visual inspection. Examples are studies of the biosynthesis of the tetracycline antibiotics (McCormick, 1967) and the carotenoid pigments (Cerda-Olmedo & Torres-Martinez, 1979). No such parameters can be used for studying the GA pathway in *G. fujikuroi*. The problem of discovering whether each colony still produces GA was elegantly solved by Phinney (see below) using the amylase halo half-seed assay (Fukuyama, 1971; Bearder, MacMillan, & Phinney, 1979a; Bukovac et al., 1979) which is a variation of the barley half-seed assay of Jones and Varner (1967).

The origin of the mutant strains used for the study of the GA pathway will now be discussed. The genetics of the fungus *G. fujikuroi* were first investigated by Gordon, who had acquired a large collection of strains from widely differing hosts and geographical locations (Gordon, 1960). Gordon was occasionally successful in crossing some of the strains and was able to show that the fungus was heterothallic and that mating type was controlled by a pair of alleles (Gordon, cited in Phinney & Spector, 1967). Significant progress could not be made because production of perithecia was erratic and often took up to a year to appear. This stumbling block was removed by Phinney (Spector & Phinney, 1966; Phinney & Spector, 1967) who found that a medium containing stems of *Citrus medica* supported the development of the sexual stage of the fungus. When strains of opposite mating type were grown in this medium, perithecia were regularly produced in 3 to 6 weeks. A large number of strains were available for study because Phinney had acquired Gordon's collection and also personally collected a large number of strains in the field in Japan.

Genetic studies with these strains resulted in the identification of two genes, g_1 and g_2, which controlled GA production (Spector & Phinney, 1968). The gene g_1 controlled the whole pathway, and $g_1 -$ strains neither produced GAs, nor metabolized added precursors such as GA_{12} aldehyde (Phinney, unpublished work). The gene g_2 controlled 13-hydroxylation since $g_2 +$ strains produced the whole spectrum of GAs including GA_1 and GA_3, whereas $g_2 -$ strains synthesized no GA_1 or GA_3 but accumulated GA_4 and GA_7. This interpretation was based on the previously discussed work by Geissman et al. (1966) which indicated that GA_4 and GA_7 were precursors of GA_1 and GA_3. The identification of $g_2 -$ strains was originally made by TLC examination of culture filtrates. However, one $g_2 -$ strain, mutant R-9, was studied in more detail. This was obtained from a wild-type strain N-3844 isolated in the field in Japan. During subculturing, perithecia fortuitously appeared and the ascospores were separated and germinated to give 500 different strains, of which the mutant R-9 was one.

The strain REC-193A was obtained in a similar manner to R-9. It had characteristics that made it a suitable strain to use for the generation of GA-less mutants by using Phinney's selection procedure. First, it was a wild-type for GA production. Second, it had a nonspreading style of growth on agar, which allowed many distinct colonies to be grown on the same plate. Third, it produced low levels of extra-mycelial α-amylase. The fact that this strain was slow growing made possible a number of feeding studies in the trophophase (see Bu'Lock, 1967) before normal GA production had begun.

The use of REC-193A as a wild-type parent for the production of mutants became redundant when Phinney discovered that colonial-type growth could be induced by incorporation of griseofulvin into the agar plating medium. Thus a high-producing strain of *G. fujikuroi* could be substituted. The strain chosen was a vigorous wild-type, GF-1a, which was a good GA producer.

The mutant B1-41a was derived by UV irradiation of the wild-type GF-1a and selected as a GA-deficient mutant from 50,000 GA-producing strains by the barley halo half-seed assay. A typical mutation/selection experiment was performed as follows. Micro-conidia from a wild-type GA-producing strain were irradiated with UV light to *ca*. 85% killing. The microconidia were then plated out onto potato-dextrose agar (PDA) plates, and incubated in the dark for 24 hr. After incubation in light for a further 48 hr, each colony was assayed for gibberellin activity in the amylase halo half-seed assay. The assay is performed by placing an embryo-less half of a barley seed on top of each colony. After 2 days the half-seeds are removed and are placed onto PDA in petri dishes. After a further 2 days an aqueous solution of potassium iodide and iodine is added to the petri dishes, and a colorless halo appears around half-seeds that have been in contact with GA-producing colonies. The size of the halo that appears is proportional to the quantity of GA produced by the corresponding colony. Thus non-GA-producing colonies may be identified. The basis of the assay is as follows. GAs from the colony stimulate the release of α-amylase from the aleurone layer in the half-seed. The α-amylase diffuses radially from the half-seed into the PDA and hydrolyzes the starch present. The iodine that is added forms the characteristic blue color only with the starch, so that areas of α-amylase activity stand out as colorless haloes. This assay is sensitive only to C_{19}-GAs.

The biosynthetic studies using strains R-9, REC-193A, GF-1a, and B1-41a will now be presented.

The GA phenotype of strain R-9 was confirmed by combined gas chromatography-mass spectrometry (GC-MS) analysis of culture filtrates (Bearder et al., 1973e). GA_1 and GA_3 were absent but otherwise the normal complement of diterpenoids were observed, including GA_4, GA_7, and GA_{16} (Bearder et al., 1973e; Bearder, 1973).

A more definitive demonstration that mutant R-9 did not produce GA_1 and GA_3 was provided by growing the mutant in the presence of $[6-^3H]GA_{12}$ alcohol (*38*) (Bearder et al., 1973e), which although not a true intermediate had been shown to be an efficient precursor of GA_3 in mutant B1-41a (see later) and in wild-type ACC-917 (Hanson & Hawker, 1973). After 11 days the cultures were worked up in the usual way. No radioactivity was detected at the Rf of GA_{12} alcohol (*38*) methyl ester and only a trace at the Rf of GA_3 methyl ester where GA_1 and GA_{16} methyl esters would also occur. Most of the radioactivity was located at the Rf corresponding to GA_4 and GA_7 methyl esters. This band was recovered and shown by GC-MS and combined gas chromatography-radio-counting (GC-RC) to contain radio-labeled GA_4 and GA_7 methyl esters. It was notable that the GC-RC results indicated that the GA_4 had a much higher specific activity than the GA_7. This indicated that GA_7 was synthesized quicker via endogenous substrates than from the added GA_{12} alcohol. This is probably a consequence of the relatively short lifetime of the 1,2-dehydrogenating enzyme noted by McInnes et al. (1977), enhanced by the slow incorporation of GA_{12} alcohol (*38*) into C_{19}-GAs. It is suggested that GA_{12} alcohol is metabolized rapidly via GA_{14} alcohol (*54*) and GA_{14} aldehyde to GA_{14}, which temporarily accumulates and causes diffusion of the GA_{14} into the culture medium (see Fig. 5.21). The polar GA_{14} molecule can diffuse back into the mycelium only relatively slowly, effectively limiting the rate of incorporation of GA_{12} alcohol into C_{19}-GAs. This idea will be returned to in relation to the metabolism of C_{20}-GAs in mutant B1-41a. The absence of significant quantities of radio-labeled GA_3 in the feed of $[6\text{-}^3\text{H}]GA_{12}$ alcohol to strain R-9 was established by the addition of unlabeled GA_3 methyl ester which was recovered having a ^3H incorporation of less than 0.05%. The position of the genetic lesion was thus firmly identified. The strain seemed ideal for testing the possible biosynthesis of GA_3 from GA_1, as the absence of GA_3 removed the possibility of product feed-back inhibition. $[1,2\text{-}^3\text{H}_2]$-$GA_1$ (5 Ci mmol.$^{-1}$, 22 μg) was incubated with R-9 for 11 days. The extract of the culture filtrate, which contained 94% of the added radioactivity, was supplemented with unlabeled GA_3 and fractionated by partition chromatography by the method of Pitel, Vining, and Arsenault (1971a). The GA_1 peak contained 90.5% of the radio-activity, but the GA_3 showed an incorporation of only 2.7% or 5.4% (depending on the exact location of the tritium in the substrate). In another experiment the metabolism of large amounts of (unlabeled) GA_1 could not be observed. This low incorporation of GA_1 into GA_3 was consistent with the previously discussed work by Pitel et al. (1971b), which was published at about the same time.

GC-MS analysis of an 11-day culture of REC-193A grown on a potato dextrose liquid (PDL) medium revealed the presence of the usual diterpenoid metabolites, but in low yield (Bearder, MacMillan, & Phinney, 1973d). The following compounds were detected: GA_{14}, GA_{13}, GA_9, the triacid (*9*), GA_4, GA_7, the diacid (*8*), GA_1, iso-GA_3 (*33*), GA_3, GA_{12}, and GA_{25}. In 5-day cultures of this strain virtually none of these metabolites were present despite the considerable quantity of mycelium formed. At that time, GA_{12} aldehyde had recently been shown to be an efficient precursor to GA_3 (Cross et al., 1968c), and so it was decided to investigate the metabolism of GA_{12} aldehyde at this early stage (trophophase) of fungal growth. Media containing GA_{12} aldehyde were inoculated with REC-193A, and after 5 days' growth the culture filtrate was examined by GC-MS. The GA_{12} aldehyde was completely metabolized to two compounds identified as GA_{12} and GA_{14}. Up to 5 mg of GA_{12} aldehyde could be completely transformed by 100 ml of mycelial suspension. Time-course studies showed that GA_{12} and GA_{14} were initially formed in equal amounts but that after 3 days the GA_{12} reached a maximum concentration and thereafter only GA_{14} increased until the GA_{12} aldehyde was completely consumed. This seems to suggest that the amount of GA_{12} is controlled by feedback inhibition or that the 3-hydroxylating enzyme is synthesized after the C-7 oxidizing enzyme. In a control experiment in the absence of the fungus, GA_{12} aldehyde remained unchanged. When GA_{12} was fed to the fungus under similar conditions it remained unmetabolized. To prove that the GA_{12} and GA_{14} had been derived from exogenously applied GA_{12} aldehyde, $[6-^2H]GA_{12}$ aldehyde was prepared and fed to the fungus in a similar fashion. GC-MS showed that GA_{12} and GA_{14} were formed without dilution of the ^2H-label. The degree of conversion of GA_{12} aldehyde was determined by feeding $[6-^3H]GA_{12}$ aldehyde in an identical manner. After 5 days, the metabolites were extracted and separated by liquid/liquid partition chromatography. The metabolites identified were GA_{12} (8.3%), GA_{14} (45%), GA_4 (ca. 17%), and GA_7 (ca. 6%). Other unidentified products were formed in low yield, including a metabolite tentatively identified by MS as a 1-hydroxy GA_{12} (46).

Structures (42)–(47)

Structures (48)–(53)

This work corroborated that of Cross et al. (1968c) and also demonstrated that in the early growth phase of the fungus, although gibberellins are not produced, some of the enzymes of the GA pathway are present.

The strain GF-1a is a vigorous high-yielding strain of *G. fujikuroi* which gives a wide range of metabolites similar to ACC-917 (Bearder et al., 1974). In a detailed investigation of mevalonate-derived metabolites of strain GF-1a, MacMillan and Wels (1974) detected 72 compounds of which 25 were known diterpenes, including 15 GAs. The compounds identified are listed in Table 5.11. The structures assigned to these new compounds were only tentative. Bearder et al. (1974) made a simple time-course study of gibberellin production in GF-1a on a synthetic medium. After 36 hr the mycelium had begun to pigment, signalling roughly the depletion of inorganic nitrogen and the onset of GA production (Bearder et al., 1973a; Bu'Lock et al., 1974). Analysis of the culture filtrate extract at this stage by GC-MS showed the presence of GA_9, GA_{14}, GA_{24}, GA_{25}, the diacid (8), and the triacid (9). After 60 hr these metabolites were still present but GA_1, GA_3, GA_4, GA_7, GA_{13}, GA_{15}, and GA_{16} were also present. After 7 days the only change in the metabolites in the culture filtrate was an increase in the amount of GA_3. The neutral metabolites consisted mainly of the three kaurenolides (10), (11), and (44). A methanolic mycelial extract was shown to contain *ent*-kaurene (1) as the major component, but traces of *ent*-kaurenol (2) and *ent*-kaurenal (3) were also detected. The early production of the non-3-hydroxylated GAs agrees well with the observed metabolism of GA_{12} aldehyde in REC-193A, which suggests that either the formation of non-3-

Table 5.11. Compounds in $[2\text{-}^3H_2]$mevalonate feeds to *G. fujikuroi*, strain GF-1a, identified by GC-MS in fractions from a Sephadex LH20 partition column

Fraction	dpm × 10³	Compound	Fraction	dpm × 10³	Compound
A	867	—	M	190	GA_7, (35), GA_{25}, GA_{37}
B	812	—	N	349	(46), GA_{14}
C	2,317	(7), (42), (8), (45)	O	545	GA_{14}, GA_{54}, (45), (9), (47)
D	1,217	(7), (43)	P	2104	mevalonate
E	594	(43)	Q	677	GA_{16}, GA_{36}, GA_{47}
F	810	GA_9, GA_{12}, GA_{15}, (6)	R	465	GA_{13}, GA_{16}, GA_{47}, (48)
G	264	(44), (6)	S	1341	GA_1, GA_3, GA_{13}
H	860	GA_{24}, (12), (44)	T	4981	GA_1, GA_3, GA_{13}, (49)
I	1,277	GA_4, GA_{24}, (7), (43), (8), (12), (45)			
J	3,084	GA_4, (7), (8)	U	979	GA_3, (33)
K	1,209	GA_7, (8)	V	188	(33), (50)
L	637	GA_7, GA_{25}, GA_{37}	W	86	(49)

After MacMillan & Wells (1974).

295

hydroxylated GAs is controlled by feedback inhibition or that the 3-hydroxylating enzyme is not synthesized before those that oxidize the 20-position.

When mutant B1-41a was grown on the same synthetic medium as GF-1a, it was indistinguishable from its parent in rate of growth and morphological appearance (Bearder et al., 1974). After 7 or 11 days no metabolites could be identified in culture filtrates when aliquots of derivatized extracts identical to those used for the parent strain were examined by GC. When much larger aliquots of mutant B1-41a culture filtrate were examined, traces of GA_{13}, GA_3, and the two seco-acids (8) and (9) were detected. This experiment indicated that the percentage leak of the mutant was ca. 2%. A value of 2.7% was obtained by a comparison of the incorporation of $[2\text{-}^{14}C]$-mevalonic acid into GA_3 in resuspension cultures of GF-1a and mutant B1-41a. A mycelial extract of mutant B1-41a contained ent-kaurene (1), ent-kaurenol (2), and ent-kaurenal (3). As in the case of the parent GF-1a, ent-kaurene was the principal component. Although still a minor metabolite, the aldehyde (3) accumulated to a greater extent in B1-41a than in GF-1a.

The position of the metabolic block was determined by a series of feeding studies of labeled and unlabeled substrates. When mutant B1-41a was grown in 50% N-ICI medium supplemented with $[2\text{-}^3H_2]$-mevalonate, radio-TLC showed the presence of ent-kaurene (15% incorporation). The mycelial and culture filtrate extracts from mutant B1-41a grown for 6 days on 50% N-ICI medium containing ent-$[15,17\text{-}^3H]$kaurene were examined by radio TLC. In the mycelial extract 58% of the total radioactivity fed co-chromatographed with ent-kaurene and 1.2% with ent-kaurenal (3). In the culture filtrate extract 2% of the activity ran with ent-kaurene and 4% remained on the origin. Thus only 4% of the added ent-kaurene was metabolized to substantially polar compounds. This experiment also illustrates that ent-kaurene resides predominately in the mycelium and that published incorporations of various precursors into ent-kaurene, calculated from ent-kaurene isolated from culture filtrates only, are probably gross underestimates.

For subsequent feeds to mutant B1-41a, resuspension cultures were used to ensure reproducible mycelium/substrate ratios and to reduce the amounts of interfering endogenous metabolites. This technique involved growing the fungus in 50% N-ICI medium until pigmented (3–4 days) and then harvesting the mycelium by filtration and resuspending in fresh medium containing no ammonium nitrate and normally adjusted to pH 3.5. Under these conditions unlabeled ent-kaurenol (2), ent-kaurenal (3), and ent-kaurenoic acid

(*4*) were fed to mutant B1-41a in separate experiments. From *ent*-kaurenol and *ent*-kaurenal no metabolites were detected in the culture filtrates and GC traces of the derivatized mycelial extract from each feed were identical to those of the controls, except for the intense peaks of the unmetabolized *ent*-kaurenol and *ent*-kaurenal, respectively. In contrast, *ent*-kaurenoic acid was rapidly and completely metabolized by resuspended cultures of B1-41a. After 20 hr, no *ent*-kaurenoic acid could be detected in the mycelium or culture filtrate, and the latter was shown by GC and GC-MS to contain the same metabolites in approximately the same proportions as the culture filtrates of an unsupplemented 7-day-old culture of the parent strain GF-1a. These experiments indicated that the metabolic block in the mutant was at the conversion of *ent*-kaurenal into *ent*-kaurenoic acid.

Since *ent*-kaurenol or *ent*-kaurenal neither accumulated in the mutant nor were metabolized in feeding experiments, the results could also be accomodated by a metabolic block in all oxidations at the 19-position of *ent*-kaurene.

The mutant B1-41a proved to be an excellent vehicle for the study of GA biosynthesis in the pathway after *ent*-kaurenoic acid. The use of the mutant B1-41a had some important advantages over studies with wild-type strains. Added substrates did not compete with endogenously produced metabolites for enzyme sites and were therefore metabolized more efficiently. The virtual absence of endogenous diterpenoid metabolites in culture filtrates allowed the use of unlabeled substrates, metabolites being identified semi-quantitatively by GC-MS. Resuspension cultures of the mutant were employed to increase reproducibility of feeding studies and to eliminate traces of endogenous metabolites from the original culture medium. The validity of the procedure was confirmed by the similarity of results with mutant B1-41a and some of those previously published using conventional techniques (e.g., Pitel et al., 1971b).

This technique greatly simplified biosynthetic studies, and using these methods Bearder et al. (1975b) reported the metabolic fate in the mutant of 22 kaurenoids and GAs. The results of these experiments are shown in Table 5.12 and interpreted in Fig. 5.20. These results are most conveniently discussed in a retrograde sequence starting at GA_3, the major terminal GA in wild-type strains and one that remained unmetabolized when fed to mutant B1-41a.

GA_4 was almost completely metabolized to GA_3, but minor quantities of GA_7, GA_{16}, and GA_1 were detected. GA_1 and GA_{16} were not metabolized when re-fed. Thus GA_{16} is neither a precursor nor a product of GA_7 as suggested by Katsumi and Phinney (1969). The

Table 5.12. Metabolites of substrates incubated with *G. fujikuroi* mutant B1–41a at pH 3.5 for 20 hr

Substrate	Metabolites (% total, ± 3%)[a]						
	GA_3	GA_1	GA_{16}	GA_{13}	(7)	(9)	GA_9
Ent-kaurenoic acid (*4*)	28	2	1	7	6	4	1
Ent-7α-hydroxy-kaurenoic acid (*5*)	41	1		4		4	3
Ent-7-oxokaurenoic acid (*51*)						100	
GA_{12} alcohol (*38*)[b]	17			5			
GA_{12} alcohol (*38*)	28			6			
GA_{12} aldehyde[b]	4						4
GA_{12} aldehyde	20	2	1	4			6
GA_{12} aldehyde[c]	52	1	1	9			9
GA_{12}[b]							37
GA_{12}	7			2			58
GA_{14} aldehyde	34			8			
GA_{14}	32	2	1	4			
GA_{24}							
GA_{25}[c]							
GA_{15}[c]							
GA_{36}[c]							
GA_{37}							
GA_4	79	2	2				
GA_4[b]							
GA_7	71						
GA_7[b]							
GA_{16}			100				
GA_1		100					
GA_3[b]	100						
GA_{42}							
GA_{13} anhydride (*39*)				100			

After Bearder et al. (1975).
[a] Determined by GC on 2% QFl as methyl esters or methyl ester

Metabolites (% total, ± 3%)[a]												
GA$_{24}$	GA$_{25}$	(10)	GA$_{42}$	GA$_4$	GA$_7$	(8)	GA$_{14}$	GA$_{15}$	GA$_{12}$	(52)	GA$_{36}$	GA$_{37}$
1	1	2	1	4	1	21	22					
1	8					28	10					
				2	2		73					
				2	2		62					
	1			1			85	3				
2	2			2	3		54	2	2			
1	10		7	2	1		5	2				
7	11								40			
14	4						8	5	2			
			2	2			54					
		1	4	4		50					1	
45	51									4		
	100											
							100					
											100	
												100
				2	15							
				100								
					29							
					100							
			100									

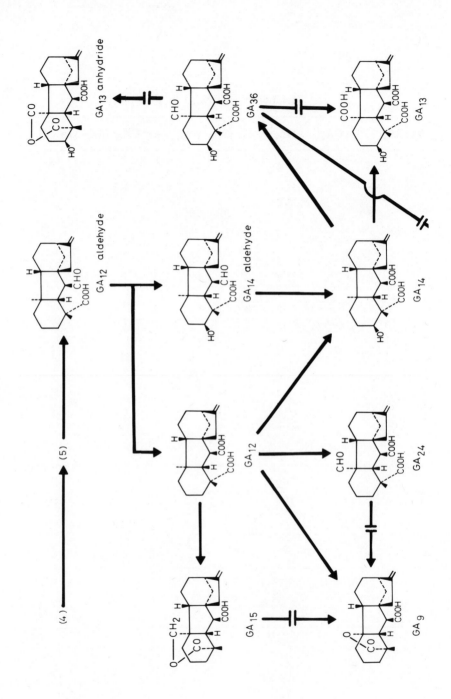

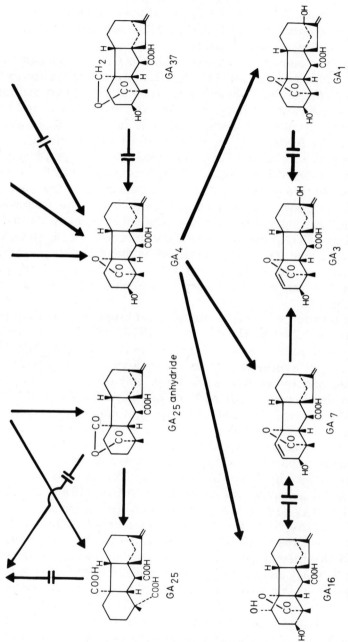

Figure 5.20 A scheme illustrating the feeding studies with mutant B1-41a. (Bearder et al., 1975b.)

non-metabolism of GA_1 agrees with the previously mentioned feeds of GA_1 to mutant R-9. These results are almost identical to those that Vining and co-workers (Pitel et al., 1971b) achieved with the wild-type ACC-917 on synthetic media.

GA_{14} was converted in good yield into GA_3, in contrast to the result of Evans and Hanson (1975). This may be explained by the absence of endogenous competing substrate in the mutant. Although substantial quantities of GA_{14} remained unmetabolized after 20 hr it was almost completely consumed after 5 days. [It should also be noted that in feeds of GA_{14} to mutant B1-41a by Hedden et al. (1974), when 40% of the GA_{14} remained after 5 days a higher pH (4.5) was employed.] The relatively slow metabolism of GA_{14} seems most likely to be a result of inefficient translocation to the site of the C-20 oxidizing enzyme(s) due to its high polarity. This is supported by the finding that at pH 7.0 when GAs are completely ionized the metabolism of GA_{14} was completely suppressed. Under these conditions GA_4 and GA_7 also remained unmetabolized, presumably for similar reasons. Precursors of GA_{14} such as GA_{12} [and GA_{12} alcohol (*38*)] also gave temporary accumulations of GA_{14} as noted for mutant R-9.

The position of GA_{14} aldehyde as an intermediate in the pathway was established by Hedden et al. (1974). [6-^{3}H]GA_{12} aldehyde was fed to a resuspension culture of mutant B1-41a. After 2 hr, GA_{14} aldehyde synthesized from $3\beta,7\beta$-dihydroxykaurenolide (*44*), was added and the culture worked up. Isolation of the GA_{14} aldehyde revealed that it contained 0.5% of the radioactivity of [6-^{3}H]-GA_{12} aldehyde fed. A parallel culture that was not supplemented with GA_{14} aldehyde was also worked up after 2 hr and analyzed by GC-MS. Almost all the GA_{12} aldehyde had been converted into GA_{14}. In the same publication, GA_{14} aldehyde was shown to be a good precursor of the 3-hydroxy GAs. In a 5-day fermentation with mutant B1-41a, GA_{14} aldehyde was metabolized to GA_3 (15%), GA_7 (17%), GA_1 (5%), GA_4 (19%), GA_{13} (5%), GA_{14} (9%), and an isomer of GA_1 (15%). GA_{14} alcohol (*54*) was also fed under similar conditions and gave the same range of metabolites except for the absence of GA_1 and the presence of a substance assigned the structure corresponding to GA_{42} alcohol (*53*). The alcohol (*54*) was not claimed as an authentic intermediate in GA biosynthesis, but was assumed to be converted via a metabolic grid such as Fig. 5.21. A parallel 5-day fermentation of GA_{14} gave a similar spectrum of metabolites to GA_{14}, although 40% remained unmetabolized and traces of GA_{42} were detected.

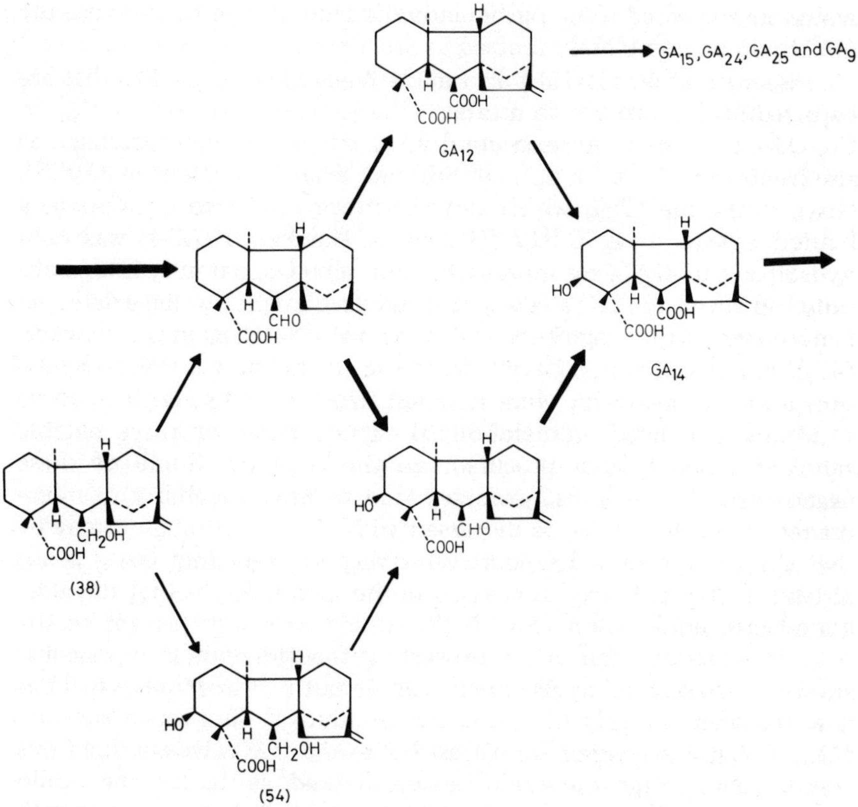

Figure 5.21 Metabolic grid involving oxidation at C-3 and C-7.

The conversion of GA_{12} aldehyde into GA_{14} in B1-41a is particularly efficient, as indeed it is in REC-193A. Hedden et al. (1974) found that in only 2 hr a feed of GA_{12} aldehyde was almost completely metabolized to GA_{14} by mutant B1-41a. If this conversion occurs as quickly in wild-type strains during normal GA biosynthesis— when GA_{14} does not accumulate—it follows that a previous step in the GA pathway must be rate limiting. In such a situation only low concentrations of GA_{14} will develop, and it may be metabolized before it diffuses into the culture medium. When high concentrations of a precursor such as GA_{12} aldehyde are fed, high concentrations of GA_{14} are produced in the mycelium causing rapid diffusion of GA_{14} into the medium. This extracellular GA_{14} must then be transported back into the mycelium to be further metabolized, a process that would be expected to be particularly difficult at high pH because the GA would be completely ionized.

Bearder et al. (1975b) also made feeds of the C_{20}- GAs that are oxidized at C-20 to try to establish the process involving the C_{20}- to C_{19}-GA conversion. As expected GA_{13} remained unmetabolized, as also reported Cross et al. (1968c) and Evans and Hanson (1975). GA_{13} anhydride (39), which served as a precursor to C_{19}-GAs to a limited extent in ACC-917 (Hanson & Hawker, 1972b), was only hydrolyzed to GA_{13} by mutant B1-41a, albeit apparently enzymatically. However, the GC analysis method used might not have detected conversions of the magnitude (~1%) recorded by Hanson and Hawker. GA_{37} and GA_{36}, which have C-20 at the alcohol and aldehyde oxidation level, respectively, were also fed. Unexpectedly, both of these substrates remained unmetabolized. Thus, none of these eligible candidates acted as a precursor to the C_{19}-GAs. Some of these negative results are most probably due to impermeability of membranes to the substrate, as discussed with GA_{14}. Although it may be that GA_{13} is a terminal GA, it seems very unlikely that GA_{36} is not at least on the pathway to GA_{13}. Another possibility is that the aldehyde/lactol equilibrium (Fig. 5.22), which probably lies far to the right in aqueous solution, is protecting the aldehyde group against enzymic attack. Similar arguments can be put forward for GA_{37}. Thus it would seem unlikely that the corresponding alcohol (open lactone) GA_{37} is not a precursor to at least GA_{36} and GA_{13}. However, in this case the alcohol group would remain masked, as the lactone would not be in equilibrium with the hydroxy-acid. Indeed, other experiments have indicated that GA_{37} in the lactone form cannot be a precursor to other C_{20}- or C_{19}-GAs (Bearder, MacMillan, & Phinney, 1976c). Another biosynthetic scheme that fits the facts is that intermediates between GA_{14} and GA_4 are enzyme-bound and not in equilibrium with exogenously supplied compounds. The C-20 oxidized C_{20}-GAs that appear in culture filtrates could be derived by leakage of these bound intermediates. Such bound intermediates are not

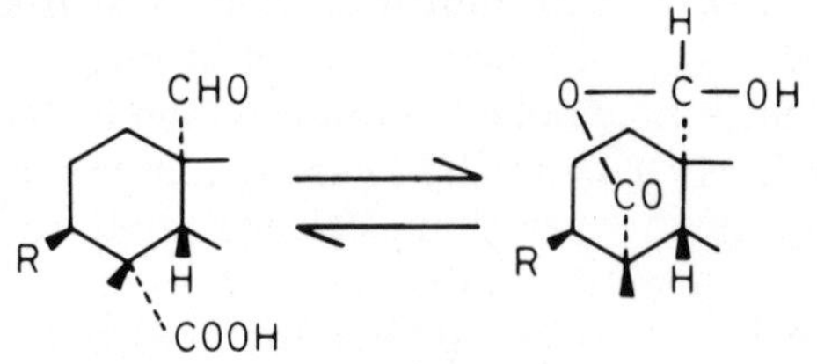

Figure 5.22   Aldehyde/lactol equilibrium in GA_{36} ($R = OH$) and GA_{24} ($R = H$).

uncommon in biosynthetic pathways (cf. fatty acid biosynthesis). However, it should be noted that experiments involving C_{20}-GAs with ^{18}O-labels in carboxyl functions at C-4 or C-6 preclude complexation by covalent links at these positions (Bearder et al., 1979a; Bearder and MacMillan, unpublished work).

Finally, an alternative pathway has been mooted which involves the 3-hydroxylated precursors to the C_{19}-GAs remaining at the aldehyde oxidation level at C-7. This suggestion originated from Cross' observation that GA_{12} aldehyde was a better precursor of GA_3 than was GA_{14} and took some support from the more efficient metabolism of GA_{14} aldehyde versus GA_{14} reported by Hedden et al. (1974). However, Bearder et al. (1975b) noticed no significant difference in the efficiency of metabolism of GA_{14} aldehyde compared to GA_{14}. The previously discussed tentative identification of GA_{13} aldehyde as a naturally occurring precursor of C_{19}-GAs by Dockerill et al. (1977) also gave weight to this theory. However, feeding studies by Bearder and Phinney (1979) suggests that GA_{13} aldehyde (*41*) is not an efficient precursor to C_{19}-GAs. They partially synthesized GA_{13} alcohol anhydride (*55*) and GA_{13} aldehyde anhydride (*56*) and fed each to mutant B1-41a for 5 days. In each case hydrolysis of the anhydride occurred to give as major products the aldehyde (*41*) and alcohol (*57*), respectively. In the case of the GA_{13} aldehyde anhydride (*56*) feed, approximately 20% GA_{13} was also formed. The GA_{13} alcohol anhydride (*55*) was hydrolyzed almost exclusively; only traces of GA_{13} aldehyde (*41*) and GA_{13} were detected. In neither case were any C_{19}-GAs detected. This implies that the "GA_{13} aldehyde" (Fig. 5.17) isolated by Dockerill et al. (1977) from a cell-free preparation of *G. fujikuroi*, and subsequently incorporated into GA_3 (13%) by an intact culture, was misidentified.

(55)

(56)

(57)

Structures (*55*)–(*57*)

These results—taken with the fact that no C_{19}-GA aldehydes have ever been identified as products of *G. fujikuroi*—would seem to make this theory rather contrived. Furthermore, as will now be discussed, GA_{12} is an excellent precursor to non-3-hydroxylated C_{19}-GAs (GA_{15}, GA_{24}, GA_{25}, and GA_9), and this pathway obviously does not proceed through 7-aldehyde intermediates (see below). It would seem unlikely that both pathways operate by a different mechanism, and so it would be more logical to invoke 7-carboxylic acid intermediates for both 3-hydroxylated and non-3-hydroxylated GA pathways and look elsewhere for the reasons behind the negative feeding results. It is also worth noting that there is no evidence for the participation of 20-functionalized C_{20}-GA-7-aldehydes in the pathway in higher plants (see Graebe et al., 1980).

Returning to the feeding studies by Bearder et al. (1975b), feeds of GA_{12} alcohol (*38*), GA_{12} aldehyde, and GA_{12} to mutant B1-41a pinpointed the branching point where the non-3-hydroxylated GA pathway diverges from the main (3-hydroxylated) pathway. These feeds were also done with [6-^{3}H]-labeled substrates to validate the general approach of being able to dispense with radiotracers when using the mutant B1-41a. As expected the metabolites of these substrates showed similar specific activities to their precursors (by GC-RC), proving the absence of endogenous unlabeled metabolites. At pH 3.5 the 20-hr feed of GA_{12} aldehyde gave a complete conversion into all the common fungal GAs. Non-3-hydroxylated GAs accounted for only about 5% of the total metabolites. The major GA was GA_{14}, but this was shown to be only a temporary accumulation as 5-day feeds under the same conditions gave low concentrations of GA_{14} and correspondingly higher quantities of GA_3. The pH 3.5 20-hr feed of GA_{12} alcohol (*38*) gave an almost identical range of GAs except that no non-3-hydroxylated GAs could be detected. In contrast, under the same culture conditions, GA_{12} gave only 17% 3-hydroxylated GAs, the balance of the metabolites being the non-3-hydroxylated GAs. This contrast is in accord with the results of Cross et al. (1968c) who recorded lower incorporations of GA_{12} compared to GA_{12} aldehyde into GA_3. Thus, for the first time it was shown that GA_{12} was an efficient precursor for the non-3-hydroxylated GAs.

These results are most readily interpreted in terms of the metabolic grid illustrated in Fig. 5.21. The complete absence of non-3-hydroxylated GAs from GA_{12} alcohol (*38*) feeds are due to preferential 3-hydroxylation to GA_{14} alcohol (*54*) rather than to conversion via GA_{12} aldehyde. It should be remembered that despite their efficient metabolism there was no evidence for the intermediacy of C-7 alcohols in the normal biosynthesis of GA_3 (Hanson & Hawker, 1973). At pH 7.0 a similar set of results was obtained except that for GA_{12} alcohol (38) and GA_{12} aldehyde feeds GA_{14} was even more prominent and other 3-hydroxylated GAs were present in only trace quantity. Under these conditions, GA_{12} was incompletely metabolized (40% remained) and only non-3-hydroxylated gibberellins (GA_9, GA_{15}, GA_{24}, and GA_{25}) were formed. Feeds of GA_{12} under these conditions would seem to be ideal for the preparation of the relatively rare non-3-hydroxylated C_{20}-GAs. The pH effect may be adequately explained in terms of increased ionization of substrates at high pH hindering transport across membranes. Thus the GA_{14} formed from GA_{12} alcohol (38) and GA_{12} aldehyde is expelled into the culture medium and cannot now be efficiently transported back into the mycelium to be further metabolized. In the GA_{12} feed, the slower migration of the substrate into the mycelium at high pH allows the preferential specificity of the 20-hydroxylating enzyme for GA_{12} compared with the 3-hydroxylating enzyme to be expressed more distinctly. [At pH 7.0 neither GA_4 nor GA_7 is metabolized, probably for a similar reason (Bearder et al., 1975b).] It would be expected that a 3-hydroxylated $GA(GA_{14})$ would be more highly ionized than its non-3-hydroxylated counterpart (GA_{12}) from inspection of their approximate partition coefficients between ethyl acetate and buffer (Durley & Pharis, 1972).

The effect of inorganic nitrogen on the progress of the GA pathway has been superficially investigated by Bearder et al. (1979a). Preliminary experiments suggested that the dehydrogenation of GA_4 to GA_7 was readily inhibited by the presence of inorganic nitrogen. This fact fitted the data of McInnes et al. (1977) who found that cultures of strain ACC-917 grown on a complex nitrogen source produce relatively larger amounts of GA_1, GA_{16}, and GA_{47} than when grown on a synthetic medium. It was noted that this finding could be explained by the fact that the complex medium slowly releases nitrogen, which inhibits the GA_4 1,2-dehydrogenase preferentially and allows 1α-, 2α-, and 13-hydroxylases to act slowly on the accumulating GA_4.

Pursuing the biogenesis of the non-3-hydroxylated GAs, GA_{15}, GA_{24}, GA_{25}, and GA_{25} anhydride (52) were fed to resuspension cultures of *G. fujikuroi* mutant B1-41a (Bearder et al., 1975b). In accord with the feeds of the corresponding 3-hydroxylated GAs (see above), no C_{19}- GA (GA_9) was produced; but in contrast to the non-metabolism of GA_{36}, the corresponding GA, GA_{24}, was metabolized to GA_{25} anhydride (52) and GA_{25}. This suggests the existence of the pathway $GA_{24} \rightarrow GA_{25}$ anhydride $\rightarrow GA_{25}$. Feeds of GA_{25} and GA_{25}

anhydride (52) were consistent with this pathway. This sequence points to GA_{24} being in the lactol form when it is oxidized (see Fig. 5.22). If a free 20-aldehyde group is required for the oxidative loss of C-20 to give the C_{19}-GAs then an unfavorable aldehyde/lactol equilibrium could be responsible for the lack of formation of GA_9 from exogenously applied GA_{24}. It would be particularly contrived to suggest that these negative results were due to the wrong oxidation state at C-7, as GA_{12} is a better precursor to GA_9 than GA_{12} aldehyde. The nonmetabolism of GA_{15} has been shown not to be due to permeation problems by the following experiment (Bearder et al., 1976c), which also provided some insight into the mechanism of the C_{20}- to C_{19}-GA process. GA_{12} alcohol (38) was labeled in the carboxyl group with ^{18}O by hydrolysis of GA_{12} alcohol methyl ester with $K^{18}OH$. The labeled material (56 atom% ^{18}O) was oxidized to GA_{12} containing 56 atom% ^{18}O at C-19 (Fig. 5.23). Both the labeled alcohol and the acid were fed to resuspension cultures of mutant B1-41a, and the derivatized metabolites were analyzed by GC-MS in the usual way. $[19-^{18}O_1]GA_{12}$ was metabolized as expected to, inter alia, GA_{15}, GA_{24}, GA_{25}, and GA_9 which contained 32, 53, 55, and 53 atom% ^{18}O, respectively. Thus, GA_{15} that had lost half of the ^{18}O-label could not be a precursor to any of the other three nonhydroxylated GAs as these had retained the full complement of ^{18}O. The loss of label in going to GA_{15} confirmed that GA_{15} is derived by lactonization of the corresponding hydroxy acid and not by attack of the 19-oic acid group at C-20 (Fig. 5.23). The C_{19}-GA GA_9 had also retained all the label, and this provided some important information on the mechanism of the C_{20}- to C_{19}-GA conversion process. Formation of GA_9 by lactonization of the 10-hydroxy 19-oic acid can be excluded as this would result in the loss of half the label. Therefore, a Baeyer-Villiger type mechanism leading to GA_9 via a 10-alcohol can be ruled out. The formation of the lactone ring in GA_9 must involve an intermediate with an electrophilic center at C-10 which is attacked by the 19-oic acid. The feed of the $[19-^{18}O]$-labeled GA_{12} alcohol gave, inter alia, GA_{36}, GA_{13}, GA_4, GA_1, and GA_3 containing 59, 54, 54, 56, and 55 atom% ^{18}O. The retention of ^{18}O-label in the 3-hydroxylated C_{19}-GAs means that the same conclusion can be drawn concerning C_{20}- to C_{19}-GA conversion in the main pathway to GA_3. In this experiment, GA_{37} could not be detected but it can be reasonably inferred that GA_{37} like GA_{15} is a terminal gibberellin and not on the pathway to GA_{36}, GA_{13}, or the 3-hydroxylated C_{19}-GAs. These ^{18}O-labeling experiments also showed that the formation of the C_{19}-GAs does not involve intermediates in which the 19-oic acid group is covalently bound to an enzyme as this would also incur loss of ^{18}O.

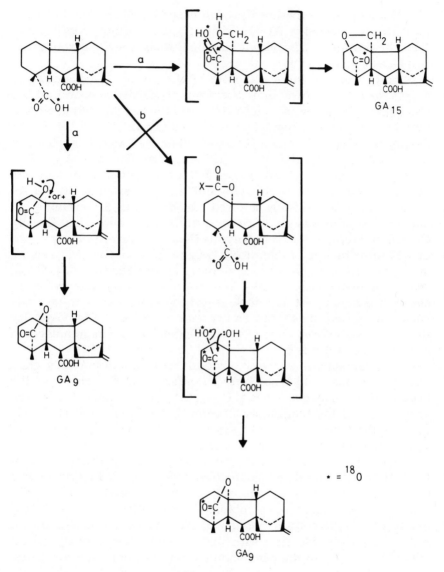

Figure 5.23   The fate of $[19\text{-}^{18}O_1]GA_{12}$ in *G. fujikuroi.*   Pathway *a* is observed. The Baeyer-Villiger (pathway *b*) does not occur. (Bearder et al., 1976c.)

In preliminary experiments (Bearder and MacMillan, unpublished) $[7\text{-}^{18}O]GA_{14}$ was fed to B1-41a, giving C_{19}-GAs without loss of label. Thus the possibility of covalent binding through the C-6 carboxyl can also be excluded.

Early work on the metabolism of GA_9 by wild-type cultures of *G. fujikuroi* had shown that it was not efficiently metabolized but was converted in low yield into GA_{10} and the lactone (59) identified by TLC (and autoradiography) (Cross et al., 1964, 1968b). In the mutant B1-41a, 30% of the GA_9 in a 20-hr feed was metabolized to three products (Bearder et al., 1976b), identified as GA_{10} (2%), GA_{20} (7%), and GA_{40} (20%) (Fig. 5.24). The extent of metabolism was increased to 75% by carrying out the incubation at pH 4.8 for 7 days. Large-scale feeds allowed the detection of several other metabolites produced in trace quantity. These were identified by GC-MS as GA_{11}, the two didehydro GA_9 derivatives (58) and (59), $16\alpha,17$-dihydro-$16\alpha,17$-dihydroxy-GA_9 (60), 3-*epi*-GA_4 (61), and a monohydroxy GA_9, which was tentatively identified as (62). GA_{40}, the major metabolite of GA_9, was not metabolized when re-fed to cultures of mutant B1-41a; it does not therefore appear to be a precursor of the two didehydro compounds or to GA_{11}. The possibility that GA_{11} was derived from GA_9 via the didehydro compounds could not be tested as it was impossible to obtain them free of GA_9.

The metabolism of GA_9 obviously does not occur to a great extent in normal wild-type *G. fujikuroi* fermentations. However, the occurrence of small quantities of GA_{10}, GA_{11}, GA_{20}, and GA_{40} in wild-type strains of the fungus (see Bearder, 1980; McInnes et al., 1977) suggests that these are not an expression of nonspecific "microbiological transformation" in the mutant.

Returning to the mutant studies on the main pathway (Bearder et al., 1975b), the results of incubations with *ent*-7α-hydroxykaurenoic acid (5) and *ent*-kaurenoic acid will now be discussed (see Fig. 5.25). The total ion current traces obtained by GC-MS of the derivatized extracts from the culture medium of 20-hr feeds of each compound were quantatively identical with that obtained from normally grown wild-type cultures of the parent strain GF-1a except for the concentration of GA_{14}.

Although equal concentrations of *ent*-kaurenoic acid (4) and *ent*-7α-hydroxykaurenoic acid (5) were fed to the same amount of resuspended mycelium of the mutant, the absolute concentration of the metabolites from *ent*-7α-hydroxykaurenoic acid was approximately 10 times less than that from *ent*-kaurenoic acid. Although little or no *ent*-7α-hydroxykaurenoic acid remained in the culture medium, considerable amounts were extracted from the mycelium

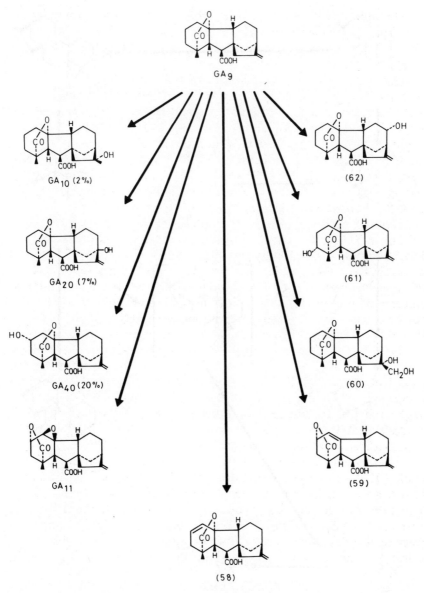

Figure 5.24   The metabolism of GA$_9$ in *G. fujikuroi* mutant B1-41a. (Bearder et al., 1976b.)

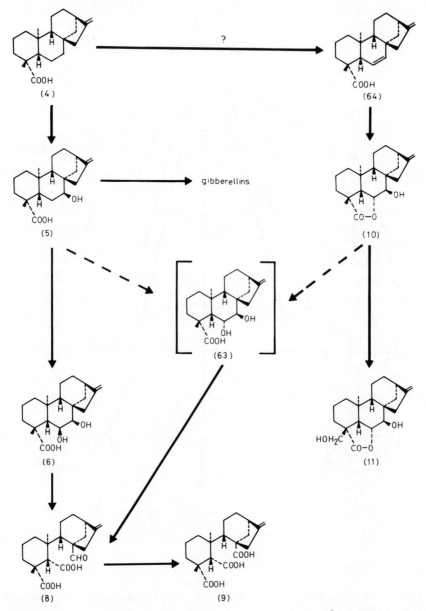

Figure 5.25 A scheme for the biogenesis of the kaurenolides and *seco*-ring B metabolites in *G. fujikuroi.*

with methanol. These observations corroborated the results of Lew and West (1971), who found that *ent*-kaurenoic acid was converted into GA_3 by a wild-type strain of *G. fujikuroi* (~12%) but the *ent*-7α-hydroxy derivative (5) was only incorporated to the extent of *ca.* 4%. In the work with the mutant, the *ent*-7α-hydroxylation of kaurenoic acid and the ring contraction of *ent*-7α-hydroxykaurenoic acid to GA_{12} aldehyde could not be demonstrated as discrete steps. This was not unexpected, as *ent*-7α-hydroxykaurenoic acid and GA_{12} aldehyde do not accumulate in the fungus and the existence of these compounds as metabolites of labeled precursors in *G. fujikuroi* has only been established by isotope dilution analysis (Hanson et al., 1972).

Ent-7-oxokaurenoic acid (51) was also fed to mutant B1-41a, since this compound had been suggested as a possible intermediate as a substrate for the ring contraction process (West, 1973). After 20 hr the *ent*-7-oxokaurenoic acid had been completely metabolized to a single compound identified as the triacid (9). Thus evidence for the intermediacy of the 7-oxoacid (51) was not forthcoming.

A number of feeds to the mutant B1-41a have provided evidence relevant to the biogenesis of the kaurenolides or 6,7-secokaurenoids. A conversion of *ent*-7α-hydroxykaurenoic acid into 7β-hydroxy-kaurenolide was observed, but this is now thought to be in error. The low levels of metabolism (see above) may well have been confused by slightly high control levels of endogenous metabolites. It seems from recent work (Beale et al., 1982) that *ent*-7α-hydroxykaurenoic acid is *not* a precursor of 7β-hydroxykaurenolide. When 7β-hydroxy-kaurenolide (10) was fed to B1-41a for 20 hr, 45% remained unmetabolized and *ca.* 35% was converted into 7β,18-dihydroxy-kaurenolide (11). This was expected, as the 18-hydroxylation of 7β-hydroxykaurenolide had been reported by Cross et al. (1968b) using a wild-type strain of *G. fujikuroi*. However, in the present case two other minor metabolites were also formed. These were identified as the seco-ring B diacid (8) and triacid (9), which were formed in 10% and 3% yields. These products may well have been derived from *ent*-6β,7α-dihydroxykaurenoic acid (63) for the reason outlined in the following paragraphs.

When *ent*-6β,7α-dihydroxykaurenoic acid (63) was fed to mutant B1-41a for 20 hr, it was completely consumed, giving as products the two seco-ring B acids (8) and (9) in 80% and 20% yield, respectively. Thus 7β-hydroxykaurenolide may act as a minor precursor to the seco-ring B metabolites. However, there is no reason to doubt that the major route is via the previously discussed *ent*-6α,7α-dihydroxy-kaurenoic acid (6). In fact, the feed of the 7β-hydroxykaurenolide hydrolysis product (63) had been planned to investigate the formation of the kaurenolides in the fungus, as Hanson et al. (1972) had suggested that this occurred by *ent*-6β-hydroxylation of *ent*-7α-hydroxykaurenoic acid. *Ent*-6β,7α-dihydroxykaurenoic acid (*63*) was therefore the expected precursor. The failure of this compound to act as a precursor of 7β-hydroxykaurenolide suggested that the kaurenolide might be formed directly by attack of the 19-oic acid group on an electrophilic center generated at C-6 by abstraction of the *ent*-6β-hydrogen atom. The possibility of a reaction of this nature was shown by the result of a feed of *ent*-kaur-6,16-dienoic acid (*64*) to mutant B1-41a (Bearder, 1973). GC examination of the culture filtrate of the feed revealed only a small quantity of the substrate but also a modest amount of a single metabolite. The balance of the substrate was found in methanolic extracts of the mycelium. The single product detected in the culture filtrate was identified by GC-MS as 7β-hydroxykaurenolide (*10*).

The most likely mechanism for its formation is via *ent*-6α,7α-epoxykaurenoic acid (*65*), which could undergo nucleophilic attack by the 19-oic acid group at C-6 leading directly to 7β-hydroxykaurenolide (Fig. 5.26). An epoxidation of this type would not be unprecedented as it is known that fungi that can introduce axial hydroxyl groups into saturated steroids are capable of transform-

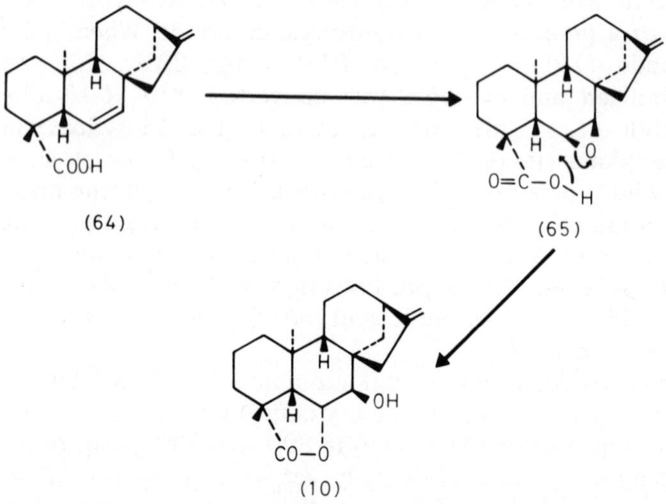

Figure 5.26 A scheme for the biosynthesis of 7β-hydroxykaurenolide from *ent*-kaur-6,16-dienoic acid in *G. fujikuroi* mutant B1-41a. (Bearder, 1973; Beale et al., in press.)

ing the unsaturated analogues into epoxides (Bloom & Shull, 1955; Talalay, 1957). A microbiological transformation in *G. fujikuroi* reported by Jefferies and co-workers (Bakker et al., 1974) probably involves a similar mechanism (see pp. 328 et seq.). These workers found that *ent*-kaur-2,16,dien-19-ol (*104*) was converted into the *ent*-gibberellin hydroxylactone (*109*). This compound was probably formed from the epoxide (*119*) by intramolecular attack of the 20-carboxylic acid group on the 2α-position. The likelihood of the involvement of a 2,3-epoxide in the formation of this compound was increased by the co-occurrence of the epoxide (*111*) in the feed.

In 1973 when this work was first done there was no suggestion that the *ent*-kaur-6,16-dienoic acid (*64*) was in fact the natural precursor to 7β-hydroxykaurenolide. However, recently Hedden and Graebe (1981) have shown that in a cell-free preparation of *Cucurbita maxima* endosperm the dienoic acid (*64*) is the authentic precursor of the kaurenolides. The acid (*64*) is derived from *ent*-kaurenoic acid (*4*) by a *trans*-elimination of hydrogen. In this system *ent*-7α-hydroxykaurenoic acid (*5*) is only a precursor of the GAs *not* the kaurenolides.

In the fungus, evidence has been presented by Hanson (see p. 272) and Bearder et al. (1973b) that *ent*-7α-hydroxykaurenoic acid (*5*) was a precursor to the kaurenolides. However, a recent re-examination of kaurenolide biosynthesis in mutant fungus B1-41a by Beale et al. (1982) has provided the following facts. Radio-labeled *ent*-7α-hydroxykaurenoic acid is not incorporated into the kaurenolides. Kaurenolides are formed with both oxygen atoms from the 19-oic acid group of *ent*-[19-$^{18}O_2$]kaurenoic acid retained in the lactone of the kaurenolides. Beale et al. also repeated the feed of *ent*-kaur-6, 16-dienoic acid (*64*) and recorded an incorporation of 52% into 7β-hydroxykaurenolide (*10*) in 2 hr. *Ent*-6α,7α-epoxykaurenoic acid (*65*) was also converted by the fungus into 7β-hydroxykaurenolide (*10*). This conversion was also shown to occur readily in the absence of the fungus in buffered solutions in the pH range 2.5–6.5. Kaurenolide (*30*) was also tested as a substrate in incubations with the fungus. After 5 days no hydroxykaurenolides could be detected, and the single product formed was tentatively identified as the triacid (*9*).

These studies suggest that kaurenolide biosynthesis in the fungus occurs by a similar pathway to that found in *C. maxima*, namely *ent*-kaurenoic acid (*4*) undergoes a *trans*-6α,7β-dehydrogenation to give *ent*-kaur-6,16-dienoic acid (*64*) and thence to 7β-hydroxykaurenolide (*10*) via the epoxyacid (*65*) (see Fig. 5.26). Unequivocal evidence for this pathway would be provided by the detection of *ent*-kaur-6, 16-dienoic acid *(64)* in the fungus. The instability of the presumed intermediate epoxyacid *(65)* militates against it ever being detected as a fungal metabolite.

This pathway gains some support from the metabolism of various *ent*-kaurenoid analogues (see Section 5.3). Many of these analogues are metabolized to analogues of 7β-hydroxykaurenolide, whereas these metabolites are never observed in feeds of *ent*-7α-hydroxykaurenoides.

5.2.3 Studies with wild-type strains of *Gibberella fujikuroi* using growth retardants

A large number of synthetic compounds are known that can cause reduced growth when applied to plants. The action of these dwarfing agents (or growth retardants) like many cases of genetic dwarfing can often be ameliorated by the application of gibberellins (Tolbert, 1961; Lockhart, 1962; Zeevaart & Lang, 1963; Zeevaart, 1964; Cathey, 1964). The possibility that these compounds might act by inhibiting GA biosynthesis in the plants was first suggested by Lang and co-workers (Kende, Ninneman & Lang, 1963). This theory was most easily investigated by examining the effect of the growth retardants on GA biosynthesis in the fungus. Indeed, the majority of the growth retardants tested showed some inhibition of GA production in *G. fujikuroi*. The literature on this subject has been admirably reviewed by Lang (1970) and Cross (1968). Growth retardants and related compounds that show this effect in intact cultures of *G. fujikuroi* include AMO-1618 *(66)*, CCC *(67)*, the quaternary ammonium salt *(68)* (Hedden et al., 1977) and related compounds (Cho et al., 1979), triadimefon *(69)* (Buchenauer & Grossmann, 1977), geranylimidazole *(70)*, *ent*-17-nor-16-dimethyl-aminokaurane methiodide *(71)* and decylimidazole *(72)* (Wada, 1978a, 1978b), ancymidol *(73)* (Shive & Sisler, 1976), triarimol *(74)* (Coolbaugh, Heil & West, 1982), and various secondary and tertiary amines and ammonium salts (Echols et al., 1981). The site of the inhibition in *G. fujikuroi* has been located (prior to *ent*-kaurene between geranylgeranyl pyrophosphate and copalyl pyrophosphate for CCC *(67)*, and AMO-1618 *(66)* (Cross & Myers, 1969; Barnes et al., 1969), and *(68)* (Hedden et al., 1977). Whereas triarimol *(69)* (Coolbaugh et al., 1981) and 1-decylimidazole *(70)* (Wada et al., 1978b) have been shown to inhibit the oxidation of *ent*-kaurene, the sterol biosynthesis inhibitor SKF 525 *(75)* also inhibits the incorporation

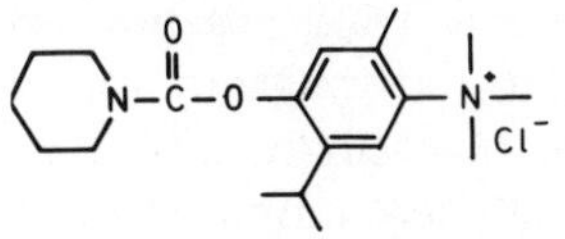

(66) AMO - 1618

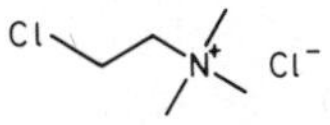

(67) CCC

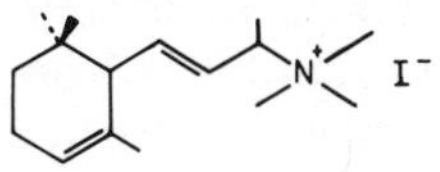

(68)

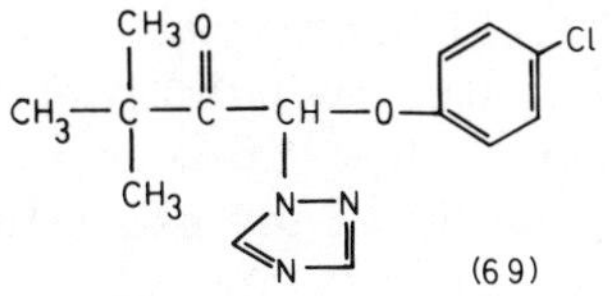

(69)

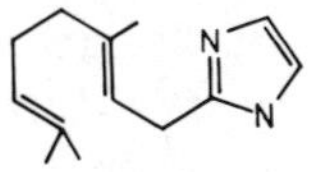

(70)

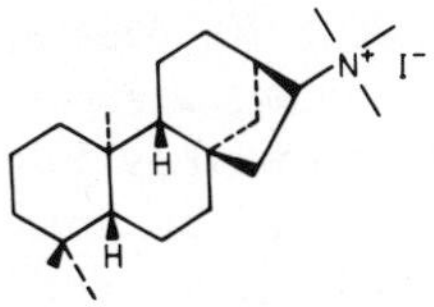

(71)

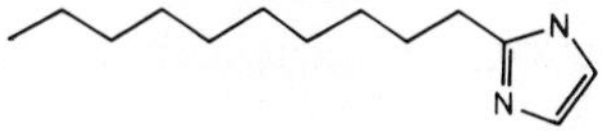

(72)

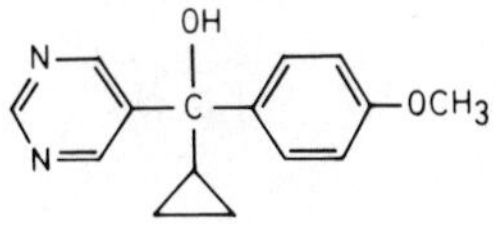

(73)

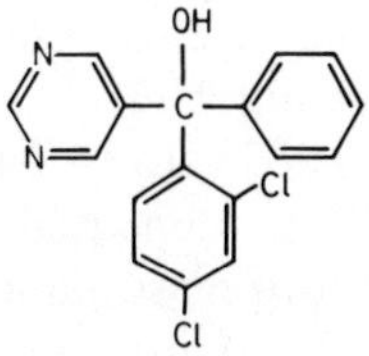

(74)

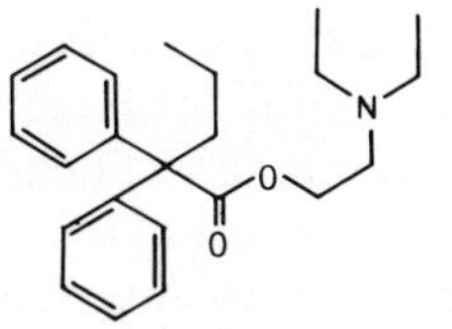

(75)

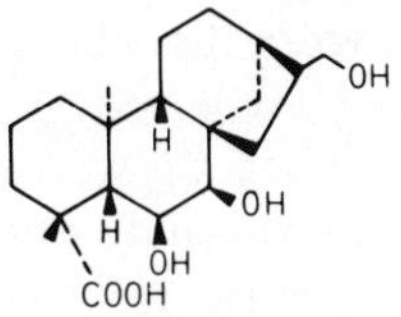

(76)

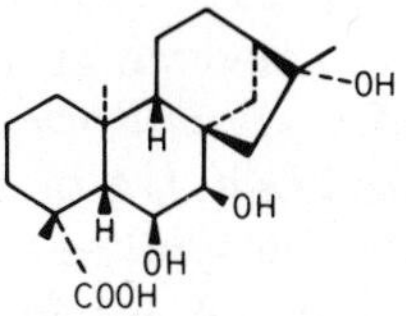

(77)

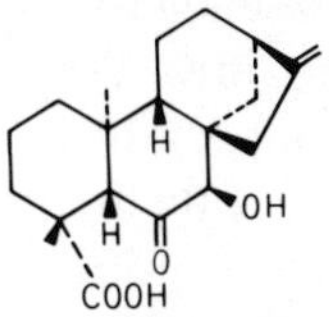

(78)

Structures (66)–(78)

of mevalonate into GAs in cultures of G. *fujikuroi*; in this case the position of inhibition seems to be between *ent*-kaurenal and *ent*-kaurenoic acid (Reid, 1969).

Until recently, there was no evidence of any compound inhibiting any step after *ent*-kaurenoic acid. However, Echols et al. (1981) have reported that diethyloctylamine hydrochloride and octyltriethyl-ammonium iodide caused the accumulation of the 6,7,17-triol (*76*) in addition to inhibiting the formation of other diterpenoid metabolites.

This is a particularly interesting result as this triol has not been detected as a normal metabolite of the fungus, although it is a constituent of higher plants (Beeley, Gaskin & MacMillan, 1975). The accumulation of this compound indicates that these ammonium salts interfere with the oxidation of the 6,7-diol moiety to the seco-ring B diacid (8). This inhibition could lead to the build-up of the 6,7-diol (6) which then undergoes a "microbiological transformation" by anti-Markownikov hydration of the 16,17-olefin. 17-Hydroxylated *ent*-kauranes are unknown in the fungus, although 16-hydroxylated *ent*-kauranes and GAs are often encountered. These 16-hydroxylated metabolites, caused by orthodox hydration of the exocyclic methylene group, seem to occur when relatively large quantities of the olefinic precursor are present (e.g., Hedden et al., 1974; Bearder et al., 1975b; Fraga et al., 1981). The data presented by Echols et al. (1981) do not allow one to reject the possibility that the compound they identified was in fact the $6\beta,7\beta,16\alpha$-triol (77), a metabolically more reasonable product.

A number of diterpenoid substrate analogues that act as inhibitors of GA biosynthesis will be featured in Section 5.3.2.

Hedden et al. (1977) were the first to demonstrate that cultures of *G. fujikuroi* containing growth retardants were useful for feeding studies in GA biosynthesis. They showed that the iodide (68) blocked GA biosynthesis between mevalonate and *ent*-kaurene, probably by inhibiting *ent*-kaurene synthetase A activity. The growth retardant (68) caused the incorporation of [2-^{14}C]mevalonate into GA_3 to be inhibited by 98%. In the presence of (68), cultures incorporated 26% of added *ent*-[^{14}C]kaurene into GA_3. Thus, chemically inhibited cultures can be used as an alternative to the mutant B1-41a and, as with the mutant, unlabeled precursors (after *ent*-kaurene) can be employed in feeding studies.

The suppression of endogenous precursors of *ent*-kaurene was used to good effect by Dockerill and Hanson (1978) in their previously discussed investigation into the metabolic fate of C-20 in the conversion of C_{20}-GAs into C_{19}-GAs. Later work by Hanson and Sarah (1979) into the biogenesis of the kaurenolides and seco-ring B metabolites in the fungus also involved AMO-1618-inhibited cultures, allowing the use of unlabeled substrates. 7β-Hydroxykaurenolide (10) (140 mg) was incubated with *G. fujikuroi* in the presence of AMO-1618 for 6 days. Extraction of the metabolites yielded the diacid (8) isolated as the diester (12 mg) 7,18-dihydroxykaurenolide (11) (43 mg), and the unchanged substrate (64 mg). This result is in good agreement with a previously mentioned feed of 7β-hydroxykaurenolide to mutant B1-41a, when the formation of the fujenal diacid (8) was attributed to hydrolysis of the lactone ring (see Fig. 5.25). Hanson suggested an alternative route in which the 7β-hydroxykaurenolide is 6β-hydroxylated [presumably leading to *ent*-6-oxo-7α-hydroxy-kaurenoic acid (*78*)] followed by cleavage to give the required 7-formyl-6-oic acid (*8*). The relatively low yields of the fujenal derivative from this experiment and the corresponding one with B1-41a points to this being only a minor route to the seco-ring B compounds. The major route is more likely to be through the 6,7-diol (*6*) (see Fig. 5.25). In the same paper, Hanson drew attention to a formal relationship between the ring contraction reaction in GA biosynthesis and the glycol cleavage reaction in fujenal biosynthesis and suggested that the enzymes might be related. Ring contraction to form the GAs involves the oxidative loss of an *ent*-6α-hydrogen and the *ent*-7α-hydroxy hydrogen. Dehydrogenation of the *ent*-6α- and *ent*-7α-hydroxyl groups in a similar manner could afford a mechanism for the cleavage of ring B. This theory gains some support from the results of the feeding studies of many kaurenoid analogues, in which it is often observed that when ring contraction is retarded the *ent*-6α, 7α-diol accumulates (see Section 5.3).

In a further feeding experiment by Hanson and Sarah (1979), 7,18-dihydroxykaurenolide (500 mg) gave 7β-hydroxykaurenolide 18-carboxylic acid (*79*) (21 mg). As this acid was readily decarboxylated, affording 7β-hydroxy-18-norkaurenolide (*43*), this substance was tested as a substrate for the fungus. The norkaurenolide (*43*) (200 mg) underwent hydroxylation at the 4-position to give 4β,7β-dihydroxy-18-norkaurenolide (*12*) (57 mg). The compounds (*43*) and (*12*) have been detected as metabolites of *G. fujikuroi* (MacMillan & Wels, 1974; Yamane, Murofushi & Takahashi, 1974), and so the biosynthetic sequence mapped out by these feeding experiments (Fig. 5.27) probably represents their normal biosynthetic origin.

The feed of the synthetic 18-hydroxykaurenolide (*80*) described in the same paper will also be considered here for convenience. When this compound (*80*) (243 mg) was fed, 11α-hydroxy-18-norkaurenolide (*81*) (18 mg) was formed (Fig. 5.28). However, (*81*) was shown to be an artifact as a repeat experiment, using a milder extraction procedure involving methylation of the crude product, furnished the methyl 11α-hydroxykaurenolide 18-carboxylate (*82*) (18 mg). A second metabolite (5 mg) was tentatively identified as 1β-hydroxykaurenolide 18-carboxylic acid (*83*). The formation of these two hydroxykaurenolides is entirely consistent with the known occurrence of 1β,7β-dihydroxykaurenolide (*84*) and 11α,7β-dihydroxykaurenolide (*85*) in culture filtrates of *G. fujikuroi* (Hedden, MacMillan, & Grinstead, 1973).

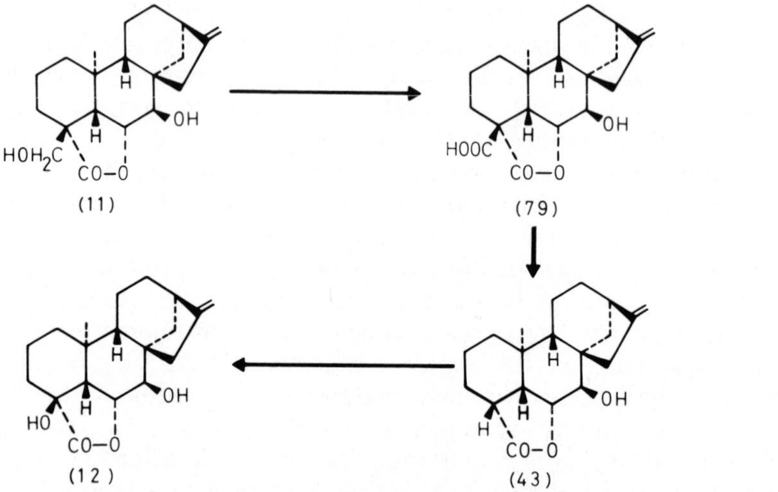

Figure 5.27 A possible biogenesis of $4\beta,7\beta$-dihydroxy-18-norkaurenolide (*12*) based on the work of Hanson and Sarah (1979).

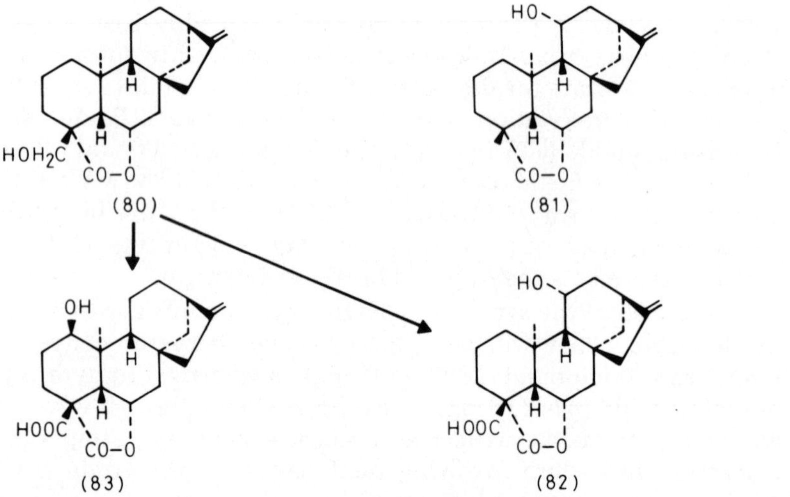

Figure 5.28 The fate of 18-hydroxykaurenolide in *G. fujikuroi*. (Hanson & Sarah, 1979.)

Structures (84)–(85)

5.2.4 Summary of the diterpenoid pathway in *Gibberella fujikuroi*

Figure 5.29 illustrates the commonly agreed biosynthetic pathway in the fungus from *ent*-kaurene to GA_{14} which can act as a precursor to all the common 3-hydroxylated GAs. The sequential oxidation of *ent*-kaurene to *ent*-7α-hydroxykaurenoic acid (5) has been particularly well investigated in cell-free systems of the higher plants (see Hedden et al., 1978), and results from *in vivo* studies in the fungus support this pathway. The ring contraction of *ent*-7α-hydroxykaurenoic acid (5) to give GA_{12} aldehyde has been shown to occur *in vivo*, but the extreme experimental difficulties involved in studying the process have not allowed the exact mechanism to be unambiguously defined. It is most likely, however, that the process is the same as that established in cell-free extracts of *C. maxima* (Graebe et al., 1980), namely the abstraction of the *ent*-6α-hydrogen concerted with or followed by the migration of C-8 to C-6 and with extrusion of C-7 and loss of the *ent*-7α-hydroxylic proton (Fig. 5.29). GA_{12} aldehyde lies at the branch point between the 3-hydroxy GAs and the non-3-hydroxy GAs. Oxidation at C-7 to GA_{12} leads mainly to the non-3-hydroxylated GAs, whereas oxidation at C-3, which is the principal reaction, leads to the 3-hydroxylated GAs.

Despite extensive study, the intermediates are not known between the C_{20}-GAs that are nonfunctionalized at C-20 (i.e., GA_{12} and GA_{14}) and their C_{19}-GA counterparts (GA_9 and GA_4). However, the following constraints of the overall reaction have been elucidated. Intermediates that have double bonds at the 1-, 5-, or 9-positions are ruled out, excluding many processes known in steroid biosynthesis. The formation of the γ-lactone of the C_{19}-GAs involves attack of the 19-oic acid group on C-10, excluding a Baeyer-Villiger-type mechanism, or a mechanism involving a 19-peracid such as that suggested by McCorkindale (1976). Carbon-20 is apparently lost as carbon dioxide, unlike the analogous steroidal C-19 which is lost as formate in an

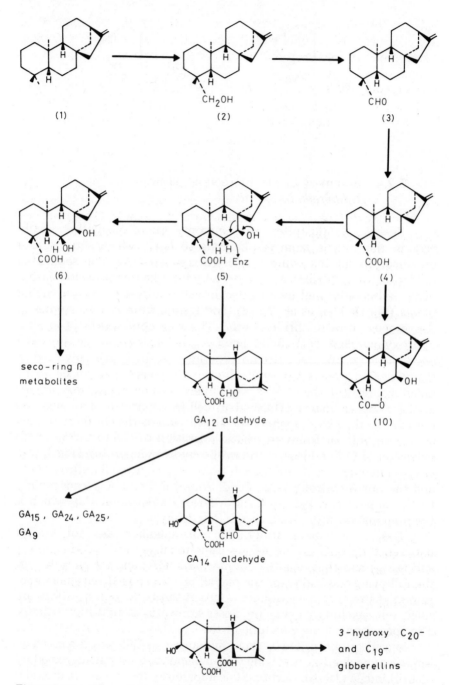

(1)

(2)

(3)

(4)

(5)

(6)

CH₂OH

CHO

COOH

COOH

Enz

OH

OH

H OH

COOH

seco-ring β
metabolites

GA₁₂ aldehyde

CHO

COOH

(10)

CO—O

OH

GA₁₅ , GA₂₄ , GA₂₅ ,
GA₉

GA₁₄ aldehyde

HO

CHO

COOH

3-hydroxy C₂₀⁻
and C₁₉⁻
gibberellins

HO

COOH

COOH

Figure 5.29 The pathway from *ent*-kaurene to GA_{14} in *G. fujikuroi.*

estrogen biosynthesis (Akhtar et al., 1981) or the steroidal C-14 which appears as formic acid (Pascal, Chang, & Schroepfer, 1980). Hanson has proposed a mechanism for the reaction involving the decomposition of a C-20 peracid (Fig. 5.19). This mechanism could also be compatible with the incorporation of GA_{13} anhydride (*39*) into C_{19}-GAs observed by Hanson and Hawker (1972b). It is difficult to postulate a sequence of intermediates that is entirely consistent with the experimental results and also compatible with our knowledge of this sequence in higher plants (Graebe et al., 1980). To come to a conclusion about the pathway in the fungus, the plant pathway cannot be ignored—it would seem to be highly improbable for the major steps in the biosynthesis of such complex molecules as GAs to be significantly different in mechanism. In higher plants, GA_{36}, but not GA_{13}, acts as precursor of the C_{19}-GA, GA_4 (see Graebe et al., 1980; Chapter 3, this volume). This fact must carry considerable weight when considering the probable biosynthetic sequence in the fungus.

The proposal that the sequential oxidation of C-20 occurs with C-7 at the aldehyde oxidation level seems to be an artificial contrivance to account for the relatively slow metabolism of GA_{14} and the observed nonincorporation of other C_{20}-GAs. However, these results can be adequately explained in terms of membrane permeability problems. Furthermore, 7-carboxy-C_{20}-GAs are metabolized to C_{19}-GAs in the higher plant system (e.g., $GA_{36} \rightarrow GA_4$) (Graebe et al., 1980) and in the non-3-hydroxy GA pathway in the fungus (e.g., $GA_{12} \rightarrow GA_9$). It seems to be no coincidence that the metabolism of no GA more polar than GA_{14} has ever been observed in *G. fujikuroi*. In conclusion, either of the two pathways illustrated in Fig. 5.30 would seem to adequately contain the experimental results without recourse to numerous constraints. A similar scheme could be constructed for the biosynthesis of GA_9 from GA_{12}.

The pathways among the C_{19}-GAs are uncontroversial and are summarized in Figs. 5.16 and 5.24. The enzymes in the fungal diterpenoid pathway have variable substrate specificities (as will be elaborated in the next sections), which means that under certain circumstances alternative pathways may be involved to a minor extent. These are indicated in Figs. 5.16 and 5.24 by broken lines. The fungal GAs that have not yet been biosynthetically studied are probably due to the operation of these enzymes on more plentiful substrates.

The C_{20}- to C_{19}-GA pathway in the fungus will not be understood with any degree of confidence until successful cell-free systems have been developed.

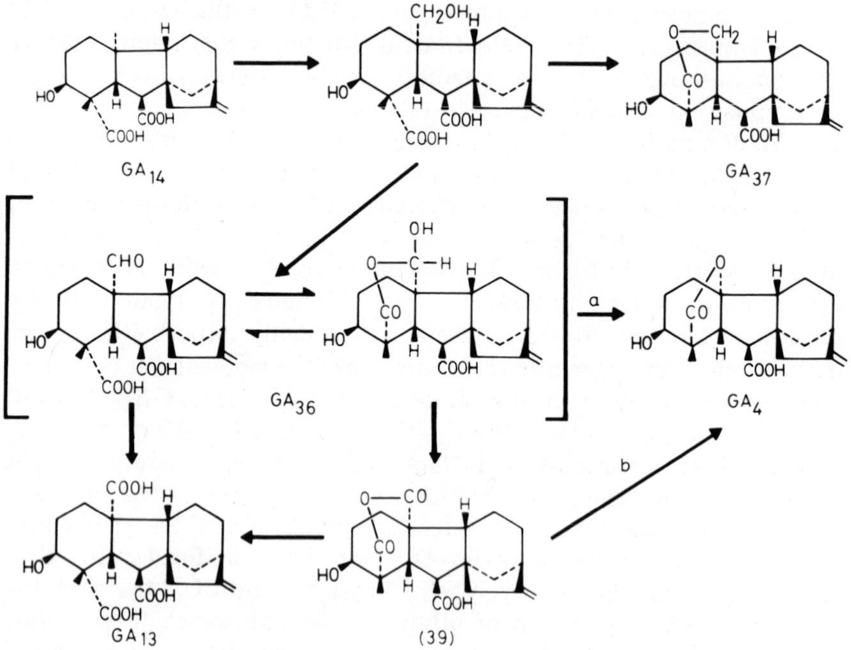

Figure 5.30 A scheme illustrating two possible pathways (*a* and *b*) to GA_4.

The biosynthesis of the seco-ring B metabolites is summarized in Fig. 5.25. The main pathway seems to be via *ent*-6α,7α-dihydroxy-kaurenoic acid (*6*), although the route through 7β-hydroxykaurenolide (*10*) may make a contribution.

In the light of recent work by Beale et al. (1982), kaurenolide biosynthesis in *G. fujikuroi* probably parallels that found in *C. maxima* by Hedden and Graebe (1981); namely, that *ent*-kaurenoic acid (*4*) undergoes a *trans*-6α,7β-dehydrogenation to *ent*-kaur-6,16-dienoic acid (*64*) which is oxidized to *ent*-6α,7β-epoxykaurenoic acid (*65*) which undergoes reaction to give 7β-hydroxykaurenolide (*10*).

5.3 METABOLISM OF KAURENOID AND GIBBERELLIN ANALOGUES IN *GIBBERELLA FUJIKUROI*

5.3.1 Introduction

The fungus *Gibberella fujikuroi* has been used to hydroxylate a wide range of exogenously applied structural types—for example, steroids and flavanoids (e.g., Charney & Herzog, 1967; Kieslich, 1976). However, the only compounds discussed here are those that are structurally related to normal metabolites of *G. fujikuroi*.

There are a number of reasons for performing these analogue feeds:

1. To help elucidate the normal pathway of GA biosynthesis in the fungus.
2. To investigate the substrate specifities of the enzymes in the fungal pathway.
3. To prepare analogues of fungal metabolites
   (a) such as rare radio-labeled plant GAs
   (b) as potential inhibitors of GA biosynthesis or action
   (c) as tools to investigate the structure-activity relationship of GA activity.

5.3.2   With wild-type strains of *Gibberella fujikuroi*

A number of radioactive GAs and kaurenoid analogues have been fed to *G. fujikuroi* for investigating the normal diterpenoid pathways in the fungus. These have already been discussed in Section 5.2.1. It is, however, worth mentioning that of all the unnatural GA analogues fed the metabolism of GA_{12} alcohol (*38*) to GA_3 has been shown to be so efficient that this route has been used to prepare $[17\text{-}^{14}C]GA_3$ (Hanson & Hawker, 1973). This procedure could now of course be improved by the application of "growth retardant blocked cultures" or by the use of mutant B1-41a. This is also true of most of the work described in the next section. The major disadvantages of the older methods using wild-type strains are that only principal metabolites of the applied substrates are detected or isolated because of the high "background" of normal metabolites. Also, the compound fed is often incompletely metabolized because of competition with the endogenous substrate.

The first recorded example of the metabolism of an unnatural diterpenoid substrate by *G. fujikuroi* was the transformation of steviol (*86*). Steviol is the aglycone of stevioside (*87*), an intensely sweet-tasting glycoside from the South American shrub *Stevia rebaudiana*. As early as 1963 the GA-related structure had prompted Ruddat, Lang, and Mossettig (1963) to test its biological activity. The significant response found in the highly specific dwarf maize assay (*d-5*) lead Ruddat and co-workers (Ruddat, Heftmann, & Lang, 1965; Ruddat, 1968) to test steviol (*86*) as a potential precursor to GA_3 in the fungus.

Structures (86)–(90)

[^{14}C]Steviol (biosynthesized from [^{14}C]acetate in *S. rebaudiana*) was incubated with cultures of *G. fujikuroi* for 6 days. The culture filtrate was then extracted and examined by radio-TLC. Only a small amount of activity, if any, was associated with GA$_3$. The major radioactive peak co-chromatographed with GA$_4$ and GA$_7$ in one solvent system but was quite distinct from these GAs in an alternative system. The metabolite was shown to have GA-like biological activity in the *d-5* and *d-1* dwarf maize assays and in a dwarf pea assay. It was concluded that steviol had been converted into a GA-like compound distinct from the nine GAs then known. It was suggested by virtue of its chromatographic mobility that the metabolite might be an isomer of

GA_4 in which the hydroxyl group was at C-13 instead of C-3 (the GA now known as GA_{20}). No attempts were made to chemically characterize this major product. Subsequently, the transformation of steviol (*86*) was studied by Hanson and White (1968). A 7-day incubation of steviol with *G. fujikuroi* was carried out, and the resulting metabolites were separated into acidic and neutral fractions. TLC analysis of the neutral extract showed two spots not present in the control. Extensive chromatography led to the purification of the more polar metabolite. This compound was assigned the structure (*88*) on the basis of extensive NMR analysis. However, the new metabolite did not have TLC properties similar to those of the active substance reported by Ruddat et al. (1965). Further progress in the metabolism of steviol in the fungus was not made until the introduction of the mutant B1-41a (see next section).

The widespread occurrence in plants of diterpenes having a carbon skeleton grossly enantiomeric with that of the known GAs raised the possibility of the existence of an enantiomeric series of GAs in nature. Despite the admittedly remote chance of success, Cross et al. (1968a) tested [17-^{14}C]kaurene (*89*) as a precursor of (−)-gibberellic acid in *G. fujikuroi*. Not unexpectedly, the gibberellic acid isolated from this experiment was found not to be radioactive. This experiment was later repeated with a similar result in growth retardant-inhibited cultures of the fungus (Hedden et al., 1977).

The metabolism of ring-A-modified analogues of normal kaurenoid precursors was first investigated by the Jefferies group, stimulated no doubt by the plentiful supply of kaurenoid diterpenes from the local West Australian flora. Prompted by current evidence that *ent*-6α,7α-dihydroxykaurenoic acid (*6*) might be a key intermediate in GA biosynthesis, Jefferies et al. (1970, 1974) fed *ent*-kaurenol-19-hemisuccinate (*90*) to resuspension cultures of *G. fujikuroi* for 2 days with a view to blocking GA biosynthesis at some step. Hydrolysis of the acidic fraction from the fermentation gave the two metabolites (*91*) and (*92*) in 12% and 20% yields, respectively; GA_3 was also detected. The structures of the metabolites were firmly established by a combination of spectroscopic and chemical methods. As these substances were isolated by hydrolysis of the acidic fraction of the culture filtrate, the succinate group had obviously remained intact in the formation of these products. Backfeeding experiments with labeled samples of the diol (*91*) and triol (*92*) reinforced the suspicion that *ent*-6α,7α-hydroxylated intermediates were not involved in GA biosynthesis (Fig. 5.31). Whereas the [^{14}C]diol (*91*) showed a 1.8% incorporation into GA_3, the triol (*92*) yielded less than 0.2%.

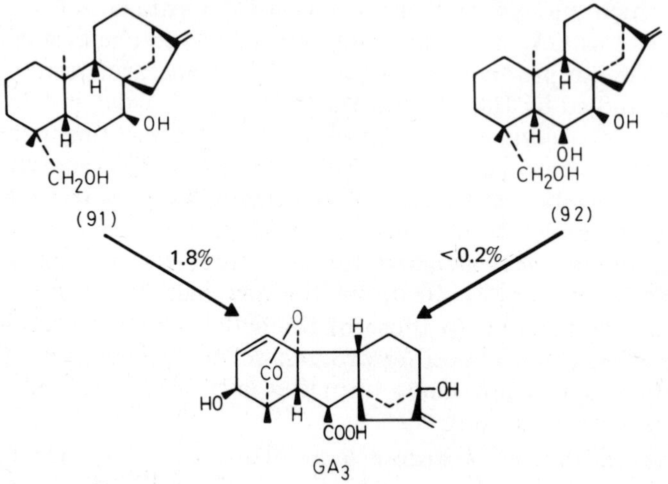

Figure 5.31 Incubation of the diol (91) and triol (92) with *G. fujikuroi*. (Jefferies et al., 1974.)

Incubations of *ent*-kaurenyl succinate (90) gave substantial quantities of GA_3, indicating enzymic hydrolysis of the succinate group occurs to some extent.

The observed efficient hydroxylation of *ent*-kaurenyl succinate led the Jefferies group to develop a microbiological method for the facile production of reasonable quantities of the true fungal metabolites *ent*-7β-hydroxykaurenoic acid (5) and *ent*-6α,7α-dihydroxykaurenoic acid (6). The 2-carboxyethyl ester (93) of *ent*-kaurenoic acid was prepared, which is an isostere of *ent*-kaurenol-19-hemisuccinate (90) (Croft et al., 1974). Incubation of this ester with resuspension cultures of *G. fujikuroi* for 7 days furnished *ent*-7α-hydroxykaurenoic acid (5) (30%) and *ent*-6α,7α-dihydroxykaurenoic acid (6) (12%), isolated by alkaline hydrolysis of the acidic extract

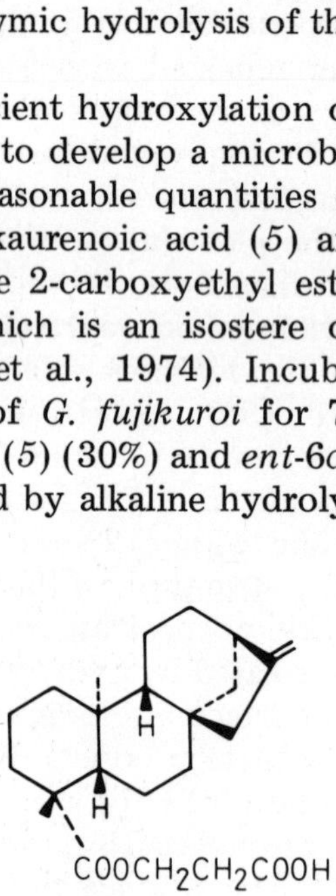

(93)

Structure (93)

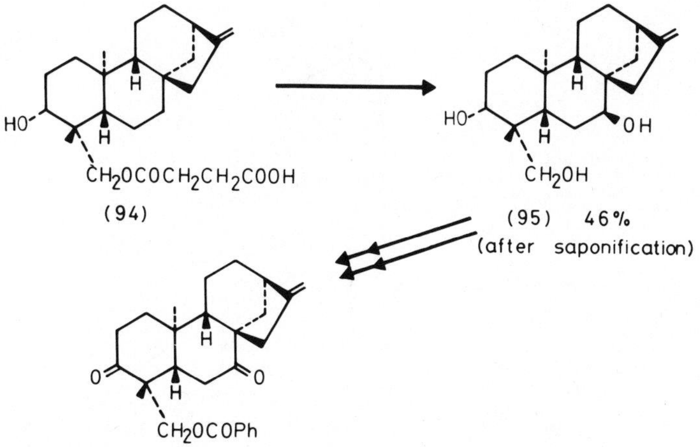

Figure 5.32 Metabolism of *ent*-kaur-3β,19-diol hemisuccinate (*94*) in *G. fujikuroi.* (Jefferies et al., 1974.)

of the culture filtrate. The presence of the 19-succinoyloxy group or 2-carboxyethyl ester apparently acts as an efficient block for both ring contraction of the *ent*-7α-hydroxy derivative and the cleavage of the diol to give seco-ring B derivatives.

Jefferies, Knox, and Ratajczak (1974) also investigated the metabolism of two ring-A-hydroxylated *ent*-kaurenyl succinates. *Ent*-3β,19-dihydroxykaurene-19-succinate (*94*) is a major constituent of the leaves of *Goodenia strophiolata* (Middleton & Jefferies, 1968). Fermentation of *G. fujikuroi* mycelium with the succinate (*94*) for 7 days gave a major acidic product (Fig. 5.32), which was purified as the triol (*95*) in 46% yield after hydrolysis of the crude acidic extract. The structure was defined by proton magnetic resonance (PMR) spectroscopy and MS of the triol (*95*) and the derived diketobenzoate (*96*). The epimeric *ent*-kauren-3α,19-diol-19-hemisuccinate (*97*) was exposed to *G. fujikuroi* under similar conditions (Fig. 5.33). After saponification the acidic extract gave the triols (*98*) and (*99*) in 38% and 11% yield, respectively. The major triol was chemically correlated with the previous diketobenzoate (*96*), and the minor triol (*99*) underwent ring closure to give the ether (*100*).

A feed of *ent*-3-oxokauren-19-ol succinate (*101*) was also reported by Jefferies et al. (1974). The primary products from this substrate proved unstable, and so the acidic fraction was reduced with LiAlH₄ to furnish a small quantity of *ent*-3β,7α,19-triol (*95*). As *ent*-3α-hydroxylation is a feature of the diterpenoid biosynthesis in *G. fujikuroi*

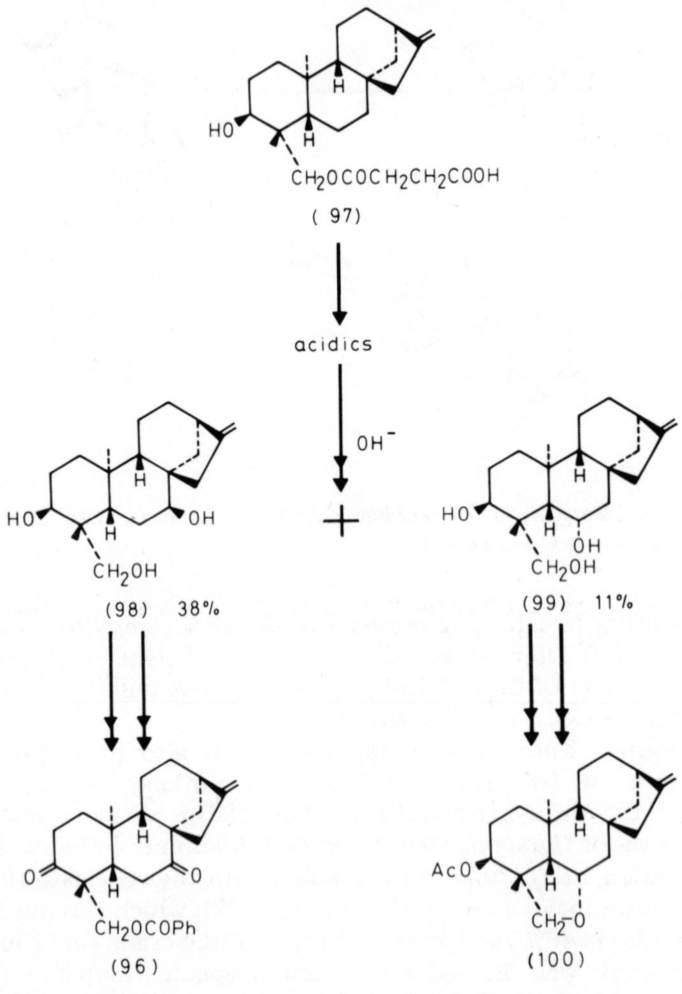

Figure 5.33 Metabolism of *ent*-kaur-3α,19-diol-19 hemisuccinate (*97*) in *G. fujikuroi.* (Jefferies et al., 1974).

(albeit at a later stage in the pathway) the metabolic fate was investigated of the [17-^{14}C]-derivatives of the three *ent*-3α-hydroxykaurenoids (*98*), (*99*), and (*102*) featured in this chapter. Neither of the triols (*98*) and (*99*) showed significant alteration on separate incubation with the fungus. The diol (*102*) was metabolized to a small extent, giving a compound with the TLC mobility of the *ent*-3α,6β,19-triol (*99*).

In separate publications, the Jefferies group described their results from feeds of *ent*-kaura-2,16-dien-19-ol (*104*) and its hemi-

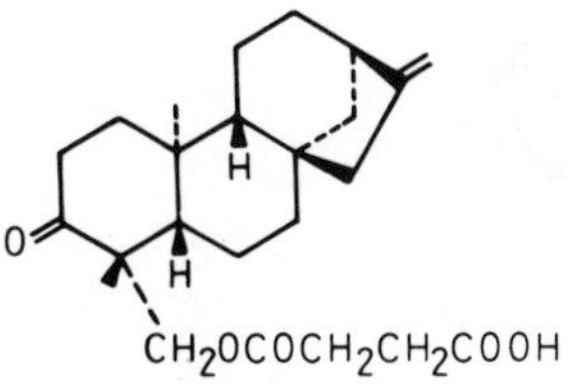

CH₂OCOCH₂CH₂COOH
(101)

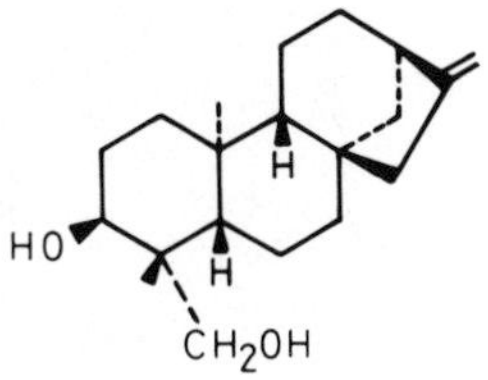

CH₂OH
(102)

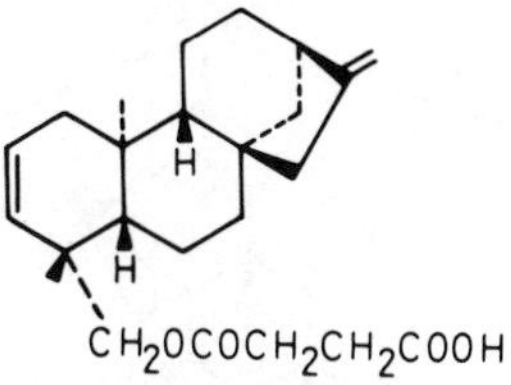

CH₂OCOCH₂CH₂COOH
(103)

Structures (101)-(103)

succinate (103) with resuspension cultures of G. fujikuroi (Bakker et al., 1974; Cook, Jefferies, & Knox, 1975). Ent-kaura-2,19-dienol ranks with steviol (86) as one of the few kaurenoids not native to the fungus that show GA-like bioactivity in Phinney's dwarf maize assays (Katsumi et al., 1964). As this activity could be due to its metabolism to GAs in the bioassay plant, it was an attractive idea to study the metabolism of the compound in the fungus G. fujikuroi. In early experiments ent-[17-^{14}C]kaur-2,16-dienol (104) was fed to resuspended cultures of the fungus, but the rate of metabolism was poor. The succinate (103) was prepared and proved to be converted to a similar range of products, but in much higher yield. Thus in contrast to the 3-hydroxy-19-succinates (94) and (97) but as with ent-kaurenyl succinate (90), the ester group was quite labile in the fungus. From large-scale feeds the three kaurenoid metabolites (105), (106), and (107) were isolated from the culture filtrate (<0.3%) (Cook et al., 1975) (Fig. 5.34). The structures were assigned on the basis of detailed MS and PMR studies and the chemical cleavage of the vic-diol group.

The formation of the ent-1α,6α,7α,16β-tetraol (106) in this case would seem to be further evidence for the reassignment of the structure of the triol (76) which accumulates in growth retardant-inhibited cultures of the fungus (Echols et al. 1981) (see p. 317).

The GA metabolites of ent-[17-^{3}H$_2$]-kaur-2,16-dienol hemisuccinate (103) were much more plentiful (Bakker et al., 1974). These are illustrated in Fig. 5.35 with their yields. The structures of these compounds were rigorously proven and with the exception of the 13-deoxy-GA$_5$ (108) were shown to be metabolites of the labeled precursor.

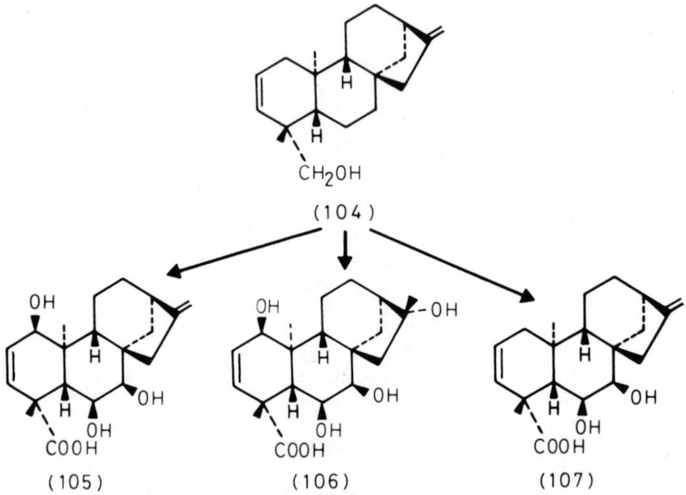

Figure 5.34 Three kaurenoid metabolites of *ent*-kaur-2,16-dien-19-ol (*104*) in *G. fujikuroi* isolated as their methyl esters. (Cook et al., 1975.)

It will be noted that these metabolites consist of both C_{20}- and C_{19}-GA analogues and that they fall into two groups with respect to the metabolic fate of the ring A double bond. Four of the compounds have retained this functionality, and in the other three it has been oxygenated. This partition in the pathway corresponds to the formation of non-3-hydroxylated and 3-hydroxylated GAs in the normal GA pathway. The microbiological epoxidation of the 2,3-double bond was not completely unexpected as microbes are known to epoxidize olefins located where axial hydroxyl groups are introduced in the corresponding saturated compounds (Bloom & Shull, 1955; Talalay, 1957). To clarify the biosynthetic relationship between these metabolites, backfeeding experiments were undertaken with [17-^{14}C]-labeled samples of these GAs obtained by feeding [17-^{14}C]dienylsuccinate (*103*).

The results from feeding the labeled epoxy-diacid (*111*) to resuspension cultures are summarized in Fig. 5.36. After 4 days, 50% of the substrate was metabolized to two other 2,3-oxidized metabolites originally isolated from the dienyl succinate feed, (*109*) and (*110*), and GA_6 (*115*). Although the large amount of unmetabolized epoxy-diacid (*111*) could have been due to transport problems it was suggested that the epoxy-diacid was probably a terminal metabolite of the dienyl succinate and that it provided only a minor route to the lactone diacid (*109*) which was the major metabolite of (*103*). It was

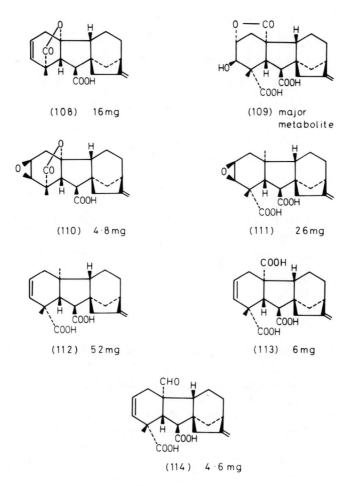

Figure 5.35   Metabolites of *ent*-[17-³H₂]kaur-2,16-dienol hemisuccinate (*103*) (6.82 g) in *G. fujikuroi*. (Bakker et al., 1974.)

thus argued that the epoxy-diacid might be the precursor of the minor quantities of deoxy-GA₆ (*110*) obtained from (*103*) but that a pathway involving a late-stage 7-aldehyde intermediate could not be excluded. In considering the observed metabolism of the epoxy diacid (*111*), it is instructive to consider this process in terms of the normal metabolites of the fungus which are analogous to these compounds. Thus the metabolism of the epoxy-diacid (*111*) is equivalent to the conversion of GA₁₄ into GA₄, GA₇, GA₃, GA₁, and GA₁₃. [The epoxy acid (*115*) may be considered analogous to both GA₁ and GA₃, as the normal introduction of the 1,2-olefin is blocked by the epoxide

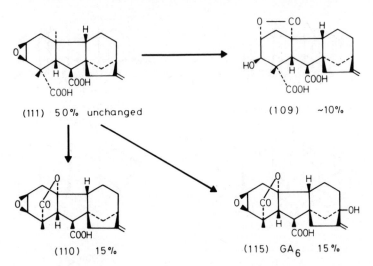

Figure 5.36 The metabolism of the [17-^{14}C]epoxy-diacid (*111*) in *G. fujikuroi.*
(Bakker et al., 1974.)

group.] This conversion, therefore, correlates well with that expected by analogy with the normal GA pathway in the fungus. It is interesting that the epoxy diacid (*111*) is metabolized only slowly, as is its analogue GA$_{14}$ in similar feeds.

A feed of [17-^{14}C]-labeled diene diacid (*112*) gave two major metabolites, identified as diene aldehyde (*114*) and diene triacid (*113*); minor amounts of the C$_{19}$-GAs GA$_5$ (*116*) and GA$_6$ (*115*) were tentatively identified (Fig. 5.37). This conversion equates to the metabolism of GA$_{12}$ to GA$_{24}$ and GA$_{25}$, with minor quantities of GA$_9$ and GA$_{20}$, and thus it is qualitatively almost identical to feeds of GA$_{12}$ to mutant B1-41a (Bearder et al., 1975b). Noticeably almost no 2,3-epoxy-GAs (3β-hydroxy analogues) are formed, in similarity to the metabolism of GA$_{12}$. A quantitative difference is that in the diene diacid (*112*) feed C$_{19}$-GA formation is apparently not favored. Thus, as in the equivalent case of the biosynthesis of native GA in the fungus, it seems that the presumed precursor of all the GA analogues—the diene acid aldehyde (*117*)—is at a branch point in the pathways to the 2,3-diene GAs and the 2,3-epoxy GAs. The oxidation of C-7 to the diene diacid (*112*) leads to the 2,3-diene GA while prior oxidation at the 2,3-double bond is the major branch, giving 2,3-epoxy GAs and their derivatives. This similar substrate specificity between GA$_{12}$ aldehyde and the putative 2,3-ene GA$_{12}$ aldehyde (*117*) corroborates

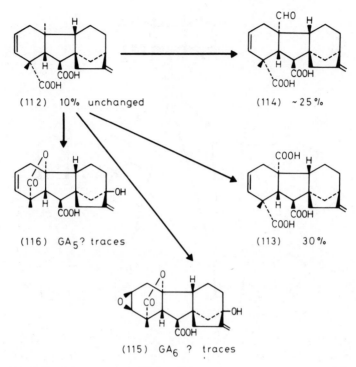

Figure 5.37 Metabolism of [17-^{14}C]diacid (*112*) in *G. fujikuroi.* (Bakker et al., 1974.)

the theory that the same enzyme system is responsible for the epoxidation of the 2,3-ene as for the 3β-hydroxylation of GA$_{12}$ aldehyde.

Further evidence for the aldehyde branch point was sought by feeding the chemically derived [17-^{14}C]diene diol (*118*). The metabolites were analyzed qualitatively and quantitatively by radio-TLC of their methyl esters, and the results are summarized in Fig. 5.38. It

(117)

Structure (*117*)

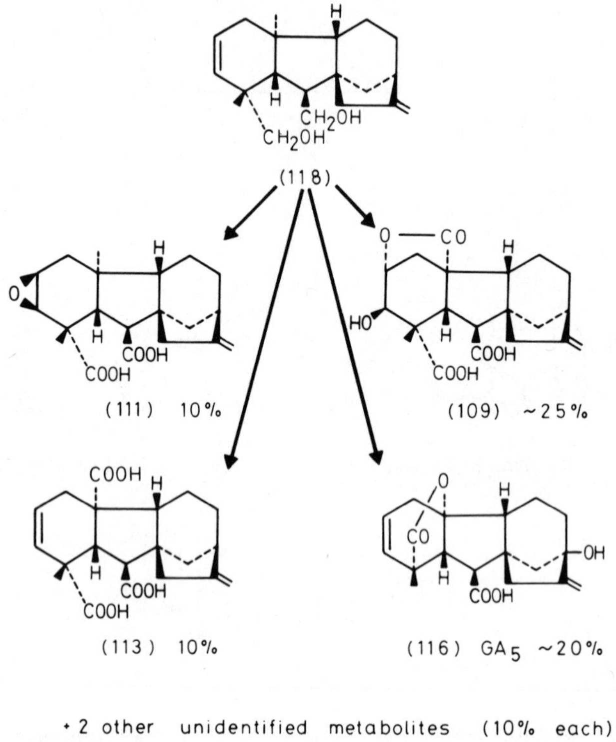

Figure 5.38 The metabolism of [17-^{14}C]diol (*118*) in *G. fujikuroi.* (Bakker et al., 1974.)

can be seen that in this experiment substantial conversion into 2,3-epoxy GAs is observed (~35%) compared with only traces of the 2,3-epoxy GAs from the diene diacid (*112*) feed. The efficient production of the lactone (*109*) in this feed and the absence of significant quantities of GA$_6$ (*115*) contrasted with the relative proportions of these amounts in the epoxy diacid (*111*) feed. This was taken as further evidence for an alternative pathway to the lactone (*109*) not involving the diacid (*111*), such as a 7-aldehyde. However, the diene diol (*118*) feed produced large quantities of GA$_5$ (*116*), in sharp contrast to the small quantities of C$_{19}$-GAs observed in the diene diacid (*112*) feed. In this case, an alternative pathway to the 2,3-ene C$_{19}$-GAs not involving the diene diacid (*112*) [such as the diene aldehyde (*117*)] would be inconsistent with the previously discussed branch point at the diene aldehyde (*117*). It is noteworthy that the 13-hydroxy GAs were not observed in the feeds of the dienylsuccinate (*103*), only in the backfeeding experiments. This is consistent with a relatively high degree of substrate specificity for the 13-hydroxylating enzyme and is reminiscent of the absence of GA_{20} from feeds of early GA intermediates but its formation in feeds of GA_9.

The feeding studies by Bakker et al. (1974) were completed by the demonstration by isotope dilution analysis that the dienol (104) did not act as a precursor to GA_4 or GA_3 in the fungus. The transformations described above may be relevant to facets of GA biosynthesis in higher plants. Considering the obscurity of the biosynthesis of GA_5 (116) and GA_6 (115) in higher plants, these studies showed that precursors containing 2,3-double bonds early in the pathway cannot be excluded. Metabolism to GA_5 and GA_6 may account for the activity of the dienol (104) in the dwarf maize bioassay.

Although qualitative differences in the range of metabolites formed in feeding experiments are probably significant and may be used with caution to give insight into the normal GA pathway, it should be noted that quantitative changes will be much less reliable. It should be remembered that the presence in ring A of 2,3-olefin or 2,3-epoxy groups will considerably distort the shape of the A ring. In particular, any transannular interactions between C-19 and C-20 will be less favorable. This could possibly account for the apparent absence of C_{20}-GA lactones (C-19 $\rightarrow$ C-20) in these feeds. In addition, if such a transannular mechanism is required for C-20 loss, this could also account for the relatively small quantities of C_{19}-GA produced. It must also be remembered that the applied analogues are competing with endogenous substrates for enzyme sites, and thus small differences in substrate specificity may be reflected by large changes in the proportions that the metabolites formed. Undoubtedly, the microbiological production of GA_5 and GA_6 would be realized more efficiently by dienol (104) feeds plus growth retardant. The absence of any $2\beta,3\beta$-epoxy GA_{25} (119) from these feeds suggests that as this compound is formed it quickly undergoes intramolecular reaction to give the lactone (109). Thus, indirectly this observation provides further evidence that the pathway to the C_{19}-GA does not go via an intermediate containing a free 20-oic acid.

(119)   (120)

Structures (119)-(120)

The introduction of a fluorine atom into a biologically active molecule is known to have marked effects on biological properties (Schlosser, 1978). The reason for this is believed to be that the substitution of fluorine for hydrogen would not be expected to alter the size or shape of a molecule but would have marked effects on its polarity and hydrogen-bonding ability. Fluoro-substituted GAs were first prepared using a microbiological route by Bateson and Cross (1974). The readily available 7β,18-dihydroxykaurenolide (11) was fluorinated at the 18-position to give 7β-hydroxy-18-fluorkaurenolide (120). This compound was then converted chemically into 18-fluoro-GA_{12} aldehyde (121) by a route analogous to that used for the preparation of GA_{12} aldehyde (Cross et al., 1968c). The fluoroaldehyde (121) was then added to a fermentation of *G. fujikuroi* at the onset of GA_3 production. After 7 days, the fermentation was worked up in the usual way. Careful chromatography of the acidic extract gave 18-fluoro GA_9 (122) (11%) and an inseparable mixture of 18-fluoro GA_3 (123) (12%) and GA_3 (Fig. 5.39). The biological activities of these substances have been measured by Stoddart (1972) (see Volume II, Chapter 2).

Structural analogues of known GA precursors have been fed to *G. fujikuroi* with the intention that these might act as enzyme inhibitors. *Ent*-15β-fluorokaurenol (124) was prepared from xylopic acid (125) by Cross and Erasmuson (1978). Xylopic acid is a constituent of the fruit of the Nigerian shrub *Xylopia aethiopica* (Ekong

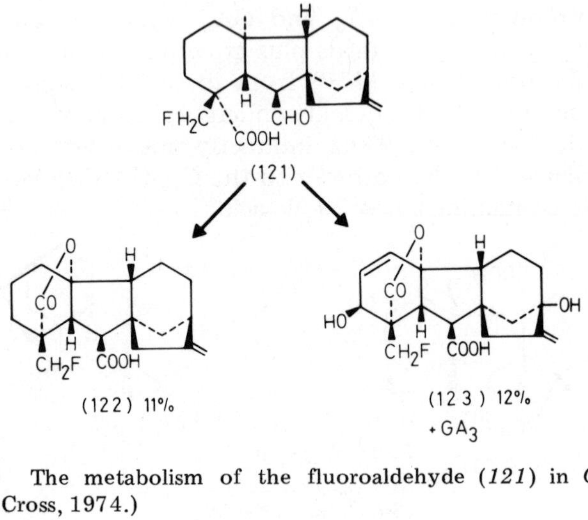

Figure 5.39   The metabolism of the fluoroaldehyde (121) in *G. fujikuroi*. (Bateson & Cross, 1974.)

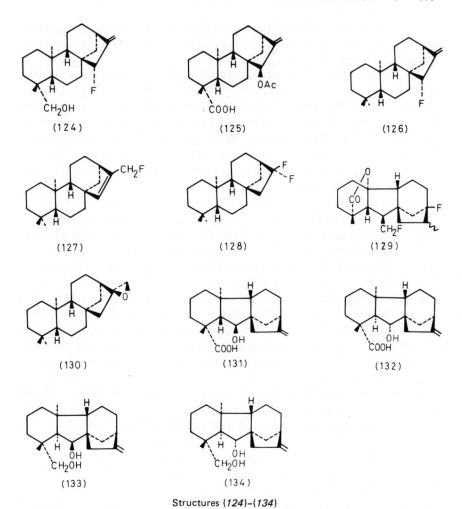

CH₂OH
(124)

·COOH
(125)

(126)

(127)

(128)

(129)

(130)

(131)

(132)

(133)

(134)

Structures (124)–(134)

& Ogan, 1968). However, when this compound was fed to a fermentation of *G. fujikuroi* it did not significantly affect the yield of GA_3. Banks et al. (1980) synthesized the fluorokaurene derivatives (126), (127), and (128) from *ent*-kaurene. However, addition of each of these compounds to cultures of *G. fujikuroi* neither affected the production of normal diterpenoid metabolites nor caused the formation of fluorinated metabolites. Boulton and Cross (1981) prepared the difluoro compound (129) from GA_3 and investigated its effect on normal GA production in the fungus. Experiments were performed with concentrations of (129) of 10 mg liter^{-1} and 25 mg liter^{-1}, and significant increases in the production of *ent*-kaurene, GA_{13}, and

GA_7 were claimed. However, large differences between the quantities of metabolites observed in two control fermentations make the recorded results difficult to assess.

Work has been published recently by Hanson and co-workers in the area of diterpene analogues as potential enzyme inhibitors. *Ent*-kauran-16β,17-epoxide *(130)* was shown to inhibit the biosynthesis of *ent*-kaurene from mevalonic acid in *G. fujikuroi* (Hanson, Willis, & Parry, 1980). Furthermore, the epoxide *(130)* also inhibited the conversion of *ent*-[^{14}C]kaurene in GA_3 in the fungus. Five days after the beginning of the feed, the control incorporation *ent*-[^{14}C]-kaurene into GA_3 (5.7%) was cut to 2.7% by 20 ppm of epoxide and to 1.7% by 40 ppm epoxide. In a further experiment the epoxide *(130)* was shown to be reversibly binding to a protein that is involved in *ent*-kaurene metabolism. The epoxide *(130)* also caused a modest (10–20%) retardation to the growth of young rice seedlings, and it inhibited the characteristic "bakanae effect" in rice seedlings infected with *G. fujikuroi*. The same workers have recently reported the remarkably specific inhibition of the ring contraction step in the fungus by two synthetic analogues of *ent*-7α-hydroxykaurenoic acid (Hanson, Parry, & Willis, 1981). The two epimeric hydroxy acids *(131)* and *(132)* were prepared in three steps from the naturally occurring fujenal *(8)*. These hydroxy acids were then reduced to the corresponding diols *(133)* and *(134)*. The structures of these derivatives were deduced from their PMR spectra, but that of the hydroxy acid *(131)* was confirmed by X-ray analysis. The four compounds were separately incubated at 40 mg liter^{-1} with *G. fujikuroi* in the presence of [^{14}C]mevalonate at a concentration of 40 mg liter^{-1} (typical of GA_3 production under the chosen conditions). The effect on the ^{14}C-incorporation into GA_3 was compared with that in control cultures. Whereas the *ent*-6β-alcohols *(132)* and *(134)* had little effect on GA_3 biosynthesis, the *ent*-6α-alcohols *(131)* and *(133)* completely blocked its formation. Even at a concentration of 4 mg liter^{-1}, the hydroxy *(131)* acid reduced by 10-fold the [^{14}C]mevalonate incorporation into GA_3 in a 10-day-old fermentation. Furthermore, *ent*-[^{14}C]-7α-hydroxykaurenoic acid *(5)* accumulated in the blocked fermentations. However, when [6-^{3}H]GA_{12} aldehyde was incubated with the fungus in the presence of the *ent*-6α-hydroxy acid *(131)*, there was a 31% incorporation of radioactivity into GA_3 compared with only 2.2% in a control. This increased incorporation demonstrates that the hydroxyacid does not block a later step in GA biosynthesis. The high incorporation figure is presumably due to a more efficient metabolism in the absence of endogenous precursors, as is found in the mutant B1-41a and in *G. fujikuroi* inhibited by conventional growth retardants. This is the first definite demonstration of an inhibitor acting at this stage in the pathway. The differential activity of the *ent*-6α- and *ent*-6β-alcohols was also observed in a bioassay. The *ent*-α-diol causes a 30% inhibition of growth of young rice plants when applied at a level of 40 μg per plant, whereas the *ent*-β-diol was inactive.

5.3.3 With *Gibberella fujikuroi* mutant B1-41a

The mutant B1-41a has been shown to be an excellent instrument for the study of the fungal GA pathway after *ent*-kaurenoic acid (*4*). The facility of the mutant in metabolizing *ent*-kaurenoic acid stimulated the study of the fate of structural analogues of *ent*-kaurenoic acid in the fungus. A particularly prominent candidate was steviol (*86*), which was considered a potential intermediate in higher plant GA biosynthesis (Lang, 1970). It was pointed out in Section 5.3.2 that steviol metabolism in the fungus had been investigated with limited success in wild-type strains by Ruddat et al. (1965), Ruddat (1968), and Hanson and White (1968).

Steviol (*86*) was prepared by enzymic hydrolysis of the naturally occurring glycoside, stevioside (*87*), using a crude pectolytic enzyme preparation (Bearder et al., 1975c). In initial studies steviol was incubated with resuspended cultures of the mutant B1-41a for 20 hr and for 5 days. The metabolites were extracted from the culture filtrate, derivatized, and analyzed by combined GC-MS. The identity and proportion of each metabolite found is shown in Fig. 5.40 and Table 5.13. Traces of the native diterpenoid metabolites GA_{13}, the diacid (*8*), and GA_3 were also detected from leakage of the metabolic block in B1-41a. However, the small amount of GA_3 was no greater than that found in controls. This furnishes additional evidence that the normal pathway to GA_3 in wild-type strains does not involve GA_1. In this experiment questions of substrate transport cannot detract from its failure to give GA_3 as the GA_1 is being formed *in situ*. The known GAs produced from steviol—GA_1, GA_{18} (*139*), GA_{19} (*141*), and GA_{20}—were identified from their published mass spectra. Of the other metabolites, those asterisked in Table 5.13 were identified by isolation and characterization in a traditional manner. The other metabolites were identified by correlation of their mass spectra with analogous known compounds. The time course of the metabolism of steviol at pH 4.8 was examined. After only 15 min, 250 μg of the substrate in 10 ml resuspended mycelium was completely converted into the *ent*-7α-hydroxy derivative (*135*). After 1 hr, traces of GA_{53}

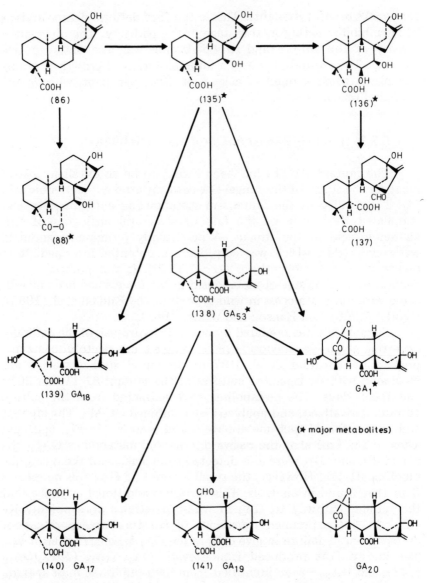

Figure 5.40   The metabolic fate of steviol (*86*) in *G. fujikuroi* mutant B1-41a. (Bearder et al., 1975c.)

Table 5.13. The metabolism of steviol (86) in G. fujikuroi mutant
B1–41a for 20 hr and for 5 days

| Metabolite | Percent of total | |
	20 hr	5 days
GA$_{53}$ (138)*	11.0	18.5
Ent-7α-dihydroxykaurenoic		
acid (135)*	47.0	0
GA$_{19}$ (141)	0.5	3.5
Ent-6α,7α,13-trihydroxy		
kaurenoic acid (136)*		
and GA$_{20}$	9.0	11.0
GA$_{18}$ (139)	10.0	15.5
Seco-ring B diacid (137)	0.5	6.0
GA$_1$ *	11.0	26.0
7β,13-dihydroxykaurenolide		
(88)	6.0	7.5
Other minor peaks	5.0	12.0

After Bearder et al. (1975c).

(138), ent-6α,7α,13-trihydroxykaurenoic acid (136), and GA$_{18}$ (139)
appeared. Five hours later GA$_1$ and 7β,13-dihydroxykaurenolide (88)
were also present. GA$_{19}$ (141) and the diacid (137) were detected
after 1 day. All the metabolites other than GA$_{18}$ (139) and the
kaurenolide increased in concentration at the expense of ent-7α,13-
dihydroxykaurenoic acid (135).

There are two major differences between the metabolism of
steviol and that of ent-kaurenoic acid in mutant B1-41a. First, no
1,2-ene GAs are formed from steviol. Second, in the steviol feed the
ent-7α- and ent-6α,7α-dihydroxy-derivatives- accumulate in short-time
incubations, whereas they can be detected only by isotope dilution
analysis in feeds of endogenous precursors. The accumulation of these
metabolites is reminiscent of that observed by Jefferies in feeds of
ent-kaurenol hemisuccinate (90) and ent-kaurenoic acid 2'-carboxy-
ethyl ester (93) (Jefferies et al., 1974; Croft et al., 1974).

The effect of varying pH on the rate of metabolism of steviol (86)
and its ent-7α-hydroxy-derivative (135) was determined. Steviol was
completely metabolized between pH 1–11, whereas ent-7α,13-
dihydroxykaurenoic acid (135) was only fully metabolized between

pH 3–5. The effect of steviol concentration on the proportions of metabolites produced was remarkably slight, except that a large increase in the amount of GA_{19} was observed at low steviol loadings.

The major metabolites (asterisked in Fig. 5.40), apart from GA_1, were isolated and re-fed to the fungus at pH 4.8. The results are summarized in Fig. 5.40. The *ent*-7α,13-dihydroxykaurenoic acid (*135*) gave a spectrum of metabolites almost identical to that of steviol; the only difference was that no 7β,13-dihydroxykaurenolide (*88*) was detected. This contrasts with the production of 7β-hydroxy-kaurenolide (*10*) from *ent*-7α-hydroxykaurenoic acid (*5*) in cultures from a wild-type strain of the fungus (Hanson et al., 1972) and of the mutant B1-41a (Bearder et al., 1975b). However, this result is in line with the recently discovered biogenesis of 7β-hydroxykaurenolide in *Cucurbita maxima* (Hedden & Graebe, 1981) and may be considered as evidence for this pathway in the fungus.

GA_{53} (*138*) was metabolized inefficiently to one main product, identified as GA_{19} (*141*) by GC-MS; traces of GA_1, GA_{17} (*140*), GA_{18} (*139*), and GA_{20} were also formed (Fig. 5.40). The slow rate of metabolism may be ascribed to transport problems, as was the slow utilization of GA_{14} (Bearder et al., 1975b) which is at an identical oxidation level. Qualitatively, the range of metabolites formed is what one would expect of an analogue of GA_{12}.

GA_{18} (*139*) was not metabolized when re-fed to B1-41a cultures, a result that was considered surprising in view of the conversion of GA_{14} into GA_3 by the mutant (Bearder et al., 1975b). However, this result falls into place if it is assumed that the polarity of exogenously applied substrate is a significant factor in the observed inefficiency of metabolism in *G. fujikuroi*.

When *ent*-6α,7α,13-trihydroxykaurenoic acid (*136*) was fed to the mutant, it was converted in 26% yield into the diacid (*137*) identified from the MS of its methyl ester and trimethylsilyl ether (TMSi) derivatives. This result is analogous to that observed in the wild-type strain ACC-917 with the native substrate *ent*-6α,7α-dihydroxykaurenoic acid (*6*) (Hanson et al., 1972). A minor *ent*-X,6,7,13-tetrahydroxykaurenoic acid was also detected by GC-MS.

The results of the steviol feed in the mutant are obviously substantially different from the metabolism in the wild-type LM-45-399 observed by Ruddat et al. (1965). The fact that GA_1 is the major product in the mutant but that another GA (presumably GA_{20}) predominates in the wild-type is probably because in the wild-type the steviol and its metabolites must compete for the enzymes with the endogenous intermediates. This would quite likely radically alter the nature of the products.

The metabolism of steviol in the mutant demonstrates the potential for the production of inaccessible GAs. For example, 290 mg of GA_{18} (*139*) was originally obtained by extracting 162 kg of immature seeds of *Lupinus luteus* (Koshimizu et al., 1968); the same quantity of GA_{18} could now be obtained by incubation with mutant B1-41a of the steviol derived from 130 g of the dried leaves of *Stevia rebaudiana*.

Encouraged by the success of the steviol feed, derivatives of steviol were then tested with the mutant (Bearder et al., 1976a). Isosteviol (*142*), obtained by acid-catalyzed ring-*C/D*-rearrangement of steviol (*86*), was fed to resuspension cultures of mutant B1-41a for 1 and 5 days. The metabolites extracted from the culture medium were analyzed by combined GC-MS as their methyl esters and methyl ester TMS ethers. From the 5-day incubation, the metabolites illustrated in Fig. 5.41 were identified by direct comparison with the derivatives of products obtained by rearrangement of the appropriate GA or *ent*-13-hydroxykaurenoid with dilute hydrochloric acid. Notably, no seco-ring B metabolites were detected. Thus, although the *ent*-6α,7α-dihydroxy derivatives (*144*) were formed in high yield they do not appear to be converted into seco-ring B compounds, unlike the corresponding *ent*-6α,7α,13-trihydroxykaurenoic acid (*136*) in the steviol feed.

The fact that no 3-hydroxylated metabolites were formed from isosteviol (*142*) is also of particular interest. A further example of this substrate specificity was provided by the metabolism of steviol acetate (*150*) (Bearder et al., 1976a). Small-scale feeds of steviol acetate were performed under the conditions used for steviol and isosteviol feeds. After 5 days the metabolites shown in the Fig. 5.42 were identified by GC-MS of the methyl ester and methyl ester trimethylsilyl ether derivatives of the acetates extracted from the culture filtrates and the free alcohols obtained by alkaline hydrolysis of this extract. Traces of nonacetylated GA_{18} (*139*) and 7β,13-dihydroxykaurenolide (*88*) were also detected and were attributed to enzymatic hydrolysis of a fraction of the steviol acetate fed. As in the metabolism of steviol (*86*) the principal metabolite after 1 day was the *ent*-7α-hydroxy derivative, which was almost completely metabolized after 5 days. A large-scale incubation (281 mg in 3 liter) was performed under similar conditions to those of the small-scale experiment. After 7 days the major metabolites were isolated in crystalline form and characterized in the usual way. Appreciable amounts of unmetabolized steviol acetate (*150*) were found in mycelial extracts, and the yields of metabolites were lower than in the small-scale feeds. The yields of metabolites were as follows: *ent*-13-acetoxy-6α,7α-dihydroxykaurenoic acid (*153*)

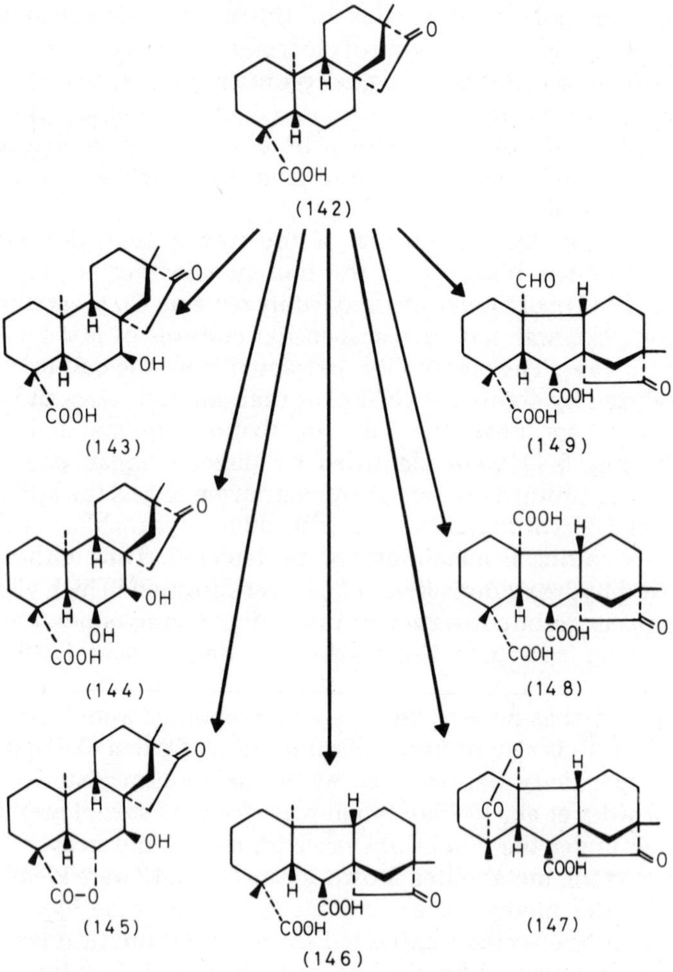

Figure 5.41 The metabolism of isosteviol (*142*) in *G. fujikuroi*. (Bearder et al., 1976a.)

10%, GA_{17} acetate (*154*) 4%, GA_{20} acetate (*156*) 20%. These three compounds were each re-fed to the mutant. *Ent*-13-acetoxy-6α, 7α-dihydroxykaurenoic acid (*153*) was slowly metabolized to the seco-ring B compound (*155*) which was obtained in 25% yield after an incubation of 5 days. As expected, the GA_{25} analogue, GA_{17} acetate (*154*), was not metabolized by the mutant. GA_{20} acetate (*156*) was mainly unmetabolized but gave a trace of a compound which was tentatively identified from MS as the 2-epimer (*157*) of GA_{29} acetate.

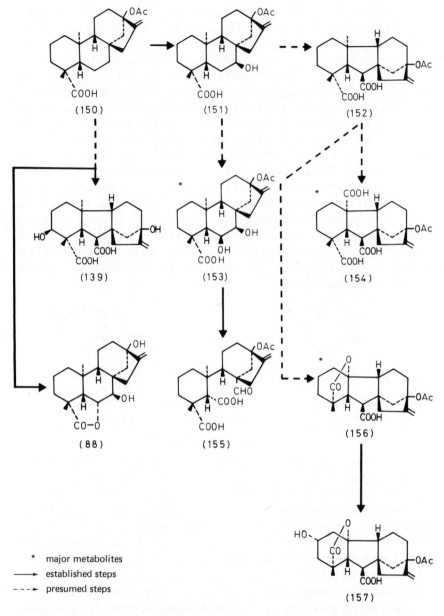

*   major metabolites
——→ established steps
- - -→ presumed steps

Figure 5.42   The metabolic fate of steviol acetate (*150*) in *G. fujikuroi* mutant B1-41a. (Bearder et al., 1976a.)

2α-Hydroxylation of GA_{20} acetate would be analogous to the observed 2α-hydroxylation of GA_{20} to GA_{40} by the mutant (Bearder et al., 1976b) (Fig. 5.24).

Comparison of the metabolites from steviol acetate (150) with those from steviol (86) and ent-kaurenoic acid (4) showed that the presence of a 13-acetoxy group completely suppressed 3β-hydroxylation. This observation provided the possibility of using the mutant to obtain quantities of the rare plant GAs GA_{17} (140) and GA_{20}. For example, the amount of steviol acetate (150) derived via stevioside (87) from 5 g of dried leaves of S. rebaudiana would be converted by the mutant via GA_{20} acetate (156) into the same quantity (7 mg) of GA_{20} isolated from 60 kg of immature seed of Pharbitis nil (Murofushi et al., 1968). This preparative route was used by Bearder et al. (1976a) to prepare radioactivity labeled GA_{20}. [14,15,17-^{3}H]Steviol acetate was prepared via steviol norketone (158) by treatment with $Ph_3P{=}C^3H_2$. The derived acetate was then fed to mutant B1-41a as before to give [^{3}H]GA_{20} (acetate) for use in studies of GA biosynthesis in higher plants.

Structures (158)–(159)

The fate in mutant B1-41a of one other related derivative was briefly examined (Bearder et al., 1973c). Steviol methyl ester (159) was metabolized to several mono- and dihydroxy-steviol methyl esters not hydroxylated in ring A and so far unidentified except for the methyl ester of ent-7α,13-dihydroxykaurenoic acid (135). No GAs were formed, indicating that methylation of the carboxyl group in steviol prevents ring B contraction. Acetylation of ent-7α-hydroxykaurenoic acid (5) likewise prevented ring B contraction to GAs, and the acetate was metabolized to a monohydroxy derivative.

The remarkable degree of nonspecificity of many of the enzymes in the fungal pathway to ring C/D analogues of ent-kaurenoic acid was shown to extend to the metabolism of trachylobanic acid (160) which occurs in flower heads of the sunflower (Helianthus annuus) (St. Pyrek, 1970). Bearder et al. (1979b) fed trachylobanic acid (160) to resuspension cultures of mutant B1-41a in the usual way. GC-MS analysis of the acidic metabolites from the culture filtrate indicated the presence of the 12,16-cyclo-analogues of GA_4, GA_9, GA_{12}, GA_{13}, GA_{14}, GA_{15}, GA_{24}, GA_{25}, GA_{37}, and GA_{47}. The proportions of the metabolites varied with the length of fermentation, pH, and concentration of the substrate. On a preparative scale trachylobanic acid (220 mg) yielded 12,16-cyclo GA_9 (*161*) (26 mg), 12,16-cyclo GA_{12} (*162*) (35 mg), and a mixture (19 mg) of the 12,16-cyclo-analogues of GA_4 (*163*) and GA_{14} (*164*) from a 5-day fermentation at pH 7.0. (Fig. 5.43). The structures of the isolated metabolites were supported by their PMR and CMR spectra. Interestingly, the 12,16-cyclo GA_9 (*161*) retained biological activity in the lettuce hypocotyl, cucumber hypocotyl, and dwarf rice bioassays.

The remaining feeds discussed in this section were analyzed only by combined GC-MS of the derivatized metabolites. These assignments were not confirmed by isolation and full characterization. Although the results are not as well documented as those previously discussed, the extensive experience of MS of GAs accumulated by MacMillan and colleagues has allowed the assignment of structures to

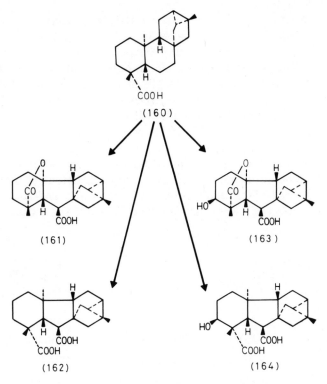

Figure 5.43 Major metabolites from a feed of trachylobanic acid (*160*) to *G. fujikuroi*. (Bearder et al., 1979b.)

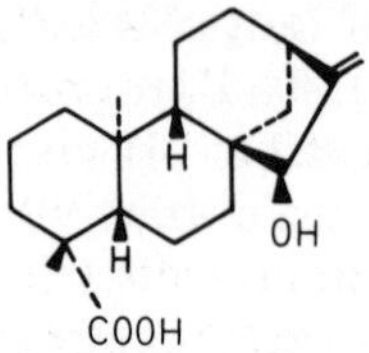

(165)

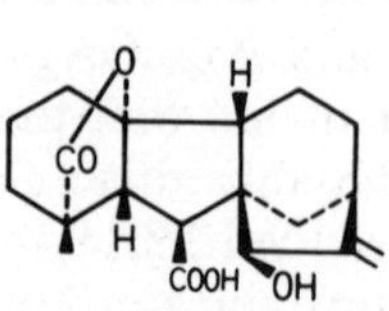

(166) GA$_{45}$

Figure 5.44   The production of GA$_{45}$ (166) by the metabolism of ent-15α-hydroxykaurenoic acid (165) in G. fujikuroi. (Bearder et al., 1975a.)

many metabolites with some confidence. This technique was used to advantage in assigning the structure to a new GA isolated from seed of pear (*Pyrus communis* L.) (Bearder et al., 1975a). The new GA GA$_{45}$ (166) was found to be identical (MS of the methyl ester TMSi ether derivative) to one of the metabolites from an incubation of ent-15α-hydroxykaurenoic acid (165) (Fig. 5.44). This substrate was obtained by saponification of the corresponding acetate (125) (xylopic acid).

The mutant B1-41a converted ent-15α-hydroxykaurenoic acid (165) into 15β-hydroxy analogues of GA$_1$, GA$_3$, GA$_4$, GA$_7$, GA$_9$, (GA$_{45}$), GA$_{12}$, GA$_{13}$, GA$_{14}$, and GA$_{15}$ (Bearder and Kybird, unpublished). The 15-hydroxylated metabolites were characterized by an intense ion at *m/e* 156 in the MS of the methyl ester TMSi ether derivatives. The 15β-hydroxy GA$_9$ (GA$_{45}$) (166) was identical to the GA from *P. communis* identified by MS.

When xylopic acid (125) itself was incubated with the mutant B1-41a, no GA metabolites were identified but a single product was formed that was tentatively identified as the ent-3α,7α-dihydroxy-derivative. The more bulky acetate group apparently blocks ring contraction. Ent-15-oxokaurenoic acid (167) and ent-15β-hydroxy-kaurenoic acid (168) were also hydroxylated by the mutant but not

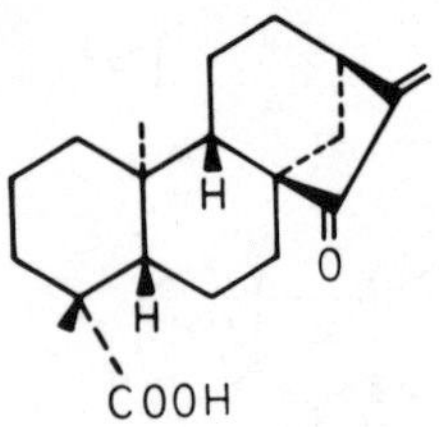

(167)

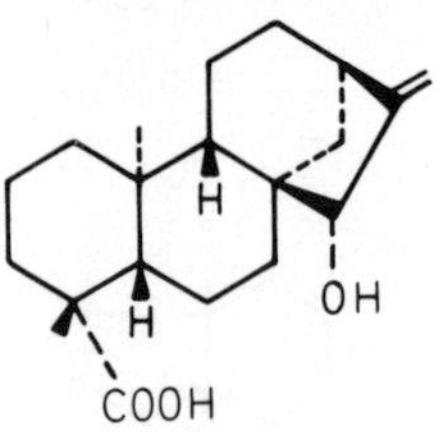

(168)

Structures (167)–(168)

converted into GAs (Bearder and MacMillan, unpublished). Lunnon, MacMillan, and Phinney (1977) used the mutant B1-41a to study the metabolism of *ent*-3α-hydroxykaurenoic acid (*169*) (Fig. 5.45) and *ent*-2α,3α-dihydroxykaurenoic acid (*172*) (Fig. 5.46), both of which were prepared from the diol (*104*). These studies were undertaken to further investigate the lack of substrate specificity of the enzymes of the GA pathway in the fungus. The feeds were carried out for 5 days in the usual way, and the analysis of metabolites was undertaken

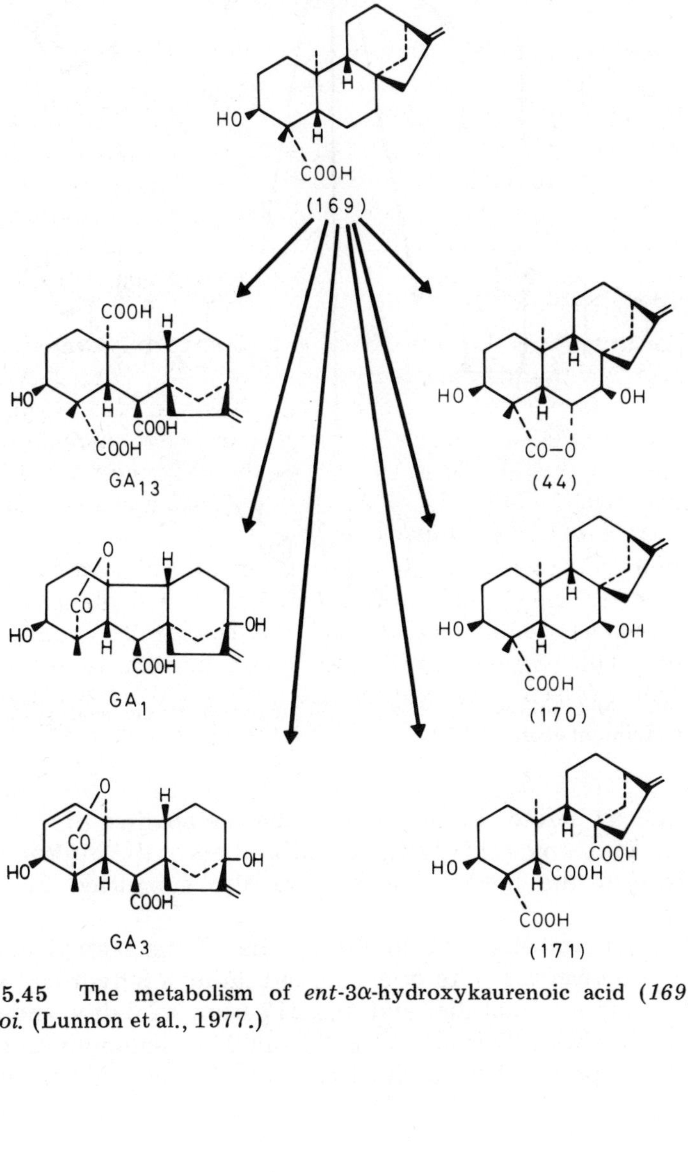

Figure 5.45 The metabolism of *ent*-3α-hydroxykaurenoic acid (*169*) in *G. fujikuroi.* (Lunnon et al., 1977.)

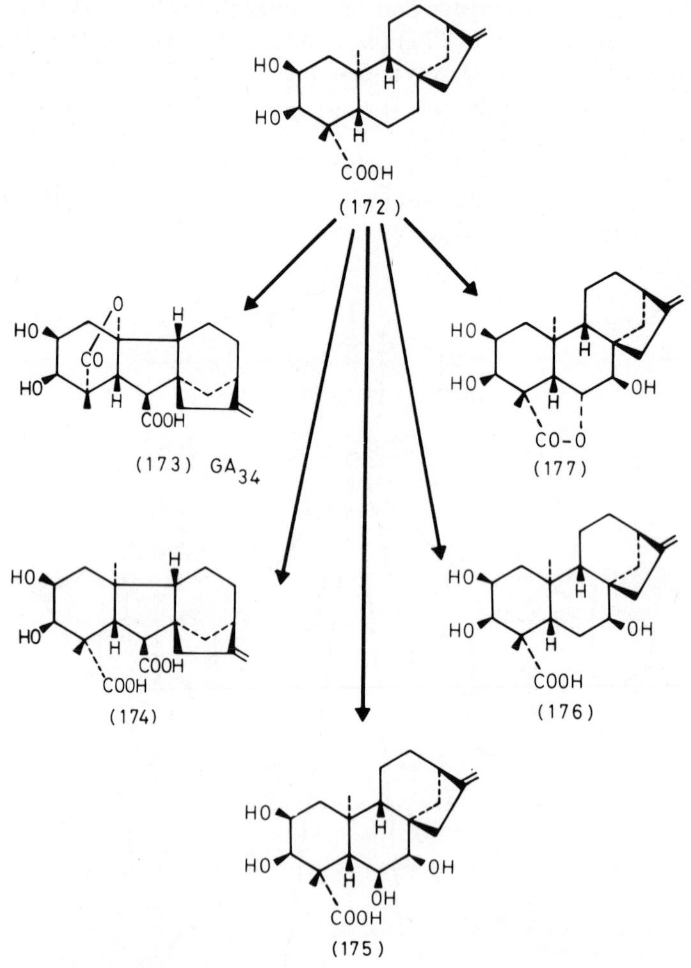

Figure 5.46 Metabolites of *ent*-2α,3α-dihydroxykaurenoic acid (*172*) in *G. fujikuroi*. (Lunnon et al., 1977.)

by GC-MS as before. In the case of the metabolites of *ent*-2α,3α-dihydroxykaurenoic acid (*172*), the TMSi ethers of the isopropylidene derivatives of the methyl esters were also submitted to GC-MS analysis.

Useful diagnostic ions in the spectra of the methyl ester trimethylsilyl derivatives were m/e 129, indicating a 3-hydroxyl group; m/e 147 from a vicinal diol and m/e 217 from a 2,3-diol; and M^+-49 from 2,3-dihydroxy GAs and M^+-150 from 2- or 3-monohydroxy-C_{20} GAs. The isopropylidene derivatives were characterized by intense ions at m/e 43 and 58; kaurenolides unsubstituted in ring A characteristic m/e 135 and 107 due to the elimination of trimethylsilanol. *Ent*-6α,7α-dihydroxykaurenoic acid (*6*) shows base peak of m/e 269, but when ring A is hydroxylated this is replaced by strong ions at m/e 357 and 267.

The metabolites of *ent*-3α-hydroxykaurenoic acid (*169*) (Fig. 5.45) included the known substrates GA_1, GA_3, GA_{13}, and 3β,7β-dihydroxykaurenolide (*44*). Two new compounds were tentatively identified as *ent*-3α,7α-dihydroxykaurenoic acid (*170*) and the *ent*-3α-hydroxy seco-ring B compound (*171*). The *ent*-2α,3α-dihydroxykaurenoic acid (*172*) was also efficiently consumed by the mutant fungus (Fig. 5.46), giving the known higher plant GA GA_{34} (*173*) and four other metabolites tentatively identified as 2β,3β-dihydroxy GA_{12} (*174*), 2β,3β,7β-trihydroxykaurenolide (*177*), *ent*-2α,3α,7α-trihydroxykaurenoic acid (*176*) (major product), and *ent*-2α,3α,6α,7α-tetrahydroxykaurenoic acid (*175*). These results showed that the hydroxylation of *ent*-kaurenoic acid at the *ent*-2α and *ent*-3α positions did not prevent normal transformation to GAs, although ring contraction was retarded judging by the accumulation of the *ent*-7α-hydroxykaurenoic acid analogues. *Ent*-2α-hydroxylation apparently also slowed the C_{20}- into C_{19}-GA conversion.

Finally, some preliminary results of other feeds to mutant B1-41a will be discussed which bear on the question of substrate specificity. (Bearder and MacMillan, unpublished). A number of other ring-D-modified *ent*-kaurenoic acid analogues are apparently metabolized by mutant B1-41a all the way through to C_{19}-GA analogues; these include stachenoic acid (*178*), *ent*-kauranoic acid (*179*) and its 16-epimer, and *ent*-kaur-15-en-19-oic acid (*180*). A free 19-oic acid group is apparently essential for the C_{20}- to C_{19}-GA conversion as GA_{12} alcohol (*38*) methyl ester is not converted into C_{19}-GA but does undergo oxidation at the 3β- and 20-positions, giving inter alia, the methyl esters of GA_{14} alcohol (*54*) and GA_{13} alcohol (*57*). An unencumbered C-7 oxygen function is likewise essential. GA_{12} alcohol

(178)

Structure (*178*)

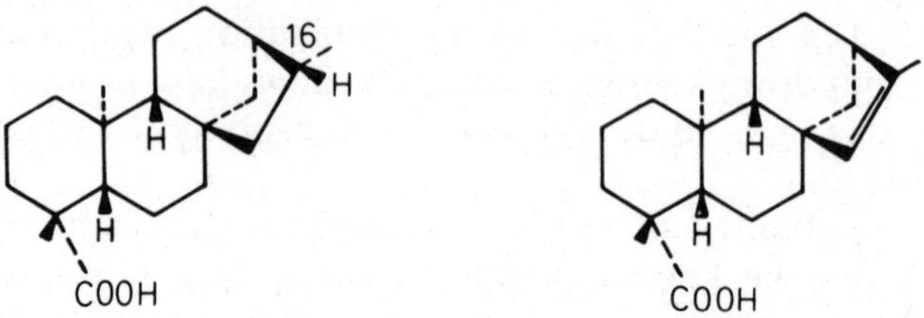

Structures (*179*)–(*180*)

(*38*) 7-acetate gave no C_{19}-GAs, although GA_{14} alcohol (*54*) 7-acetate was among the products. The main product was apparently 16,17-hydrated GA_{14} alcohol 7-acetate (*53*).

The C_{20}- to C_{19}-GA conversion was also blocked by acetylation at the 3-position. GA_{14} acetate (*181*) was completely metabolized to the C_{20}-GA analogues GA_{13} acetate (*184*), GA_{37} acetate (*182*), and GA_{36} acetate (*183*) (Fig. 5.47). These assignments were confirmed by alkaline hydrolysis of the extract to give GA_{13}, GA_{36}, and GA_{37}

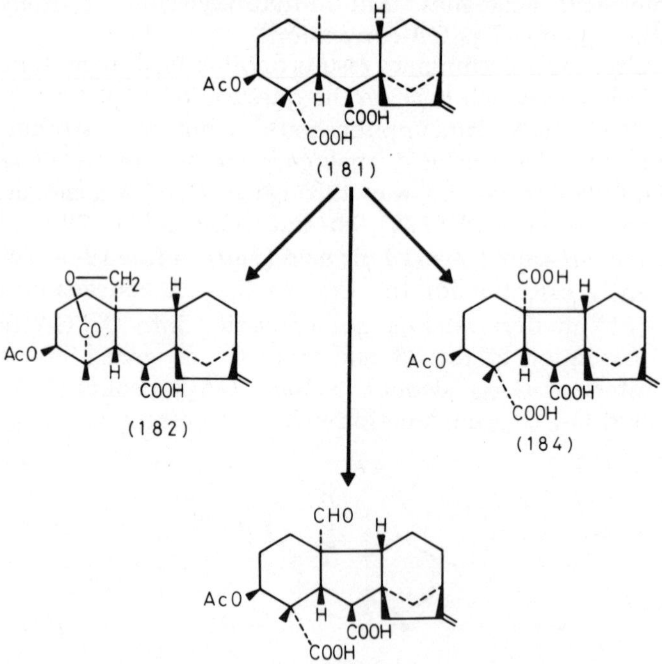

Figure 5.47 The metabolites of GA_{14} acetate (*181*) in *G. fujikuroi*. (Bearder & MacMillan, unpublished.)

identical to authentic samples by MS. This transformation could be used as a useful route to the rare GA_{36} and GA_{37} from the relatively more accessible GA_{14}. Epimerization of the 3-hydroxyl in GA_{14} does not prevent the C_{20}- into C_{19}-GA conversion, since 3-*epi*-GA_4 and 3-*epi*-GA_1 were detected in incubations with 3-*epi*-GA_{14}. When 3-*epi*-GA_4 and the GA_4 isomer (*185*) were separately fed to the mutant, 2,3-olefin formation did not occur and a small degree of 13-hydroxylation was the only metabolic transformation observed in each case.

(1 85)

Structure (*185*)

In summary, the mutant B1-41a has been and remains a very useful tool for the transformation of kaurenoid and GA analogues. Disadvantages of its use are its restricted availability and the requirement for substrates to have C-19 at the carboxylic acid oxidation state. The use of a suitable growth retardant in conjunction with a wild-type strain can metabolize analogues of *ent*-kaurene onwards, but it must be borne in mind that the added growth retardant might also influence subsequent reactions in the pathway (see Echols et al., 1981).

5.3.4 With wild-type strains of *Gibberella fujikuroi* using growth retardants

Hedden et al. (1977) were the first to use the blockage of GA biosynthesis with a growth retardant as a tool for the microbiological conversion of analogues of normal intermediates. Resuspension cultures of the wild-type strain GF-1a containing $100 \, mg \, liter^{-1}$ (*68*) were used for the study of metabolism of three isomers of the natural intermediate *ent*-kaur-16-ene (*1*).

The compounds kaur-16-ene (*89*) (the enantiomer of the natural precursor), 13β-kaurene (*186*) (phyllocladene), and *ent*-kaur-15-ene (*187*) were separately incubated under these conditions. However, in none of the feeds were any metabolites detected (GC) in the culture filtrate, and the mycelial extracts showed only the presence of the

(186)   (187)

Structures (186)–(187)

applied substrate. The negative results in the case of the kaur-16-ene
(89) and 13β-kaurene (186) feeds might be expected owing to the
gross stereochemical differences between these and *ent*-kaur-16-ene
(1); more so in the case of kaur-16-ene (89) as this ^{14}C-labeled
substrate had previously been shown not to be incorporated into the
enantiomer of GA_3 in cultures of the wild-type strain ACC-917 (Cross,
Norton & Stewart, 1968a). In retrospect, the nonmetabolism of *ent*-
kaur-15-ene (187) is notable, as *ent*-kaur-15-ene-19-oic acid (180) is
converted into GAs in mutant B1-41a (Bearder and MacMillan
unpublished). In the light of Hanson's recent work (Hanson et al.,
1980), this result is most easily accommodated by assuming a high
degree of specificity in the *ent*-kaurene carrier protein.

An advantage of the wild-type plus growth-retardant technique
over the mutant B1-41a was illustrated in the work of Lunnon et al.
(1977). They were able to study the metabolism of ring A hydroxy
analogues of *ent*-kaurenol (2), which were much more accessible than
the corresponding *ent*-kaurenoic acid (4) derivatives that would be
required for feeding studies with mutant B1-41a. The two epimeric
ent-2-hydroxykaurenoids (188) and (193) were separately fed to
resuspension cultures of GF-1a containing the growth retardant (68).
In 5 days each substrate had been completely metabolized. The *ent*-
kaur-2α,19-diol (188) gave metabolites that were identified by GC-MS
as (in order of elution) *ent*-2α,6α,7α-trihydroxykaurenoic acid (189),
2β-hydroxy GA_{14} (174), GA_{43} (192), 2β,7β-dihydroxykaurenolide
(191), and GA_{34} (173) (Fig. 5.48). On another column 2β-hydroxy
GA_{12} (190) was also detected by GC-MS. *Ent*-kauren-2β,19-diol (193)
gave (in order of elution)*ent*-2β,6α,7α-trihydroxykaurenoic acid (194),
2α-hydroxy GA_{14} (195), a 2α-dihydroxy GA_{14}, 2α,7β-dihydroxy-
kaurenolide (196), and GA_3 (Fig. 5.49). The quantity of GA_3
produced was at least 25 times greater than that in control cultures
containing the inhibitor, and so it was construed that GA_3 was a
genuine metabolite of the diol (193). In a third experiment, *ent*-kaur-
16-en-3α,19-diol (102) was incubated with the wild-type GF-1a in the

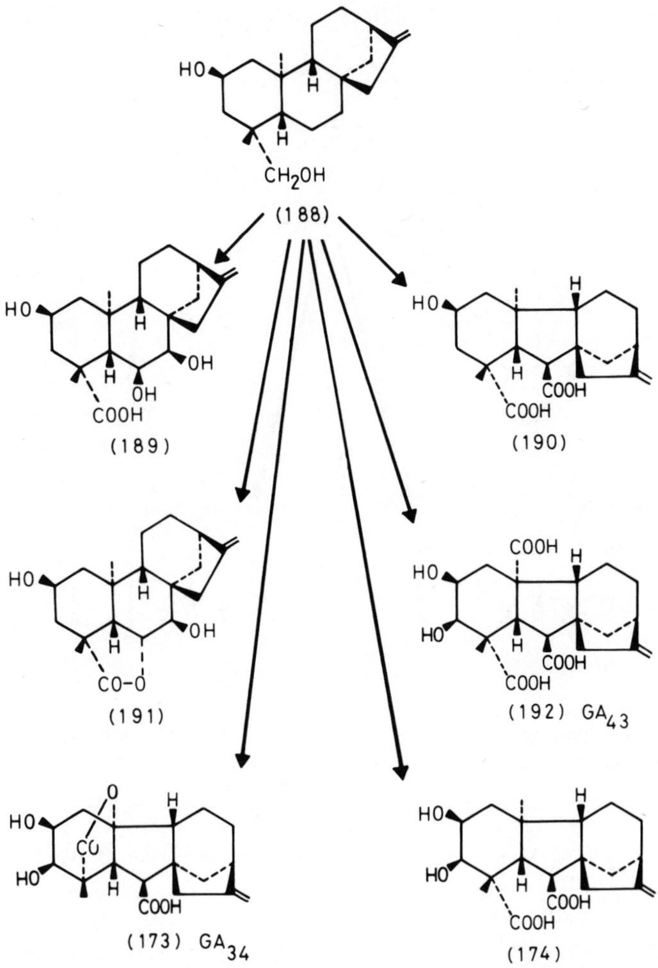

Figure 5.48 The metabolites of *ent*-kaur,2α,19-diol (*188*) in *G. fujikuroi.* (Lunnon et al., 1977.)

presence of (*68*). This compound, as expected, gave the same range of metabolites as that derived from *ent*-3α-hydroxykaurenoic acid (*169*) with mutant B1-41a (Fig. 5.45). In the feeds of the *ent*-2-hydroxy-kauren-19-ols (*188*) and (*193*), no *ent*-7α-hydroxykaurenoic acid analogues were detected. This suggests that presence of a 2-hydroxyl group in *ent*-7α-hydroxykaurenoic acid does not affect the rate of ring contraction, compared with the *ent*-3α-hydroxy and *ent*-2α,3α-dihydroxy-analogues. However, a 2-hydroxyl group did seem to inhibit

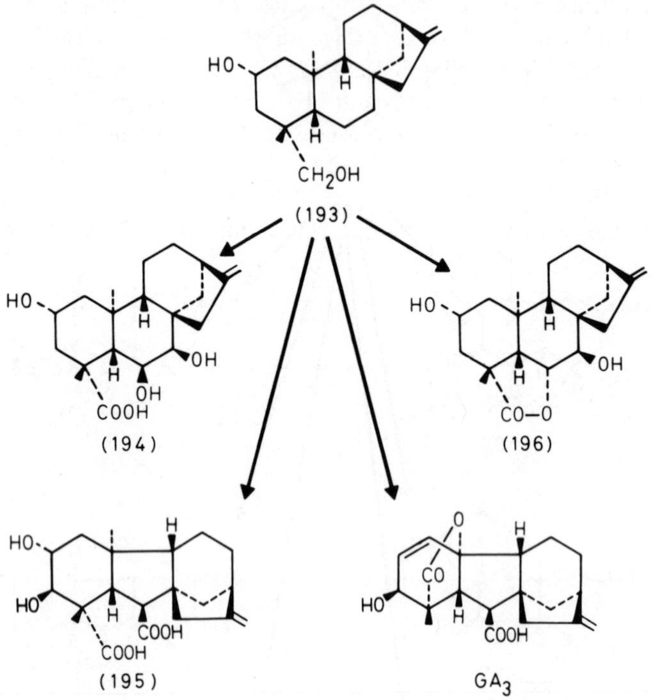

Figure 5.49 The metabolites of *ent*-kaur-2β,19-diol (*193*) in *G. fujikuroi.* (Lunnon et al., 1977.)

conversion into C_{19}-GAs. The formation of GA_3 from the *ent*-2β-hydroxykaurenol (*193*) is noteworthy. It was presumed that this group underwent dehydration subsequent to ring contraction to give the 1,2-double bond of GA_3. Although such an intermediate cannot be ruled out in the normal biosynthesis of GA_3 (from Hanson's feeds of doubly labeled mevalonates), there is no evidence for an intermediate of this type in the normal pathway.

Cross and co-workers have used the wild-type strain ACC-917 inhibited with the growth retardant AMO-1618 for the study of the metabolism of fluorinated analogues of *ent*-kaurenoic acid.

Cross and Erasmuson (1978, 1981) prepared *ent*-15β-fluoro-kaurenoic acid (*197*) from desacetoxy xylopic acid (*165*) by a sequence involving fluorination with Et_2NSF_3. The fluoro acid (*197*) was incubated with *G. fujikuroi* ACC-917 (nonreplacement cultures), and the metabolites were isolated as methyl esters and characterized by infra-red spectroscopy (IR) and by comparison of PMR and MS with proton analogues (Fig. 5.50). One fermentation using a dilute culture medium yielded from the acids the seco-ring B metabolite (203) and 15α-fluoro-GA$_{14}$ (198). In the second fermentation of the fluoro acid (695 mg) in a stronger medium the acidic metabolites were chromatographed on silica gel to give the anhydride form of the metabolite (203) (presumably an artifact of chromatography), the novel lactone (202) (23 mg), and 15α-fluoro GA$_4$ (199) (20 mg). The neutral fraction yielded 15α-fluoro-7β-hydroxykaurenolide (201) and a metabolite tentatively assigned the structure 15α-fluoro-1β,7β-

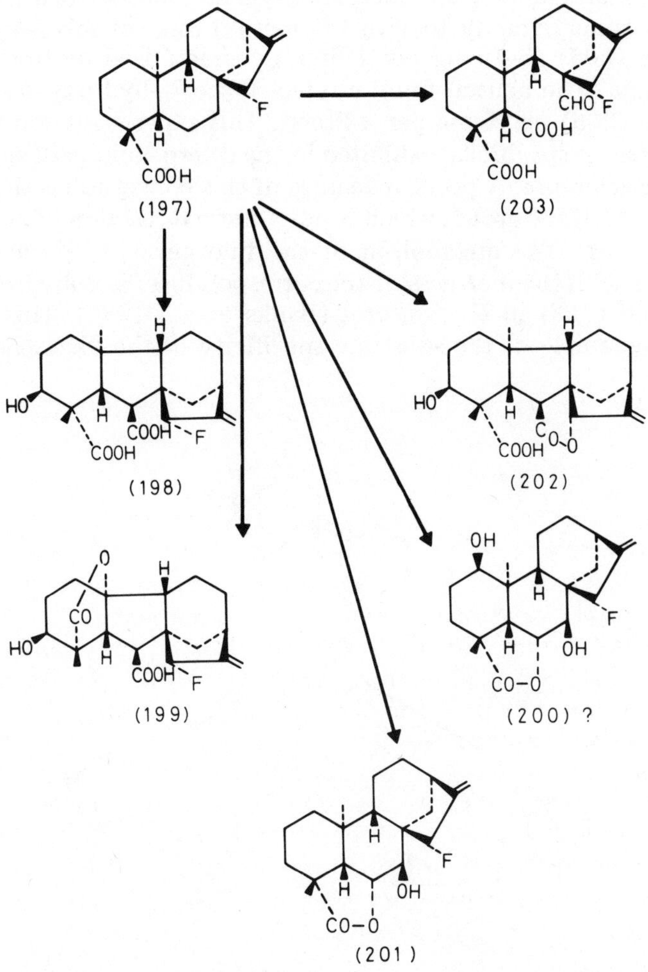

Figure 5.50 Metabolites of *ent*-15β-fluorokaurenoic acid (197) in *G. fujikuroi.* (Cross & Erasmuson, 1978, 1981.)

dihydroxykaurenolide (200). The origin of the new lactone (202) was not pinpointed. Three possible derivations seem possible: (a) as an artifact of the isolation procedure, (b) as a product of a microbially derived substitution of fluoride ion at the 15-position, or (c) as a new GA—the AMO-1618 was added after GA production had begun. In further work Cross and Filippone (1980) examined the fate of ent-16, 16-difluoro-17-nor-kaurenoic acid (204) in the fungus. Addition of the fluoro acid (204) (40 mg liter^{-1}) to a stirred fermentation of G. fujikuroi ACC-917 in the presence of AMO-1618 (10 mg liter^{-1}) gave crude acidic and neutral fractions (Fig. 5.51). The acid extract was chromatographed on silica, methylated, and purified by liquid-liquid chromatography to give the methyl ester of the GA$_7$ ketone analogue (205) (5–10 mg per 4 liters) characterized by full spectroscopic data. The neutral fraction yielded the 7β-hydroxykaurenolide analogue (206) (ca. 5 mg per 4 liters). This experiment confirms the low degree of specificity exhibited in the diterpenoid pathway of the C/D ring structure. A peculiar feature of this conversion is the formation of a 3-ketone group, which is not a normal reaction of the fungal GA pathway. The metabolism of the fluoroacids (197) and (204) compares with the inertness of the corresponding fluorohydrocarbons (126) and (128) in G. fujikuroi (Banks et al., 1980). This may be another example of the substrate specificity of the "kaurene carrier protein."

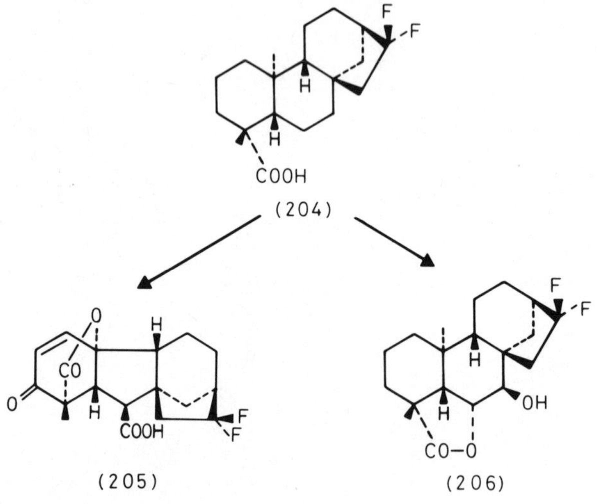

Figure 5.51   Metabolites of ent-16,16-difluoro-17-norkaurenoic acid (204) in G. fujikuroi. (Cross & Filippone, 1980.)

In a series of publications Hanson and co-workers have reported the use of AMO-1618 inhibited cultures of *G. fujikuroi* to observe the metabolism of various diterpene analogues of natural kaurenoid intermediates.

Sideritis spp. contain 7- and 18-hydroxylated *ent*-kaurenes suitable for feeding to the fungus (Breton et al., 1969; Piozzi et al., 1968, 1971; de Quesada, Rodriguez, & Valverde, 1973, 1974; Gonzalez et al., 1973). The naturally occurring *epi*-candicandiol (*207*) incubated with *G. fujikuroi* ACC-917 for 4 days gave a triol shown by chemical and spectroscopic means to have the structure (*208*) (Fraga, Hanson, & Hernandez, 1978) (Fig. 5.52). When the incubation was repeated incorporating AMO-1618, an acidic metabolite was isolated as its methyl ester. This was identified as the methyl ester of *ent*-7α,18-dihydroxykaur-16-en-19-oic acid (*209*) by NMR spectroscopy and chemical correlation with the triol (*208*). The ester had been over-looked in the first fermentation (without AMO-1618) as its TLC *Rf* was identical to that of GA$_3$ methyl ester. [18-^{3}H]Epi-candicandiol (*207*) was incorporated by the fungus plus AMO-1618 into the triol (*208*) (16.5%) and acid (*209*) (14.8%). Dilution analysis failed to detect a 7,18-dihydroxykaurenolide, and TLC evidence could not be found for other acidic and possibly *ent*-gibberellane metabolites. Thus it was noted that the oxidation at C-19 was not substantially affected by 18- and *ent*-7α-hydroxylation but that the 18-hydroxylation prevented microbial attack at the 6-position (both lactonization and ring contraction).

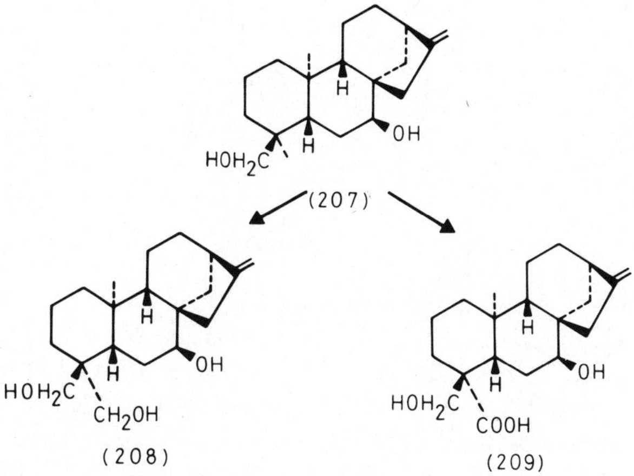

Figure 5.52 The metabolism of epi-candicandiol (*207*) in *G. fujikuroi.* (Fraga et al., 1978.)

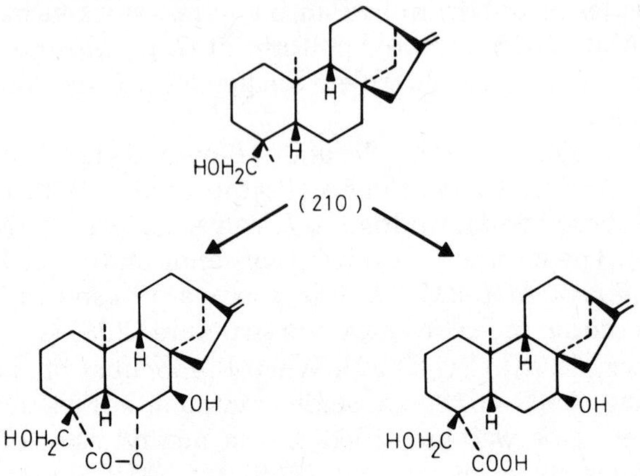

Figure 5.53 Metabolites of candol B (*210*) in *G. fujikuroi.* (Fraga et al., 1980.)

This effect was further examined in a subsequent publication (Fraga et al., 1980). Candol B (*210*) (*ent*-18-hydroxy-kaur-16-ene, 28 mg) was incubated for 5 days with an AMO-1618-inhibited fermentation of *G. fujikuroi* ACC-917. Two metabolites were isolated (Fig. 5.53) and shown to be 7,18-dihydroxykaurenolide (*11*) (18 mg) and *ent*-7α,18-dihydroxykaurenoic acid (*208*) (10 mg) by comparison with authentic samples (Cross et al., 1963; Fraga et al., 1978). Candol A (*ent*-7α-hydroxykaur-16-ene) (*211*) (75 mg) under similar fermenta-

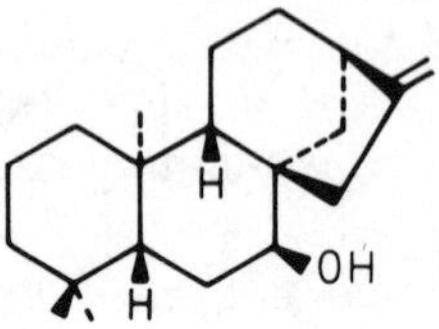

(211)

Structure (*211*)

tion conditions yielded the seco-ring B diacid (*8*) (5 mg), GA$_4$ mixed with GA$_7$ (5 mg), and GA$_3$ (15 mg), identified from the PMR spectra of their methyl esters. Although these compounds are normal metabolites of the fungus, they are not detected in the AMO-1618-inhibited control cultures. Sideradiol (*212*) (*ent*-7α,18-dihydroxykaur-15-ene), which is the double-bond isomer of *epi*-candicandiol (*207*), was also incubated with fungus (Fig. 5.54). Incubation of 200 mg of (*212*) yielded two metabolites, which were isolated and characterized as

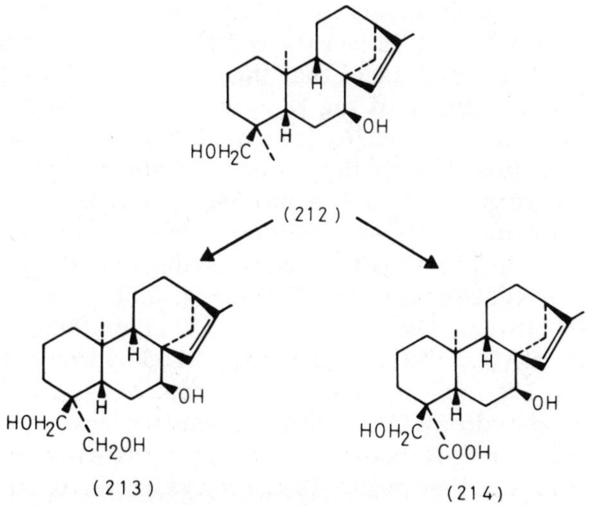

Figure 5.54   Metabolites of sideradiol (*212*) in *G. fujikuroi*. (Fraga et al., 1980.)

ent-7α,18,19-trihydroxykaur-15-ene (*213*) (10 mg) and *ent*-7α-18-dihydroxykaur-15-ene-19-oic acid (*214*) (8 mg). No GA or fujenal analogues were detected. Thus the fate of (*212*) with *G. fujikuroi* is analogous to that of its $\Delta^{16,17}$-isomer (*207*) (Fraga et al., 1978). The results of these feeds led to the following conclusions. *Ent*-7α-hydroxylation does not inhibit the sequential oxidation of C-19. *Ent*-18-hydroxylation also does not effect 19-oxidation, but it does inhibit functionalization at C-6 which would lead to ring contraction and kaurenolide and fujenal formation. Furthermore, the isomerization of the exocyclic olefin to the endocyclic position does not affect oxidation at C-19.

To further probe the mechanism of the inhibition of enzymic attack at C-6 in *ent*-18-hydroxykaurenes, Fraga et al. (1980) prepared *ent*-18-chloro-7α-hydroxykaur-16-ene (*215*). However, repeated

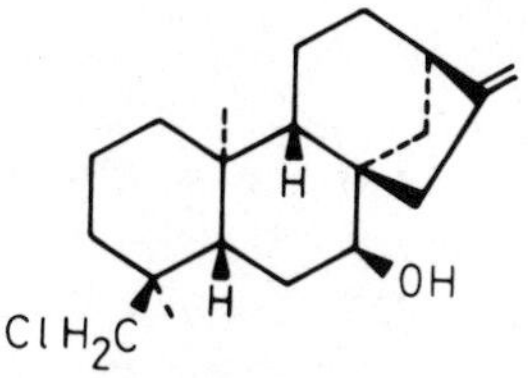

(*215*)

Structure (*215*)

attempts failed to elicit metabolism of this compound in the presence or absence of AMO-1618. In fact the compound appeared to be toxic to the fungus. It seemed likely that the inhibition of reaction at C-6 was due to the proximity of the 18-hydroxyl group, and so ent-15α, 19-dihydroxy-kaur-16-ene (216) [prepared from xylopic acid (215)] was fed to the fungus with the purpose of observing the effect on ent-7α,15α-hydroxylation by the proximal 15-hydroxyl group. The diol (216) (400 mg) gave two products, which were isolated and characterized as 7β,15β-dihydroxykaurenolide (217) (20 mg) and ent-7α,15α-dihydroxykaurenoic acid (218) (Fig. 5.55). However, a more detailed (GC-MS) analysis in MacMillan's laboratory showed the presence of GA_{45}, 15β-hydroxy-GA_{12}, 15β-hydroxy-GA_{14}, 15β-hydroxy-GA_4, 15β-hydroxy-GA_{15}, 15β-hydroxy-GA_{13}, 15β-hydroxy-GA_1, and 15β-hydroxy-GA_3. While discussing the effect of 18-hydroxylation on ring B oxidation, it is appropriate to draw attention to the observation that while 18-hydroxykaurene is subjected to ent-7α-hydroxylation, when 18-hydroxykaurenolide (80) is fed in the presence of AMO-1618 no 7,18-dihydroxykaurenolide (11) is formed although 11α-hydroxylation occurs and the 18-position undergoes further oxidation (Fig. 5.28) (Hanson & Sarah, 1979).

The idea that oxidation at a certain position in the ent-kaurene skeleton can be hindered by a closely situated hydroxyl group led Fraga et al. (1981) to study the effect of ent-3β-hydroxylation on

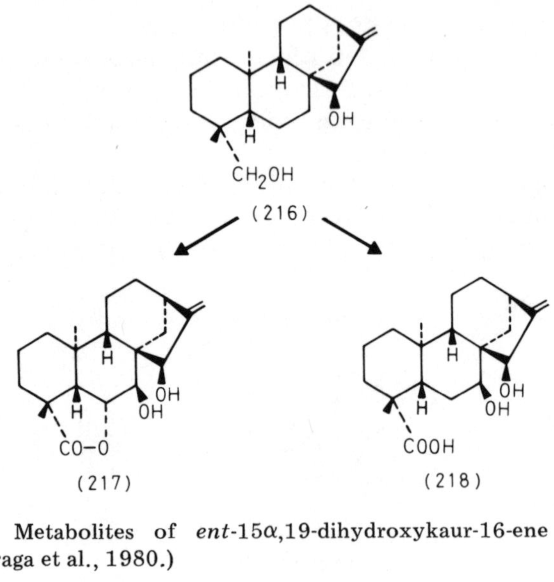

Figure 5.55 Metabolites of ent-15α,19-dihydroxykaur-16-ene (216) in G. fujikuroi. (Fraga et al., 1980.)

oxidation at C-19. As in other previous papers the substrates were prepared from *Sideritis* diterpenes. *Ent*-3β-hydroxykaur-16-ene (*219*) was obtained by deoxygenation of the 7- and 18-functions of linearol (*220*). When (*219*) was incubated with the fungus, no metabolism could be detected and the bulk of the feed was recovered unchanged.

(219)   (220)

Structures (*219*)–(*220*)

Ent-3β,18-dihydroxykaur-16-ene (*221*) was prepared from foliol (*222*). In this case, (*221*) (200 mg) was almost totally consumed by the fungus to give one major metabolite, which was isolated and shown to be *ent*-3β,7α,18-trihydroxykaurene (*222*, foliol) (48 mg) (Fig. 5.56). When foliol (140 mg) was backfed to *G. fujikuroi* it was substantially metabolized, undergoing hydration of the exocyclic methylene group to give the tetraol (*223*) (90 mg). This latter transformation is one that is frequently observed when non-13-hydroxy-

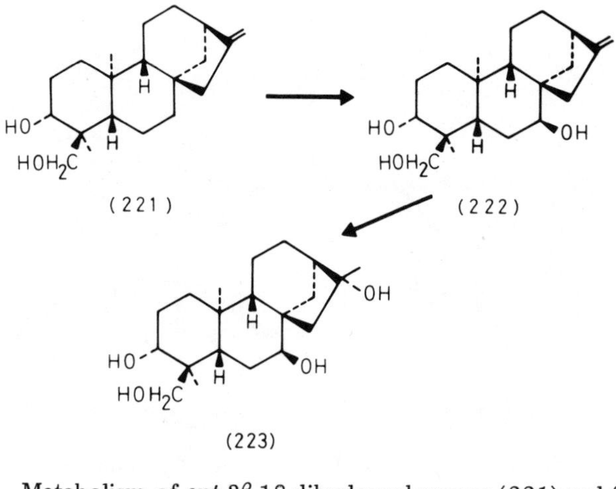

(221)   (222)

(223)

Figure 5.56 Metabolism of *ent*-3β,18-dihydroxykaurene (*221*) and foliol (*222*) in *G. fujikuroi*. (Fraga et al., 1981.)

lated kaurenoids and GAs accumulate in the fungus, and such products are almost always metabolically inert (Hanson & Ball, unpublished; Bearder, unpublished; Hedden et al., 1974). This series of experiments showed that ent-3β-hydroxylation blocks oxidation at C-19. Thus, it was suggested that the ent-3β-hydroxyl could be interacting with the enzyme responsible for oxidation of C-19 in a similar way in which the 18-hydroxyl might bind to the 6-functionalizing enzyme. The lack of ent-7α-hydroxylation of ent-3β-hydroxykaurene (219) compared with that observed with ent-3β-18-dihydroxykaurene (221) suggested that although 19-oxidation was not essential for ent-7α-hydroxylation to occur, some polar functionality at C-4 was required.

Hanson et al. (1979) have also studied the metabolism of the ent-atisirene analogue of the natural intermediate ent-7α-hydroxy-kaurenoic acid (5). This substance (224) was obtained by saponification of the corresponding angelate, gummiferolic acid, a constituent of Margotia gummifera (Pinar, Rodriguez & Alenary, 1978). A 5-day incubation of the hydroxy acid (224) (300 mg) in the presence of AMO-1618 gave two metabolites not detected in control fermentations. These substrates were isolated and characterized as the GA_{12} analogue (225) (15 mg) and the GA_{14} analogue (226) (58 mg) (Fig. 5.57). More highly oxygenated GAs were not detected. Although possibly a consequence of the sensitivity of the analytical method used, this apparent lack of oxidation at C-20 was suggested to reflect the increased steric crowding at C-20 caused by the two-carbon

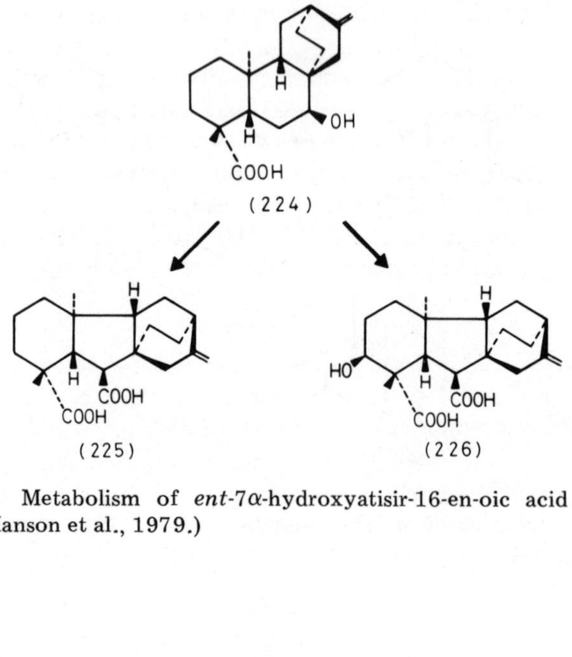

Figure 5.57 Metabolism of ent-7α-hydroxyatisir-16-en-oic acid (224) in G. fujikuroi. (Hanson et al., 1979.)

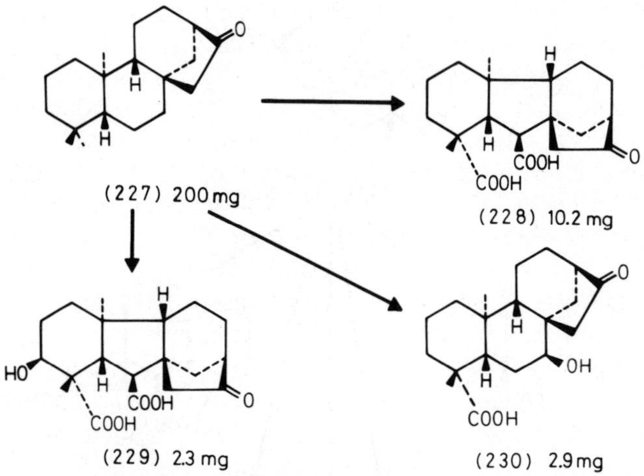

Figure 5.58   Metabolites of *ent*-16-oxo-17-norkaurene (*227*) in *G. fujikuroi.* (Wada et al., 1979.)

bridge (C-13 and C-14) as compared to the methylene (C-14) of the GAs. In the previously discussed metabolism of stachenoic acid (*178*), this bridge would not be expected to cause as much hinderance.

Japanese chemists have also been active in the study of the metabolism of *ent*-kaurene analogues in chemically blocked fermentations of *G. fujikuroi*. Wada, Imai, and Shibata (1979) examined the fate of *ent*-16-oxo-17-norkaurane (*227*), *ent*-17-hydroxykaur-15-ene (*231*), and *ent*-15α-hydroxykaur-16-en-19-oic acid (*165*) with mycelial suspensions of the fungus containing 1-decylimidazole (*70*) or CCC (*67*). The norketone (*227*) (200 mg) was incubated with 540 ml of resuspended mycelium containing CCC. After 2 days the acidic metabolites were ρ-bromophenacylated and isolated by liquid-liquid chromatography. After dephenacylation and methylation, the methyl esters of 16-oxo-17-norGA$_{12}$ (*228*) (10.2 mg), 16-oxo-17-norGA$_{14}$ (*229*) (2.3 mg), and 7β-hydroxy-16-oxo-17-norkauran-19-oic acid (*230*) (2.9 mg) were identified by IR, PMR, and MS of authentic samples (Fig. 5.58).

The alcohol (*231*) (500 mg) was metabolized under similar conditions, and methyl esters of metabolic products were isolated in like fashion. From spectral data the metabolites were identified as the GA$_{12}$ analogue (*232*) (38 mg), the GA$_{24}$ analogue (*233*) (11 mg), the GA$_{15}$ analogue (5.4 mg) (*235*), the GA$_4$ and the GA$_7$ analogues (*236*) and (*237*) (88 mg), and the GA$_{14}$ analogue (*234*) (3.1 mg) (Fig. 5.59).

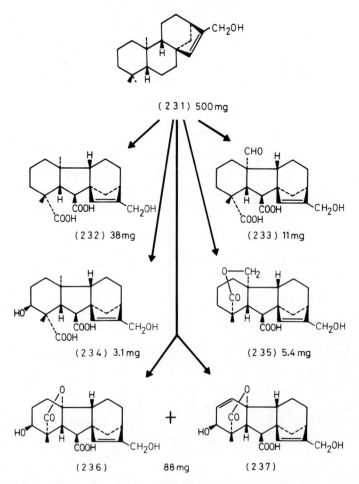

Figure 5.59 Metabolism of *ent*-17-hydroxykaur-15-ene (*231*) in *G. fujikuroi.*
(Wada et al., 1979.)

The 15-hydroxy-acid (*165*) was slowly metabolized in the mycelial suspension to yield *ent*-7α,15α-dihydroxykaur-16-en-19-oic acid (*218*) after 2 days. After 7 days GA analogues were formed, but endogenous GAs were also produced. When the experiment was repeated using 1-decylimidazole (*70*) as blocking agent (50 ppm) the major metabolites were tentatively identified by GC-MS as 15β-hydroxy-GA$_{12}$, 15β-hydroxy-GA$_{25}$ and 15β-hydroxy-GA$_7$. This range of metabolites compares with the 15β-hydroxy-analogues of GA$_9$ (GA$_{45}$), GA$_{25}$, GA$_{24}$, and GA$_4$ detected in a similar experiment with the mutant B1-41a [Bearder et al., 1975a; see also the feed of *ent*-kauran-15α,19-diol (*216*), p. 364]. In general, feeding results fit the theory that ring D modification does not prevent "GA" formation although 3-hydroxylation may be impaired.

The metabolism of steviol (*86*) by *G. fujikuroi* in the presence of plant growth retardants was studied by Murofushi, Nagura, and Takahashi (1979). The growth retardants used were CCC (*67*), AMO-1618, and the quaternary ammonium salt (*238*). The latter

(238)

Structure (*238*)

compound was most successful in blocking normal GA biosynthesis. Metabolites were quantitated by GC and identified by GC-MS by comparison with spectra published by Bearder et al. (1975c). The metabolites identified were *ent*-7α-13-dihydroxykaurenoic acid (*135*), (*138*), *ent*-6α,7α,13-trihydroxykaurenoic acid (*136*), GA_1, GA_{18} (*139*), and GA_{19} (*141*). The only significant difference from the results of the feeds of steviol with mutant B1-41a (Bearder et al., 1975c) was that the seco-ring B diacid (*137*), GA_{17} (*140*), and also GA_{20} were not detected. To examine the potential of the method for the preparation of 13-hydroxy GAs, a 2-day incubation of 200 mg of steviol (*86*) was undertaken. This gave an ethyl acetate-soluble acidic fraction which was purified by partition chromatography on Sephadex LH20 by the method of MacMillan and Wels (1973). Recrystallization or TLC gave *ent*-7α,13-dihydroxykaurenoic acid (*135*) (5.7 mg), *ent*-6α, 7α-trihydroxykaurenoic acid (*136*) (25.2 mg), GA_{53} (*138*) (3.5 mg), GA_{18} (*159*) (12.2 mg), GA_1 (35.7 mg), and GA_{19} (*141*) (2.3 mg). It was pointed out that conditions for optimal metabolism were not sought.

Wada and Yamashita (1980) demonstrated the transformation of *ent*-12β-hydroxykaurene (*239*) into 12α-hydroxyGAs in mycelial suspensions of *G. fujikuroi* blocked by the presence of CCC (*67*) (Fig. 5.60). The alcohol (*239*) was prepared chemically from *ent*-kaurene (*1*), and the acidic metabolites were methylated and examined by GC-MS. Of the seven peaks detected, five were tentatively identified as 12α-hydroxyGA_{12} (*240*), 12α-hydroxyGA_{25} (*241*), *ent*-7α,12β-dihydroxykaurenoic acid (*244*), *ent*-6α,7α,12β-trihydroxy-kaurenoic acid (*243*), and GA_{39} (*242*). It is notable that with this substrate no C_{19}-GA analogues were formed.

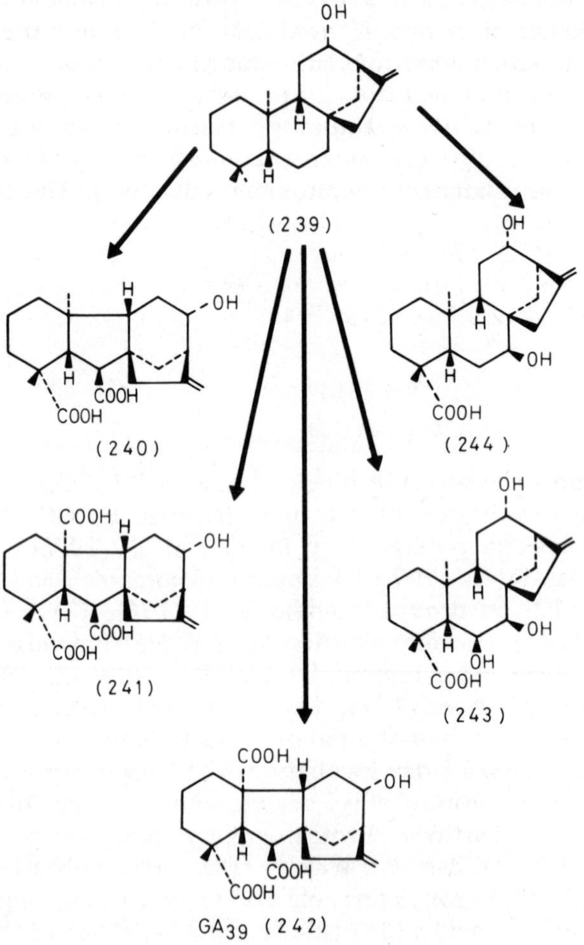

Figure 5.60 Metabolites of *ent*-12β-hydroxykaurene (*239*) in *G. fujikuroi.* (Wada & Yamashita, 1980.)

5.3.5 Summary of substrate specificity of enzymes in the diterpene pathway

In this section the main steps in the GA pathway are considered with reference to the ability of the enzymes to cope with variations in the structure of their substrates. The data are taken from the three previous sections. In many of the feeding studies published, it is impossible to judge the significance of the absence of particular metabolites because the sensitivity of experimental procedures varies between laboratories. In cases where analogue metabolites are recognized by TLC (versus a control containing normal GAs) the low resolution of the technique may obscure the presence of certain metabolites. However, all the metabolites should appear on a TLC plate. GC-MS is a very powerful technique in this kind of work, but it must be recognized that more complex metabolites such as conjugates may not be detected. Quantitation of the metabolites has only been achieved satisfactorily in the feeds using radioactive substrates or in preparative scale experiments where losses incurred during workup are expected to be minimal. Even in these experiments the fate of all the substrate added is seldom traced. It is with this heterogeneous pool of data that the following generalizations are made.

Oxidation at C-19. The oxidation of the *ent*-kaurene isomers kaurene (*89*), *ent*-kaur-15-ene (*187*), and 13β-kaurene (*186*) does not occur. In the light of the work described below on the oxidation of more polar analogues at C-19, it seems that the failure of these hydrocarbons to be hydroxylated may be due to a high substrate specificity in the kaurene carrier protein. The metabolic inertness of the fluoro compounds (*126*) and (*128*) versus the corresponding 19-oic-acids (*197*) and (*204*) may also be due to the specificity of the kaurene carrier protein (Banks et al., 1980; Cross & Erasmuson, 1981; Cross & Filippone, 1980).

Hydroxylation at C-19 is not prevented by various changes in the C and D rings—for example, *ent*-12β-hydroxylation, a 17-hydroxyl and endocyclic double bond, and a 16-norketone (Wada, 1979; Wada & Yamashita, 1980). *Ent*-7α-hydroxylation also does not prevent 19-hydroxylation (Fraga et al. 1980). In ring A, hydroxylation at C-19 is not halted by 18-hydroxylation but *ent*-3β-hydroxylation completely suppresses it.

Significant inhibition of the further oxidation at C-19 of an *ent*-19-hydroxykaurene analogue has not been observed. These oxidation steps are only marginally affected by numerous modifications; even *ent*-gibberellane 19-alcohols are oxidized [e.g., the diol (*19*)].

Oxidation at C-7. Oxidation at C-7 is seldom affected by structural modification. *Ent*-7α-hydroxylation occurs even when the 19-oic acid group is absent—for example, with *ent*-3β,18-dihydroxykaurene (*221*) (Fraga et al., 1981), steviol (*86*) methyl ester (Bearder et al., 1973c), and *ent*-kaurenyl succinate (*90*) (Jefferies et al., 1974). However, some polar functionality at C-4 is apparently required (Fraga et al., 1981).

Oxidation at C-6.   It should be noted that the *ent*-6β-oxidation in general refers to formation of kaurenolides that are thought to be derived via 6,7-ene and *ent*-6α,7α-epoxy intermediates. Therefore, an inhibition of kaurenolide formation might be attributed to inhibition of the formation of either the olefin or the epoxide. Blockage of the 19-oic acid group would also prevent kaurenolide formation since internal nucleophilic attack at the *ent*-6β-position would be impossible.

Oxidation at C-6 (α or β) apparently occurs with considerable facility in numerous substrate analogues, including some in which C-19 is blocked (Jefferies et al., 1974). However, Fraga et al. (1981) have shown that 18-hydroxylation inhibits hydroxylation at the *ent*-6α-position and that *ent*-3β-hydroxylation inhibits *ent*-6β-oxidation. It is likely that the enzymes responsible for the *ent*-7α-hydroxylation of *ent*-kaurenoic acid and the *ent*-6α,7α-epoxidation of *ent*-kaur-6, 16-dienoic acid are identical or very similar (Bloom & Shull, 1955; Talalay, 1957). Therefore, the inhibition of kaurenolide formation by an *ent*-3β-hydroxyl group is probably due to interference in the formation of the 6,7-ene. In many feeds the occurrence of oxidation at the *ent*-6β-position giving 7β-hydroxykaurenolide analogues cannot be determined, as often only acidic fractions have been analyzed. Oxidation at the *ent*-6α position is apparently unaffected by the oxygen function at C-7 since both the 7-epimer of *ent*-7β-hydroxy-kaurenoic acid (*5*) and *ent*-7-oxo-kaurenoic acid (*51*) give seco-ring B derivatives (Bearder, 1973; Bearder et al., 1975b).

Ring B contraction.   Ring   contraction   of   *ent*-7α-hydroxy-kaurenoic acid analogues does not occur unless C-19 is present as a free carboxylic acid and the *ent*-7α-hydroxyl is not esterified. However, ring contraction seems to occur slower with substrate analogues since, in many feeds of *ent*-kaurenoic acid analogues, *ent*-7α-hydroxy-derivatives are observed in contrast to the almost undetectable quantities of *ent*-7α-hydroxykaurenoic acid present in *ent*-kaurenoic acid feeds. Ring contraction is, however, prevented by *ent*-15α-acetoxylation (Bearder & MacMillan, unpublished) and by 18-hydroxylation (Fraga et al., 1978). Hanson has noted a formal resemblance between ring contraction and the cleavage of the glycol in *ent*-6α,7α-dihydroxykaurenoic acid (*6*) (Hanson & Sarah, 1979). It is also observed that in analogues where ring contraction is slow, formation of seco-ring B metabolites is similarly retarded.

Oxidation at C-3.   3-Hydroxylation or its equivalent occurs in most *ent*-gibberellane analogues. However, it is notable that certain changes in ring *D* completely inhibit 3-hydroxylation. Thus in feeds of isosteviol (*142*) and steviol acetate (*150*), no 3-hydroxy analogues were observed. In many analogue feeds, 3-hydroxylation is obviously retarded and non-3-hydroxylated GA analogues become the major products. 3-Hydroxylation is also not prevented by hydroxylation at the adjacent 2-position (Lunnon et al., 1977) or by fluorination at the 18-position (Bateson & Cross, 1974). A free 19-oic acid is not necessary, because GA_{12} alcohol (*38*) methyl ester is 3-hydroxylated.

Oxidation at C-20.   Oxidation at C-20 and formation of C_{19}-GAs rarely occurs as efficiently in analogues as with the normal precursors, although in most cases C_{19}-GAs are formed to some extent. A number of structural factors completely prevent C-20 oxidation. *Ent*-7α-hydroxyatisirenoic acid (*224*) is only metabolized to 20-methyl GAs (Hanson et al., 1979). Hanson suggests that this is due to steric interference by the D-bridge. *Ent*-16-oxo-17-norkaurane (*227*) is likewise only metabolized to GA_{12} and GA_{14} analogues. GA_{12} alcohol (*38*) acetate is 3-hydroxylated but does not apparently undergo oxidation at C-20 (Bearder & MacMillan, unpublished). Other structural features allow C-20 oxidation but not the formation of C_{19}-GAs. A free 19-oic acid group seems to be essential for C_{19}-GA formation but not for oxidation at C-20 because GA_{12} alcohol (*38*) methyl ester is 3-hydroxylated and oxidized at C-20 giving, inter alia, GA_{13} alcohol (*57*) 7-methyl ester, but no C_{19}-GA analogues are formed.

Likewise, *ent*-12β-hydroxykaurene (*239*) gives several C_{20}-GAs but no C_{19}-GAs. The bulk of the β-face of ring A has a marked effect on C_{19}-GA formation since GA_{14} acetate (*181*) is completely converted into a mixture of the acetates of the three C_{20}-GAs GA_{37}, GA_{36}, and GA_{13} (Bearder & MacMillan, unpublished). Lunnon et al. (1977) have observed that in comparison with other ring-A-hydroxylated precursors the presence of an *ent*-2α-hydroxyl group slows the C_{20}- to C_{19}-GA step. A 2,3-double bond also seems to have this effect (Bakker et al., 1974).

Introduction of the 1,2-double bond.   The desaturation of ring A of GA_4 analogues seems to involve the enzyme in the GA pathway most sensitive to structural variations. Only in relatively rare cases are GA_7 or GA_3 analogues formed. GA_7 analogues have been detected in feeds of 17β-hydroxykaur-15-ene, (*231*), and *ent*-15α-hydroxykaurenoic acid (*165*) (Wada et al., 1979). A GA_7 ketone analogue (*205*) was formed on incubation with *ent*-16,16-difluoro-17-norkaurenoic acid (*204*) (Cross & Filippone, 1980). 3-*Epi*-GA_4 and the isomeric GA_4 (*185*) did not undergo desaturation, and it has been noted in Sections 5.2.1 and 5.2.2 that GA_1 is a very poor precursor to GA_3. Time-course studies by McInnes et al. (1977) have illustrated the sensitive nature of this enzyme.

13-Hydroxylation.   13-Hydroxylation is also relatively rare in suitable analogue feeds. It has, however, been observed with GA_9 in high concentration (Bearder et al., 1976b) and in 3-*epi*-GA_4 and the isomer (*185*) (Bearder and MacMillan, unpublished).

5.4 DITERPENOID BIOSYNTHESIS IN *SPHACELOMA MANIHOTICOLA*

Recently, one other fungus has been shown to produce GA. In 1972 J. C. Lozano of the Centro International de Agricultura Tropical (CIAT) in Columbia described a previously unreported disease of cassava (*Manihot esculenta*). Among other symptoms, the disease caused extensive elongation of the internodes of young infected cassava plants (CIAT Annual Report, 1972). The causative agent of this "superelongation disease" was shown to be the Deuteromycete *Sphaceloma manihoticola* (CIAT Annual Report, 1975). In a study of the disease, Krausz (1976) showed that the fungus produced GA-like substances in liquid culture, but unreproducible results precluded further investigation. Rademacher and Graebe (1979) first identified the active substance as GA_4, and Zeigler, Powell and Thurston (1980) subsequently also identified GA_4 in culture filtrates of this fungus. More recent studies suggest the biosynthetic pathway in *Sphaceloma manihoticola* to be similar to that in *G. fujikuroi*, apart from the absence of 1,2-dehydro- and 13-hydroxy-GAs (Graebe et al., 1980; Bearder and MacMillan, unpublished work).

REFERENCES

Akhtar, M., Calder, M. R., Corina, D. L., & Wright, J. N. (1981). The status of oxygen atoms in the removal of C-19 in oestrogen biosynthesis. J. Chem. Soc., Chem. Commun., 129–130.

Bakker, H. J., Cook, I. F., Jefferies, P. R., Knox, J. R. (1974). Gibberellin metabolites from *ent*-kaura-2,16-dien-19-ol and its succinate in *Gibberella fujikuroi*. Tetrahedron *30*, 3631–3640.

Banks, R. E., Bateson, J. H., Cross, B. E., & Erasmuson, A. (1980). Fluorinated kaurenoids. Part 1. Synthesis and biological activity of *ent*-15β-fluorokaur-16-ene, *ent*-17-fluorokaur-16-ene and *ent*-17,17-difluorokaur-16-ene. J. Chem. Res. (S), 46.

Barnes, M. F., Light, E. N., & Lang, A. (1969). The action of plant growth retardants on terpenoid biosynthesis. Inhibition of gibberellic-acid production in *Fusarium moniliforme* by CCC and AMO-1618: Action of these retardants on sterol biosynthesis. Planta *88*, 172–182.

Bateson, J. H., & Cross, B. E. (1972). Production of fluorogibberellins by *Gibberella fujikuroi* from a fluorinated analogue of a gibberellin precursor. J. Chem. Soc., Chem. Comm., 649–650.

Bateson, J. H., & Cross, B. E. (1974). The microbiological production of analogues of mould metabolites. Part 1. Production of fluorogibberellic acid and fluorogibberellin A_9 by *Gibberella fujikuori*. J. Chem. Soc., Perkin Trans. I, 1131–1136.

Beadle, G. W., & Tatum, E. L. (1941). Genetic control of biochemical reactions in *Neurospora*. Proc. Nat. Acad. Sci. (U.S.A.) *27*, 499–506.

Beale, M. H., Bearder, J. R., Down, G. H., Hutchison, M., MacMillan, J., & Phinney, B. O. (1982). The biosynthesis of kaurenolide diterpenoids by *Gibberella fujikuroi*. Phytochem. *21*, 1279–1287.

Bearder, J. R. (1973). Diterpenoid biosynthesis in mutants of *Fusarium moniliforme*. Ph.D. Thesis, University of Bristol.

Bearder, J. R. (1980). Plant hormones and other growth substances—their background, structures and occurrence. In: Hormonal Regulation of Development I. Molecular Aspects of Plant Hormones. Encyclopedia of Plant Physiology New Series Vol 9, pp. 9–112. MacMillan, J., ed. Springer-Verlag, Berlin, Heidelberg, New York.

Bearder, J. R., Dennis, F. G., MacMillan, J., Martin, G. C., & Phinney, B. O. (1975a). A new gibberellin (A_{45}) from seed of *Pyrus communis* L. Tetrahedron Lett., 669–670.

Bearder, J. R., Frydman, V. M., Gaskin, P., Harvey, W. E., Hedden, P., MacMillan, J., Phinney, B. O., & Wels, C. M. (1973a). Gibberellin biosynthesis in *Gibberella fujikuroi* using the mutant B1-41a. In: Plant Growth Substances, pp. 241–253. Hirokawa Press, Tokyo.

Bearder, J. R., Frydman, V. M., Gaskin, P., Hatton, I. K., Harvey, W. E., Macmillan, J., & Phinney, B. O. (1976b). Fungal products. Part XVII. Microbiological hydroxylation of gibberellin A_9 and its methyl ester. J. Chem. Soc., Perkin Trans. I, 173–183.

Bearder, J. R., Frydman, V. M., Gaskin, P., MacMillan, J., Wels, C. M., & Phinney, B. O. (1976a). Fungal products. Part XVI. Conversion of isosteviol and steviol acetate into gibberellin analogues by mutant B1-41a of *Gibberella fujikuroi* and the preparation of [3H]-gibberellin A_{20}. J. Chem. Soc., Perkin Trans. I, 173–178.

Bearder, J. R., Hedden, P., MacMillan, J., Wels, C. M., & Phinney, B. O. (1973b). Gibberellin biosynthesis in the mutant B1-41a of *Gibberella fujikuroi*. J. Chem. Soc., Chem. Comm., 777–778.

Bearder, J. R., MacMillan, J., Matsuo, A., & Phinney, B. O. (1979b). Conversion of trachylobanic acid into novel pentacyclic analogues of gibberellins by *Gibberella fujikuroi* mutant B1-41a. J. Chem. Soc., Chem. Comm., 649–650.

Bearder, J. R., MacMillan, J., & Phinney, B. O. (1973d). 3-Hydroxylation of gibberellin A_{12}-aldehyde in *Gibberella fujikuroi* strain Rec-193A. Phytochem. *12*, 2173–2179.

Bearder, J. R., MacMillan, J., & Phinney, B. O. (1973e). Conversion of gibberellin A_1 into gibberellin A_3 by the mutant R-9 of *Gibberella fujikuroi*. Phytochem. *12*, 2655–2659.

Bearder, J. R., MacMillan, J., & Phinney, B. O. (1975b). Fungal products. Part XIV. Metabolic pathways from *ent*-kaurenoic acid to the fungal gibberellins in mutant B1-41a of *Gibberella fujikuroi*. J. Chem. Soc., Perkin Trans. I, 721–726.

Bearder, J. R., MacMillan, J., & Phinney, B. O. (1976c). Origin of oxygen atoms in the lactone bridge of C_{19}-gibberellins. J. Chem. Soc., Chem. Comm., 834–835.

Bearder, J. R., MacMillan, J., & Phinney, B. O. (1979a). The use of fungal mutants in the elucidation of gibberellin biosynthesis. Proc. FEBS Meet. V. 55 (Regul. Second. Prod. Plant Horm. Metab.), 25–35.

Bearder, J. R., MacMillan, J., Phinney, B. O., Hanson, J. R., Rivett, D. E. A., & Willis, C. L. (1982). Gibberellin A_{13} T-aldehyde: A proposed intermediate in the fungal biosynthesis of gibberellin A_3. Phytochem. *21*, 2225–2230.

Bearder, J. R., MacMillan, J., Wels, C. M., Chaffey, M. B., & Phinney, B. O. (1974). Position of the metabolic block for gibberellin biosynthesis in mutant B1-41a of *Gibberella fujikuroi*. Phytochem. *13*, 911–917.

Bearder, J. R., MacMillan, J., Wels, C. M., & Phinney, B. O. (1973c). Metabolism of steviol and its derivatives by *Gibberella fujikuroi*, mutant B1-41a. J. Chem. Soc., Chem. Comm., 778-779.

Bearder, J. R., MacMillan, J., Wels, C. M., & Phinney, B. O. (1975c). The metabolism of steviol to 13-hydroxylated *ent*-gibberellanes and *ent*-kauranes. Phytochem. *14*, 1741-1748.

Bearder, J. R. & Phinney, B. O. (1979). Attempts to identify the C_{20}- precursors of C_{19}-gibberellins. Abstracts of the Tenth International Conference on Plant Growth Substances. Madison, Wisconsin.

Bearder, J. R., & Sponsel, V. M. (1977). Selected topics in gibberellin metabolism. Biochem. Soc. Trans. *5*, 569-582.

Beeley, L. J., Gaskin, P., & MacMillan, J. (1975). Gibberellin A_{43} and other terpenes in endosperm of *Echinocystis macrocarpa*. Phytochem. *14*, 779-783.

Birch, A. J., Rickards, R. W., Smith, H., Harris, A., & Whalley, W. B. (1959). Studies in relation to biosynthesis—XXI. Rosenanolactone and gibberellic acid. Tetrahedron 7, 241-251.

Birch, A. J., Rickards, R. W., Smith, H., & Winter, J. (1960). The allogibberic-gibberic acid rearrangement. Chem. Ind., 401-402.

Bloom, B. M., & Shull, G. M. (1955). Epoxidation of unsaturated steroids by microorganisms. J. Amer. Chem. Soc. 77, 5767-5768.

Boulton, K., & Cross, B. E. (1981). Inhibitors of the biosynthesis of gibberellins. Part I. 7-Fluoro-10β-fluoromethyl-1β,8-dimethylgibbane-1α,4aα-carbolactone. J. Chem. Soc., Perkin Trans. I, 427-432.

Breton, J. L., Gonzalez, A. G., Rocha, J. M., Panizo, F. M., Rodriguez, B., Valverde, S. (1969). Candicandiol, a new diterpene from *Sideritis candicans* Ait., var. 'eriocephala' Webb (Labiatae), Tetrahedron Lett., 599-602.

Brown, S. A., (1972). Methodology. Specialist Periodical Reports. Biosynthesis *1*, 1-40.

Brown, S. A., & Wetter, L. R. (1972). Methods for investigation of biosynthesis in higher plants. Progr. Phytochem., *3*, 1-45.

Buchenauer, H., & Grossmann, F. (1977). Triadimefon: Mode of action in plants and fungi. Neth. J. Plant Pathol. *83* (Suppl. 1), 93-103.

Bukovac, M. J., Yuda, E., Murofushi, N., & Takahashi, N. (1979). Endogenous plant growth substances in developing fruit of *Prunus cerasus* L. Part VII. Isolation of gibberellin A_{32}. Plant Physiol. *63*, 129–132.

Bu'Lock, J. D. (1967). Essays in Biosynthesis and Microbial Development. Wiley, London.

Bu'Lock, J. D., Detroy, R. W., Mostatek, Z., & Munim-Al-Shakarchi, A. (1974). Regulation of secondary biosynthesis in *Gibberella fujikuroi*. Trans. Brit. Mycol. Soc. *64*(2), 377–389.

Cathey, H. M. (1964). Physiology of growth retarding chemicals. Ann. Rev. Plant Physiol., *15*, 271–302.

Cavell, B. D., & MacMillan, J. (1967). Isolation of (−)-kaur-16-en-19-oic acid from the mycelium of *Gibberella fujikuroi*. Phytochem. *6*, 1151–1154.

Cerda-Olmedo, E., & Torres-Martinez, S. (1979). Genetics and regulation of carotenoid biosynthesis. Pure Appl. Chem. *51*, 631–637.

Charney, W., & Herzog, H. L. (1967). Microbiol Transformations of Steriods. A Handbook. Academic Press, New York, London.

Cho, K. Y., Sakurai, A., Kamija, Y., Takahashi, N., & Tamura, S. (1979). Effects of the new plant growth retardants of quaternary ammonium iodides on gibberellin biosynthesis in *Gibberella fujikuroi*. Plant Cell Physiol. *20*, 75–81.

CIAT Annual Report: Centri Internacional de Agricultura Tropical, Cali, Columbia (1972).

CIAT Annual Report: Centro Internacional de Agricultura Tropical, Cali, Columbia (1975).

Coates, R. M., & Cavender, P. L. (1980). Stereochemistry of the enzymatic cyclization of copalyl pyrophosphate to kaurene in enzyme preparations from *Marah macrocarpus*. J. Amer. Chem. Soc. *102*, 6359–6361.

Cook, I. F., Jefferies, P. F., & Knox, J. R. (1975). Acidic *ent*-kauranoids from the metabolism of *ent*-kaura-2,16-dien-19-ol in *Gibberella fujikuroi*. Tetrahedron *31*, 251–255.

Coolbaugh, R. C., Heil, D. R., & West, C. A. (1981). Comparative effects of substituted pyrimidines on growth and gibberellin biosynthesis in *Fusarium moniliforme*. Plant Physiol. (in press).

Croft, K. D., Ghisalberti, E. L., Jefferies, P. R., Knox, J. R., Mahoney, T. J., &

Sheppard, P. N. (1974). Chemical and microbiological syntheses of intermediates in gibberellin biosynthesis. Tetrahedron *30*, 3663-3667.

Cross, B. E. (1968). Biosynthesis of the gibberellins. Progr. Phytochem. *1*, 195-222.

Cross, B. E., & Erasmuson, A. (1978). Microbiological production of 9α-fluorogibberellin A_4, 9α-fluorogibberellin A_{14} and other fluoroterpenoids. J. Chem. Soc., Chem. Comm., 1013-1015.

Cross, B. E., & Erasmuson, A. (1981). The microbiological production of analogues of mould metabolites. Part 2. Production of 9α-fluorogibberellin A_4, 9α-fluorogibberellin A_{14}, and other fluoroterpenoids by *Gibberella fujikuroi*. J. Chem. Soc., Perkin Trans. I, 1918-1922.

Cross, B. E., Erasmuson, A., & Filippone, P. (1981). Fluorinated kaurenoids. Part 2. Preparation of methyl *ent*-17,17,17-trifluorokaur-15-en-19-oate and *ent*-16,16-difluoro-17-norkauran-19-oic acid from xylopic acid. J. Chem. Soc., Perkin Trans. I, 1293-1297.

Cross, B. E., & Filippone, P. (1980). Production of *gem*-difluoronor-derivatives of gibberellin A_7 and 7-hydroxykaurenolide by *Gibberella fujikuroi*. J. Chem. Soc., Chem. Comm., 1097-1098.

Cross, B. E., Galt, R. H. B., & Hanson, J. R. (1964). The biosynthesis of the gibberellins. Part I. (−)-Kaurene as a precursor of gibberellic acid. J. Chem. Soc., 295-300.

Cross, B. E., Galt, R. H. B., Hanson, J. R., Curtis, P. J., Grove, J. F., & Morrison, A. (1963). New metabolites of *Gibberella fujikuroi*. Part II. The isolation of fourteen new metabolites. J. Chem. Soc., 2937-2943.

Cross, B. E., Galt, R. H. B., & Norton, K. (1968b). The biosynthesis of the gibberellins—II. Tetrahedron *24*, 231-237.

Cross, B. E., & Myers, P. L. (1969). The effect of plant growth retardants on the biosynthesis of diterpenes by *Gibberella fujikuroi*. Phytochem. *8*, 79-83.

Cross, B. E., & Norton, K. (1965). New metabolites of *Gibberella fujikuroi*. Part VIII. Gibberellin A_{12}. J. Chem. Soc., 1570-1572.

Cross, B. E., Norton, K., & Stewart, J. C. (1968a). An attempt to find evidence for the existence of (+)-gibberellic acid. Phytochem. 7, 83-84.

Cross, B. E., Norton, K., & Stewart, J. C. (1968c). The biosynthesis of the gibberellins—III. J. Chem. Soc. (C), 1054-1063.

Cross, B. E., Stewart, J. C., & Stoddart, J. L. (1970). 6β,7-Dihydroxykaurenoic acid: Its biological activity and possible role in the biosynthesis of gibberellic acid. Phytochem. 9, 1065-1071.

Davis, B. D. (1955). Intermediates in amino acid biosynthesis. Adv. Enzymol. 16, 247-312.

Dawson, R. M., Jefferies, P. R., & Knox, J. R. (1975). Cyclization and hydroxylation stereochemistry in the biosynthesis of gibberellic acid. Phytochem. 14, 2593-2597.

Dennis, D. T., & West, C. A. (1967). Biosynthesis of gibberellins. The conversion of (−)-kaurene to (−)-kauren-19-oic acid in endosperm of Echinocystis macrocarpa Greene. J. Biol. Chem. 242, 3293-3300.

Dockerill, B., Evans, R., & Hanson, J. R. (1977). Removal of C-20 in gibberellin biosynthesis. J. Chem. Soc., Chem. Comm., 919-921.

Dockerill, B., & Hanson, J. R. (1978). The fate of C-20 in C_{19} gibberellin biosynthesis. Phytochem. 17, 701-704.

Durley, R. C., & Pharis, R. P. (1972). Partition coefficients of 27 gibberellins. Phytochem. 11, 317-326.

Echols, L. C., Maier, V. P., Poling, S. M., & Sterling, P. R. (1981). New bioregulators of gibberellin biosynthesis in Gibberella fujikuroi. Phytochem. 20, 433-437.

Ekong, D. E. U., & Ogan, A. U. (1968). Chemistry of the constituents of Xylopia aethiopica. The structure of xylopic acid a new diterpene acid. J. Chem. Soc. (C), 311-312.

Evans, R., & Hanson, J. R. (1975). Studies in terpenoid biosynthesis. Part XIII. The biosynthetic relationship of the gibberellins in Gibberella fujikuroi. J. Chem. Soc., Perkin Trans. I, 663-666.

Evans, R., Hanson, J. R., & Mulheirn, L. J. (1973). Studies in terpenoid biosynthesis. Part X. Incorporation of (5S)-[5-^{3}H] mevalonic acid into gibberellic acid. J. Chem. Soc., Perkin Trans. I, 753-756.

Evans, R., Hanson, J. R., & White, A. F. (1970). Studies in terpenoid biosynthesis. Part VI. The stereochemistry of some stages in tetracyclic diterpene biosynthesis. J. Chem. Soc. (C), 2601-2603.

Fraga, B. M., Gonzalez, A. G., Hanson, J. R., & Hernandez, M. G. (1981). The microbiological transformation of some ent-3β-hydroxykaur-16-enes by Gibberella fujikuroi. Phytochem. 20, 57-61.

Fraga, B. M., Hanson, J. R., & Hernandez, M. G. (1978). The microbiological transformation of epicandicandiol *ent*-7α,18-dihydroxykaur-16-ene, by *Gibberella fujikuroi*. Phytochem. *17*, 812–814.

Fraga, B. M., Hanson, J. R., Hernandez, M. G., & Sarah, F. Y. (1980). The microbiological transformation of some *ent*-kaur-16-ene 7-, 15- and 18-alcohols by *Gibberella fujikuroi*. Phytochem. *19*, 1087–1091.

Fukuyama, M. (1971). Gibberellin mutants in the fungus, *Gibberella fujikuroi*. Ph.D. Thesis, University of California, Los Angeles, U.S.A., pp. 1–122. (Diss. Abs., 1971–1972, *32*, 3219-B).

Galt, R. H. B. (1965). New metabolites of *Gibberella fujikuroi*. Part IX. Gibberellin A_{13}. J. Chem. Soc., 3143–3151.

Geissman, T. A., Verbiscar, A. J., Phinney, B. O., Cragg, G. (1966). Studies on the biosynthesis of gibberellins from (−)-kaurenoic acid in cultures of *Gibberella fujikuroi*. Phytochem. *5*, 933–947.

Gonzalez, A. G., Fraga, B. M., Hernandez, M. G., & Luis, J. G. (1973). New diterpenes from *Sideritis candicans*. Phytochem. *12*, 2721–2723.

Gordon, W. L. (1960). Distribution and prevalence of *Fusarium moniliforme* Sheld. (*Gibberella fujikuroi* (SAW.) WR) producing substances with gibberellin-like biological properties. Nature *186*, 698–700.

Graebe, J. E., Dennis, D. T., Upper, C. D., & West, C. A. (1965). Biosynthesis of gibberellins. I. The biosynthesis of (−)-kaurene, (−)-kauren-19-ol, and *trans*-geranyl-geraniol in endosperm nucleus of *Echinocystis macrocarpa* Greene. J. Biol. Chem. *240*, 1847–1853.

Graebe, J. E., Hedden, P., & Rademacher, W. (1980). Gibberellin biosynthesis. In: Gibberellins—Chemistry, Physiology and Use. British Plant Growth Regulator Group Monograph No. 5, pp. 31–47, Lenton, J. R., ed. British Plant Growth Regulator Group, Wantage.

Graebe, J. E., & Ropers, H. J. (1978). The gibberellins. In: Phytohormones and Related Compounds: A Comprehensive Treatise, Vol. 1. The Biochemistry of Phytohormones and Related Compounds, pp. 107–204, Letham, D. S., Goodwin, P. B., Higgins, T. J. V. eds. Elsevier/North-Holland, Amsterdam, Oxford, New York.

Hanson, J. R. (1966). The chemistry of tetracyclic diterpenoids III. The partial synthesis of kaurenolide. Tetrahedron *22*, 2877–2882.

Hanson, J. R. (1971). The biosynthesis of the diterpenes. Forschr. Chem. Org. Nat. *29*, 395–415.

Hanson, J. R., & Hawker, J. (1972a). The chemistry of the tetracyclic diterpenes—XII. The labelling and functionalisation of the kauranoid 6β-position. Tetrahedron *28*, 2521-2526.

Hanson, J. R., & Hawker, J. (1972b). The formation of the C_{19}-gibberellins from gibberellin A_{13} anhydride. Tetrahedron Lett., 4299-4302.

Hanson, J. R., & Hawker, J. (1973). Preparation of [^{14}C]-gibberellic acid. Phytochem. *12*, 1973-1975.

Hanson, J. R., Hawker, J., & White, A. F. (1972). Studies in terpenoid biosynthesis. Part IX. The sequence of oxidation on ring B in kaurene-gibberellin biosynthesis. J. Chem. Soc., Perkin Trans. I, 1892-1896.

Hanson, J. R., Parry, K. P., & Willis, C. L. (1981). Mimics of intermediates in gibberellin biosynthesis as plant growth regulators. J. Chem. Soc., Chem. Comm., 285-286.

Hanson, J. R., & Sarah, F. V. (1979). Studies in terpenoid biosynthesis. Part 23. Relationships between the kaurenolides and the seco-ring B metabolites of *Gibberella fujikuroi*. J. Chem. Soc., Perkin Trans. I, 3151-3154.

Hanson, J. R., Sarah, F. Y., Fraga, B. M., & Hernandez, M. G. (1979). The microbiological preparation of two "atisagibberellins." Phytochem. *18*, 1875-1876.

Hanson, J. R., & White, A. F. (1968). The transformation of steviol by *Gibberella fujikuroi*. Tetrahedron *24*, 6291-6293.

Hanson, J. R., & White, A. F. (1969). Studies in terpenoid biosynthesis. Part IV. Biosynthesis of the kaurenolides and gibberellic acid. J. Chem. Soc. (C), 981-984.

Hanson, J. R., Willis, C. J., & Parry, K. P. (1980). The inhibition of gibberellic acid biosynthesis by *ent*-kauran-16β,17-epoxide. Phytochem. *19*, 2323-2325.

Haslam, E. (1974). The Shikimate Pathway. Halstead, New York.

Hedden, P., & Graebe, J. E. (1981). Kaurenolide biosynthesis in a cell-free system from *Cucurbita maxima* seeds. Phytochem. *20*, 1011-1015.

Hedden, P., MacMillan, J., & Grinstead, M. J. (1973). Fungal products. Part VIII. New kaurenolides from *Gibberella fujikuroi*. J. Chem. Soc., Perkin Trans. I, 2773-2778.

Hedden, P., MacMillan, J., & Phinney, B. O. (1974). Fungal products. Part XII.

Gibberellin A_{14}-aldehyde, an intermediate in gibberellin biosynthesis in *Gibberella fujikuroi*. J. Chem. Soc., Perkin Trans. I, 587–592.

Hedden, P., MacMillan, J., & Phinney, B. O. (1978). The metabolism of the gibberellins. Ann. Rev. Plant Physiol. *29*, 149–192.

Hedden, P., Phinney, B. O., MacMillan, J., & Sponsel, V. M. (1977). Metabolism of kaurenoids by *Gibberella fujikuroi* in the presence of the plant growth retardant, N,N,N-trimethyl-1-methyl-(2′,6′,6′-trimethylcyclohex-2′-en-1-yl) prop-2-enylammonium iodide. Phytochem. *16*, 1913–1917.

Honda, K., Shishibori, T., & Suga, T. (1980). Biosynthesis of (−)-kaurene. ^{13}C n.m.r. spectroscopic evidence for the mechanism of formation of ring D. J. Chem. Res. (S), 218–219.

Jefferies, P. R., Knox, J. R., & Ratajczak, J. (1970). Metabolites from the succinate ester of (−)-kaurenol. Tetrahedron Lett., 3229–3231.

Jefferies, P. R., Knox, J. R., & Ratajczak, J. (1974). Metabolic transformations of some *ent*-kaurenes in *Gibberella fujikuroi*. Phytochem. *13*, 1423–1431.

Jones, R. L., & Varner, J. E. (1967). The bioassay of gibberellins. Planta *72*, 155–161.

Jones, K. C., West, C. A., & Phinney, B. O. (1968). Isolation, identification and biological properties of gibberellin A_{14} from *Gibberella fujikuroi*. Phytochem. *8*, 283–291.

Katsumi, M., & Phinney, B. O. (1969). The biosynthesis of gibberellins. In: Gibberellin: Chemistry, Biochemistry, Physiology, pp. 195–219, Tamura, S., ed. Tokyo University Press, Tokyo.

Katsumi, M., Phinney, B. O., Jefferies, P. R., & Henrick, C. A. (1964). Growth response of the d-5 and an-1 mutants of maize to some kaurene derivatives. Science *144*, 849–850.

Kende, H., Ninnemann, H., & Lang, A. (1963). Inhibition of gibberellic acid biosynthesis in *Fusarium moniliforme* by AMO-1618 and CCC. Naturwissenschaften *50*, 599–600.

Kieslich, K. (1976). Microbial Transformations of Non-steriodal Cyclic Compounds. Georg Thieme Verlag., Stuttgart.

Koshimizu, K., Fukui, H., Kusaki, T., Ogawa, Y., & Mitsui, T. (1968). Isolation and structure of gibberellin A_{18} from immature seeds of *Lupinus luteus*. Agr. Biol. Chem. *32*, 1135–1140.

Krausz, J. P. (1976). Ph.D. Dissertation, Cornell University.

Lang, A. (1970). Gibberellins: Structure and metabolism. Ann. Rev. Plant Physiol. *21*, 537–570.

Lew, F. T., & West, C. A. (1971). (−)-Kaur-16-en-7β-ol-19-oic acid, an intermediate in gibberellin biosynthesis. Phytochem. *10*, 2065–2076.

Lockhart, J. A. (1962). Kinetic studies of certain anti-gibberellins. Plant Physiol. *37*, 759–764.

Lunnon, M. W., MacMillan, J., & Phinney, B. O. (1977). Fungal products. Part 20. Transformation of 2- and 3-hydroxylated kaurenoids by *Gibberella fujikuroi*. J. Chem. Soc., Perkin Trans. I, 2308–2316.

McCorkindale, N. J. (1976). The biosynthesis of terpenes and steroids. In: The Filamentous Fungi, pp. 369–422, Smith, J. E., & Berry, D. R., eds. Wiley, New York.

McCormick, J. R. D. (1967). In: Antibiotics. Vol. II. Biosynthesis, Gottlieb, D., & Shaw, P. D., eds. Springer-Verlag, New York.

McInnes, A. G., Smith, D. G., Arsenault, G. P., & Vining, L. C. (1973). Biosynthesis of gibberellins in *Gibberella fujikuroi*. Gibberellin A$_{16}$. Can. J. Biochem. *51*, 1470–1474.

McInnes, A. G., Smith, D. G., Durley, R. C., Pharis, R. P., Arsenault, G. P., MacMillan, J., Gaskin, P., & Vining, L. C. (1977). Biosynthesis of gibberellins in *Gibberella fujikuroi*. Gibberellin A$_{47}$. Can. J. Biochem. *55*, 728–735.

MacMillan, J. (1971). Diterpenes—the gibberellins. In: Aspects of Terpenoid Chemistry and Biochemistry, pp. 153–180, Goodwin, T. V., ed. Academic Press, New York.

MacMillan, J., & Wels, C. M. (1973). Partition chromatography of gibberellins and related diterpenes on columns of Sephadex LH-20. J. Chromatogr. *87*, 271–276.

MacMillan, J., & Wels, C. M. (1974). Detailed analysis of metabolites from mevalonic lactone in *Gibberella fujikuroi*. Phytochem. *13*, 1413–1417.

Middleton, E. J., & Jefferies, P. R. (1968). Substances derived from *Goodenia strophiolata* F. Muell. Aust. J. Chem. *21*, 2349–2351.

Murofushi, N., Nagura, S., & Takahashi, N. (1979). Metabolism of steviol by *Gibberella fujikuroi* in the presence of plant growth retardant. Agr. Biol. Chem. *43*, 1159–1161.

Murofushi, N., Takahashi, N., Yokota, T., & Tamura, S. (1968). Gibberellins in immature seeds of *Pharbitis nil*. Part 1. Isolation and structure of a novel gibberellin. Gibberellin A_{20}. Agr. Biol. Chem. *32*, 1239-1245.

Pascal, R. A., Chang, P., & Schroepfer, G. J. (1980). Possible mechanisms of demethylation of 14α-methyl sterols in cholesterol biosynthesis. J. Amer. Chem. Soc. *102*, 6599-6601.

Phinney, B. O., & Spector, C. (1967). Genetics and gibberellin production in the fungus *Gibberella fujikuroi*. Ann. N.Y. Acad. Sci. *144*, 204-210.

Pinar, M., Rodriguez, B., & Alenary, A. (1978). Gummiferolic acid, a new *ent*-atis-16-ene diterpenoid from *Margotia gummifera*. Phytochem. *17*, 1637-1640.

Piozzi, F., Venturella, P., Bellino, A., & Mondelli, R. (1968). Diterpenes from *Sideritis sicula* ucria. Tetrahedron *24*, 4073-4081.

Piozzi, F., Venturella, P., Bellino, A., Paternostro, M. P., Rodriguez, B., & Valverde, S. (1971). Revised structures for candicandiol and epicandicandiol. Chem. Ind., 926.

Pitel, D. W., Vining, L. C., & Arsenault, G. P. (1971a). Improved methods for preparing pure gibberellins from cultures of *Gibberella fujikuroi*. Isolation by adsorption or partition chromatography on silicic acid and by partition chromatography on sephadex columns. Can. J. Biochem. *49*, 185-193.

Pitel, D. W., Vining, L. C., & Arsenault, G. P. (1971b). Biosynthesis of gibberellins in *Gibberella fujikuroi*. The sequence after gibberellin A_4. Can. J. Biochem. *49*, 194-200.

Popjak, G., & Cornforth, J. W. (1966). Substrate stereochemistry in squalene biosynthesis. Biochem. J. *101*, 553-568.

de Quesada, T. G., Rodriguez, B., & Valverde, S. (1973). New diterpenes of *Sideritis leucantha* and *Sideritis linearifolia*. Ann. Quimica *69*, 757-769.

de Quesada, T. G., Rodriguez, B., & Valverde, S. (1974). Diterpenes from *Sideritis lagascana* and *Sideritis valverdei*. Phytochem. *13*, 2008.

Rademacher, W., & Graebe, J. E. (1979). Gibberellin A_4 produced by *Sphaceloma manihoticola*, the cause of the superelongation disease of cassava (*Manihot esculenta*). Biochem. Biophys. Res. Commun. *91*, 35-40.

Reid, W. W. (1969). Effect of SKF7997 and SKF525 on diterpene and sterol biosynthesis in *Gibberella fujikuroi* from (2-^{14}C) mevalonate. Biochem. J. *113*, 37P-38P.

Rowe, J. W. (1968). The Common and Systematic Nomenclature of Cyclic Diterpenes. 3rd Rev. Forest Products Laboratory, U.S. Department of Agriculture, Madison, Wisconsin.

Ruddat, M. (1968). Biosynthesis and metabolism of steviol. In: Biochemistry and Physiology of Plant Growth Substances, pp. 341–346, Wightman, F., & Setterfield, G., eds. The Runge Press, Ottawa.

Ruddat, M., Heftmann, E., & Lang, A. (1965). Conversion of steviol to a gibberellin-like compound by *Fusarium monoliforme*. Arch. Biochem. Biophys. *111*, 187–190.

Ruddat, M., Lang, A., & Mossettig, E. (1963). Gibberellin activity of steviol, a plant terpenoid. Naturwissenschaften *50*, 23.

Schlosser, M. (1978). Introduction of fluorine into organic molecules: Why and how. Tetrahedron *34*, 3–17.

Schmidt, S. (1961). Der Einfluss der Kulturbedingungen auf die Bildung von keimungsford-ernden Staffen (Gibberellinen) durch *Gibberella fujikuroi* und die Isolierung eines bisher unbekannten Gibberellins. Flora *151*, 455–486.

Schmidt, S. (1962). Zusammensetzung von gibberellin D. Flora *152*, 527–529.

Shive, J. B., & Sisler, H. D. (1976). Effects of ancymidol (a growth retardant) and triarimol (a fungicide) on the growth, sterols and gibberellins of *Phaseolus vulgaris* (L.). Plant Physiol. *57*, 640–644.

Spector, C., & Phinney, B. O. (1966). Gibberellin production: Genetic control in the fungus *Gibberella fujikuroi*. Science *153*, 1397–1398.

Spector, C., & Phinney, B. O. (1968). Gibberellin biosynthesis: Genetic studies in *Gibberella fujikuroi*. Physiol. Plant *21*, 127–136.

Sprinson, D. B. (1960). Biosynthesis of aromatic compounds from D-glucose. Adv. Carbohydrate Chem. *15*, 235–270.

Stoddart, J. L. (1972). The biological activity of fluorogibberellins. Planta *107*, 81–88.

St. Pyrek, J. (1970). New pentacyclic diterpene acid. Trachyloban-19-oic acid from sunflower. Tetrahedron *26*, 5029–5032.

Talalay, P. (1957). Enzymic mechanisms in steroid metabolism. Physiol. Rev. *37*, 362–389.

Tolbert, N. E. (1961). Antigibberellins. Advan. Chem. Ser. 28, 145–151.

Verbiscar, A. J., Cragg, G., Geissman, T. A., & Phinney, B. O. (1967). Studies on the biosynthesis of gibberellins—II. The biosynthesis of gibberellins from (—)-kaurenol, and the conversion of gibberellins ^{14}C-GA-4 and ^{14}C-GA-7 into ^{14}C-GA-3 by Gibberella fujikuroi. Phytochem. 6, 807—814.

Wada, K. (1978a). Inhibition of gibberellin biosynthesis by geraniol derivatives and 17-nor-16-azakauranes. Agr. Biol. Chem. 42, 787—791.

Wada, K. (1978b). New gibberellin biosynthesis inhibitors, 1-n-decyl- and 1-geranylimidazole: Inhibitors of (—)-kaurene 19-oxidation. Agr. Biol. Chem. 42, 2411–2413.

Wada, K., Imai, T., & Shibata, K. (1979). Microbial productions of unnatural gibberellins from (—)-kaurene derivatives from Gibberella fujikuroi. Agric. Biol. Chem. 43, 1157–1158.

Wada, K., & Yamashita, H. (1980). Synthesis and microbial transformation of ent-12β-hydroxykaur-16-ene. Agr. Biol. Chem. 44, 2249–2250.

West, C. A. (1973). Biosynthesis of gibberellins. In: Biosynthesis and Its Control in Plants, pp. 143–169, Milborrow, B. V., ed. Academic Press, London.

Yamane, H., Murofushi, N., & Takahashi, N. (1974). Structure of a new nor-kaurenolide from Gibberella fujikuroi. Agr. Biol. Chem. 38, 207–210.

Zeevart, J. A. D. (1964). Effects of the growth retardant CCC on floral initiation and growth in Pharbitis nil. Plant Physiol. 39, 402–408.

Zeevart, J. A. D., & Lang, A. (1963). Suppression of floral induction in Byrophyllum daigremontianum (Kalanchoe daigremontiana) by a growth retardant. Planta 59, 509–517.

Zeigler, R. S., Powell, L. E., & Thurston, H. D. (1980). Gibberellin A_4 production by Spaceloma manihoticola, causal agent of cassava superelongation disease. Phytopathol. 70, 589–593.

6

Gibberellin Conjugates

Gernot Schneider

6.1 INTRODUCTION

Investigations that led to the isolation and characterization of plant hormones have also produced evidence of the existence of so-called bound, water-soluble, or polar hormones from which free hormones can be liberated by hydrolysis treatment. These compounds are widespread in their occurrence and are occasionally present in high concentrations. It is now realized that some of them are plant hormones connected to either a sugar or an amino acid such as indole-3-acetyl aspartic acid (Andreae & Good, 1955), gibberellin A_8-2-O-glucoside (Schreiber, Weiland, & Sembdner, 1967), the glucosyl ester of abscisic acid (Koshimizu et al., 1968), and zeatin-

N(7) glucoside (Parker et al., 1972). The related feature of these derivatized plant hormones has resulted in the use of the term "conjugates," which is defined as plant hormones (or their metabolites) that are metabolically coupled to other low molecular weight compounds by covalent binding (Sembdner, 1974; Sembdner et al., 1980). This definition excludes noncovalent associations between plant hormones and high molecular weight particles such as proteins, cell particles, organelles, and membranes. These complexes should continue to be called "bound" plant hormones. This is, however, an arbitrary limitation based on structural features, and it should not be assumed that the distinction necessarily implies uniqueness of the physiological role for one or the other class of compounds.

Many papers have been published on gibberellin (GA) conjugation during the last two decades, and the subject has been discussed in varying degrees of detail in reviews by Lang (1970), Sembdner (1974), Hedden, MacMillan, & Phinney (1978), Graebe and Ropers (1978), and Sembdner et al. (1980). The purpose of this chapter is to overview the field of GA conjugates. It is hoped that a critical appraisal will initiate new ideas and approaches and thus lead to a greater understanding of the physiological role of GA conjugates in plant growth and development. The structures of the GA conjugates discussed are illustrated in Fig. 6.1 and referred to numerically in the text.

6.2 NATURALLY OCCURRING GIBBERELLIN CONJUGATES

Correlations between endogenous GA levels and physiological effects led to Radley (1958) predicting the existence of "bound" GAs. The concept was supported by the metabolic formation of a partially identified GA_3 glucoside when GA_3 and saccharose were incubated with cucumber leaf discs (Murakami, 1961a). The detection of "water-soluble" or "bound" GAs, which may be potential conjugates, has been reported in *Phaseolus coccineus* (McComb, 1961; Jones, 1964; Sembdner et al., 1964; Gaskin & MacMillan, 1975), *Phaseolus vulgaris* (Hashimoto & Rappaport, 1966), *Pharbitis nil* (Murakami, 1961b, 1962; Ogawa, 1963, 1966b; Zeevaart, 1966), *Pharbitis purpurea* (Reinhard & Sacher, 1967a, 1967b, 1967c), *Vitis vinifera* (Weaver & Pool, 1965a, 1965b), *Nicotiana tabacum* (Sembdner & Schreiber, 1965), *Pisum sativum* (Reinhard & Konopka, 1967; Musgrave & Kende, 1970), *Hordeum vulgare* (Lazer, Baumgartner,

(i)

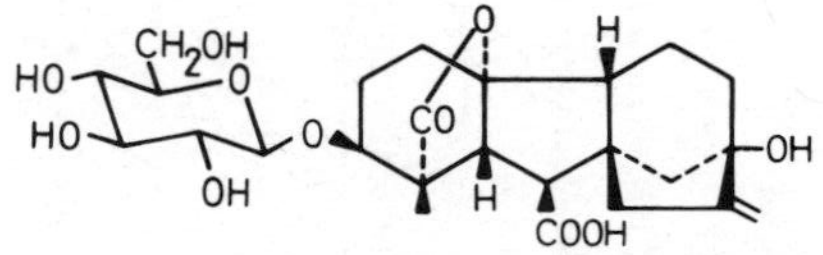

(1) GA_1-3-O-β-D-glucoside (O,S)

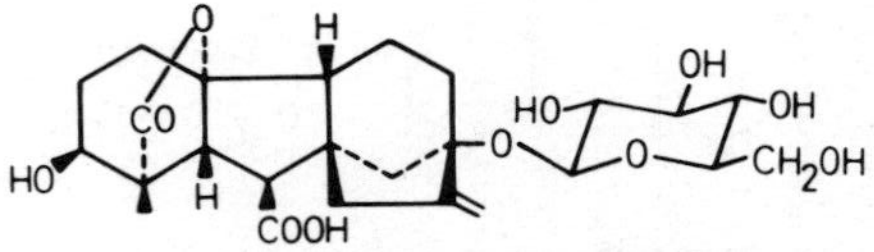

(2) GA_1 -13-O-β-D-glucoside (S)

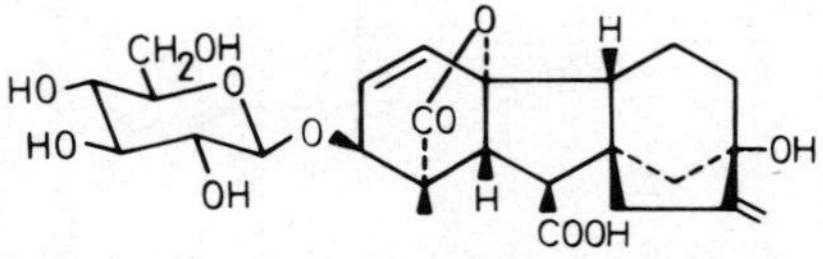

(3) GA_3 -3-O-β-D-glucoside (O,S)

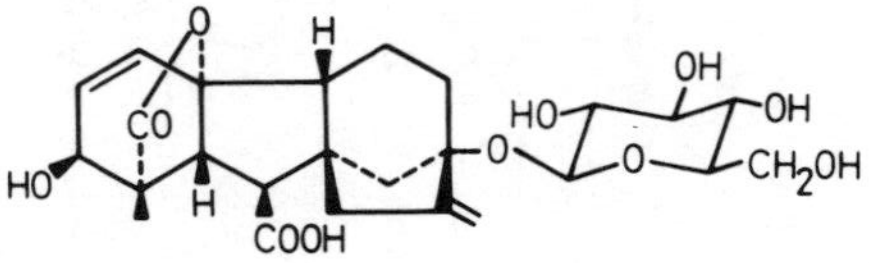

(4) GA_3 -13-O-β-D-glucoside (S)

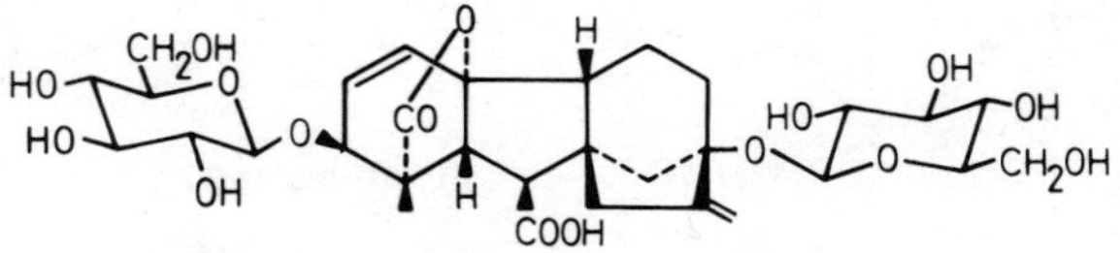

(5) GA_3 -3,13-di-O-β-D-glucoside (S)

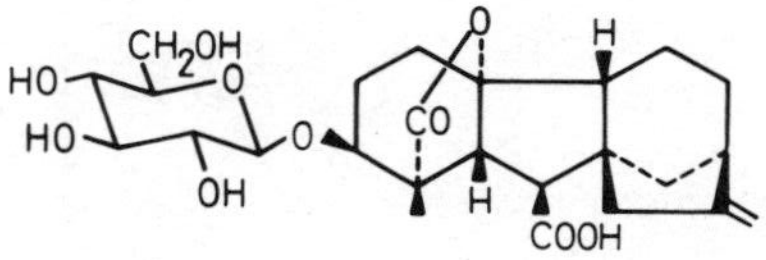

(6) GA_4 -3-O-β-D-glucoside (S)

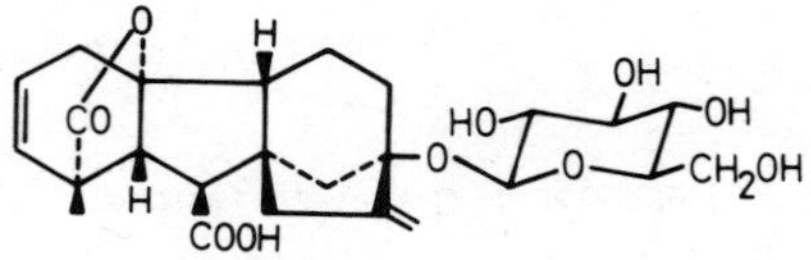

(7) GA_5 -13-O-β-D-glucoside (S)

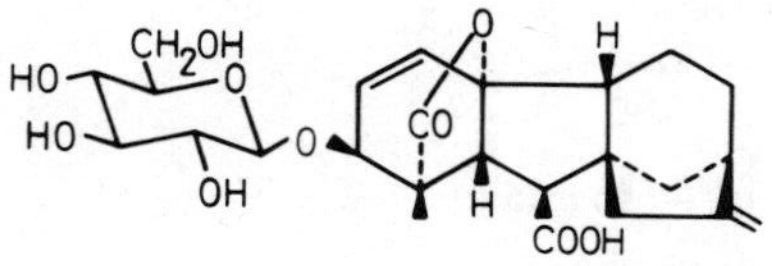

(8) GA_7 -3-O-β-D-glucoside (S)

Figure 6.1 Structures of GA conjugates [(O): naturally occurring; (s): chemically synthesized]. (i) GA glucosides; (ii) GA glucosyl esters; (iii) GA acyl and GA alkyl conjugates, gibberethiones; (iv) GA amino acid conjugates.

(9) GA_8-2-0-ß-D-glucoside (0,S)

(10) GA_8-13-0-ß-D-glucoside (S)

(11) GA_{26}-2-0-ß-D-glucoside (0)

(12) GA_{27}-2-0-ß-D-glucoside (0)

(13) GA_{29}-2-0-ß-D-glucoside (0)

(14) GA_{35}-11-0-ß-D-glucoside (0)

(15) 3-epi-GA_1-3-0-ß-D-glucoside (S)

(16) 3-epi-GA_1-13-0-ß-D-glucoside

(17) 16,17-H_2-GA_1-3-0-ß-D-glucoside (S)

(18) 16,17-H_2-GA_1-13-0-ß-D-glucoside (S)

(19) iso-GA_3-3-0-ß-D-glucoside (0,S)

(20) iso-GA_3-13-0-ß-D-glucoside (S)

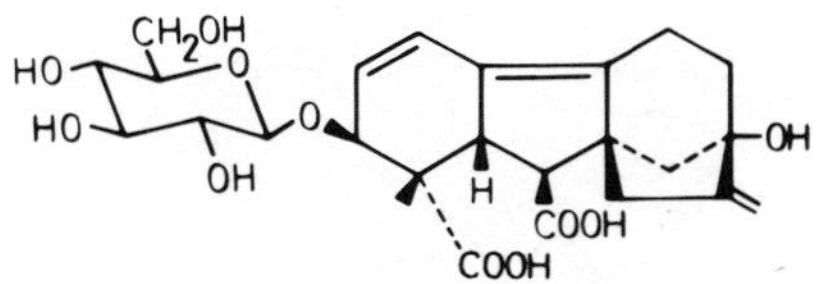

(21) Gibberellenic acid -3-0-ß-D-glucoside

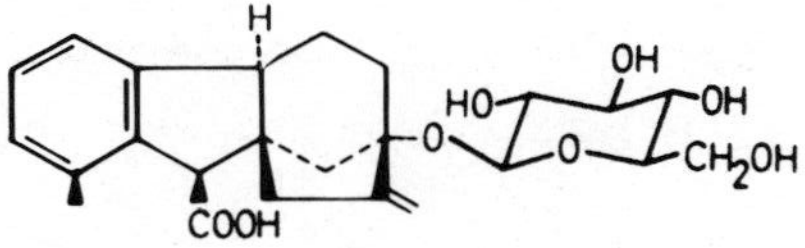

(22) allogibberic acid-13-0-0-ß-D-glucoside (S)

(ii)

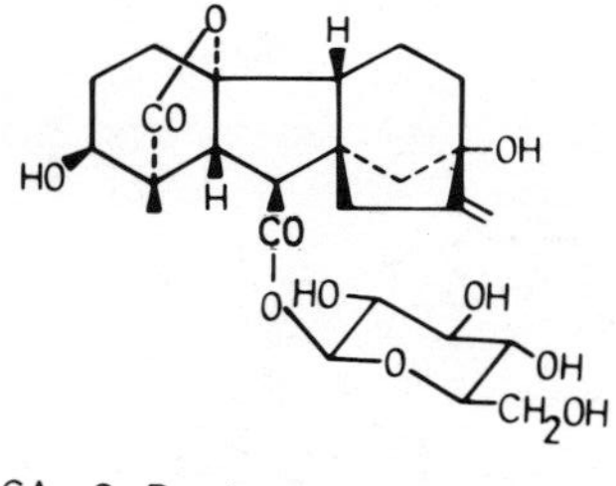

(23) GA₁ ß-D-glucosyl ester (0,S)

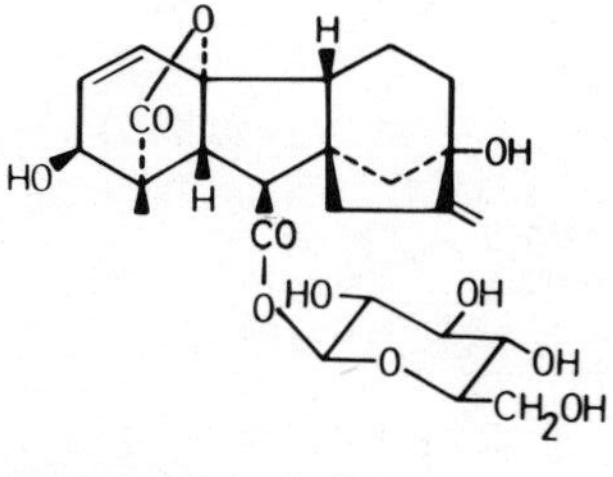

(24) GA₃ ß-D-glucosyl ester (S)

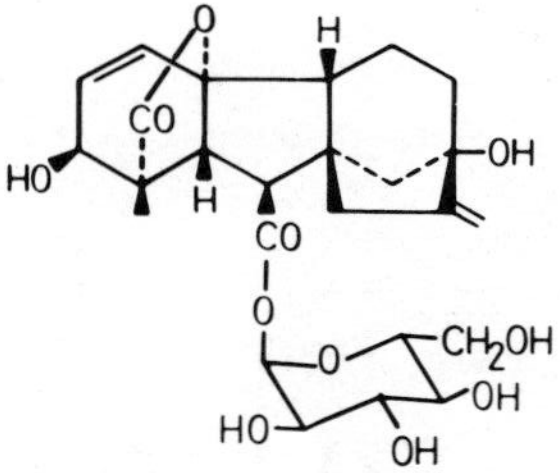

(25) GA₃ α-D-glucosyl ester (S)

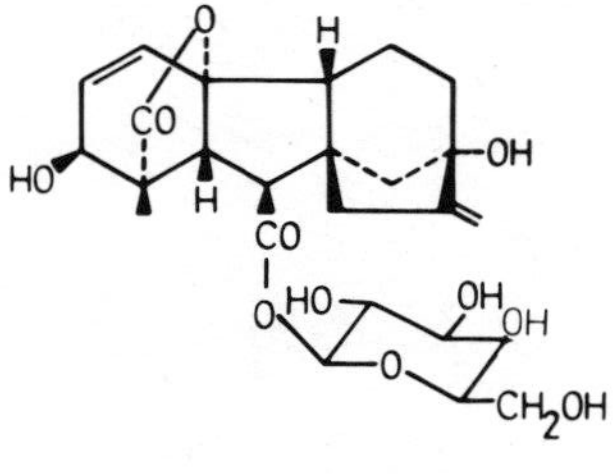

(26) GA₃ ß-D-galactosyl ester (S)

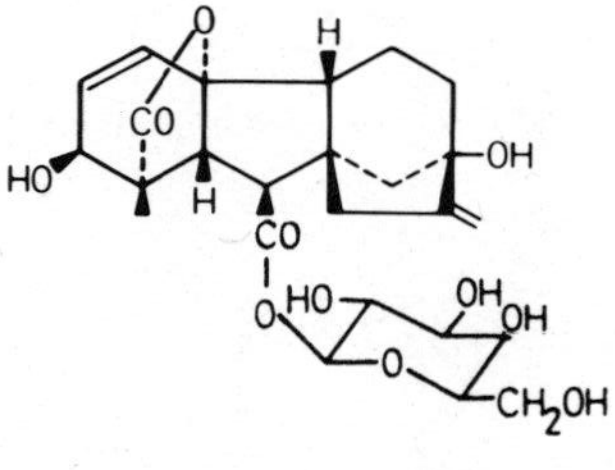

(27) GA₃ α-L-arabinosyl ester (S)

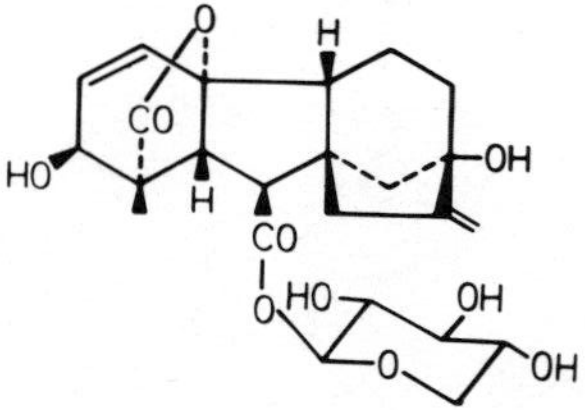

(28) GA₃ ß-D-xylosyl ester (S)

(continued)

Figure 6.1 (Continued)

(29) GA₄ ß-D-glucosyl ester (O,S)

(30) GA₅ ß-D-glucosyl ester (O,S)

(31) 13-0-ß-glucosyl-GA₅-glucosyl ester (S)

(32) GA₉ ß-D-glucosyl ester (O,S)

(33) GA₃₇ß-D-glucosyl ester (O,S)

(34) GA₃₈ ß-D-glucosyl ester (O,S)

(35) GA₄₄ ß-D-glucosyl ester (O)

(36) allogibberic acid ß-D-glucosyl ester (S)

(iii)

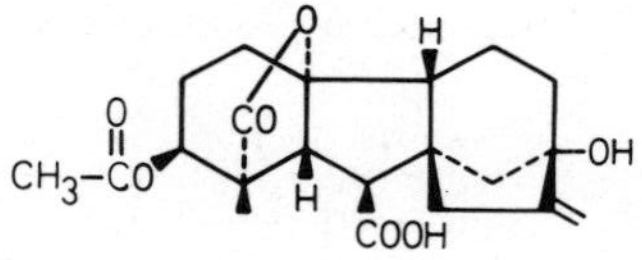

(37) 3-0-acetyl-GA₁ (O,S)

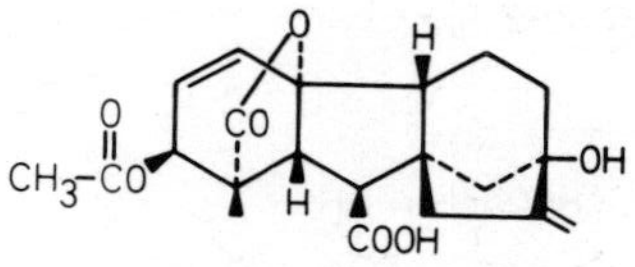

(38) 3-0-acetyl-GA₃ (O,S)

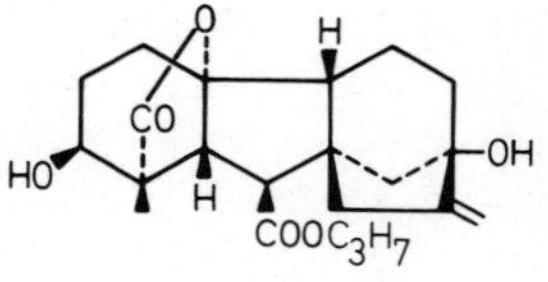

(39) GA₁-n-propyl ester (O,S)

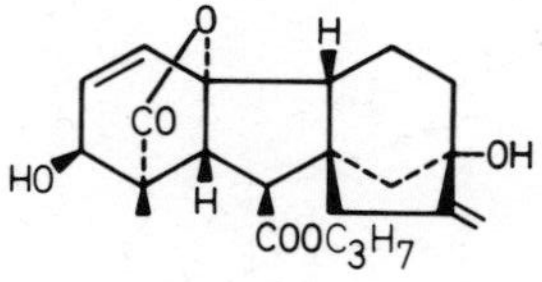

(40) GA₃-n-propyl ester (O,S)

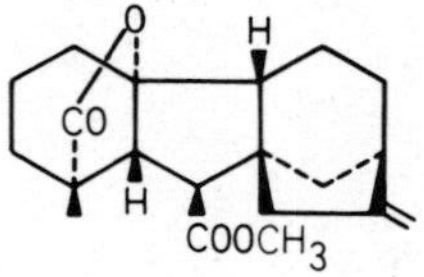

(41) GA₉-methyl ester (O,S)

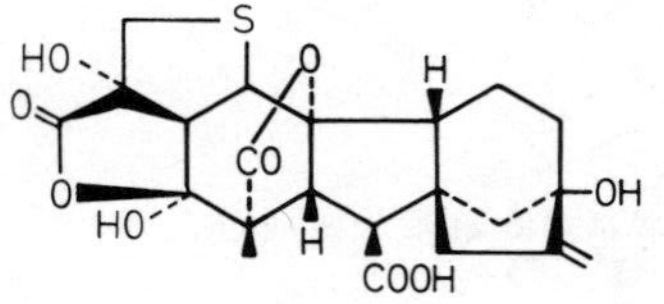

(42) gibberethione (O,S)

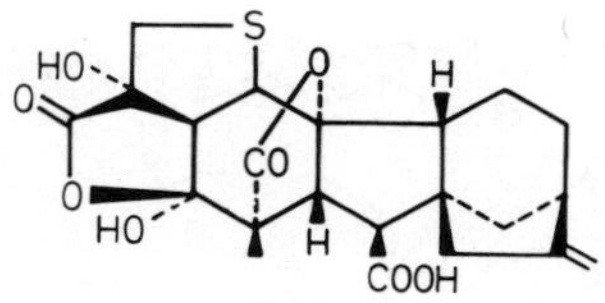

(43) 13-desoxy-gibberethione(S)

Figure 6.1 *(Continued)*

(iv)

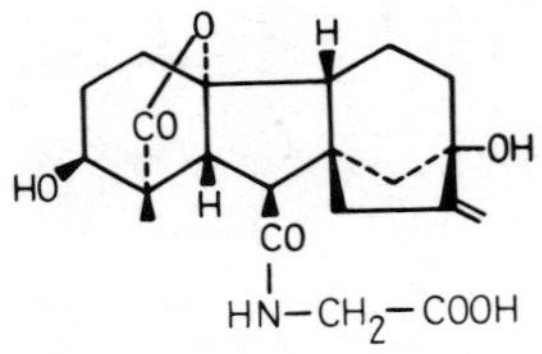

(44) GA$_1$-oyl-glycine (S)

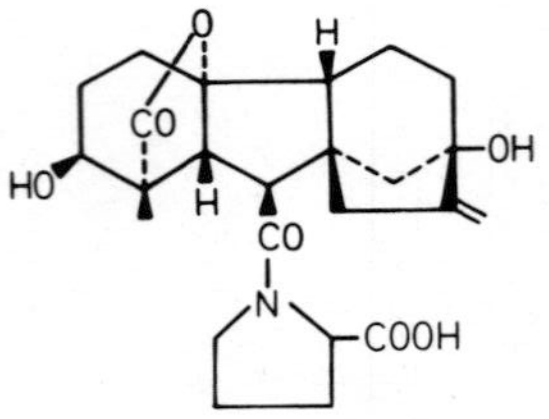

(45) GA$_1$-oyl-L-proline (S)

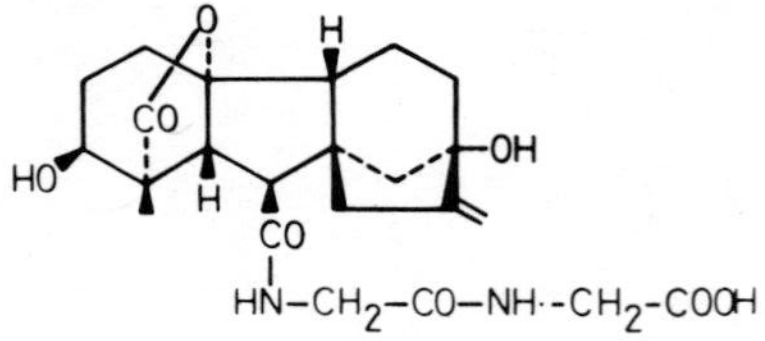

(46) GA$_1$-oyl-glycylglycine (S)

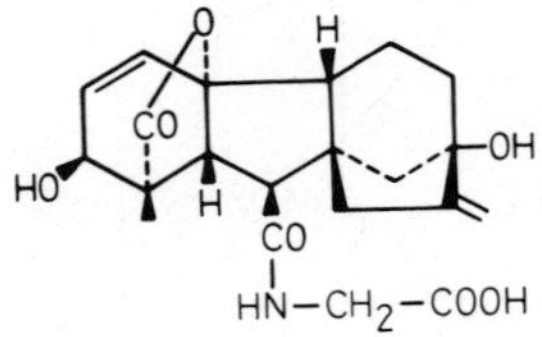

(47) GA$_3$-oyl-glycine (S)

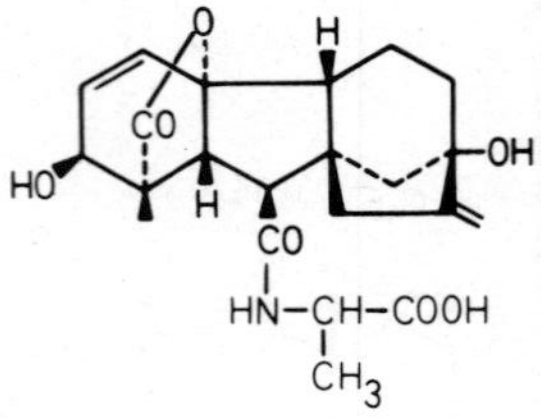

(48) GA$_3$-oyl-L-alanine (S)

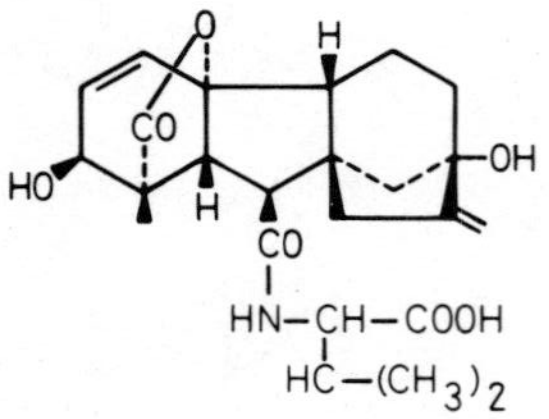

(49) GA$_3$-oyl-L-valine (S)

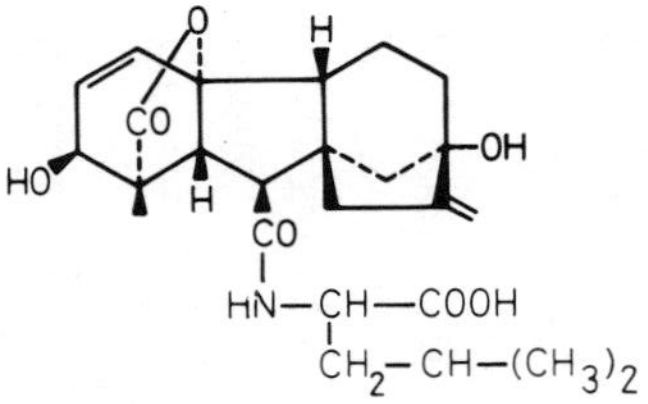

(50) GA$_3$-oyl-L-leucine (S)

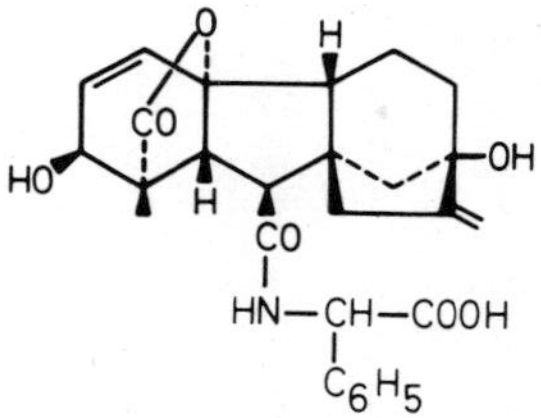

(51) GA$_3$-oyl-L-phenylalanine (S)

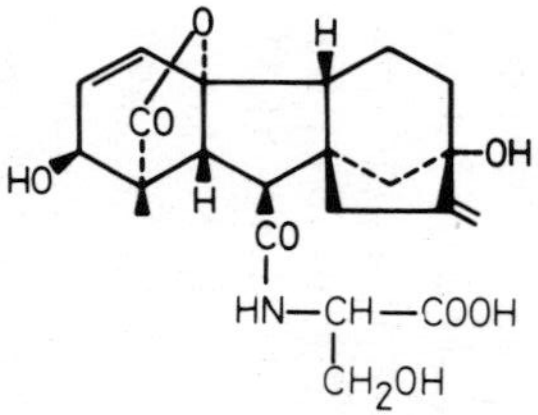

(52) GA$_3$-oyl-L-serine (S)

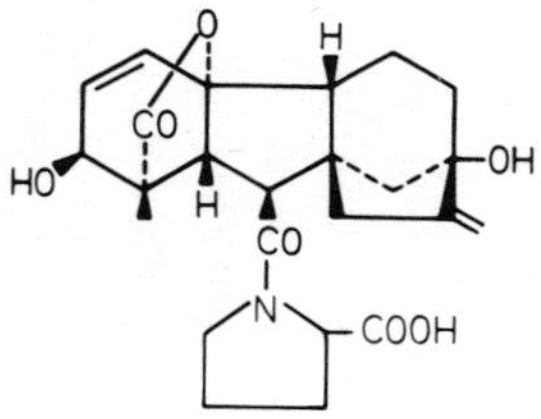

(53) GA$_3$-oyl-L-proline (S)

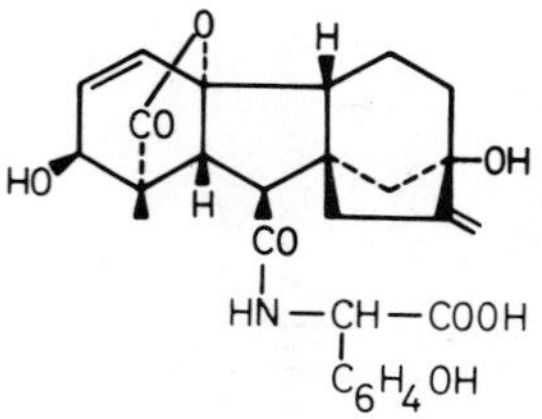

(54) GA$_3$-oyl-L-tyrosine (S)

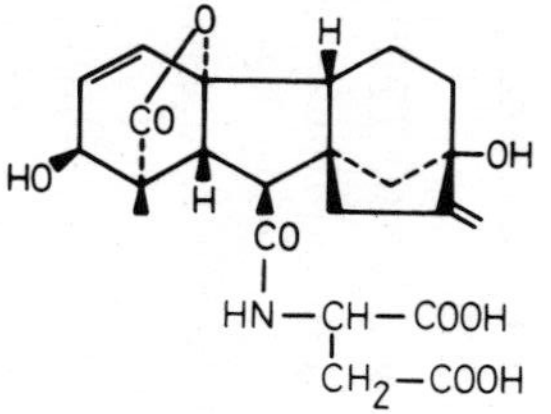

(55) GA$_3$-oyl-L-aspartic acid (S)

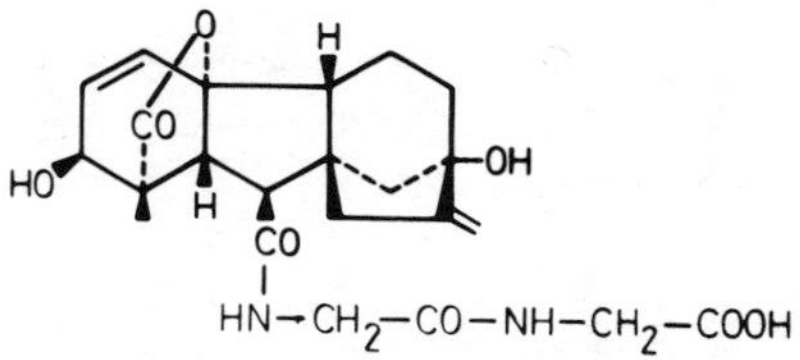

(56) GA$_3$-oyl-glycylglycine (S)

Figure 6.1 (Continued)

397

& Dahlstrom, 1961), *Lycopersicon esculentum* (Pegg, 1966), *Lupinus luteus*, *Prunus persica*, and *Sechivum edula* (Ogawa, 1966a, 1966b), and *Tulipa gesneriana* (Aung & Rees, 1974). In many cases the belief that "bound" GAs are present in extracts is based on bioassay data rather than physicochemical evidence (Murakami, 1961b; Ogawa, 1963, 1966a, 1966b; Weaver & Pool, 1965b; Hashimoto & Rappaport, 1966; Reinhard & Sacher, 1967a). The liberation of free GAs either by enzyme hydrolysis (Murakami, 1961a; McComb, 1961; Jones, 1964; Sembdner et al., 1964; Reinhard & Sacher, 1967a, 1967c; Hayashi et al., 1971; Gaskin & MacMillan, 1975; Dathe, 1977; Sponsel, Gaskin, & MacMillan, 1979) or by chemical hydrolysis (Murakami, 1961b; Lazer et al., 1961; Sembdner et al., 1964; Barendse & DeKlerk, 1975) is usually regarded as evidence of the presence of GA conjugates. Although these procedures yield only indirect evidence, they are often the only feasible way to monitor GA conjugates in physiological studies. Care should, however, be exercised because chemical hydrolysis, in particular, can lead to the production of artifacts (see Section 6.7.1).

6.2.1 Endogenous gibberellin conjugates

Endogenous GA conjugates can currently be classified as either glucosyl derivatives, alkyl esters, or acetates.

6.2.1.1 Gibberellin glucosyl conjugates

GA glucosyl conjugates represent the main group of GA conjugates. To date glucose, in the pyranose form, is the only sugar component to have been detected in GA conjugates. As far as is known the sugar unit is linked to the GA moiety either by a hydroxyl group, resulting in the formation of a GA glucoside (GA glucosyl ether), or via the C-7 carboxyl function to yield a GA glucosyl ester.

Gibberellin glucosides. The first structurally identified GA conjugate was GA_8-2-*O*-glucoside (*9*), originally called "Phaseolus ε" (Sembdner et al., 1964), which was isolated from maturing fruits of *Ph. coccineus* L. cv. Prizewinner (Schreiber et al., 1967; Weiland, 1968; Schreiber, Weiland, & Sembdner, 1970). GA_8 glucoside (*9*) has also been identified in extracts from mature seeds of *Ph. vulgaris* (Hiraga et al., 1974a, 1974b), immature seeds of *Ph. nil* (Tamura et al., 1968; Yokota et al., 1971b), and shoot apices of *Althea rosea*

(Harada & Yokota, 1970). A series of four GA glucosides has been isolated from immature seeds of *Ph. nil* (Yokota et al., 1969, 1970, 1971b). In three of these—GA_{26}-2-*O*-glucoside (*11*), GA_{27}-2-*O*-glucoside (*12*), and GA_{29}-2-*O*-glucoside (*13*)—the glucose unit is attached to the same 2-hydroxy group as in (*9*). The fourth *Pharbitis* glucoside is GA_3-3-*O*-glucoside (*3*), which carries the conjugating moiety at the physiologically important 3β-hydroxy group (see Stoddart & Venis, 1980). In addition, *Pharbitis* seed has yielded the 3-*O*-glucosides of iso-GA_3 (*19*) and gibberellenic acid (*21*). These conjugates are also found in extracts from immature *Ph. vulgaris* seed and are probably artifacts derived from GA_3-3-*O*-glucoside (*3*) during purification (Asakawa et al., 1974; Yamaguchi et al., 1979; Yokota, Murofushi, & Takahashi, 1980). More recently, GA_3-*O*-3-glucoside (*3*) has been identified in seeds of *Quamoclit pennata* (Yamaguchi et al., 1979). A further 3-*O*-glucoside, GA_1-3-*O*-glucoside (*1*), has been reported to occur in seeds of *Dolichos lablab* (Yokota et al., 1978), while GA_{35}-11-*O*-glucoside (*14*) has been isolated from immature seeds of *Cytisus scoparius* (Yamane et al., 1971, 1974). To date, no GA-13-*O*-glucosides have been detected. This is, perhaps, unexpected in view of the large number of 13-hydroxy GAs and the occurrence of steviol-13-*O*-glucoside in *Stevia rebaudiana* (Mosettig et al., 1963).

Gibberellin glucosyl esters. Following indications of the occurrence of neutral GA derivatives (Hashimoto & Rappaport, 1966), Hiraga et al. (1972, 1974a) succeeded in isolating the glucosyl esters of GA_1 (*23*), GA_4 (*29*), GA_{37} (*33*), and GA_{38} (*34*) from neutral ethyl acetate and *n*-butanol extracts of mature *Ph. vulgaris* seed. A polar fraction from needles of *Picea sitchensis* has been shown to contain GA_9 glucosyl ester (*32*). The isolated ester was identified by comparison with synthetic material (Lorenzi, Horgan, & Heald, 1976). More recently, glucosyl esters of GA_5 (*30*) and GA_{44} (*35*) originating from immature seeds of *Pharbitis purpurea* have been isolated and structurally elucidated by Yamaguchi, Kobayashi, and Takahashi (1980). It seems probable that the number of GA glucosyl esters will continue to increase as there are many indications in the literature of base-hydrolyzable fractions that potentially contain such conjugates.

6.2.1.2 Other gibberellin conjugates

In addition to the main group of GA glucosyl esters and ethers, there are only a few other types of conjugates, the occurrence and significance of which may be of limited importance. 3-*O*-Acetyl GA_3 (*38*) has been isolated as an endogenous product of the fungus *Fusarium moniliforme* (Schreiber et al., 1966). 3-*O*-Acetyl GA_1 (*37*) and 3-*O*-acetyl GA_3 (*38*) are both produced as minor metabolites of GA_4 in modified *Gibberella fujikuroi* cultures (McInnes et al., 1977). Acetylated GAs have not yet been found in higher plants. Seeds of *Cucumis sativus* are a source of GA_3 *n*-propyl ester (*40*) and a further conjugate tentatively identified as GA_1 *n*-propyl ester (*39*) (Hemphill, Baker, & Sell, 1973). Cultivated prothalia of *Lygodium japonicum* contain GA_9 methyl ester (*41*) which when exogenously applied produces antheridia formation at $10^{-10} M$ and inhibits archeogonium formation at $10^{-9} M$ (Yamane et al., 1979; Volume II, Chapter 6). A sulphur-containing GA derivative has been obtained from immature seed of *Ph. nil.* It was originally called "pharbitic acid" but is now named gibberethione (*42*). This compound could be formally considered as an amino acid conjugate (Yokota et al., 1974).

6.2.2 Metabolic formation of gibberellin conjugates

This section deals with GA conjugates that have been detected as metabolic products after the application of radioactively labeled free GAs to plants, plant organs, or tissues. It should be pointed out that not all the conjugation processes observed necessarily represent physiologically relevant reactions, as many experiments have been performed with readily available rather than endogenous GA substrates. Another restriction originates from the application of superoptimal concentrations of precursors, which may induce reactions more typical of xenobiotic attack than *in vivo* physiological conditions. An additional problem concerns the limited analytical techniques that are employed in many studies. As a consequence, identifications of many metabolite GA conjugates are incomplete and require further confirmation. This is especially true of quantitative data. Table 6.1 lists the GA conjugates that have been detected in metabolic studies and outlines the analytical evidence upon which the identifications are based.

The first classical experiment on metabolic conjugation was performed by Murakami (1961a), who incubated leaf discs of *Cucumis sativus* in an aqueous solution of GA_3 and sucrose. The resulting polar compound showed biological activity and released GA_3 and glucose after both acidic and enzymic hydrolysis. It was suggested that the putative conjugate was GA_3 glucoside. The same metabolite was formed by incubating GA_3 and glucose with tissues of *Pharbitis nil*, *Ipomea batatas*, *Diospyros kaki*, *Arachis hypogaea*, *Lupinus luteus*, and *Pyrus simonis*. Despite their preliminary nature these simple experiments by Murakami (1961a) convincingly demonstrated the general capability of plant tissues to conjugate GA_3.

The metabolism of $[^3H]GA_1$ has been extensively investigated in terms of conjugate formation. After application of $[1,2-^3H_2]GA_1$ to germinating seeds of *Ph. vulgaris*, a GA_8 glucoside-like fraction was isolated and characterized by thin-layer chromatography (TLC) and gas chromatography (GC) (Nadeau & Rappaport, 1972). This was confirmed by Yamane, Murofushi, and Takahashi (1975) who observed that in maturing and germinating *Ph. vulgaris* seed the GA_1 label was favorably incorporated into GA_8 glucoside and to a lesser extent into GA_8. $[^3H]GA_1$ glucosyl ester was also detected, although the identification was tentative and based on acid hydrolysis followed by analysis of the related GA moiety by gas chromatography-radioactivity counting (GC-RC). $[^3H]GA_8$ glucoside was also found after $[1,2-^3H_2]GA_1$ was applied to a mutant of *Hordeum vulgare* (Stoddart & Foster, 1976). In dwarf and normal *Zea mays* seedlings, $[^3H]GA_1$ was converted to GA_8 glucoside as well as to an unidentified GA_1 conjugate with properties clearly different from those of GA_1 glucoside (Davies & Rappaport, 1975a, 1975b). In barley aleurone layers, $[^3H]GA_1$ was mainly converted into GA_8 glucoside and GA_1 glucoside. Inhibition of 2β-hydroxylation with a GA_8 cold trap led to enhanced glucosylation of GA_1. In addition to these two glucosides a polar, amphoteric GA conjugate was isolated which appeared to be identical with the unidentified *Zea* conjugate (Nadeau, Rappaport, & Stolp, 1972; Rappaport et al., 1974; Nadeau & Rappaport, 1974). A similar array of metabolites was obtained with protoplast, but not vacuole preparations, from *H. vulgare* seedlings (Rappaport & Adams, 1978). The amphoteric nature of the conjugate was determined by paper electrophoresis, which showed an isoelectric point at pH 2.9. Acidic hydrolysis yielded gibberellin C, which was identified by GC-RC. Exclusion chromatography on Sephadex suggested that the compound had a molecular weight of *ca.* 700–800 daltons. The available evidence suggests that the conjugate moiety has peptide-like properties (Nadeau et al., 1972; Stolp, Nadeau, & Rappaport, 1973; Rappaport et al., 1974). The same amphoteric conjugate may also have been obtained after the application of $[^3H]GA_1$ to *P. sativum* seedlings, which also produced a second putative conjugate that was thought to be GA_1 glucosyl ester on the basis of TLC data (Durley, Railton, & Pharis, 1974).

Table 6.1. Metabolic formation of GA conjugates and their characterization

Plant material	GA applied	Conjugate formed	Basis of identification*			Reference
			Conjugate	Parent GA	Conjugating unit	
Bryophyllum daigremontianum leaves, shoots	[³H]GA$_{20}$	glucosyl ester-like C/D-rearranged GA$_{20}$	TLC	AlkH/AcH GC-RC	—	Durley et al., 1975
Hordeum vulgare aleurone layers	[³H]GA$_1$	GA$_1$ glucoside-like	TLC, GC, HVE	AcH GC	—	Nadeau et al., 1972; Rappaport et al., 1974
		GA$_8$ glucoside-like	TLC, GC, HVE	AcH GC	—	"
		GA$_1$ X ("ampho"-GA$_1$)	TLC, HVE Sephadex excl.	AcH GC	—	"
protoplasts	[³H]GA$_1$	GA$_8$ glucoside GA$_1$ X ("ampho" GA$_1$)	TLC/GC-RC	—	—	Rappaport & Adam, 1978

(continued)

Tissue	Substrate	Product				Reference
excised leaves	[³H]GA₄	GA₄ glucosyl ester-like	DEAE-Seph.	EH/TLC		Liebisch et al., 1980
	[³H]GA₁	GA₁ glucosyl ester-like	DEAE-Seph.	EH/TLC	—	"
	[³H]GA₁	GA₈ glucoside-like				Stoddart & Foster, 1976
Lactuca sativa excised hypocotyls	[³H]GA₁	GA₁ glucoside-like	TLC, HVE	EH HVE	EH HVE	Stoddart & Jones, 1977
	[³H]GA₁	GA₁ glucosyl ester-like	TLC, HVE	EH HVE	EH/HVE TLC	"
	[³H]GA₉	GA₉ glucosyl ester-like	TLC, HVE	AlkH	—	Nash et al., 1978
		GA₂₀ glucoside-like	TLC, HVE		—	"
Pharbitis nil seedlings	GA₃	GA₃ glucoside-like	PC bioassay	AcH, EH, PC	—	Murakami, 1961a
	[³H]GA₁ [¹⁴C]glucose	GA₁ glucoside-like	TLC	AcH, TLC-RC bioassay	AcH, TLC-RC	Barendse, 1971
	[¹⁴C]GA₃	GA₃ glucoside-like	DEAE Seph. TLC-RC	EH, AcH TLC		Barendse & DeKlerk, 1975; Barendse, 1974

Table 6.1. *(Continued)*

Plant material	GA applied	Conjugate formed	Basis of identification*			Reference
			Conjugate	Plant GA	Conjugating unit	
Phaseolus coccineus maturing pods	[^{14}C]GA$_3$	GA$_3$-3-*O*-glucoside	TLC	TLC		Sembdner et al., 1972
	GA$_3$	GA$_8$-like-glucoside	TLC NMR (acetates) MS	EH TLC	EH TLC	"
seedlings	[^{3}H]GA$_4$	GA$_4$ glucosyl ester-like	TLC	EH/TLC	—	Nash & Crozier, 1975
Phaseolus vulgaris germinating seeds	[^{3}H]GA$_1$	GA$_8$ glucoside	TLC-RC E GC-RC	AcH, EH TLC-GC	—	Nadeau & Rappaport, 1972
seedlings	[^{3}H]GA$_3$	GA$_3$-3-*O*-glucoside	IR NMR MS auth. mat.	EH TLC MS	PC	Asakawa et al., 1974
		iso-GA$_3$-3-*O*-glucoside	IR, NMR, MS	EH, NMR TLC MS	PC	"
		gibberellenic acid-3-*O*-glucoside	PC, TLC UV	EH, TLC	PC	"

404

	Substrate	Product			PC	Reference
immature seeds	[³H] GA₁	GA-X glucoside-like	PC, NMR IR	—	PC	"
	[³H] GA₄ }[³H] GA₂₀	GA₁ glucosyl ester	TLC-RC auth. mat.	—	—	Yamane et al., 1975
	[³H] GA₂₀	GA, glucoside-like	TLC-RC	EH, GC-RC AcH TLC	—	Takahashi et al., 1976
	[³H] GA₂₀	GA₂₀ glucoside-like	TLC-RC	EH TLC	—	"
	[³H] GA₅	GA₅ glucosyl ester-like	—	EH, TLC, GC-RC	—	"
		GA₅ glucoside-like	—	AlkH, TLC EH GC-RC	—	"
mature seeds	[³H] GA₁	GA₈ glucoside	TLC GC-RC	TLC, EH, GC-RC	—	Yamane et al., 1977
	[³H] GA₄ [³H] GA₅ [³H] GA₈ [³H] GA₂₀					
	[³H] GA₄	GA₄ glucosyl ester-like	TLC-RC	EH TLC	—	"
Pisum sativum shoots of etiolated plants	[³H] GA₄ }[³H] GA₉ [³H] GA₂₀	GA conjugates	TLC, RC	—	—	Railton et al., 1974a, 1974b

(continued)

Table 6.1. (*Continued*)

| Plant material | GA applied | Conjugate formed | Basis of identification* | | | | Reference |
			Conjugate	Parent GA	Conjugating unit		
immature seeds	[³H]GA₉	12α-OH-GA₉ glucoside-like	—	EH, GC-MS	—		Frydman & Mac-Millan, 1975; Sponsel & MacMillan, 1976, 1977; Sponsel, 1979
	[³H]GA₂₀	GA₂₀ conjugate	—	EH, GC-MS	—		Frydman & MacMillan, 1975
		GA₂₉ conjugate	—	EH, GC-MS	—		
	[2α-²H₁ -2α-³H₁]- GA₂₉	GA₂₉ conjugate	—	EH, GC-MS	—		"

406

Pseudotsuga menziesii shoots	$[^3H]GA_4$	GA_{34} glucoside-like	—	—	Wample et al., 1975
Zea mays	$[^3H]GA_1$	GA_8 glucoside-like	TLC RC methyl ester	AcH GC	Davies & Rappaport, 1975a, 1975b "
		GA_1-X(ampho-GA_1)	TLC RC	AcH GC	
Arachis hypogaea, Cucumis sativus, Diospyros kaki, Ipomea batatas, Lupinus luteus, Pyrus simonii	GA_3	GA_3 glucoside-like	PC bio-assay	AcH, EH PC	Murakami, 1961a

*Abbreviations: AcH—acidic hydrolysis; AlkH—alkaline hydrolysis; EH—enzymic hydrolysis; E—electrophoresis; GC—gas chromatography; GC-MS—combined gas chromatography/mass spectrometry; HVE—high-voltage electrophoresis; MS—mass spectrometry; NMR—nuclear magnetic resonance; PC—paper chromatography; RC—radio counting; TLC—thin layer chromatography.

The metabolism of $[^3H]GA_1$ in suspension cultures of *Lycopersicon esculentum* has been shown to result in the production of GA_8-2-O-glucoside via GA_8 (Liebisch, 1980). Excised hypocotyls of *Lactuca sativa* have been used for a detailed study of the metabolism of $[^3H]GA_1$ (Stoddart & Jones, 1977). The formation of two GA_1 glucosyl derivatives was established by high-voltage electrophoresis, and the metabolites were tentatively identified as GA_1 glucosyl ester and GA_1 glucoside. $[^3H]GA_1$ methyl ester was metabolized much more slowly than $[^3H]GA_1$ resulting exclusively in the production of the corresponding glucoside. The data of Stoddart and Jones (1977) indicate that in many instances tentatively identified conjugates may, in fact, be comprised of a mixture of glucosides and glucosyl esters.

The simultaneous feeding of $[^3H]GA_1$ and $[^{14}C]$glucose to seedlings of *Ph. nil* afforded a doubly labeled conjugate, which according to TLC and hydrolysis experiments was believed to be GA_1 glucoside (Barendse, 1971). Application of $[^{14}C]GA_3$ to maturing fruits of *Ph. coccineus* was reported to result in the accumulation of a GA_8-like glucoside and GA_3 glucoside (*3*) (Sembdner et al., 1968; Weiland, 1968). In a preparative scale experiment with ripening excised *Ph. coccineus* pods, application of GA_3 resulted in a 17% incorporation into GA_3-3-O-glucoside (*3*) (Sembdner et al., 1972). $[^3H]GA_3$ is also converted to GA_3-3-O-glucoside (*3*) by *Ph. vulgaris* seedlings (Asakawa et al., 1974). Iso-GA_3-3-O-glucoside (*19*) and the 3-O-glucoside of gibberellenic acid (*21*) were also detected, but, as mentioned earlier, these compounds are probably artifacts derived from GA_3 glucoside (*3*). In addition, a further polar GA metabolite was isolated but not structurally elucidated. The formation of a GA_3 glucoside-like conjugate was reported to occur in seedlings of *Ph. nil* (Barendse, 1974). The conjugate was shown to be identical with authentic GA_3-3-O-glucoside (*3*) on the basis of TLC and enzyme and acid hydrolysis as well as bioassay data (Barendse & DeKlerk, 1975). GA_3 is thought to be an endogenous constituent in *Pharbitis* seedlings.

$[^3H]GA_4$ administered to *d-1* mutant *Z. mays* seedlings has been shown to be converted into a glucosylated form (Sembdner, 1969). Nash and Crozier (1975) fed $[^3H]GA_4$ to seedlings of *Ph. coccineus*, which contain endogenous GA_4, and found that it was metabolized to a polar compound that released GA_4 after cellulase hydrolysis. It was suggested the metabolite was a GA_4 conjugate, possibly a glucosyl ester. Studies of $[^3H]GA_4$ metabolism in immature and mature seeds of *Ph. vulgaris* have also produced evidence for the formation of GA_4 glucosyl ester (Yamane et al., 1975, 1977; Takahashi, Murofushi, & Yamane, 1976). The main pathway, however, appears to proceed via GA_1 and GA_8 to glucosylated GA_1 and GA_8 glucoside. [3H]GA_4 is rapidly metabolized by vegetative shoots of *Pseudotsuga menziesii* to GA_{34} and a polar metabolite assumed to be GA_{34} glucoside (Wample, Durley, & Pharis, 1975).

[3H]GA_5 fed to maturing seeds of *Ph. vulgaris* was found to be metabolized to GA_8 glucoside, possibly via GA_8. The identification of GA_8 glucoside was based on enzymic liberation of GA_8 from acidic butanol extracts. The simultaneous release of putative GA_5 indicated the potential occurrence of GA_5 glucoside, while liberation of GA_5 from the neutral butanol fraction was taken to suggest the presence of GA_5 glucosyl ester (Yamane et al., 1975, 1977; Takahashi et al., 1976). GA_{20}, which is a native constituent, appeared to be converted to GA_{20} glucoside by *Ph. vulgaris* seed. Evidence for this conversion was based on characterization, by TLC and GC-RC, of GA_{20} liberated after enzymic hydrolysis of the acidic, butanol fraction (Yamane et al., 1975, 1977; Takahashi, 1974). In general, glucosylation of applied 3H-GAs occurs much more readily in mature than in immature *Ph. vulgaris* seed (Takahashi et al., 1976; Yamane et al., 1977). [3H]GA_{20} fed to mature leaves of *Bryophyllum daigremontianum* was converted into a C/D-ring-rearranged derivative and its glucosylated form from which the aglucone was liberated by alkaline hydrolysis. This indicates that the conjugate is likely to be a glucosyl ester rather than a glucoside (Durley, Pharis, & Zeevaart, 1975).

Applied 3H-GAs are rapidly metabolized in both seeds and seedlings of *P. sativum*. After application of [3H]GA_5, [3H]GA_4, and [3H]GA_{20} to dwarf pea seedlings, high levels of radioactivity were found in butanol fractions (Railton, Durley, & Pharis, 1974a, 1974b). Although this was thought to represent glycosylated GAs, no firm basis of identification was provided. [3H]GA_9 administered to developing seed of *P. sativum* cv. Progress is converted into 12α-hydroxy GA_9 and its glucoside. The tentative structure of the 12α-hydroxy GA_9 glucoside was based on GC-MS identification of the parent GA liberated by pectinase hydrolysis (Frydman & MacMillan, 1975; Sponsel & MacMillan, 1976, 1977). As formation of 12α-hydroxy GA_9 is believed to be via an artificially induced rather than an endogenous pathway, the glucosylation step is of undetermined metabolic significance (Sponsel & MacMillan, 1976; Sponsel et al., 1979; Sponsel, 1979). Although conjugation of endogenous GA_{20} and GA_{29} appears to occur in *P. sativum* seed, the pathway seems to be less favored than catabolic degradation (Frydman & MacMillan, 1975; Sponsel & MacMillan, 1977; Sponsel et al., 1979; Sponsel, 1979).

6.3 CHEMICAL SYNTHESIS OF GIBBERELLIN CONJUGATES

Chemical synthesis is important for confirmation of the structures of endogenous GA conjugates. It also helps to overcome the limited availability of natural material, which can only be isolated through the use of elaborate preparative purification procedures. Moreover, syntheses of labeled conjugates and analogues with interesting variations in structure pave the way for intricate physiological and metabolic investigations.

6.3.1 Gibberellin glucosyl conjugates

6.3.1.1 Gibberellin glucosyl esters

The first attempts to glucosylate GAs at the C-7 carboxyl group were performed by Keay, Moffatt, and Mulholland (1965). Silver gibberellate was reacted with 2,3,4,6-tetra-O-acetyl-α-D-glucopyranosyl bromide (α-acetobromoglucose) to yield the tetra-O-acetylglucosyl ester of GA_3. The reaction of free GAs with α-acetobromoglucose in the presence of catalysts such as Ag_2O, Ag_2CO_3, and Et_3N (Koenigs-Knorr–condition) has proved to be a convenient and efficient procedure for the glucosylation of the carboxyl group of both GA_3 and GA_8 (Weiland, 1968; Schreiber, Weiland, & Sembdner, 1969). Difficulties do, however, arise because deacetylation under alkaline conditions leads to transesterification and hydrolysis of the glucosyl ester linkage. Sodium methoxide treatment at low temperatures has proved to be useful for deacetylation in the synthesis of the glucosyl esters of GA_1 (*23*), GA_3 (*24*), GA_4 (*29*), GA_{37} (*33*), GA_{38} (*34*) as well as ^{3}H-labeled derivatives of (*23*) and (*29*) (Hiraga, Yokota, & Takahashi, 1974c; Chapter 7, this volume). However, this procedure is not without problems, because partial transesterification to the corresponding GA methyl ester cannot be completely excluded and has been found to occur in the synthesis of the glucosyl esters of GA_5 (*30*) and allogibberic acid (*36*) (Schneider, 1981). Another method for removing the protective acetyl groups involves the use of helicase, the digestive juice of snails (*Helix pomatia*) (Miersch & Liebisch, 1974, 1983). A tendency for the crude enzyme preparation to cleave the ester linkage can be suppressed by the addition of glucose to the incubation medium. The overall yield of $[^3H]GA_3$ glucosyl ester (24) synthesized via this procedure is about 55%.

Lorenzi et al. (1976) synthesized GA_9 glucosyl ester (32) in order to establish the structure of the endogenous compound isolated from *Picea sitchensis*. The synthesis of GA_3-α-D-glucosyl ester (25) was achieved by reacting with 3,4,6-tri-*O*-acetyl-2-*O*-trichloracetyl-β-D-glucopyranosyl chloride (Brigl β-chloride) (Miersch & Liebisch, 1974, 1983). Other derivatives that have been synthesized include the β-D-galactosyl, α-D-arabinosyl, and β-D-xylosyl esters of GA_3 (26, 27, 28) (Beneš & Liebisch, 1983).

6.3.1.2 Gibberellin glucosides

When GA hydroxyl groups are glucosylated, it is necessary to protect the other competing functions. Mild conditions must be employed for both the glucosylation reaction itself and the removal of the protective groups. The use of strong bases and acids (even strong Lewis acids) is prohibited because of the chemical lability of GAs. However, Koenigs-Knorr conditions work stereospecifically and without any apparent side effects (Schreiber et al., 1969; Schneider, 1972). Hydroxyl groups that are not involved in the glucosylation can be protected by acetylation. Sodium methoxide treatment conveniently cleaves the acetates without affecting the glucosidic bond. Carboxyl groups are best protected by methylation. Subsequent nonhydrolytic deprotection involves the use of lithium propyl thiolate (Bartlett & Johnson, 1970; Schneider, 1972, 1981). The overall glucosylation procedures are outlined in Fig. 6.2. This synthesis scheme has been used to synthesize the 3-*O*-β-D-glucosides of GA_1 (1), GA_3 (3), 3-*epi*-GA_1 (15), [16,17-H_2]GA_1 (17), iso-GA_3 (19), GA_4 (6), and GA_7 (8) as well as GA_8-2-*O*-β-D-glucoside (9) (Schneider, Sembdner, & Schreiber, 1974, 1977b; Schneider, 1980, 1981). The synthesized 3-*O*-β-D-glucosides of GA_1 (1) and GA_3 (3), and GA_8-2-*O*-β-D-glucoside (9) have been used to confirm the structure of putative conjugates isolated from plant material. Uniformly labeled GA_3-3-*O*-β-D-glucoside has been prepared by Wilzbach tritiation of the acetate followed by deacetylation (Schneider, 1978).

In contrast to most other tertiary alcohols, the 13-hydroxy group of GAs is susceptible to derivatization and can be glucosylated relatively easily. Consequently, the 13-*O*-β-D-glucosides of GA_1 (2), GA_3 (4), 3-*epi*-GA_1 (16), [16,17-H_2]GA_1 (18), iso-GA_3 (20), GA_5 (7), GA_8 (10), and allogibberic acid (22) can all be readily synthesized

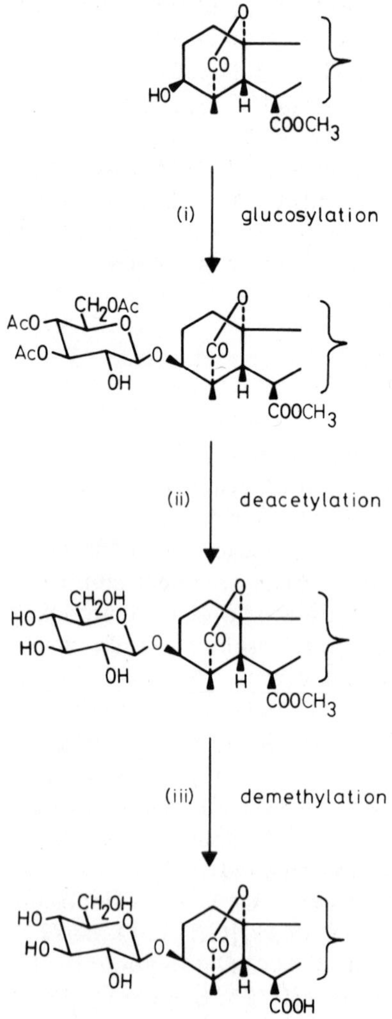

Figure 6.2 Scheme of chemical glucosylation of GAs. (i) α-Acetobromoglucose/Ag_2CO_3; (ii) CH_3ONa; (iii) $LiSC_3H_7$.

(Schneider et al., 1974, 1977b; Schneider, 1980, 1981). The endogenous occurrence of GA-13-O-glucosides has not yet been demonstrated. However, there are tentative indications of the metabolic formation of GA_5-13-O-glucoside and GA_{20}-13-O-glucoside (see Section 6.2.2).

If the substrate is a 3β,13-dihydroxy GA methyl ester, glucosylation results in a mixture of 3-O- and 13-O-monoglucosylated GAs and the 3,13-di-O-glucoside as indicated by the synthesis of GA_3-3,13-di-O-β-D-glucoside (5) (Schneider et al., 1977b). Glucosylation of

GA_5 as the free acid produces a mixture of GA_5 glucosyl ester (*30*), GA_5-13-*O*-β-D-glucoside(*7*), and in addition the 13-*O*-β-D-glucosyl-GA_5-β-D-glucosyl ester (*31*). The latter derivative represents a further potential naturally occurring conjugate (Schneider, Miersch, & Liebisch, 1977a).

The influence of structural and steric features on the glucosylation yields of different GA hydroxy groups are compared in Fig. 6.3. It is apparent that an equatorial arrangement is favored. Inhibiting effects originate from neighboring olefinic bonds. It should be pointed out that in all cases the yield of the corresponding 13-*O*-glucoside is higher than that of the 3-*O*-glucoside (Schneider, 1980). Such structural differences may affect enzymic hydrolysis and the potential physiological behavior of the glucosides in plant tissues (see Section 6.4).

6.3.2 Gibberellin amino acid conjugates and others

The important role of amino acid conjugates in indole-3-acetic acid metabolism (see Sembdner et al., 1980) makes it tempting to speculate that GAs may be conjugated in a similar manner. However, the available evidence to support this suggestion is weak and is based on the occurrence of polar fractions in *Nicotiana tabacum* extracts (Sembdner & Schreiber, 1965) and the [^{3}H]GA_1 metabolite, amphoteric GA_1-X, which was discussed in Section 6.2.2 (Rappaport et al., 1974).

By synthesis, however, a series of GA amino acid conjugates became available. Aminolysis of anhydrides of GA_1 and GA_3 with either an amino acid alkali salt or, alternatively, an amino acid methyl ester followed by ester cleavage affords a GA amino acid conjugate (Fig. 6.4). This method has been used to synthesize a series of GA_1 and GA_3 amino acid conjugates (*44–56*) and GA_3-oyl-2-[^{14}C]glycine (*47*) (Lischewski, Adam, & Sych, 1974; Lischewski & Adam, 1976; Adam et al., 1977). Aminolysis of GA anhydrides with primary or secondary amines yields an array of neutral and basic amides of GA_1 and GA_3 (Adam et al., 1976; Müller, Knöfel, & Kramell, 1976). Gibberethione (*42*) has been synthesized by Yokota, Yamane, and Takahashi (1976). 3-Keto-GA_3 was directly coupled with ammonium mercaptopyruvate to produce gibberethione (*42*) together with isogibberethione (Fig. 6.5). The same procedure can be used to synthesize 13-deoxy-gibberethione (*43*) from 3-keto GA_7. The *n*-propyl esters of GA_1 (*39*) and GA_3 (*40*) have also been

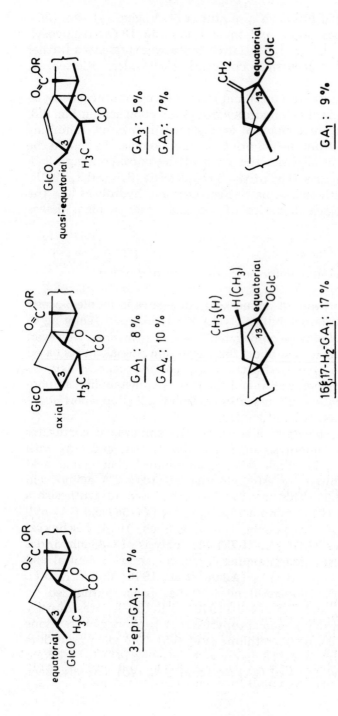

Figure 6.3 Dependence of glucosylation yield (%) on structural features of GA hydroxyl groups.

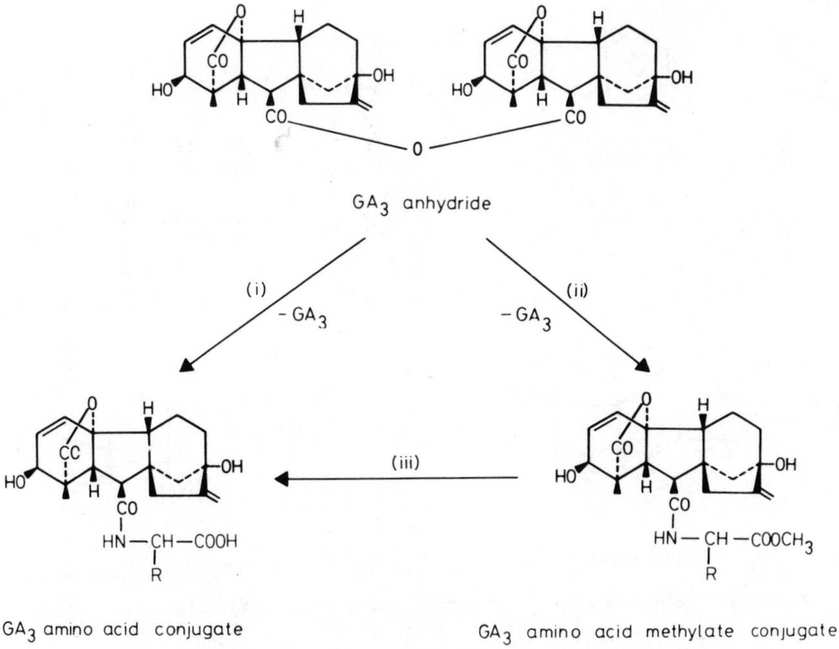

GA₃ anhydride

(i) - GA₃

(ii) - GA₃

(iii)

GA₃ amino acid conjugate

GA₃ amino acid methylate conjugate

Figure 6.4   Scheme of the synthesis of GA_3 amino acid conjugates.

(i) $R{-}CH{-}COO^-$;   (ii) $R{-}CH{-}COOCH_3$;   (iii) $LiSC_3H_7$.

$\quad\quad$ NH_2 $\quad\quad\quad\quad\quad$ NH_2

(Lischewski et al., 1974; Adam et al., 1977.)

synthesized in order to confirm their possible occurrence in *C. sativus* seed (Hemphill et al., 1973).

6.4 ENZYMIC HYDROLYSIS AND ENZYMIC FORMATION OF GIBBERELLIN CONJUGATES

There are numerous reports of investigations dealing with GA conjugates where enzymes have been used to release free GAs from polar fractions (see Section 6.2.2). In many instances this was done without any real knowledge on the nature of the conjugate involved, and—only too often—release of free GAs after enzyme treatment of crude fractions has been used as a basis for identification and quantification

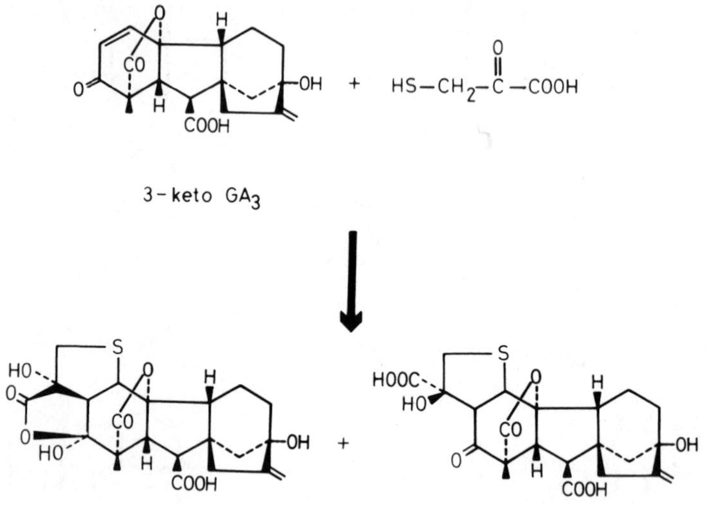

3-keto GA$_3$

(42) gibberethione

isogibberethione

Figure 6.5   Scheme of the synthesis of gibberethione (Yokota et al., 1976).

of GA conjugates. Unfortunately, this experimental approach is likely to produce less than satisfactory data.

The findings of Yokota et al. (1971b) reveal that GA$_3$-3-*O*-glucoside (*3*) cannot be completely cleaved by many commonly used enzyme preparations, including β-glucosidases. Knöfel, Müller, and Sembdner (1974) and Müller et al. (1978) have reported on the capacity of a range of enzymes to hydrolyze various GA glucosyl conjugates. Table 6.2 shows striking differences in the ability of the individual enzymes to hydrolyze GA$_1$-3-*O*-glucoside (*1*), GA$_3$-3-*O*-glucoside (*3*), and GA$_8$-2-*O*-glucoside (*9*). It appears that GA$_3$-*O*-glucoside (*3*) is most effectively hydrolyzed by cellulase and helicase, although partial isomerization to the biologically inactive iso-GA$_3$ does occur (Müller et al., 1978). Since GA$_3$ can be recovered unchanged after exposure to the incubation conditions, the formation of iso-GA$_3$ from the glucoside is associated with enzymic attack on the 3-*O*-glucosyl residue.

More detailed studies on the relationships between GA conjugate structure and hydrolytic cleavage by cellulase have been carried out (Schliemann & Schneider, 1979; Schneider & Schliemann, 1979). Hydrolysis by purified cellulase was quantitatively monitored using a glucose oxidase-peroxidase reaction. The pH optimum of cellulase-induced hydrolysis of GA_5-O-glucoside (7) was established as 3.0 (Km 0.75 ± 0.15 mM). The data illustrated in Fig. 6.6 indicate that GA_8-2-O-glucoside (9), which has an equatorially arranged glucose moiety, is the favored substrate. The decreased hydrolysis rate for the equatorial 3-O-glucoside of 3-*epi*-GA_1 (15) may be due to the close proximity of the α-orientated lactone group. The curves obtained with GA_1-3-O-glucoside (1) and GA_4-3-O-glucoside (6) indicate that a change to a 3β-axial position results in a striking decrease in the rate of hydrolysis. GA_3-3-O-glucoside (3) and GA_7-3-O-glucoside (8), which have a quasiequatorially arranged glucose moiety, show slightly enhanced sensitivity to hydrolysis due to the less-hindered steric position. Surprisingly, the spatially hindered tertiary 13-O-glucosides were cleaved as readily as the GA_8-2-O-glucoside (Fig. 6.7). Alteration of ring A substituents and hydrogenation of the 16,17-double bond did not adversely influence the rapid rate of hydrolysis of GA-13-O-glucosides. The observed differences in the ability of cellulase to hydrolyze various GA-glucosides implies that plants may possess substrate-specific GA-glucosidases.

The enzymic cleavage of GA-β-glucosyl esters is somewhat easier and less dependent on the structural features of the aglucone (Müller et al., 1978). On the other hand, enzyme specificity to the steric orientation of the glucoside bond does exist, as GA_3-α-D-glucosyl ester (25) is readily hydrolyzed by α-glucosidase but not by β-glucosidase (Liebisch, 1974). GA alkyl esters such as GA_3 propyl ester (40) have so far resisted enzyme hydrolysis as well as cleavage by microorganisms. The same applies to the synthetic GA amino acid conjugates, which resist enzymic attack by carboxypeptidase A (Serva), pronase (Calbiochem), ficin (Merck), and acylase (Schuchardt) (Borgmann, 1977).

Conversion of applied free GAs to GA glucosyl conjugates by plant tissues (Section 6.2.2) would appear to require the involvement of glucosyl transferases, and attempts have been made to isolate the enzymes involved. Most work has involved the pericarp of *Ph. coccineus* seed, which has been shown to glucosylate [^{14}C]GA_3 *in vivo* (Sembdner et al., 1972). A protein fraction was isolated that glucosylated GA_3 *in vitro* when incubated with uridine-5′-diphosphoglucose at pH 8.2. The isolated enzyme, which appears in the seed during maturation of the pods, exhibited high specificity in that it glucosylated GA_3 and GA_7 to the corresponding 3-O-glucosides (3) and (8) but did not glucosylate other GAs (Müller, Knöfel, & Sembdner, 1974). Similar preparations from seed of *Ph. nil, Ph. purpurea*, and *T. aestivum* did not contain any GA_3 or GA_7 glucosyltransferase activity (Knöfel et al., 1983).

Table 6.2. Survey of the effectiveness of enzyme preparations used for the hydrolysis of GA_1-3-O-glucoside, GA_3-3-O-glucoside, and GA_8-2-O-glucoside

Enzyme	Origin	Buffer[a]	pH	Incubation temp. (°C)	Time (hr)	3-O-Glucoside		2-O-Glucoside of GA_8[b]	Reference
						GA_1[b]	GA_3[b]		
β-Glucosidase A, B	Röhm and Haas	1	4.6	32	24	+	+	++++	Müller et al., 1978
β-Glucosidase C, D, E	Koch and Light, Fluka, Almonds	1	4.6	32		0	0	+	"
Tannase	Aspergillus niger	1	4.6	33	24	+	+	++++	"
Rhamnodiastase	Rhamnus frangula	1	4.6	33	24	0	0	0	"
Takadiastase	Ferak (Aspergillus oryzae)	1	5.5	37	48	0	0	++++	"
Elaterase	Ecballium elaterium	1	4.6	37	48	0	0	+	"
Hemicellulase	Fluka	1	5.5	37	48	0	0	++	"
Isoflavon glucoside specific glucosidase	Cicer arietinum	2	7.5	37	48	0	0	0	"
Cellulase	Serva (Aspergillus niger)	1	4.6	34	48	0	++	++++	"
Pectinase	Ferak (Aspergillus niger)	1	4.6	37	48	0	+	++++	"

Enzyme	Source	Buffer	pH	Temp	Time (h)				Reference
Esterase	Boehringer (kidney)	2	7.5/8.6	34/25	48	o	o	o	"
β-Glucanase	Bacillus sp.	1/2	4.6/7.0	37	48	o	o	o	"
Ficin	Merck (Ficus carica)	3	6.2	30	48	o	o	o	"
Strophanthobiase	Strophantus sp.	1	4.6	37	48	o	+	++++	"
Helicase	Helix pomatia	3	7.0	32	48	o	+++	++++	"
β-Glucosidase	Röhm and Haas	1 (0.5 M)	5.0	30	48	—	—	++++	Schreiber et al., 1970
Cellulase	Sigma	1 (0.2 M)	4.0	37	16	—	++	++++	Yokota et al., 1971b
Cellulase	Sigma			47	12	—	+++(+)	—	Asakawa et al., 1974
β-Glucosidase		3 (0.1 M)	5.0	37	24	—	+	—	Barendse & De-Klerk, 1975
β-Glucuronidase	Helix pomatia	3 (0.1 M)	5.1	37	24	—	+++	—	"
Cellulase		3 (0.1 M)	4.5	37	24	—	+++	—	"

[a] 1: 0.2 M acetate buffer; 2: 0.05 M Tris/HCl; 3: 0.067 M phosphate buffer.
[b] Extent of cleavage given as +: 1–20%; ++: 20–50%; +++: 80–100%.

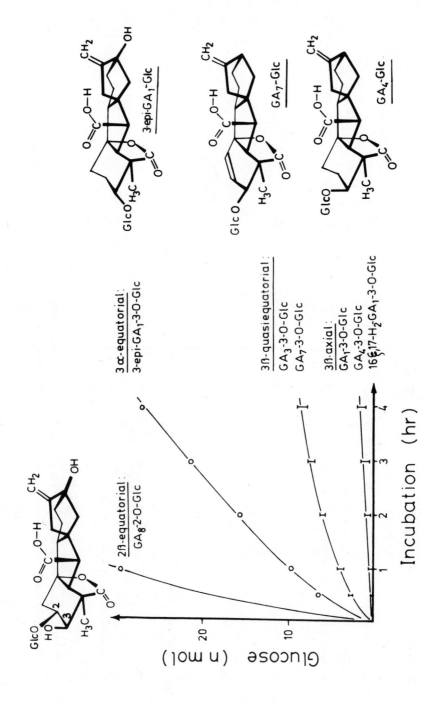

Figure 6.6 Hydrolysis of ring A glucosylated GAs by cellulase (500 µg/100 nmol glucoside) (Schneider & Schliemann, 1979).

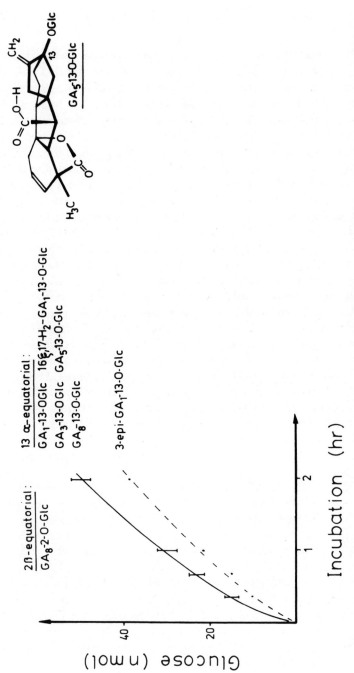

Figure 6.7 Hydrolysis of 13-*O*-glucosylated GAs by cellulase (100 μg/100 nmol glucoside) (Schliemann & Schneider, 1979).

6.5 BIOLOGICAL ACTIVITY OF GIBBERELLIN CONJUGATES

Detailed investigations are required to determine whether the response of bioassays to GA conjugates originates from the applied GA conjugate per se or from free GA released *in vivo* as a consequence of the action of hydrolyases of undetermined specificity. In dwarf pea seedlings, first indications were obtained that hydrolysis of GA conjugates such as GA_8-2-*O*-glucoside (*9*) (Sembdner, 1969; Sembdner et al., 1972, 1973) occurs during bioassay. Consequently, simultaneous feeding of condurit B epoxide, a specific inhibitor of β-glucosidase, was found to completely suppress the activity of the glucoside. In addition, correlations have been observed between the hydrolysis rate and biological activity of GA glucosyl esters in different bioassays (Hiraga et al., 1974c; Liebisch, 1974). These data have led to the view that GA conjugates per se are biologically inactive and that the growth responses they induce are dependent upon the degree of hydrolysis and the structure of the parent GA. In addition to this, the biological response may be affected by differences in uptake and transport of GA conjugates. GA_3 glucoside and GA_3 glucosyl ester are readily transported in excised barley leaves, whereas free GAs remain at the point of application in the basal zone. However, both free GAs and their conjugates move throughout the plant when applied to the root system (Liebisch, Schmidt, & Schütte, 1980).

The relative activities of GA conjugates can be calculated from log-normal dose versus response curves by comparing parallel segments of the straight curve (Sembdner et al., 1976). This process was used to compile the data on the biological activities of GA glucosides, GA glucosyl esters, and GA amino acid conjugates, as summarized in Tables 6.3 and 6.4.

GA-β-D-glucosyl esters are almost as active as the parent GAs in the dwarf rice bioassay (Table 6.3). Hydrolysis of the esters appears to take place inside the rice plant tissue since no striking differences exist between nonsterile application via the root and sterile microdrop applications via the shoot. This was confirmed by administering $[^3H]GA_1$-β-D-glucosyl ester and $[^3H]GA_4$-β-D-glucosyl ester by the microdrop method and showing that all the radioactivity moved from the neutral ester fraction into the acidic free GA fraction within 12 and 48 hr, respectively (Hiraga et al., 1974c). A similar relationship between the biological response and the rate of conjugate hydrolysis has been found with GA_3 glucosyl ester in the lettuce and dwarf pea bioassays. In the latter test system, hydrolytic activity and the induced response varied with

Table 6.3. Relative biological activity of GA glycosyl esters and GA amino acid conjugates (% of activity of parent GA)

Conjugate	Tanginbozu dwarf rice bioassay		Dwarf maize mutant bioassay		Dwarf pea bioassay	Lettuce hypocotyl bioassay	Wheat half seed amylase bioassay	Amaranthus bioassay
	Non-sterile	Microdrop (+ sterile)	d-1	d-5				
GA_1-β-D-glucosyl ester	60–70[a]	80–100[a]	5–10[a]	5–10[a]	—	—	—	—
GA_3-β-D-glucosyl ester	80–100[a] 100[b]	60–70[a]	30–60[a]	40–50[a] 1[b]	10–50[b]	20[b]	10[c]	15[c]
GA_4-β-D-glucosyl ester	100[a]	100[a]	(60–100)[a]	(80–100)[a]	—	—	—	—
GA_9-β-D-glucosyl ester	50–70[d]	—	—	—	—	30–40[d]	—	—
GA_{37}-β-D-glucosyl ester	70–80[a]	(30–100)	(30–100)[a]	(30–80)[a]	—	—	—	—
GA_{38}-β-D-glucosyl ester	100[a]	100[a]	20[a]	10–100[a]	—	—	—	—
GA_3-α-D-glucosyl ester	10[b]	—	—	0[b]	—	—	1[c]	1[c]
GA_3-oyl glycin	1[b]	1[b+]	—	—	1[b]	0[b]	1[c]	2[c]
GA_3-oyl glycylglycin	5[b]	1[b+]	—	—	1[b]	0[b]	1[c]	1[c]
GA_3-oyl prolin	1[b]	—	—	—	1[b]	0[b]	1[c]	1[c]

NOTE: ( ) = calculations of relative activity impossible or inaccurate.
[a] Hiraga et al. (1974c).
[b] Sembdner et al. (1976).
[c] Bernhardt et al. (1979).
[d] Lorenzi et al. (1976).

Table 6.4. Relative biological activity of GA-O-β-D-glucosides (% of activity of parent GA)

Conjugate	Tanginbozu (Kotake Tam = x) dwarf rice bioassay		Dwarf maize mutant bioassay			Dwarf pea bioassay	Lettuce hypocotyl bioassay	Wheat half-seed -amylase bio-assay	Amaranthus bioassay
	Non-sterile	Sterile	d-1	d-2	d-5				
GA$_1$-3-O-β-D-glucoside	30[c]	10–20[c]	—				1[c]	5–10[c]	1[e]	0[e]
GA$_1$-13-O-β-D-glucoside	50[c]	20–30[c]	—				5[c]	10–20[c]	1[e]	0[e]
GA$_3$-3-O-β-D-glucoside	90°, 40–50[c]	1[b] x 20[c]	1[b]	1–5[b]	1–10,[a] 1[b] 1[c]	1,[b] 1[c]	5[c]	2[e]	1[e]
GA$_3$-13-O-β-D-glucoside	50–100[c]	—	—	—	1[c]	1[c]	5–10[c]	1[e]	1[e]
GA$_4$-3-O-β-D-glucoside	80[c]	—	—	—	1[c]	1[c]	—	1[e]	15[e]
GA$_5$-13-O-β-D-glucoside	100[c]	—	—	—	1[c]	1[c]	5–10[c]	1[e]	7.5[e]
GA$_8$-2-O-β-D-glucoside	100[b] x	50[g]	(100)[b]	( )[b]	100[b]	70–80[b] 80–100[f]	—	0[g]	0[e]
GA$_{26}$-2-O-β-D-glucoside	( )[b]	( )[b]	(50–70)[b]	( )[b]	( )[b]	—	—	—	—
GA$_{27}$-2-O-β-D-glucoside	( )[b]	( )[b]	( )[b]	( )[b]	( )[b]	—	—	—	—
GA$_{29}$-2-O-β-D-glucoside	100[b] x	( )[b]	1–5[b]	(5–20)[b]	0[b]	—	—	—	—
GA$_{35}$-11-O-β-D-glucoside	40–80[d] 20–30[d]	—	1–5[d] 1–5[d]	5–10[d] (d-3)	(0)[d]	1[d]	—	—	—
Gibberellenic acid-3-O-β-D-glucoside	70[b] x	( )[b]	1[b]	1[b]	1–5[b]	1[b]	—	—	—

$NOTE$: ( ) = Calculation of relative activity impossible or inaccurate.

[a] Ogawa & Takahashi (1974).
[b] Yokota et al. (1971a).
[c] Sembdner et al. (1973, 1974, 1976).
[d] Yamane et al. (1973).
[e] Bernhardt et al. (1979).
[f] Sembdner et al. (1972).
[g] Crozier et al. (1970).

the physiological state of the tissue (Liebisch, 1974). In *Z. mays* low rates of hydrolysis parallel the decreased biological activity of GA_3 glucosyl ester (Hiraga et al., 1974c). GA_3-α-D-glucosyl ester does not exhibit significant activity when applied to dwarf rice seedlings under nonsterile conditions. This presumably reflects the lack of a suitable α-glucosidase (Sembdner et al., 1976).

In general, GA-*O*-β-D-glucosides show lower relative activities than GA-β-D-glucosyl esters and they are, in fact, frequently inactive (Table 6.4). The highest responses are obtained in the nonsterile dwarf rice bioassay with 2-*O*- and 13-*O*-glucosides, both of which are relatively easily hydrolyzed (see Section 6.4). Hydrolytic activity in the aqueous incubating medium seems to play a crucial role in this bioassay, since the glucosides exhibit a marked reduction in activity when applied to either the root or the shoot of rice seedlings maintained under strict sterile conditions. Further support for external hydrolysis comes from experiments with [^{3}H]GA_3-3-*O*-β-D-glucoside (*3*), which was found to be progressively split into GA_3 and iso-GA_3 in an aqueous test medium containing germinating caryopses (G. Schneider, unpublished). The hydrolytic enzymes involved in the cleavage of the conjugates are believed to originate mainly from microbes in the rhizosphere (Yokota et al., 1971a). Test systems other than the dwarf rice bioassay tend to be insensitive to GA glucosides in general, with the exception of GA_8-2-*O*-glucoside (*9*). This may be due to a lack of endogenous glucosides in plant material employed in various bioassays. The high activity of GA_8-2-*O*-glucoside (*9*) may be a consequence of hydrolysis by less specific glucosidases. In absolute terms, however, the response induced by GA_8 glucoside is low, being slightly less than that of the aglucone, GA_8.

Synthetic GA amino acid conjugates have been tested in a series of bioassays, including the nonsterile rice seedling test system, and found to be almost inactive (Table 6.3). This is in keeping with the resistance of these conjugates to enzymic hydrolysis (Section 6.4). The uptake and translocation of GA_3-oyl-glycylglycine (*56*) are quite normal, and so the plants used in the various bioassays apparently lack enzymes for cleaving GA-peptide linkages (Liebisch et al., 1980). The *n*-propyl ester of GA_3 also exhibits very low biological activity (Hemphill et al., 1973). In cucumber hypocotyl and *d-1* dwarf maize bioassays the response was 10 and 5%, respectively, of that induced by GA_3. Similarly, activities of less than 1% of that of GA_3 were observed in the barley half seed and dwarf pea bioassays. Synthetic alkyl esters of GA_3 have also been tested in the dwarf pea bioassay. They were inactive when applied via the leaves, although root application did induce a small but significant response (Moffatt & Radley, 1960).

6.6 PHYSIOLOGICAL FUNCTIONS OF GIBBERELLIN CONJUGATES

GA conjugates are generally believed to be deactivated products of GA metabolism (Lang, 1970; Sembdner, 1974; Barendse & DeKlerk, 1975; Sembdner et al., 1980). However, in contrast to other metabolic steps, conjugation can be reversed by cleavage which results in the release of the free GA. If such an event could be demonstrated to be a normal part of *in vivo* GA metabolism, GA conjugates could be structures of major physiological importance by virtue of their ability to modify rapidly the levels of biologically active free GAs. It should, however, be emphasized that as a group GA conjugates are characterized on the basis of their structural features rather than their physiological properties. The physiological functions of individual conjugates may vary widely. For instance, GA glucosyl conjugates slightly differing in both the aglucone structure and the nature of binding to the glucose may possess diverse metabolic functions and physiological roles. The possible significance of physiological functions of GA conjugates may be derived from their following characteristics: (i) ease of metabolical formation, (ii) widespread occurrence and occasional high concentrations in plant tissues, (iii) formation and hydrolysis by specific enzymes, (iv) low biological activity relative to the relevant free GA, and (v) high polarity.

It has for some time been believed that GA glucosyl conjugates act as storage products because of their preferential formation during seed ripening and the high levels in which they are found in mature seeds (Barendse, Kende, & Lang, 1968; Sembdner et al., 1968, 1972; Sembdner, 1969; Yamane et al., 1975). Indications that GA conjugates formed during seed maturation are reconverted into free GAs during germination would appear to support this assumption (Sembdner, 1969; Barendse & DeKlerk, 1975; Takahashi et al., 1976).

Recently, Stoddart and Venis (1980) have discussed the possibility that in cell membranes the glucosyl unit of GA conjugates could stabilize the GA moiety in an incorrect orientation for further interaction. Alternatively, the glucose moiety could prevent the GA from entering the active pool.

The occurrence of conjugated GAs in bleeding sap of trees has led to suggestions that these polar substances may be involved in long-distance transport (Sembdner et al., 1968; Dathe et al., 1978b). Rapidly growing and developing tissues such as germinating seeds and seedlings are also capable of conjugating GAs (Nadeau & Rappaport, 1972; Asakawa et al., 1974; Barendse, 1974; Stoddart & Jones, 1977). This finding is believed to imply that the size of endogenous GA pools may be regulated to some degree by the reversible conversion of free GAs to GA conjugates. The evidence is, however, equivocal and other attempts to correlate the ratios of free and conjugated GAs with such processes as dwarfism (Davies & Rappaport, 1975a) and the effects of light on plant growth and developments (Barendse, 1974; Lance et al., 1976) have been equally inconclusive. Much of the data are seemingly contradictory; this is, in part, due to the inadequate analytical techniques used. Moreover, in many instances, structural differences in the GA conjugates have not been considered. This is important because the properties of different types of GA conjugates can vary widely and this can affect both the experimental data obtained and the manner in which it is interpreted. This point is only too apparent when examining the three main groups of GA glucosyl conjugates.

6.6.1 Gibberellin-2β-O-glucosides

The glucose moiety is connected to a 2β-hydroxy group. There can be little doubt that 2β-hydroxylation of GAs is an important irreversible deactivation step (Reeve & Crozier, 1974; Hedden et al., 1978; Graebe & Ropers, 1978; Sembdner et al., 1980). Thus, conjugation of 2β-hydroxy GAs such as GA_8, GA_{26}, GA_{27}, and GA_{29} to form GA-2β-O-glucosides cannot play a direct role in regulating the level of active GAs. However, the linking of 2β-hydroxy GAs to glucose may be important for transport or compartmentation of these catabolites. Glucosylation and oxidation of the 2β-hydroxy GAs are competing catabolic processes. In *Vicia* and *Pisum* species conjugation appears to be the minor route, with oxidation to 2-keto compounds being the main catabolic pathway. However, in *Ph. coccineus* conjugation seems to be the favored process (see Sponsel et al., 1979; Chapter 4, this volume), although the catabolic fate of the putative GA-2β-O-glucosides that are formed remains undetermined.

6.6.2   GA-3β-*O*-glucosides

In the case of GA_1-3-*O*-glucoside (*1*) and GA_3-*O*-3-glucoside (*3*), the aglucones are highly active and glucosylation followed by hydrolysis provides a possible mechanism for the regulation of active GA pools. Some evidence for enzyme specificity in the 3-*O*-glucosylation step has been gained from studies on the *in vitro* synthesis of GA_3-3-*O*-glucoside by Müller et al. (1974) and Knöfel et al. (1983). In addition, the varying susceptibility of the individual GA-3β-*O*-glucosides to cellulase treatment, which was discussed in Section 6.4, implies that the hydrolysis of these compounds may similarly involve a high degree of enzyme specificity. The albeit limited available data on glucosylation and hydrolysis of GA-3β-*O*-glucosides thus suggests that they are well qualified to play a regulatory role in GA metabolism.

6.6.3   Gibberellin glucosyl esters

Because of their accumulation in maturing seeds and the ease with which they are cleaved, GA glucosyl esters are perhaps the favored candidate for a storage role (Hedden et al., 1978; Graebe & Ropers, 1978). However, glucosylation of acidic compounds is an extremely ubiquitous process that occurs not only to GAs but also to other endogenous and xenobiotic substances. In some of these cases, glucosylation represents a "detoxification" process. However, as discussed in Section 6.2.2, formation of GA glucosyl esters has been reported to take place in immature and germinating seeds as well as in seedlings where GA levels are seemingly limiting. It may well be that formation of GA glucosyl esters is a rapid metabolic reaction—similar to indole-3-acetic acid conjugation (Zenk, 1964)—that is followed by other metabolic or catabolic events. Both formation and hydrolysis of GA glucosyl esters seem to occur readily in plant tissues, and this could, at least in theory, be a means through which rapid changes in the levels of physiologically active GAs are mediated.

6.7   ANALYSIS OF GIBBERELLIN CONJUGATES

The difficulties of analyzing the myriad of free GAs (Gaskin & MacMillan, 1978; Yokota, Murofushi, & Takahashi, 1980; Reeve

& Crozier, 1980; Chapter 8, this volume) are further complicated by the occurrence of conjugated GA metabolites. With the exception of alkyl esters and acylated conjugates, which will not be considered in detail as they are unlikely to be the major endogenous constituents of higher plants, GA conjugates are characterized by their hydrophilic nature and high molecular weight. As a consequence an array of analytical techniques are required that are quite different from those employed when analyzing free GAs. A general outline of two procedures traditionally used for the preparative isolation of GA glucosides and GA glucosyl esters is given in Figs. 6.8 and 6.9. The scheme illustrated in Fig. 6.8 was used by Schreiber et al. (1970) to study GA conjugates in maturing pods of *Ph. coccineus*, while Hiraga et al. (1974a) used the procedures shown in Fig. 6.9 to investigate the conjugate content of *Ph. vulgaris* seed. The two flow diagrams serve as useful models when discussing problems encountered in the analysis of GA conjugates.

6.7.1 Isolation of gibberellin conjugates

In most cases, the analysis of GA conjugates starts with the extraction of plant material with methanol containing up to 20% water. After evaporation of the organic solvent, the aqueous residue is fractionated by partitioning. Vast amounts of lipids can be removed by extraction with a nonpolar solvent such as petrol ether, or benzene at pH 7.0. The neutral ethyl acetate fraction contains the less-polar GA glucosyl esters, while most of the free GAs partition into ethyl acetate at pH 2.5 (see Chapter 8). Extraction with n-butanol at pH 2.5 affords the more-polar GA glucosyl esters and GA glucosides. However, caution should be exercised when neutral and acidic n-butanol fractions are collected as in Fig. 6.9, because n-butanol is capable of absorbing up to 20% water and, therefore, accurate partition coefficients cannot be easily obtained. Furthermore, as the solubility of GA glucosyl derivatives is dominated by the hydroxyl groups on the sugar moiety, rather than the degree of ionization of the C-7 carboxyl group, pH does not control partition characteristics to the same extent as it does with free GAs. These observations explain why clear cut separations are not obtained between "neutral" and "acidic" butanol fractions and why partition characteristics cannot be used to distinguish between GA glucosides and GA glucosyl esters (Takahashi, 1974). This point is strikingly illustrated with data obtained by Stoddart and Jones (1977) when partitioning "neutral" GA_1 glucosyl ester and "acidic" GA_3-O-glucoside against

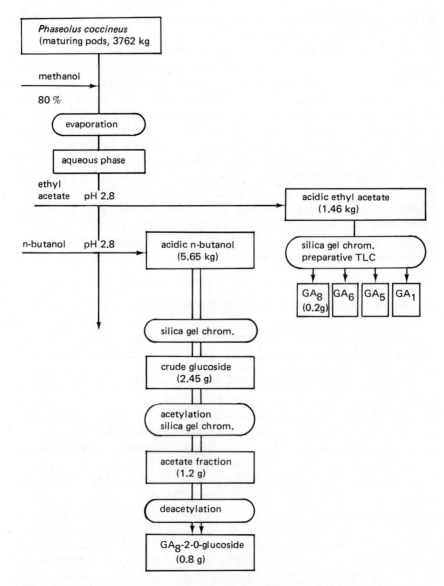

Figure 6.8 Scheme of isolation of GA$_8$-2-O-β-D-glucoside (*9*) from maturing pods of *Phaseolus coccineus* L. (Schreiber et al., 1970).

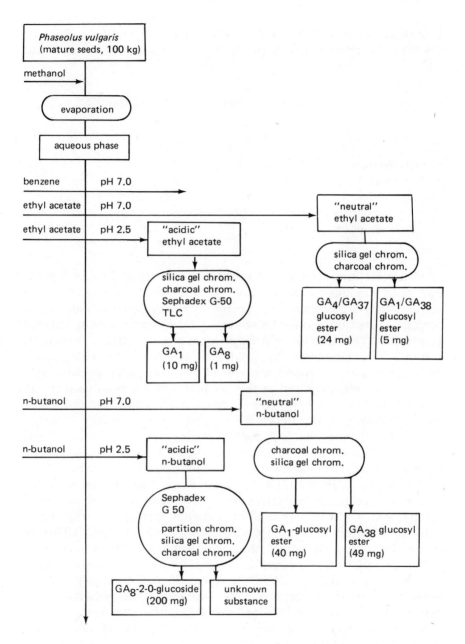

Figure 6.9 Scheme of isolation of GA glucosyl esters from mature seeds of *Phaseolus vulgaris* (according to Hiraga et al., 1974a).

Table 6.5.   Distribution of GA_1-β-D-glucosyl ester and GA_3-3-O-β-D-glucoside between aqueous phase, ethyl acetate, and n-butanol at different pH levels

	GA_1-glucosyl ester (%)	GA_3-3-O-glucoside (%)
Acidic ethyl acetate	7.1	4.9
Neutral n-butanol	38.5	33.7
Acidic n-butanol	50.2	59.0
Aqueous residue	4.2	2.4

After Stoddart & Jones (1977).

ethyl acetate and n-butanol. The data presented in Table 6.5 show that both conjugates are randomly distributed in the neutral and acidic butanol fractions. It is therefore recommended that a total conjugate fraction should be collected by extraction with butanol at pH 2.5. This fraction can subsequently be separated in neutral and acidic components by mild ion exchange chromatography. Although it has not found widespread use, ion exchange chromatography is an effective tool in this context and has been used in the isolation of GA_9 glucosyl ester (Lorenzi et al., 1976) and GA glucosides (Asakawa et al., 1974). The technique is especially effective with a DEAE-Sephadex A-25 support, as recoveries are over 90% and a whole range of neutral and acidic polar GA conjugates can be separated (Gräbner, Schneider, & Sembdner, 1976; Dathe, 1977; Dathe et al., 1978a, 1978b). Further steps that can be used for purifying GA conjugates in impure extracts include charcoal chromatography, which is routinely used by Japanese workers in conjunction with silica gel adsorption chromatography (Yokota et al., 1980). It should be noted, however, that irreversible adsorption processes and/or degradation can occur on the surface of such active supports, and this may adversely affect the recovery of GA conjugates. The same is true of preparative TLC. Partition chromatography on Sephadex G-50 is an additional procedure that has been used to purify GA conjugates in acidic extracts (Yokota et al., 1971a; Hiraga et al., 1974a).

Derivatization can be a very versatile tool through which further purification can be mediated. Acetylation and subsequent deacetylation have been used in the isolation of GA_8-2-O-glucoside from *Ph. coccineus* (see Fig. 6.8) (Schreiber et al., 1970). To be effective, derivatization reactions must produce good yields in absence of spurious side reactions. GAs are relatively fragile compounds, and so all techniques including derivatization procedures that are used in the isolation of GA conjugates must be thoroughly investigated; otherwise, artifacts and poor recoveries may prove to be recurrent problems. The particular combination of techniques used is very much a matter of trial and error and will depend upon the GA conjugate under investigation and the tissue from which it originates.

Evidence for the occurrence of GA conjugates has often been based on the liberation and identification of the parent GA after either enzymic, acid, or base hydrolysis. Unfortunately, this provides no information on the nature of the conjugating moiety and its point of attachment to the GA; in addition, the efficiency of hydrolytic cleavage is usually undetermined. Individual GA conjugates exhibit widely varying susceptibility to enzyme cleavage (see Table 6.2). Conditions for mild acid and base hydrolysis may be too weak to release the free GA, while stronger conditions can result in the formation of rearrangement and degradation products (Barendse & DeKlerk, 1975). In general, results obtained through these procedures should be considered to be of a very preliminary nature.

6.7.2 Identification of gibberellin conjugates

Reliable identification of trace quantities of endogenous plant hormones and their conjugates involves sample purification followed by chromatographic and/or spectroscopic analysis. Procedures that can be employed to verify the accuracy of such analyses have been proposed by Reeve and Crozier (1980) and are discussed in this volume by Crozier and Durley (Chapter 8). In the following section some of the most widely used chromatographic and spectroscopic procedures will be discussed. It should be noted that the amount of sample that is available has a major bearing on the procedures that can be employed. "Classical" methods such as melting point, optical rotation, elementary analyses, etc. can be used with large samples (Schreiber et al., 1970; Yokota et al., 1971b; Schneider et al., 1977b) but are of distinctly limited value when attempting to identify trace quantities of GA conjugates.

6.7.2.1 Chromatographic methods

Column and layer chromatography. Column chromatography on a silica gel support provides a group separation of GA conjugates (Yokota et al., 1980), whereas ion exchange chromatography on a DEAE Sephadex A-25 support (Gräbner et al., 1976) differentiates between neutral and acidic substances, which can be an important facet of the identification procedure. The same is true of electrophoretic separations, which resolve compounds of different charge (Schreiber et al., 1970; Stoddart & Jones, 1977; Nadeau & Rappaport, 1974; Nash, Jones, & Stoddart, 1978). TLC, in combination with a bioassay or a radioactivity detector has been widely used to analyze GA conjugates as "cold" or labeled compounds (Table 6.1). Yokota et al. (1980) recently summarized TLC and PC retention characteristics of some naturally occurring GA-O-glucosides and GA glucosyl esters. Because the individual GA glucosyl derivatives exhibit similar polarities, they do not have distinctive TLC and PC characteristics. As a consequence, neither technique is especially diagnostic and they are of limited value for identification purposes.

Gas chromatography. GC, especially in combination with MS, has proved to be one of the most powerful methods for identifying free GAs; however, it has been used only occasionally for the identification of GA conjugates because it is very difficult to transform GA conjugates to derivatives that are both volatile and stable under GC conditions.

Trimethylsilylation of GA-O-glucosides, their methyl esters, and GA glucosyl esters yields trimethylsilyl (TMSi) derivatives that chromatograph on QF-1 and OV-1 columns (Tables 6.6, 6.7) (Hiraga et al., 1974c; Schneider, Jänicke, & Sembdner, 1975). TMSi-[^{3}H]GA$_8$ glucoside methyl ester has been identified by GC-RC as a metabolite of [^{3}H]GA$_1$ in bean seeds by Nadeau and Rappaport (1972). However, because of their sheer size these derivatives are far from ideal candidates for GC, and analyses are typified by broad peaks and poor detection limits. These factors also adversely affect the discriminating power of GC-MS, and a further complicating factor is that the high molecular weight of TMSi-derivatives of GA conjugates (850–1,000 daltons) is beyond the mass range of many mass spectrometers. Recently, Rivier et al. (1981) chromatographed permethylated GA derivatives on an OV-1 capillary GC column. GA$_3$-3-O-glucoside, GA$_8$-2-O-glucoside, as well as the glucosyl ester of GA$_1$ and GA$_4$ were also permethylated and similarly analyzed. Some hydrolysis was observed when glucosyl esters were derivatized, and 3β-hydroxy

Table 6.6. GC retention times of GA glucosyl ester TMSi derivatives

	2% QF 1[a]	2% OV 1[b]
GA_1 glucosyl ester	14.8	18.8
GA_3 glucosyl ester	16.5	20.7
GA_4 glucosyl ester	13.1	14.7
GA_{37} glucosyl ester	22.0	23.0
GA_{38} glucosyl ester	25.7	28.0

After Hiraga et al. (1974).
[a] 3 mm × 1 m, 224°C, 34 ml N_2 min^{-1}.
[b] 3 mm × 1 m, 243°C, 33 ml N_2 min^{-1}.

groups were epimerized to some extent to 3α-compounds under the reaction conditions. In general, permethylated derivatives of GA conjugates appear to be superior to TMSi derivatives because of their thermostability and lower molecular weight. They should therefore be favored for GC-MS analyses.

High-performance liquid chromatography. Because of its high separating power and ability to resolve underivatized samples, high-performance liquid chromatography (HPLC) seems to be

Table 6.7. Standardized GC retentions of GA-O-glucoside TMSi derivatives and their methyl esters on 3% QF 1*

	$R_{standard}$	
	TMSi of methyl ester	TMSi of TMSi ester
GA_1-3-O-glucoside	2.45	2.49
GA_1-13-O-glucoside	2.37	2.42
GA_3-3-O-glucoside	2.20	2.01
GA_3-13-O-glucoside	1.76	1.57
Iso-GA_3-3-O-glucoside	2.01	1.94
Iso-GA_3-13-O-glucoside	1.51	1.88
GA_4-3-O-glucoside	2.00	—
Progesterone	1.00 (R_t = 6.4 min)	—

After Schneider et al. (1975); G. Schneider (unpublished).
*3 mm × 1.5m, 245°C, 175 ml N_2 min^{-1}.

the method with most potential for the analysis of hydrophilic GA conjugates. Reverse-phase columns eluted isocratically with a methanol-buffer mobile phase can readily separate very closely related conjugates such as GA_1 and GA_3 glucosides and GA-3-*O*- and GA-13-*O*-glucosides (Table 6.8). As little as 10 ng of conjugate can be detected with an absorbance monitor operating in the 200–220-nm range and, in addition, sample recoveries are high (Yamaguchi et al., 1979; G. Schneider and A. Crozier, unpublished data). It is envisaged that there are two main ways in which HPLC can be utilized in the analysis of endogenous GA conjugates. The first is as a purification procedure prior to derivatization and GC-MS, an approach adopted by Yamaguchi et al. (1979) when analyzing GA_3-3-*O*-glucoside (*3*) and gibberellenic acid glucoside (*21*) in extracts from immature seed of *Quamoclit pennata*. However, in the long term, HPLC is likely to find increased use as an analytical tool with GA conjugates being derivatized to both extend the range of chromatographic systems that can be utilized and lower the limits of detection. The use of fluorescence derivatives, for instance, offers the possibility of detection limits in the low picogram range. In due course, HPLC directly coupled to MS is likely to become a most powerful tool in the analysis of trace quantities of endogenous GA conjugates because it combines high chromatographic resolution with a sensitive and selective detector.

6.7.2.2 Spectroscopic methods

Mass spectrometry.   Despite the high molecular weight and instability of GA conjugates, especially glucosyl derivatives, MS has been successfully used to establish the structures of many natural and synthetic GA conjugates. In most cases, electron impact spectra have been obtained from either methyl esters or acetates of GA-*O*-glucosides and GA glucosyl esters (Weiland, 1968; Schreiber et al., 1970; Sembdner et al., 1980). However, anion MS can also be used, and Table 6.9 demonstrates that even methyl esters of GA glucosides exhibit a pronounced molecular ion. The spectra of GA glucoside methyl esters and their acetates without an olefinic bond in ring A contain a M-1 fragment rather than a molecular ion. This may be a useful indicator of ring A structure (Schneider et al., 1977b; Schneider, 1980, 1981). Chemical ionization and field desorption techniques have not yet been applied to GA conjugates.

The electron impact mass spectra of the TMSi derivatives of six GA glucosides and five GA glucosyl esters have been investigated

Table 6.8. Retention times (min) of GA-O-glucosides and GA glucosyl esters on reverse phase HPLC

	Wakogel LC ODS-1OH[a] (2 × 250 mm) % MeOH in NH$_4$Cl buffer				MCH-5 Varian[b] (4 × 300 mm) % MeOH in Tris/ HCl buffer	
	5%[c]	10%	20%	30%	25%	50%
GA$_1$-3-O-glucoside	—	19.0	4.7	—	21.3	3.65
GA$_1$-13-O-glucoside	—	12.4	3.4	—	16.3	3.45
GA$_3$-3-O-glucoside	19.0	16.8	4.1	2.7	19.4	3.5
GA$_3$-13-O-glucoside	—	—	—	—	13.6	3.2
Gibberellenic acid 3-O-glucoside	4.9	9.5	—	—	—	—
GA$_4$-3-O-glucoside	—	—	—	—	—	20.6
GA$_5$-13-O-glucoside	—	—	—	—	—	6.0
GA$_7$-3-O-glucoside	—	—	—	—	—	23.1
GA$_8$-2-O-glucoside	4.2	4.5	1.7	1.3	7.1	2.9
GA$_8$-13-O-glucoside	—	—	—	—	5.4	2.6
GA$_{26}$-2-O-glucoside	6.0	—	—	2.4	—	—
GA$_{29}$-2-O-glucoside	3.3	4.2	1.6	1.3	—	—
GA$_{35}$-11-O-glucoside	10.5	27.0	5.8	3.9	—	—
GA$_1$ glucosyl ester	—	—	—	3.9	—	—
GA$_3$ glucosyl ester	—	—	—	3.4	—	—
GA$_4$ glucosyl ester	—	—	—	>25.0	—	—
GA$_5$ glucosyl ester	—	—	—	—	—	8.5
GA$_{37}$ glucosyl ester	—	—	—	>25.0	—	—
GA$_{38}$ glucosyl ester	—	—	—	3.3	—	—

[a] Yamaguchi et al. (1979) pH 3.2; 1 ml min^{-1}, detection absorbance at 200–210 nm

[b] Schneider & Crozier, unpublished data; pH 3.0; 1 ml min^{-1}, detection absorbance at 205 nm

[c] pH 6.0.

by Yokota et al. (1975). Despite molecular weights of between 854 and 972 dalton, molecular ions were observed in all spectra. The fragmentation pattern of GA glucosyl esters seems to be dominated by the glucose moiety with the base peak at M-378 m/e (R-COOTMSi$^+$). This ion has a low intensity in the spectra of GA glucosides that are influenced to a much greater degree by the parent GA. This is especially noticeable in the spectra of GA$_8$-2-O-glucoside (9) and

Table 6.9. Molecular ions (m/e; in parentheses: % abundance) of synthesized GA-O-glucoside methyl esters and their peracetylated derivatives in (A) electron impact and (B) anion mass spectrometry (Mass Spectrograph Manfred v. Ardenne)

	Derivative			
	Methyl ester		Peracetylated methyl ester	
GA-glucoside	A	B	A	B
GA$_1$-3-O-glucoside	524(4)	523(15)	734(17)	733(31)
GA$_1$-13-O-glucoside	524(6)	523(30)	734(16)	—
GA$_3$-3-O-glucoside	—	522(2)	732(10)	732(20)
GA$_3$-13-O-glucoside	—	522(100)	732(5)	732(22)
GA$_4$-3-O-glucoside	508(7)	507(28)	676(2)	675(20)
GA$_5$-13-O-glucoside	506(5)	506(40)	—	673(72)
GA$_7$-3-O-glucoside	506(20)	506(5)	674(20)	674(25)
GA$_8$-2-O-glucoside	—	539(34)	—	—
Iso-GA$_3$-3-O-glucoside	—	522(100)	—	—
Iso-GA$_3$-13-O-glucoside	522(10)	522(100)	732(5)	732(55)

After Schneider (1981).

GA$_{27}$-2-O-glucoside (*12*) where the presence of a 3-O-TMSi group increases the relative abundance of the presumptive aglucone fragment to such an extent that the M-421 ion constitutes the base peak of the spectra.

Electron impact mass spectra of permethylated GA-O-glucosides and glucosyl esters have been studied by Rivier et al. (1981). Molecular ions were not obtained in all instances.

^{1}H-NMR spectroscopy. Recent technical innovations on this field have enhanced sensitivity to such an extent that the method can now be seriously considered for trace analysis. With regard to GA glucosyl conjugates, ^{1}H-NMR can help establish the position of attachment of the conjugating unit as well as structural features of both the aglucone and the sugar component. The downfield shift of 17-H$_2$ signals induced by neighboring glucose residues (see Table 6.10) may be a characteristic of GA-13-O-glucosides (Schneider, 1980, 1981).

Fluorometry.   Sulphuric acid-induced fluorescence is a well-established method for detecting GAs and GA conjugates on TLC plates. However, selection of suitable reaction conditions is of crucial importance if accurate and reproducible results are to be obtained. Maximum fluorescence is achieved with GA_3-3-O-glucoside and GA_3 glucosyl ester by treatment with 85% sulphuric acid for 20 min at 80°C, while GA_1-13-O-glucoside requires 40 min at 80°C (Fig. 6.10). The long-term stability of the fluorophor and linearity of the response down to 1–10 ng (Fig. 6.11) facilitate the use of this procedure for the quantification of submicrogram levels of GA conjugates (Geiseler, 1977; Stoddart & Jones, 1977). In addition, other fluorescent derivatives, such as methoxycoumaryl esters (see Chapter 8) may prove to be of value to HPLC-based analyses of GA glucosides.

6.7.2.3   Bioassay

GA conjugates per se are considered to be biologically inactive (Section 6.5). The induction of a response by GA conjugates appears to be strongly dependent upon cleavage by hydrolytic enzymes located in either the test medium or the plant material used in the bioassay. In practice, standardizing and controlling these factors is very complicated and restricts the applicability of bioassays especially for accurate quantification of GA conjugates. Selected test systems with a relatively high response to GA conjugates can, however, be

Table 6.10. ^{1}H-NMR δ-values (ppm) of 17-H_2 signals (2m) of GA-3-O- and GA-13-O-glucosides (acetone-D_6/D_2O TMSi intern, 100 MHz)

GA moiety	3-O-Glucoside	13-O-Glucoside
GA_1 -	4.88/5.22	5.00/5.35
GA_3	4.95/5.24	4.97/5.34
GA_4 -	4.89/4.99	—
GA_5	—	4.94/5.39
GA_7 -	4.89/5.00	—
GA_8	4.94/5.19*	5.03/5.38
Iso-GA_3 -	4.93/5.09	5.04/5.27

After Schneider (1981).
* 2-O-glucoside.

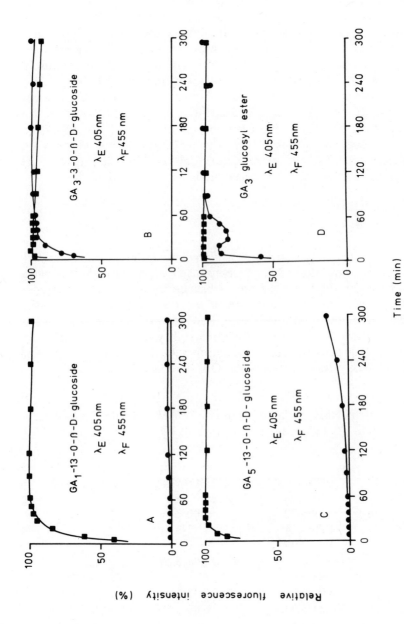

Figure 6.10 Dependence of fluorophor formation of GA conjugates upon temperature and time (85% sulphuric acid; —●— 20°C; —■— 80°C; λ_E: excitation wavelength; λ_F: fluorescence wavelength).

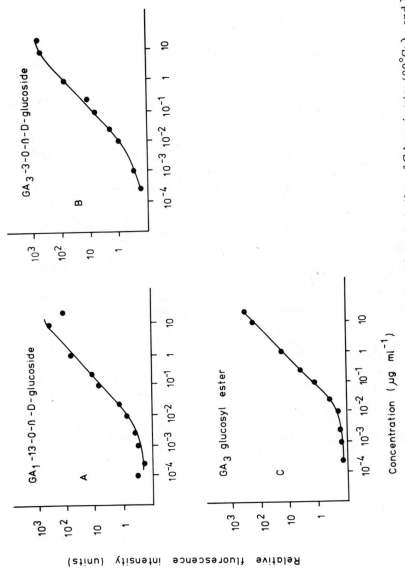

Figure 6.11 Relationship of fluorescence intensity and concentration of GA conjugates ($80°C$; λ_E and λ_F as Fig. 6.10).

441

of value in qualitative analyses. The nonsterile rice seedling bioassay, for instance, can be used to monitor activity during the purification of unknown endogenous GA glucosyl esters. It is of more limited value with GA glucosides because they exhibit much lower levels of biological activity.

6.7.3 Quantification of gibberellin conjugates

A greater understanding of the physiological role of GA conjugates is at least partially dependent upon the availability of reliable quantitative analyses of GA glucosides and GA glucosyl esters. Unfortunately, as yet, this has not been achieved and even methods for qualitative analysis leave much to be desired. Most procedures provide data that lack both precision and accuracy, primarily because of difficulties in purifying and analyzing trace quantities of polar GA conjugates in highly impure plant extracts. These problems are not insurmountable and could possibly be solved by the analytical scheme proposed in Fig. 6.12, which is based on experience obtained with preparative scale isolations and incorporates recent methodological developments. After partitioning with ethyl acetate and *n*-butanol at pH 2.5, ion exchange chromatography serves to separate acidic and neutral compounds. This is followed by steric exclusion chromatography, where the degree of purification is determined by the molecular weight distribution of components in the sample (Crozier, Zaerr, & Morris, 1980). HPLC then affords purified samples which can be either applied to capillary GC-MS after derivatization or chromatographed in further HPLC systems prior to analysis by either capillary GC-MS, fluorometry or perhaps even immunoassay. An appropriate internal marker should be used so that compensation can be made for sample losses that are encountered. In all instances, accuracy should be assessed by the procedures of Reeve and Crozier (1980).

6.8 CONCLUDING REMARKS

GA conjugation is an integral part of GA metabolism. The endogenous occurrence and biosynthesis of GA conjugates in plant tissues, especially in seeds and seedlings, have been extensively studied. Despite the apparent metabolic importance of conjugation, in

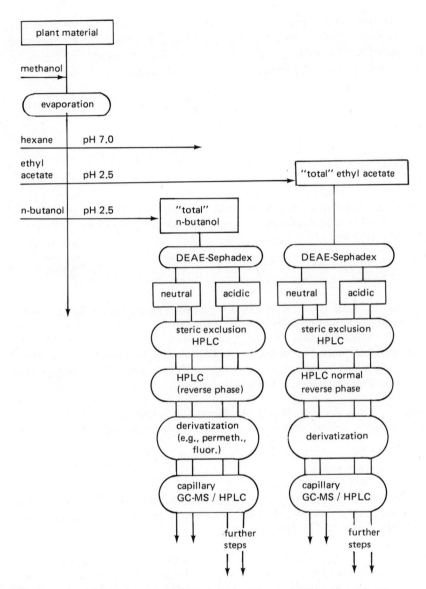

Figure 6.12 Proposed general scheme for the isolation, identification, and quantification of GA conjugates.

general the physiological role(s) of GA conjugates are still poorly understood. The development of reliable and sensitive methods for the quantitative analysis of endogenous GA conjugates is likely to be a key step toward further progress in this area.

REFERENCES

Adam, G., Lischewski, M., Sych, F.-J., & Ulrich, A. (1976). Synthese von neutralen und basischen Amiden der Gibberellin-A_3-und A_1-Reihe. J. Prakt. Chem. *318*, 105-115.

Adam, G., Lischewski, M., Sych, F.-J., & Ulrich, A. (1977). Synthese von Gibberellin-Aminosäure-Konjugaten. Tetrahedron *33*, 95-100.

Andreae, W. A., & Good, N. E. (1955). The formation of indole-acetyl aspartic acid in pea seedlings. Plant Physiol. *30*, 380-382.

Asakawa, Y., Tamari, K., Shoji, A., & Kaji, J. (1974). Metabolic products of gibberellin A_3 and their interconversion in dwarf kidney bean plants. Agr. Biol. Chem. *38*, 719-725.

Aung, L. H., & Rees, A. R. (1974). Changes in endogenous gibberellin levels in *Tulipa* bulblets during ontogeny. J. Exp. Bot. *25*, 745-751.

Barendse, G. W. M. (1971). Formation of bound gibberellins in *Pharbitis nil*. Planta *99*, 290-301.

Barendse, G. W. M. (1974). Accumulation and metabolism of radioactive gibberellic acid in seedlings of *Pharbitis nil* Chois. In: Plant Growth Substances 1973, pp. 332-341. Hirokawa, Tokyo.

Barendse, G. W. M., & DeKlerk, G. J. M. (1975). The metabolism of applied gibberellic acid in *Pharbitis nil* Choisy: tentative identification of its sole metabolite as gibberellic acid glucoside and some of its properties. Planta *126*, 25-35.

Barendse, G. W. M., Kende, H., & Lang, A. (1968). Fate of radioactive gibberellin A_1 in maturing and germinating seeds of peas and Japanese morning glory. Plant Physiol. *43*, 815-822.

Bartlett, P. A., & Johnson, W. S. (1970). An improved reagent for the O-alkyl cleavage of methyl esters by nucleophilic displacement. Tetrahedron Lett., 4459-4462.

Beneš, I., & Liebisch, H.-W. (1983). Darstellung von Glycosylestern des Gibberellin A_3. Z. Chem. (in press).

Bernhardt, D., Köhler, K.-H., & Sembdner, G. (1979). Die Aktivität von einigen freien und konjugierten Gibberellinen in *Triticum*-Halbkaryopsen- und *Amaranthus*-Gibberellintest. Biochem. Physiol. Pflanz. *174*, 607-615.

Borgmann, E. (1977). Untersuchungen über die biologische Aktivität und den Stoffwechsel konjugierter Gibberelline. Dissert.-Schrift, Martin-Luther-Universität, Halle-Wittenberg.

Crozier, A., Bowen, D. H., MacMillan, J., Reid, D. M., & Most, B. H. (1971). Characterization of gibberellins from dark-grown *Phaseolus coccineus* seedlings by gas liquid chromatography and combined gas chromatography-mass spectrometry. Planta *97*, 142-154.

Crozier, A., Kuo, C. C., Durley, R. C., & Pharis, R. P. (1970). The biological activities of 26 gibberellins in nine bioassays. Can. J. Bot. *48*, 867-877.

Crozier, A., Zaerr, J. B., & Morris, R. O. (1980). High performance steric exclusion chromatography of plant hormones. J. Chromatogr. *198*, 57-63.

Dathe, W. (1977). Untersuchungen über endogene Gibberelline und Inhibitoren während der Ontogenese des Roggens (*Secale cereale* L.). Dissert.-Schrift, Akademie der Wissenschaften der DDR.

Dathe, W., Schneider, G., & Sembdner, G. (1978a). Endogenous gibberellins and inhibitors in caryopses of rye. Phytochem. *17*, 963-966.

Dathe, W., Sembdner, G., Kefeli, V. I., & Vlasov, P. V. (1978b). Gibberellins, abscisic acid, and related inhibitors in branches and bleeding sap of birch (*Betula pubescens* Ehrh.). Biochem. Physiol. Pflanz. *173*, 238-248.

Davies, L. J., & Rappaport, L. (1975a). Metabolism of tritiated gibberellins in d-5-dwarf maize. I. In excised tissues and intact dwarf and normal plants. Plant Physiol. *55*, 620-625.

Davies, L. J., & Rappaport, L. (1975b). Metabolism of tritiated gibberellins in d-5 dwarf maize. II. [^{3}H]Gibberellin A_1, [^{3}H]gibberellin A_3 and related compounds. Plant Physiol. *56*, 60-66.

Durley, R. C., Pharis, R. P., & Zeevaart, J. A. D. (1975). Metabolism of [^{3}H]-gibberellin A_{20} by plants of *Bryophyllum daigremontianum* under long- and short-day conditions. Planta *126*, 139-149.

Durley, R. C., Railton, I. D., & Pharis, R. P. (1974). The metabolism of gibberellin A_1 and gibberellin A_{14} in seedlings of dwarf *Pisum sativum*. In: Plant Growth Substances, 1973, pp. 285-293. Hirokawa, Tokyo.

Durley, R. C., Sassa, T., & Pharis, R. P. (1979). Metabolism of tritiated gibberellin A_{20} in immature seeds of dwarf pea, cv. 'Meteor.' Plant Physiol. *64*, 214-219.

Frydman, V. M., & MacMillan, J. (1975). The metabolism of gibberellins A_9, A_{20}, and A_{29} in immature seeds of *Pisum sativum* cv. Progress No. 9. Planta *125*, 181-195.

Gaskin, P., & MacMillan, J. (1975). Polyoxygenated ent-kaurenes and water soluble conjugates in seed of *Phaseolus coccineus*. Phytochem. *14*, 1575-1578.

Gaskin, P., & MacMillan, J. (1978). GC and GC-MS techniques for gibberellins. In: Isolation of Plant Growth Substances, pp. 79-95, Hillman, J. R., ed., Cambridge University Press, Cambridge, London, New York, Melbourne.

Geiseler, G. (1977). Fluorimetrische Untersuchungen an Gibberellinen und Gibberellinkonjugaten. Ingenieurschrift, Ingenieur-Schule für Chemie, Berlin.

Gräbner, R., Schneider, G., & Sembdner, G. (1976). Fraktionierung von Gibberellinen, Gibberellinkonjugaten und anderen Phytohormonen durch DEAE-Sephadex-Chromatography. J. Chromatogr. *121*, 110-115.

Graebe, J. E., & Ropers, H. J. (1978). Gibberellins. In: Phytohormones and Related Compounds—A Comprehensive Treatise. The Biochemistry of Phytohormones and Related Compounds, Vol. 1, pp. 107-204, Letham, D. S., Goodwin, P. B., & Higgins, T. J. V., eds. Elsevier, Amsterdam.

Harada, H., & Yokota, T. (1970). Isolation of gibberellin A_8 glucoside from shoot apices of *Althaea rosea*. Planta *92*, 100-104.

Hashimoto, T., & Rappaport, L. (1966). Variations in endogenous gibberellins in developing bean seeds I. Occurrence of neutral and acidic substances. Plant Physiol. *41*, 623-628.

Hayashi, F., Boerner, D., Peterson, C. E., & Sell, H. M. (1971). The relative content of gibberellin in seedlings of gynoecious and monoecious cucumber (*Cucumis sativus*). Phytochem. *10*, 57-62.

Hedden, P., MacMillan, J., & Phinney, B. O. (1978). The metabolism of the gibberellins. Ann. Rev. Plant Physiol. *29*, 149-192.

Hemphill, D. D., Jr., Baker, L. R., & Sell, H. M. (1973). Isolation of novel conjugated gibberellins from *Cucumis sativus* seeds. Can. J. Biochem. *51*, 1647-1653.

Hiraga, K., Yokota, T., Murofushi, N., & Takahashi, N. (1972). Isolation and characterization of a free gibberellin and glucosyl esters of gibberellins in mature seeds of *Phaseolus vulgaris*. Agr. Biol. Chem. *36*, 345-347.

Hiraga, K., Yokota, T., Murofushi, N., & Takahashi, N. (1974a). Isolation and characterization of gibberellins in mature seeds of *Phaseolus vulgaris*. Agr. Biol. Chem. *38*, 2511-2520.

Hiraga, K., Yokota, T., Murofushi, N., & Takahashi, N. (1974b). Plant growth regulators in immature and mature seeds of *Phaseolus vulgaris*. In: Plant Growth Substances 1973, pp. 75-85. Hirokawa, Tokyo.

Hiraga, K., Yokota, T., & Takahashi, N. (1974c). Biological activity of some synthetic gibberellin glucosyl esters. Phytochem. *13*, 2317-2376.

Jones, D. F. (1964). Examination of the gibberellins of *Zea mays* and *Phaseolus multiflorus* using thin-layer chromatography. Nature (London) *202*, 1309-1310.

Keay, P. J., Moffatt, J. S., & Mulholland, T. P. C. (1965). Some functional derivatives and transformation products of gibberellic acid. J. Chem. Soc. 1605-1615.

Knöfel, H. D., Müller, P., & Sembdner, G. (1974). Studies on the enzymatical hydrolysis of gibberellin-*O*-glucosides. In: Biochemistry and Chemistry of Plant Growth Regulators, pp. 121-124, Schreiber, K., Schütte, H. R., & Sembdner, G., eds. Inst. Plant Biochem. Acad. Sci. GDR, Halle/S.

Knöfel, H.-D., Schwarzkopf, E., Müller, P., & Sembdner, G. (1983). Enzymic glucosylation of gibberellins. Planta (in preparation).

Koshimizu, K., Inui, M., Fukui, H., & Mitsui, T. (1968). Isolation of (+)-abscisyl-β-D-glucopyranoside from immature fruit of *Lupinus luteus*. Agr. Biol. Chem. *32*, 789-791.

Lance, B., Durley, R. C., Reid, D. M., Thorpe, T. A., & Pharis, R. P. (1976). Metabolism of [^{3}H]gibberellin A_{20} in light- and dark-grown tobacco callus cultures. Plant Physiol. *58*, 387-392.

Lang, A. (1970). Gibberellins: structure and metabolism. Ann. Rev. Plant Physiol. *21*, 537-570.

Lazer, L., Baumgartner, W. E., & Dahlstrom, R. V. (1961). Determination of endogenous gibberellins in green malt by isotopic derivative dilution procedures. J. Agr. Food Chem. *9*, 24–26.

Liebisch, H.-W. (1974). Uptake, translocation and metabolism of labelled GA_3 glucosyl ester. In: Biochemistry and Chemistry of Plant Growth Regulators, pp. 109–113, Schreiber, K., Schütte, H. R., & Sembdner, G., eds. Inst. Plant Biochem. Acad. Sci. GDR, Halle/S.

Liebisch, H.-W. (1980). Vergleichende Untersuchungen über den Stoffwechsel von GA_1, GA_3 und GA_9 in Zellsuspensionskulturen von *Lycopersicon esculentum* und verschiedenen intakten Pflanzen. Biochem. Physiol. Pflanz. *175*, 797–805.

Liebisch, H.-W., Schmidt, E., & Schütte, H. R. (1980). Verteilung radioaktiv markierter freier und konjugierter Gibberelline in abgeschnittenen Gerstenblättern. Biochem. Physiol. Pflanz. *175*, 148–153.

Lischewski, M., & Adam, G. (1976). Darstellung von Gibberellin A_3-oyl-2-^{14}C-glycin. Z. Chem. *16*, 357.

Lischewski, M., Adam, G., & Sych, F.-J. (1974). Synthesis of gibberellin amino acid conjugates. In: Biochemistry and Chemistry of Plant Growth Regulators, pp. 161–164, Schreiber, K., Schütte, H. R., & Sembdner, G., eds. Inst. Plant Biochem. Acad. Sci. GDR, Halle/S.

Lorenzi, R., Horgan, R., & Heald, J. K. (1976). Gibberellin A_9 glucosyl ester in needles of *Picea sitchensis*. Phytochem. *15*, 789–790.

McComb, A. J., (1961). "Bound" gibberellin in mature runner bean seeds. Nature (London) *192*, 575–576.

McInnes, A. G., Smith, D. G., Durley, R. C., Pharis, R. P., Arsenault, G. P., MacMillan, J., Gaskin, P., & Vining, L. C. (1977). Biosynthesis of gibberellins in *Gibberella fujikuroi*. Gibberellin A_{47}. Can. J. Biochem. *55*, 728–735.

Miersch, O., & Liebisch, H.-W. (1974). Utilization of enzyme preparations in the course of synthesis of radioactive labelled gibberellin A_3 and some derivatives. In: Biochemistry and Chemistry of Plant Growth Regulators, pp. 125–131, Schreiber, K., Schütte, H. R., & Sembdner, G., eds. Inst. Plant Biochem. Acad. Sci. GDR, Halle/S.

Miersch, O., & Liebisch, H.-W. (1983). Synthese von Gibberellinglucosylestern. Z. Chem. (in press).

Moffatt, J. S., & Radley, M. (1960). Gibberellic acid XI. The growth promoting activities of some functional derivatives of gibberellic acid. J. Sci. Food Agr. *11*, 386-390.

Mosettig, E., Beglinger, U., Dolder, F., Lichti, H., Quitt, P., & Waters, J. A. (1963). The absolute configuration of steviol and isosteviol. J. Amer. Chem. Soc. *85*, 2305-2309.

Müller, P., Knöfel, H.-D., & Kramell, R. (1976). Darstellung von Gibberellinsäure-[U-^{3}H]-amino-*n*-alkylamiden. Z. Chem. *16*, 105-106.

Müller, P., Knöfel, H.-D., Liebisch, H.-W., Miersch, O., & Sembdner, G. (1978). Untersuchungen zur Spaltung von Gibberellinglucosiden. Biochem. Physiol. Pflanz. *173*, 396-409.

Müller, P., Knöfel, H.-D., & Sembdner, G. (1974). Studies on the enzymatical synthesis of gibberellin-*O*-glucosides. In: Biochemistry and Chemistry of Plant Growth Regulators, pp. 115-119, Schreiber, K., Schütte, H. R., & Sembdner, G., eds. Inst. Plant Biochem. Acad. Sci. GDR, Halle/S.

Murakami, Y. (1961a). Formation of gibberellin A_3 glucoside in plant tissues. Bot. Mag. (Tokyo) *74*, 424-425.

Murakami, Y. (1961b). Paper-chromatographic studies on change in gibberellins during seed development and germination in *Pharbitis nil.* Bot. Mag. (Tokyo) *74*, 241-247.

Murakami, Y. (1962). Occurrence of "water-soluble" gibberellin in higher plants. Bot. Mag. (Tokyo) *75*, 451-452.

Musgrave, A., & Kende, H. (1970). Radioactive gibberellin A_5 and its metabolism in dwarf peas. Plant Physiol. *45*, 50-61.

Nadeau, R., & Rappaport, L. (1972). Metabolism of gibberellin A_1 in germinating bean seeds. Phytochem. *11*, 1611-1616.

Nadeau, R., & Rappaport, L. (1974). An amphoteric conjugate of [^{3}H]-gibberellin A_1 from barley aleurone layers. Plant Physiol. *54*, 809-812.

Nadeau, R., Rappaport, L., & Stolp, C. F. (1972). Uptake and metabolism of [^{3}H]-gibberellin A_1 by barley aleurone layers: response to abscisic acid. Planta *107*, 315-324.

Nash, L., & Crozier, A. (1975). Translocation and metabolism of [^{3}H]gibberellins by light-grown *Phaseolus coccineus* seedlings. Planta *127*, 221-231.

Nash, L. J., Jones, R. L., & Stoddart, J. L. (1978). Gibberellin metabolism in excised lettuce hypocotyls. Response to GA_9 and the conversion of [^{3}H]GA_9. Planta *140*, 143-150.

Ogawa, Y. (1963). Gibberellin-like substances occurring in the seed of *Pharbitis nil* Chois and their changes in contents during the seed development. Plant Cell Physiol. (Tokyo) *4*, 217-225.

Ogawa, Y. (1966a). Acid, neutral, and "water-soluble" gibberellin-like substances occurring in developing seed of *Sechium edule*. Bot. Mag. (Tokyo) *79*, 1-6.

Ogawa, Y. (1966b). Ethyl acetate-soluble, and "water-soluble" gibberellin-like substances in the seeds of *Pharbitis nil, Lupinus luteus* and *Prunus persica*. Bot. Mag. (Tokyo) *79*, 69-76.

Ogawa, Y., & Takahashi, N. (1974). Comparative biological effectiveness of gibberellin A_3 glucoside and gibberellin A_3. Bull. Fac. Agr., 261-267.

Parker, C. W., Letham, D. S., Cowley, D. E., & McLeod, J. K. (1972). Raphanatin, an unusual purine derivative and a metabolite of zeatin. Biochem. Biophys. Res. Commun. *49*, 460-466.

Pegg, G. F. (1966). Changes in levels of naturally occurring gibberellin-like substances during germination of seed of *Lycopersicon esculentum* Mill. J. Exp. Bot. *17*, 214-230.

Radley, M. (1958). The distribution of substances similar to gibberellic acid in higher plants. Ann. Bot. *22*, 297-307.

Railton, I. D., Durley, R. C., & Pharis, R. P. (1974a). Metabolism of tritiated gibberellin A_9 by shoots of dark-grown dwarf pea, cv. Meteor. Plant Physiol. *54*, 6-12.

Railton, I. D., Durley, R. C., & Pharis, R. P. (1974b). Studies on gibberellin biosynthesis in etiolated shoot of dwarf pea cv. Meteor. In: Plant Growth Substances 1973, pp. 294-304. Hirokawa, Tokyo.

Rappaport, L., & Adams, D. (1978). Gibberellins: Synthesis, compartmentation and physiological process. Phil. Trans. Roy. Soc. Lond. Ser. B *284*, 521-539.

Rappaport, L., Davies, L., Lavee, S., Nadeau, R., Patterson, R., & Stolp, C. F. (1974). Significance of metabolism of [^{3}H]GA_1 for plant regulation. In: Plant Growth Substances 1973, pp. 314-324. Hirokawa, Tokyo.

Reeve, D., & Crozier, A. (1974). An assessment of gibberellin structure activity relationships. J. Exp. Bot. *25*, 431-445.

Reeve, D., & Crozier, A. (1980). Quantitative analysis of plant hormones. In: Hormonal Regulation of Development I. Molecular Aspects of Plant Hormones. Encyclopedia of Plant Physiology New Series, Vol. 9, pp. 203–280, MacMillan, J., ed. Springer-Verlag, Berlin, Heidelberg, New York.

Reinhard, E., & Konopka, W. (1967). Ein Beitrag zur Analyse der Gibberelline im Samen von *Pisum sativum* L. Planta 77, 58–76.

Reinhard, E., & Sacher, R. (1967a). Über die Gibberelline in Samen und Kapseln von *Pharbitis purpurea*. I. Analyse der Gibberelline. Z. Pflanz. 58, 138–150.

Reinhard, E., & Sacher, R. (1967b). Über die Gibberelline in Samen und Kapseln von *Pharbitis purpurea*. II. Die Gibberelline in verschiedenen Entwicklungsstadien. Z. Pflanz. 58, 151–164.

Reinhard, E., & Sacher, R. (1967c). Versuche zum enzymatischen Abbau der "gebundenen" Gibberelline von *Pharbitis purpurea*. Experientia 23, 415–416.

Rivier, L., Gaskin, P., Albone, K. S., & MacMillan, J. (1981). GC-MS identification of endogenous gibberellins and gibberellin conjugates as their permethylated derivatives. Phytochem. 20, 687–692.

Schliemann, W., & Schneider, G. (1979). Untersuchungen zur enzymatischen Hydrolyse von Gibberellin-O-glucosiden. I. Hydrolysegeschwindigkeiten von Gibberellin-13-O-glucosiden. Biochem. Physiol. Pflanz. 174, 738–745.

Schneider, G. (1972). Partialsynthese von Gibberellin-A_1-O(3)-β-D-glucopyranosid. Tetrahedron Lett., 4053–4054.

Schneider, G. (1978). Synthesen radioaktiv markierter Verbindungen. Darstellung von GA_3-O(3)-β-D-glucopyranosid-U-^{3}H. Z. Chem. 18, 217.

Schneider, G. (1980). Über strukturelle Einflüsse bei der Glucosylierung von Gibberellinen. Tetrahedron 37, 545–549.

Schneider, G. (1981). Synthese von Gibberellinglucosiden. Dissert.-Schrift B, Akad. Wiss. der DDR.

Schneider, G., Jänicke, S., & Sembdner, G. (1975). Beitrag zur Gaschromatographie von Gibberellinen und Gibberellin-O-glucosiden–N,O-Bis(trimethylsilyl)acetamid als Silylierungsreagens. J. Chromatogr. 109, 409–412.

Schneider, G., Miersch, O., & Liebisch, H.-W. (1977a). Synthese von O-β-D-

Glucopyranosyl-gibberellin-*O*-β-D-glucopyranosylestern. Tetrahedron Lett., 405–406.

Schneider, G., & Schliemann, W. (1979). Untersuchungen zur enzymatischen Hydrolyse von Gibberellin-*O*-glucosiden. II. Hydrolysegeschwindigkeiten von Gibberellin-2-*O*- und Gibberellin-3-*O*-glucosiden. Biochem. Physiol. Pflanz. *174*, 746–751.

Schneider, G., Sembdner, G., & Schreiber, K. (1974). Zur Synthese von Gibberellin-A₃-β-D-glucopyranosiden. Z. Chem. *14*, 474–475.

Schneider, G., Sembdner, G., & Schreiber, K. (1977b). Synthese von *O*(3)- und *O*(13)-glucosylierten Gibberellinen. Tetrahedron *33*, 1391–1397.

Schreiber, K., Schneider, G., Sembdner, G., & Focke, I. (1966). Isolierung von *O*(2)-Acetyl-gibberellinsäure als Stoffwechsel-produkt von *Fusarium moniliforme* Sheld. Phytochem. *5*, 1221–1225.

Schreiber, K., Weiland, J., & Sembdner, G. (1967). Isolierung und Struktur eines Gibberellinglucosids. Tetrahedron Lett., 4285–4288.

Schreiber, K., Weiland, J., & Sembdner, G. (1969). Synthese von *O*(2)-β-D-Glucopyranosylgibberellin-A₃-methylester. Tetrahedron *25*, 5541–5545.

Schreiber, K., Weiland, J., & Sembdner, G. (1970). Isolierung von Gibberellin-A₈-*O*(3)-β-D-glucopyranosid aus Früchten von *Phaseolus coccineus*. Phytochem. *9*, 189–198.

Sembdner, G. (1969). Untersuchungen über Vorkommen, Stoffwechsel und biologische Wirksamkeit von Gibberellinen. Habilitationsschrift, Martin-Luther-Universität, Halle-Wittenberg.

Sembdner, G. (1974). Conjugation of plant hormones. In: Biochemistry and Chemistry of Plant Growth Regulators, pp. 283–302, Schreiber, K., Schütte, H. R., & Sembdner, G., eds. Inst. Plant Biochem. Acad. Sci. GDR, Halle/S.

Sembdner, G., Adam, G., Lischewski, M., Sych, F.-J., Schulze, C., Knöfel, H.-D., Müller, P., Schneider, G., Liebisch, H.-W., & Schreiber, K. (1974). Biological activity and metabolism of conjugated gibberellins. In: Plant Growth Substances 1973, pp. 349–355. Hirokawa, Tokyo.

Sembdner, G., Borgmann, E., Schneider, G., Liebisch, H.-W., Miersch, O., Adam, G., Lischewski, M., & Schreiber, K. (1976). Biological activity of some conjugated gibberellins. Planta *132*, 249–257.

Sembdner, G., Gross, D., Liebisch, H.-W., & Schneider, G. (1980). Biosynthesis and metabolism of plant hormones. In: Hormonal Regulation of Development I. Molecular Aspects of Plant Hormones. Encyclopaedia of Plant Physiology New Series, Vol. 9, pp. 281-444, MacMillan, J., ed. Springer-Verlag, Berlin, Heidelberg, New York.

Sembdner, G., Schneider, G., Weiland, J., & Schreiber, K. (1964). Über ein gebundenes Gibberellin aus *Phaseolus coccineus* L. Experientia *20*, 89-90.

Sembdner, G., & Schreiber, K. (1965). Gibberelline. IV. Über die Gibberelline in *Nicotiana tabacum* L. Phytochem. *4*, 49-56.

Sembdner, G., Schulze, C., Borgmann, E., Adam, G., Lischewski, M., Schneider, G., Liebisch, H.-W., Miersch, O., & Schreiber, K. (1973). Biologische Aktivität von Gibberellin-Derivaten. In: Wirkungsmechanismen von Herbiziden und synthetischen Wachstums-regulatoren. Part 10, pp. 206-212, Barth, A., Jacob, F., & Feyerabend, G., eds. Wissenschaftliche Beiträge der Martin-Luther-Universität, Halle.

Sembdner, G., Weiland, J., Aurich, O., & Schreiber, K. (1968). Isolation, structure and metabolism of a gibberellin glucoside. In: Plant Growth Regulators, pp. 70-86. S.C.I. Monograph, Vol. 31, London.

Sembdner, G., Weiland, J., Schneider, G., Schreiber, K., & Focke, I. (1972). Recent advances in the metabolism of gibberellins. In: Plant Growth Substances 1970, pp. 145-150, Carr, D. J., ed. Springer-Verlag, Berlin, Heidelberg, New York.

Sponsel, V. M. (1979). Metabolism of gibberellins in immature seeds of *Pisum sativum*. In: Plant Growth Substances 1979, pp. 170-179, Skoog, F., ed. Springer-Verlag, Berlin, Heidelberg, New York.

Sponsel, V. M., Gaskin, P., & MacMillan, J. (1979). The identification of gibberellins in immature seeds of *Vicia faba*, and some chemotaxonomic considerations. Planta *146*, 101-105.

Sponsel, V. M., & MacMillan, J. (1976). The metabolism of gibberellin in immature seeds of *Pisum sativum* cv. Progress No. 9. Plant Growth Substances 1976, pp. 366-368, Lausanne Abstracts.

Sponsel, V. M., & MacMillan, J. (1977). Further studies on the metabolism of gibberellins (GAs) A_9, A_{20} and A_{29} in immature seeds of *Pisum sativum* cv. Progress No. 9. Planta *135*, 129-136.

Stoddart, J. L., & Foster, C. A. (1976). "Bakanae barley"—a new mutant of *Hordeum vulgare* L. with an accelerated growth rate. Plant Growth Substances 1976, pp. 374-375, Lausanne Abstracts.

Stoddart, J. L., & Jones, R. L. (1977). Gibberellin metabolism in excised lettuce hypocotyls: evidence for the formation of gibberellin A_1 glucosyl conjugates. Planta 136, 261–269.

Stoddart, J. L., & Venis, M. A. (1980). Molecular and subcellular aspects of hormone action. In: Hormonal Regulation of Development I. Molecular Aspects of Plant Hormones. Encyclopedia of Plant Physiology New Series Vol. 9, pp. 445–510, MacMillan, J., ed. Springer-Verlag, Berlin, Heidelberg, New York.

Stolp, C. F., Nadeau, R., & Rappaport, L. (1973). Effect of abscisic acid on uptake and metabolism of [^{3}H]gibberellin A_1 and [^{3}H]pseudogibberellin A_1 by barley half seeds. Plant Physiol. 52, 546–548.

Stolp, C. F., Nadeau, R., & Rappaport, L. (1977). Abscisic acid and the accumulation, biological activity and metabolism of four derivatives of [^{3}H]-gibberellin A_1 in barley aleuron layers. Plant Cell Physiol. 18, 721–728.

Takahashi, N. (1974). Recent progress in the chemistry of gibberellins. In: Plant Growth Substances 1973, pp. 228–240. Hirokawa, Tokyo.

Takahashi, N., Murofushi, N., & Yamane, H. (1976). Metabolism of gibberellins in maturing and germinating bean seeds. Plant Growth Substances 1976, pp. 383–385, Lausanne Abstracts.

Tamura, S., Takahashi, N., Yokota, T., Murofushi, N., & Ogawa, T. (1968). Isolation of water-soluble gibberellins from immature seeds of Pharbitis nil. Planta 78, 208–212.

Wample, R. L., Durley, R. C., & Pharis, R. P. (1975). Metabolism of gibberellin A_4 by vegetative shoots of Douglas fir at three stages of ontogeny. Physiol. Plant. 35, 273–278.

Weaver, R. J., & Pool, R. M. (1965a). Gibberellin-like substances in grape berries as affected by seededness and ringing. Plant Physiol. 40, LXXV.

Weaver, R. J., & Pool, R. M. (1965b). Relation of seededness and ringing to gibberellin-like activity in berries of Vitis vinifera. Plant Physiol. 40, 770–776.

Weiland, J. (1968). Isolierung, Struktur und Stoffwechsel eines Gibberellin-glucosids von Phaseolus coccineus L. sowie Untersuchungen zur Glucosylierung von Gibberellinen. Dissert.-Schrift Martin-Luther-Universität, Halle-Wittenberg.

Yamaguchi, I., Kobayashi, M., & Takahashi, N. (1980). Isolation and characterization of glucosyl esters of gibberellin A_5 and A_{44} from immature seeds of *Pharbitis purpurea.* Agr. Biol. Chem. *44*, 1975-1977.

Yamaguchi, I., Yokota, T., Yoshita, S., & Takahashi, N. (1979). High pressure liquid chromatography of conjugated gibberellins. Phytochem. *18*, 1699-1702.

Yamane, H., Murofushi, N., Osada, H., & Takahashi, N. (1977). Metabolism of gibberellins in early immature bean seeds. Phytochem. *16*, 831-835.

Yamane, H., Murofushi, N., & Takahashi, N. (1975). Metabolism of gibberellins in maturing and germinating bean seeds. Phytochem. *14*, 1195-1200.

Yamane, H., Takahashi, N., Takeno, K., & Furuya, M. (1979). Identification of gibberellin A_9 methyl ester as a natural substance regulating formation of reproductive organs in *Lygodium japonicum.* Planta *147*, 251-256.

Yamane, H., Yamaguchi, I., Murofushi, N., & Takahashi, N. (1971). Isolation and structure of gibberellin A_{35} and its glucoside from immature seed of *Cytisus scoparius.* Agr. Biol. Chem. *35*, 1144-1146.

Yamane, H., Yamaguchi, I., Murofushi, N., & Takahashi, N. (1974). Isolation and structures of gibberellin A_{35} and its glucoside from immature seed of *Cytisus scoparius.* Agr. Biol. Chem. *38*, 649-655.

Yamane, H., Yamaguchi, I., Yokota, T., Murofushi, N., Takahashi, N., & Katsumi, M. (1973). Biological activities of new gibberellins A_{30}-A_{31} and A_{35}-glucoside. Phytochem. *12*, 255-261.

Yokota, T., Hiraga, K., Yamane, H., & Takahashi, N. (1975). Mass spectrometry of trimethylsilyl derivatives of gibberellin glucosides and glucosyl esters. Phytochem. *14*, 1569-1574.

Yokota, T., Kobayashi, S., Yamane, H., & Takahashi, N. (1978). Isolation of a novel gibberellin glucoside, 3-*O*-β-D-glucopyranosylgibberellin A_1 from *Dolichos lablab* seed. Agr. Biol. Chem. *42*, 1811-1812.

Yokota, T., Murofushi, N., & Takahashi, N. (1970). Structure of new gibberellin glucoside in immature seeds of *Pharbitis nil.* Tetrahedron Lett. 1489-1491.

Yokota, T., Murofushi, N., & Takahashi, N. (1980). Extraction, purification, and identification. In: Hormonal Regulation of Development I. Molecular Aspects of Plant Hormones. Encyclopedia of Plant Physiology New Series, Vol. 9, pp. 113-201, MacMillan, J., ed. Springer-Verlag, Berlin, Heidelberg, New York.

Yokota, T., Murofushi, N., Takahashi, N., & Katsumi, M. (1971a). Biological activities of gibberellins and their glucosides in *Pharbitis nil.* Phytochem. *10*, 2943-2949.

Yokota, T., Murofushi, N., Takahashi, N., & Tamura, S. (1971b). Gibberellins in immature seeds of *Pharbitis nil.* Part III. Isolation and structures of gibberellin glucosides. Agr. Biol. Chem. *35*, 583-595.

Yokota, T., Takahashi, N., Murofushi, N., & Tamura, S. (1969). Isolation of gibberellins A_{26} and A_{27} and their glucosides from immature seeds of *Pharbitis nil.* Planta *87*, 180-184.

Yokota, T., Yamane, H., & Takahashi, N. (1976). The synthesis of gibbere-thiones, gibberellin related diterpenoides. Agr. Biol. Chem. (Tokyo) *40*, 2507-2508.

Yokota, T., Yamazaki, S., Takahashi, N., & Itaka, Y. (1974). Structure of pharbitic acid, a gibberellin-related diterpenoid. Tetrahedron Lett., 2957-2960.

Zeevaart, J. A. D. (1966). Reduction of the gibberellin contents of *Pharbitis nil* seeds by CCC and after effects in the progeny. Plant Physiol. *41*, 856-862.

Zenk, M. H. (1964). Isolation, biosynthesis and function of indole acetic acid conjugates. In: Regulateurs naturels de la croissance végétale, pp. 241-249, Nitsch, J. P., ed. C.N.R.S., Paris.

7

Preparation and Isotopic Labeling of Gibberellins

Nobutaka Takahashi & Isomaro Yamaguchi

7.1 INTRODUCTION

As many as 66 nonconjugated gibberellins (GAs) have been reported as metabolites of microorganisms and higher plants. Structures have been assigned mainly on the basis of spectroscopic properties and, in some cases, have been confirmed by chemical correlation with known GAs. Recently, several new naturally occurring GAs have been characterized on the basis of mass spectrometric correlations with GA analogues produced by partial synthesis. GAs prepared in this manner have also been widely used in attempts to elucidate GA bioassay structure-activity relationships (see Volume II, Chapter 2). It is also possible to produce a range of isotopically labeled GAs and GA precursors, and these have been of value in biosynthesis and metabolism studies (see Chapter 2 to 6, this volume) as well as investigations into the mode of action of GAs (see Volume II, Chapter 1). The present chapter describes the diverse chemical, biochemical, and microbiological methods that can be employed to prepare rarer GAs and their isotopically labeled derivatives from readily available GA and *ent*-kaurenoid substrates.

7.2 CHEMICAL PREPARATION OF RARER GIBBERELLINS FROM READILY AVAILABLE SUBSTRATES

7.2.1 Gibberellins A_5, A_8, A_9, A_{20}, A_{29}, A_{55}, A_{56}, and A_{57} from A_1 and/or A_3

GA$_3$ (*1*) is commercially produced in large amounts by the fermentation of *Gibberella fujikuroi*. Under certain culture conditions, GA$_1$ (*2*) is also obtained as a fungal metabolite. However, GA$_1$ can also be prepared by selective catalytic reduction of GA$_3$; protection of the exocyclic methylene group is achieved by the addition of pyridine to the reaction mixture (Jones & MacCloskey, 1963). The reaction is catalyzed by Pd on $CaCO_3$, $BaCO_3$, or charcoal, and hydrogenation is accompanied by formation of the hydrogenolysis product (*3*) (see Fig. 7.1). Pitel and Vining (1970) found that the ratio of GA$_1$ and (*3*) is determined by the amount of pyridine in the mixture. In addition, Pd/$CaCO_3$ is more selective than Pd/$BaCO_3$ because it induces less hydrogenolysis and because smaller amounts of pyridine are required to protect the exocyclic methylene group. Subsequently, Murofushi et al. (1976) reported that the hydrogenolysis product (*3*) can be efficiently converted to GA$_1$ via the iodinated intermediate (*4*). GA$_1$ and GA$_3$ have been used as starting materials in the preparation of a range of 13-hydroxy GAs. The synthetic sequences that have been utilized are illustrated in Figs. 7.1 to 7.4.

GA$_5$ (*6*) has been prepared via several paths using either intermediate (*5*), (*7*), or (*14*) (MacMillan, Seaton, & Suter, 1960; MacMillan, & Pryce, 1967; Murofushi et al., 1974; Beale et al., 1980; Hanson, 1980). The route via the intermediate (*5*) is the simplest and gives good yields (Murofushi, Durley, & Pharis, 1977). However, MacMillan (1980) has reported that even better yields are obtained when GA$_5$ is synthesized via intermediate (*12*), which can be obtained by chlorination of the methyl ester of GA$_3$. GA$_5$ (*6*) serves as an intermediate in the synthesis of GA$_8$ (*8*) and GA$_{20}$ (*11*) from GA$_3$. Oxidation of GA$_5$ with equimolar OsO_4 in pyridine affords GA$_8$ as the major pro-

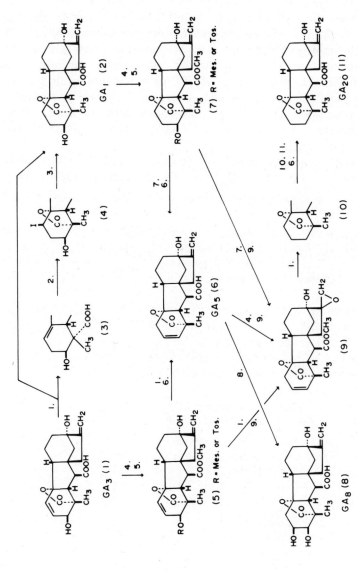

Figure 7.1 Preparation of GA5, GA8, and GA20 from GA1 and GA3. 1: H2-Pd/CaCO3-C5H5N; 2: I2, NaHCO3; 3: NaBH4; 4: CH2N2; 5: methanesulfonyl chloride or *p*-toluenesulfonyl chloride-C5H5N; 6: OH⁻ or *n*-C3H7SLi; 7: reflux in collidine; 8: OsO4; 9: *m*-chloroperbenzoic acid; 10: H2-Pd/C; 11: NaI, NaOAc, Zn.

duct, along with small amounts of a by-product that is hydroxylated at the C-16 and C-17 positions. OsO_4 oxidation of $\Delta^{2,3}C_{19}$-GAs produces $2\beta,3\beta$-dihydroxy GAs as the major product regardless of the presence or absence of a 13-hydroxy group (Murofushi et al., 1973, 1977). Preparation of GA_{20} (*11*) from GA_5 (*6*) requires selective hydrogenation. In this instance, pyridine does not adequately protect the exocyclic methylene group. GA_5 is therefore converted to the epoxide intermediate (*9*) by treatment with equimolar *m*-chloroperbenzoic acid. Hydrogenation of (*9*) produces compound (*10*), which then yields GA_{20} (Murofushi et al., 1977). These procedures, which are illustrated in Fig. 7.1, have been used to prepare high specific activity $[1\text{-}^3H]GA_5$, $[1\text{-}^3H]GA_8$, and $[2,3\text{-}^3H_2]GA_{20}$ (Murofushi et al., 1977).

Beale et al. (1980) prepared GA_5 (*6*), GA_{20} (*11*), and isotopically labeled species such as $[1\beta,3\text{-}^2H_2]GA_5$, $[1\beta,3\text{-}^3H_2]GA_5$, $[1\beta,3\alpha\text{-}^2H_2]$-

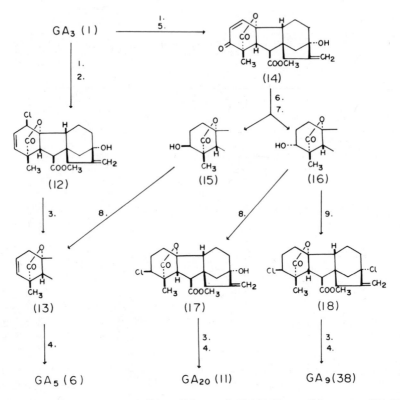

Figure 7.2 Preparation of GA_5, GA_9, and GA_{20} from GA_3. 1: CH_2N_2; 2: $SOCl_2$; 3: $(n\text{-Bu})_3SnH$; 4: OH^-; 5: MnO_2; 6: $NaBH_4$; 7: H_2O; 8: $(C_6H_5)_3P$, CCl_4; 9: $(C_6H_5)_3P$, CCl_4, C_5H_5N.

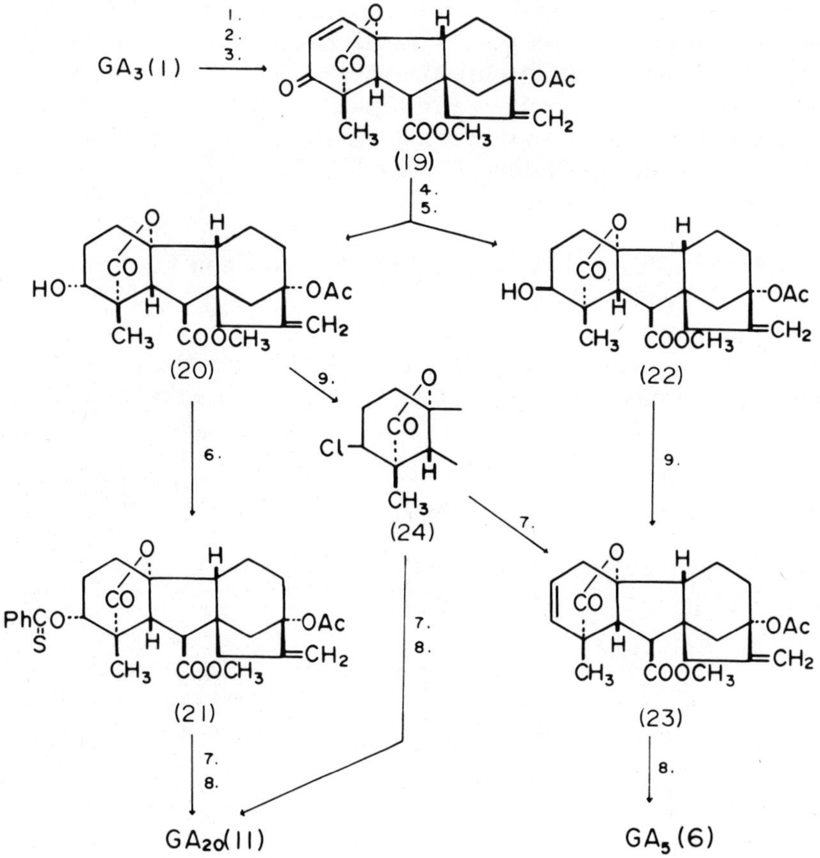

Figure 7.3 Preparation of GA_5 and GA_{20} from GA_3. 1: CH_2N_2; 2: MnO_2; 3: $(CH_3CO)_2O$, C_5H_5N; 4: $NaBH_4$; 5: H_2O; 6: $C_6H_5C(Cl)=N^+(CH_3)_2 \cdot Cl^-$, C_5H_5N, H_2S; 7: $(n\text{-}Bu)_3SnH$; 8: OH^-; 9: $(C_6H_5)_3P$, CCl_4.

GA_{20}, and $[1\beta,3\alpha\text{-}^3H_2]GA_{20}$ from GA_3 via the intermediates (19), (20), (21), and (24) (Fig. 7.3). This route provides better yields of GA_{20} from GA_3 than alternative procedures, and from the safety viewpoint it has an added advantage in that when $[^3H]GA_{20}$ is being prepared NaB^3H_4 is used instead of gaseous $^3H_2$. Hanson (1980) reported the preparation of GA_5, GA_{20}, and GA_9 (38) and their labeled species by treatment of GA_1 methyl ester (15) or 3-*epi*-GA_1 methyl ester (16) with $(C_6H_5)_3P$ and CCl_4 in both the presence and absence of pyridine (Fig. 7.2).

The α,β-unsaturated 3-keto compound (25) serves as an intermediate in the synthesis of GA_{55} (27) and GA_{57} (29). GA_{55} can also be prepared from the hydrogenolysis product (*3*), while GA_{56} (*32*) is produced from GA_3 via the intermediates (*30*) and (*31*) (Fig. 7.4) (Murofushi et al., 1979b, 1980). MacMillan (1978, 1980) has discussed procedures used to synthesize isotopically labeled GA_{29} (*69*) from GA_3 via either the α,β-unsaturated 3-keto compound (*19*) and (*23*) or the chlorinated derivative (*73*) (see Fig. 7.8).

7.2.2 Gibberellins A_9, A_{34}, A_{40}, A_{47}, A_{51}, and A_{54} from A_4 and/or A_7

Varying ratios of GA_4 (*33*) and GA_7 (*34*) are also produced by large-scale fermentation of *G. fujikuroi*. These GAs, either individually

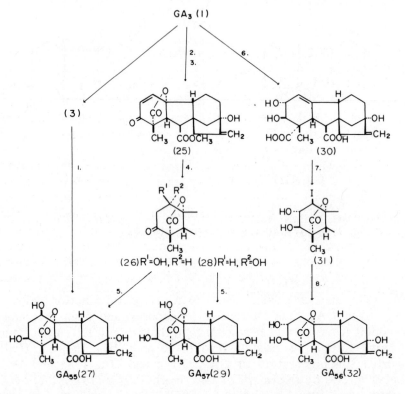

Figure 7.4 Preparation of GA_{55}, GA_{56}, and GA_{57} from GA_3. 1: *m*-chloroperbenzoic acid; 2: CH_2N_2; 3: MnO_2; 4: dilute H_2SO_4; 5: $NaBH_4$; 6: NaOH-$NaHCO_3$ (pH 13); 7: I_2, $NaHCO_3$; 8: (*n*-Bu)$_3$SnH.

or on occasion as a mixture, can be used as substrates for the production of several 13-deoxy GAs.

Yokota, Reeve, and Crozier (1976) prepared GA_9 (*38*) and $[2,3-^3H_2]GA_9$ from GA_4 via the intermediates (*35*), (*36*), and (*37*) (Fig. 7.5). Beeley and MacMillan (1976) synthesized GA_{34} (*44*), GA_{40} (*51*), and GA_{47} (*56*) from GA_4 via the key intermediate (*47*) (Fig. 7.6). GA_{40} can be easily converted to GA_{51} (*53*) (Yamaguchi et al., 1975a). Beale et al. (1980) and Hanson (1980) prepared GA_9, $[3\alpha-^2H]GA_9$, and $[3\beta-^2H]GA_9$ from a mixture of GA_4 and GA_7 via the key intermediates (*39*), (*40*), and (*41*) (Fig. 7.5). Murofushi et al. (1973, 1979b) prepared GA_{34} from GA_4 via (*36*), and GA_{47} (*56*) from GA_7 via the intermediates (*59*) and (*60*) (Fig. 7.6). GA_{54} (*58*) can be produced from GA_7 via the hydrogenolysis product (*57*) (Fig. 7.7) (Murofushi et al., 1979b). MacMillan (1978) has reported the synthesis of GA_{51} (*53*) from GA_7 (Fig. 7.8).

7.2.3 Gibberellins A_4, A_9, A_{15}, A_{25}, A_{37}, A_{43}, and A_{46} from A_{13} or A_{36}, and A_{20} from A_{17}

GA_{13} (*76*) and GA_{36} (*95*) are present in relatively high concentrations in the mother liquors after crystallization of GA_3 from industrial fermentations of *G. fujikuroi* (Bearder & MacMillan, 1973; Yamuguchi et al., 1975a). These GAs are valuable starting materials for the partial synthesis of several C_{20}-GAs.

Harrison and MacMillan (1971) and Hanson (1980) prepared GA_{25} (*83*) from GA_{13} via the key intermediate (*79*). GA_{43} methyl ester (*81*) and GA_{46} methyl ester (*88*) have also been produced from GA_{13} via the common intermediate (*79*) (Beeley, Gaskin, & MacMillan, 1975; Beeley & MacMillan, 1976). In the synthesis of both GA_{43} methyl ester and GA_{46} methyl ester, it is necessary to protect hydroxyl group(s) in the ring A of the precursor GA prior to carrying out the Wittig reaction. Pondorff reduction of the ketone (*88*) affords a higher yield of GA_{46} methyl ester than $NaBH_4$ reduction (Fig. 7.9).

Bowen et al. (1975) synthesized GA_{37} (*94*) from GA_{13} via the intermediates (*90*), (*91*), (*92*), and (*93*). Pondorff reduction of the ketone (*93*) to yield GA_{37} again proved superior to $NaBH_4$ reduction. However, it should be noted that GA_{37} is more easily prepared from GA_{36} (*95*) (Fig. 7.10). It may be possible to convert GA_{37} to GA_{15} via a chlorinated intermediate analogous to (*42*), which was used in the preparation of GA_9 from GA_4 illustrated in Fig. 7.5.

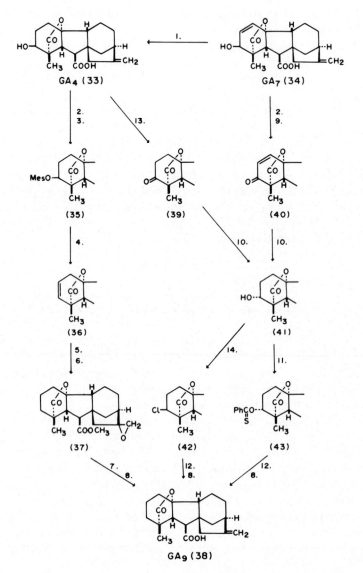

Figure 7.5 Preparation of GA_9 from GA_4 and GA_7. 1: H_2-Pd/CaCO$_3$-C$_5$H$_5$N; 2: CH_2N_2; 3: methanesulfonyl chloride-C_5H_5N; 4: reflux in collidine; 5: m-chloroperbenzoic acid; 6: H_2-Pd/C; 7: NaI, NaOAc, Zn; 8: OH$^-$; 9: MnO_2; 10: NaBH$_4$; 11: $C_6H_5C(Cl)=N^+(CH_3)_2 \cdot Cl^-$, C_5H_5N, H_2S; 12: (n-Bu)$_3$SnH; 13: Jones oxidation; 14: POCl$_3$.

464

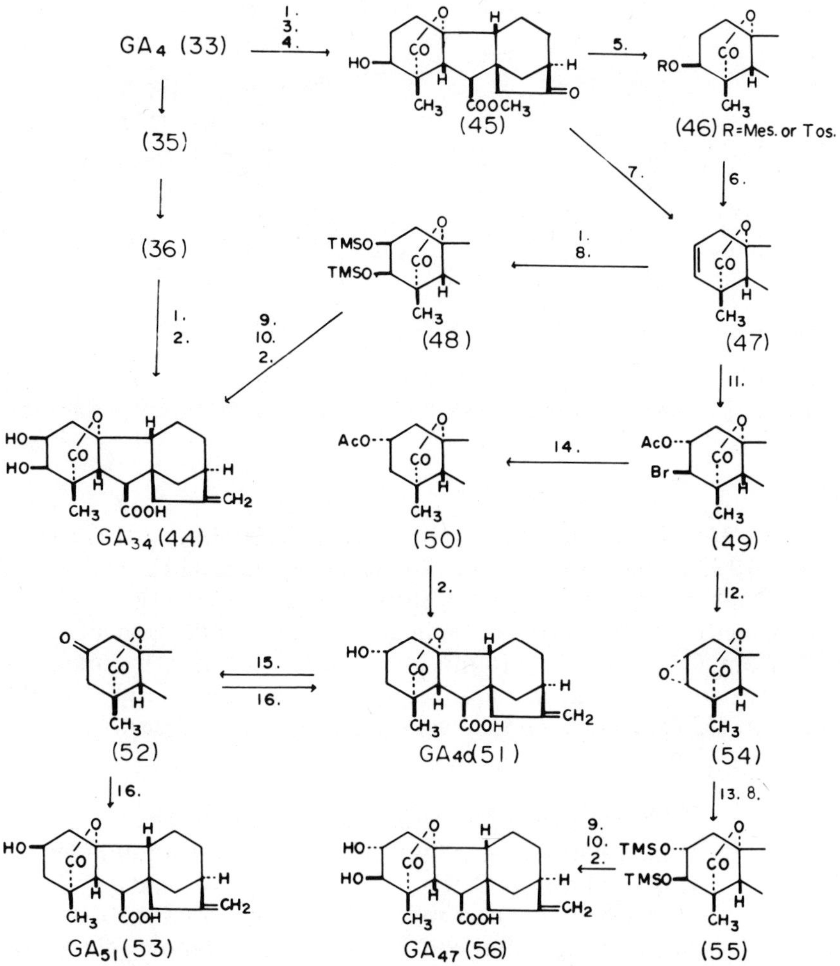

Figure 7.6 Preparation of GA_{34}, GA_{40}, GA_{47}, and GA_{51} from GA_4. 1: OsO_4; 2: OH^-; 3: $NaIO_4$; 4: CH_2N_2; 5: methanesulfonyl chloride or p-toluenesulfonyl chloride-C_5H_5N; 6: reflux in collidine; 7: $POCl_3$; 8: $[(CH_3)_3Si]_2NH$, $(CH_3)_3SiCl$; 9: $(C_6H_5)_3P=CH_2$; 10: H_3O^+; 11: $CH_3CONHBr$, LiOAc, HOAc; 12: KOH in CH_3OH; 13: 2 N H_2SO_4; 14: $(n$-Bu$)_3SnH$; 15: Jones oxidation; 16: $NaBH_4$.

To date, the chemical conversion of C_{19}-GAs to C_{20}-GAs has not been achieved. However, some C_{20}-GAs have been converted to C_{19}-GAs, thereby enabling direct chemical correlations to be made between C_{20}- and C_{19}-GA structures. Treatment of GA_{13} (*76*) with Pb(OAc)$_4$ yielded GA_4 (*33*) as well as (*96*) and its 20 → 4α lactone

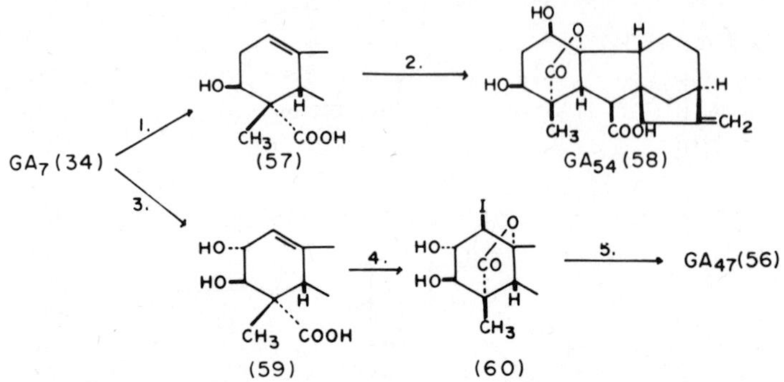

Figure 7.7 Preparation of GA_{47} and GA_{54} from GA_7. 1: H_2-Pd/$BaCO_3$-C_5H_5N; 2: m-chloroperbenzoic acid; 3: 2 N NaOH ($100°C$); 4: I_2, $NaHCO_3$; 5: $NaBH_4$.

isomer. GA_{25} (83) can similarly be converted to GA_9 (38) (Bearder et al., 1979a). Murofushi et al. (1976) synthesized GA_4 from GA_{13} via the intermediates (97), (98), (99), (100), and (101) (Fig. 7.11). These methods may also be applicable to the production of GA_{20} (11) from GA_{17}, which is found in high concentration in immature seeds of *Calonyction aculeatum* (Murofushi et al., 1973) and *Pharbitis purpurea* (I. Yamaguchi and co-workers, unpublished data).

7.2.4 Gibberellin A_{12}, Gibberellin A_{12} aldehyde, and Gibberellin A_{12} alcohol from 7β-hydroxykaurenolide; Gibberellin A_{14}, Gibberellin A_{14} aldehyde, and Gibberellin A_{14} alcohol from 3β,7β-dihydroxykaurenolide; and *ent*-11β-, 12α- and 12β-hydroxykaurenoic acid from grandiflorenic acid

GA_{12} aldehyde (110) is the first compound possessing an *ent*-gibberellane skeleton in the GA biosynthetic pathway of both higher plants and *G. fujikuroi* (Graebe, Bowen, & MacMillan, 1972; Hanson, Hawker, & White, 1972; see also Chapters 3 and 5, this volume). GA_{12} aldehyde was first prepared by Galt and Hanson (1965) from 7β-hydroxykaurenolide (102) via the *p*-toluenesulfonate (108), which was treated with 10% KOH in CH_3OH and methylated to afford the methyl ester of GA_{12} aldehyde. It was later found that treatment of the *p*-toluenesulfonate (108) with KOH in *t*-BuOH—H_2O gave better results than KOH in CH_3OH (Cross, Norton, & Stewart, 1968). The aldehyde (110) is readily converted into GA_{12} (112) by Jones oxidation and into GA_{12} alcohol (*114*) by $NaBH_4$ reduction (Fig. 7.12) (Cross et al., 1968).

GA_{14} aldehyde (*111*), which is also an important intermediate in GA biosynthesis (Hedden, MacMillan, & Phinney, 1974), has been prepared from $3\beta,7\beta$-dihydroxykaurenolide (*103*) by a similar procedure to that used for the synthesis of GA_{12} aldehyde. The 3-hydroxyl group was protected by conversion to the tetrahydro-

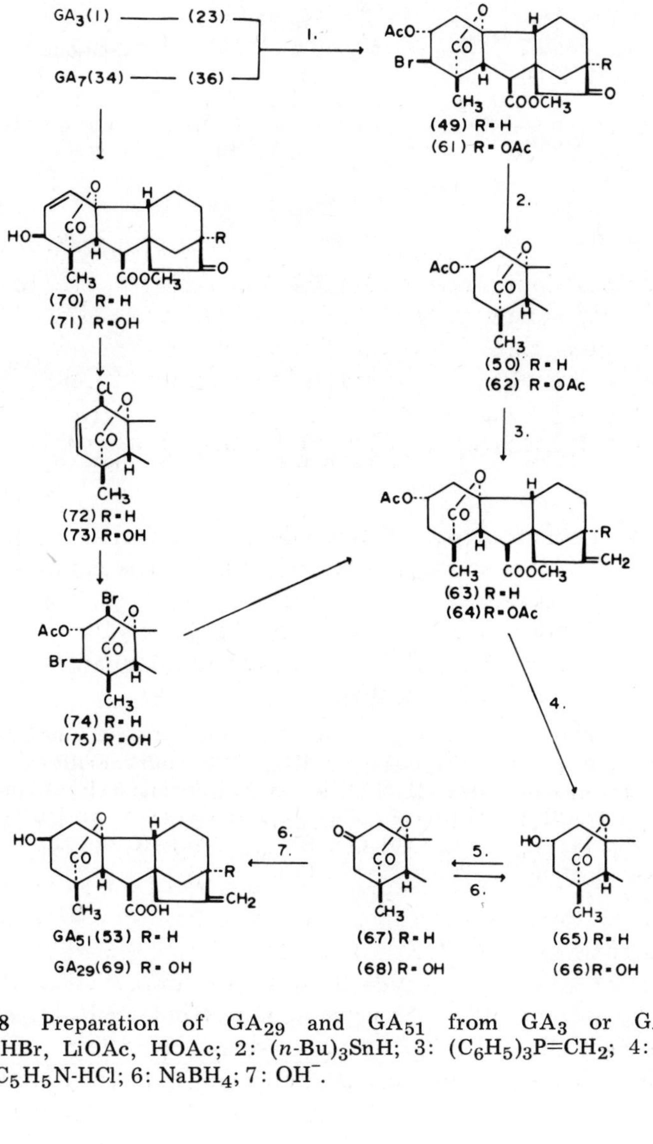

Figure 7.8   Preparation of GA_{29} and GA_{51} from GA_3 or GA_7.   1: $CH_3CONHBr$, LiOAc, HOAc; 2: $(n\text{-}Bu)_3SnH$; 3: $(C_6H_5)_3P{=}CH_2$; 4: K_2CO_3; 5: $CrO_3\text{-}C_5H_5N\text{-}HCl$; 6: $NaBH_4$; 7: OH^-.

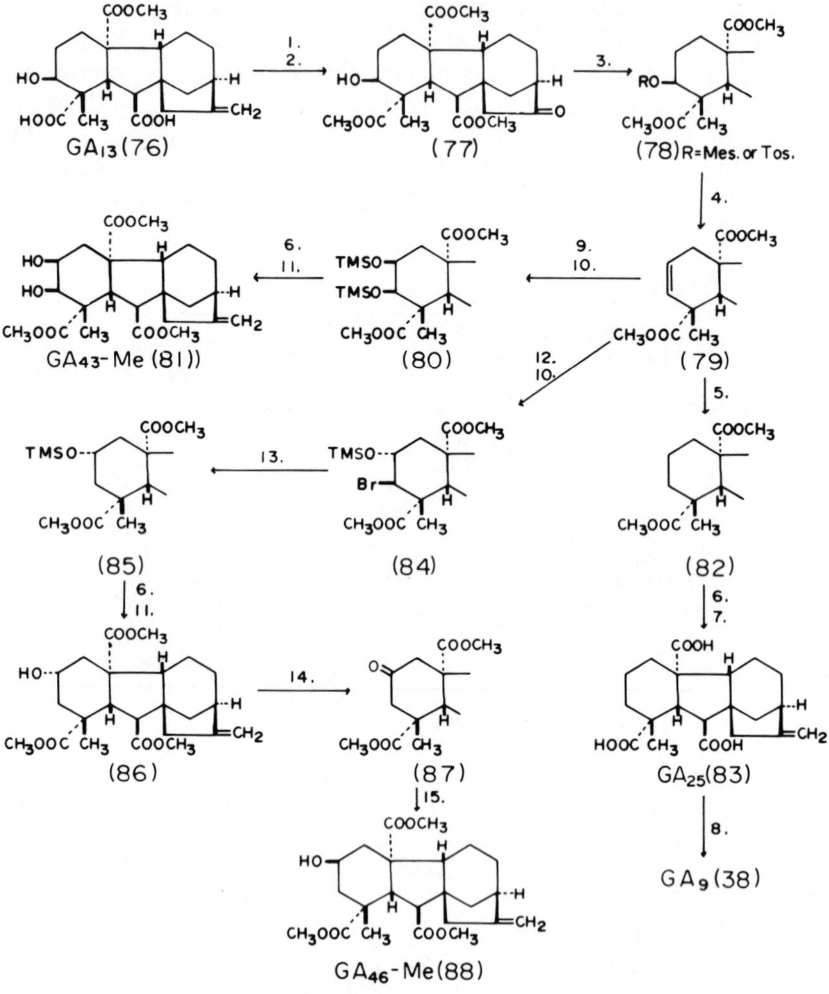

Figure 7.9 Preparation of GA_9, GA_{25}, GA_{43} methyl ester, and GA_{46} methyl ester from GA_{13}. 1: OsO_4, $NaIO_4$; 2: CH_2N_2; 3: methanesulfonyl chloride or p-toluenesulfonyl chloride-C_5H_5N; 4: reflux in collidine; 5: H_2-Adams catalyst; 6: $(C_6H_5)_3P{=}CH_2$; 7: OH^-; 8: $Pb(OAc)_4$; 9: OsO_4; 10: $[(CH_3)_3Si]_2NH$, $(CH_3)_3SiCl$; 11: H_3O^+; 12: $CH_3CONHBr$, LiOAc, HOAc; 13: $(n\text{-}Bu)_3SnH$; 14: Jones oxidation; 15: Pondorff reduction.

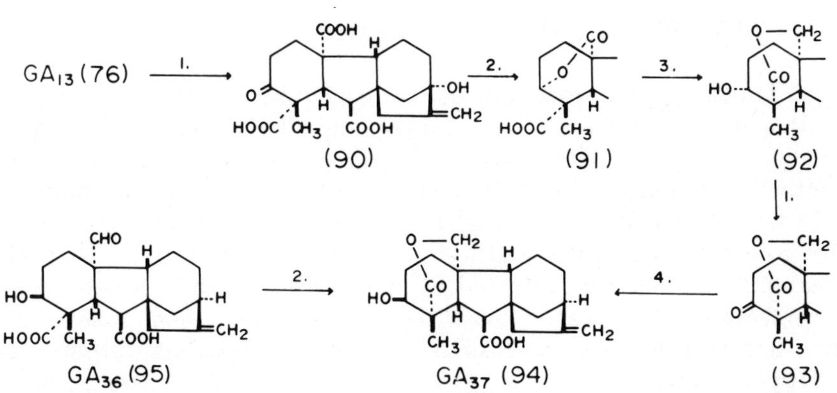

Figure 7.10 Preparation of GA_{37} from GA_{13} and GA_{36}. 1: Jones oxidation; 2: $NaBH_4$; 3: $LiBH_4$; 4: Pondorff reduction.

pyranyl ether (*105*) (Fig. 7.12) (Hedden, MacMillan, & Grinsted, 1973).

Lewis & MacMillan (1980) reported the preparation of *ent*-12α- and 12β-hydroxykaurenoic acids (*120, 121*) as well as *ent*-11β-hydroxykaurenoic acid from grandiflorenic acid (*116*) via the intermediates shown in Fig. 7.13. It may be possible to feed these kaurenoic acid derivatives to either GA deficient *G. fujikuroi* mutants or normal cultures in which *ent*-kaurene production is inhibited by growth retardants and thus obtain a range of 12-oxygenated GA metabolites such as GA_{26}, GA_{39}, GA_{48}, and GA_{49}.

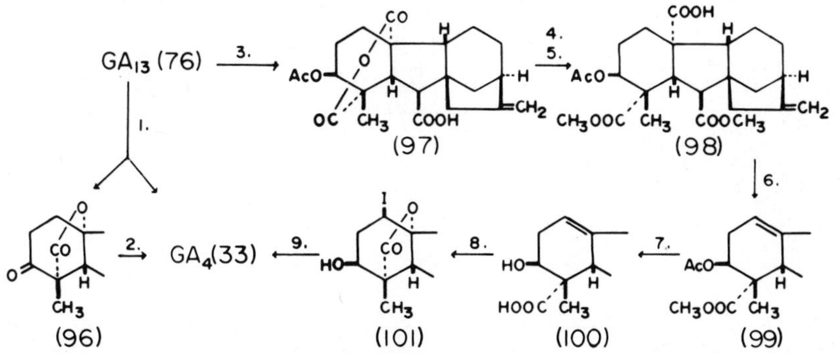

Figure 7.11 Preparation of GA_4 from GA_{13}. 1: $Pb(OAc)_4$; 2: $Cr(OAc)_2$; 3: Ac_2O, C_5H_5N; 4: CH_2N_2; 5: CH_3ONa; 6: $Cu(OAc)_2$, $Pb(OAc)_4$; 7: 1.5 N NaOH in C_2H_5OH; 8: I_2, $NaHCO_3$; 9: $NaBH_4$.

7.2.5 Removal of protecting groups

Unwanted side reactions often occur during partial synthesis of GAs and kaurenoids, and as a consequence it is necessary to protect certain functional groups. Hydroxyl groups are protected as acetates, tetrahydropyranyl ethers, or trimethylsilyl ethers. Acetate moieties can be subsequently hydrolyzed by sodium methoxide, while tetrahydropyranyl and trimethylsilyl ethers are cleaved with dilute acid. Carboxyl groups are protected by methylation, although hydrolysis to release the free acid is not without its problems since many GAs are unstable in basic and acidic conditions. However, lithium n-propylmercaptide (Bartlett & Johnson, 1970) has satisfactorily hydrolyzed the methyl esters of GA_1, GA_3, and the acetate of GA_3 glucosyl ether (Yamaguchi, Takahashi, & Fujita, 1975b; Corey, Brennan, & Carney, 1971; Schneider et al., 1977). These compounds are unstable in a traditional basic hydrolytic environment.

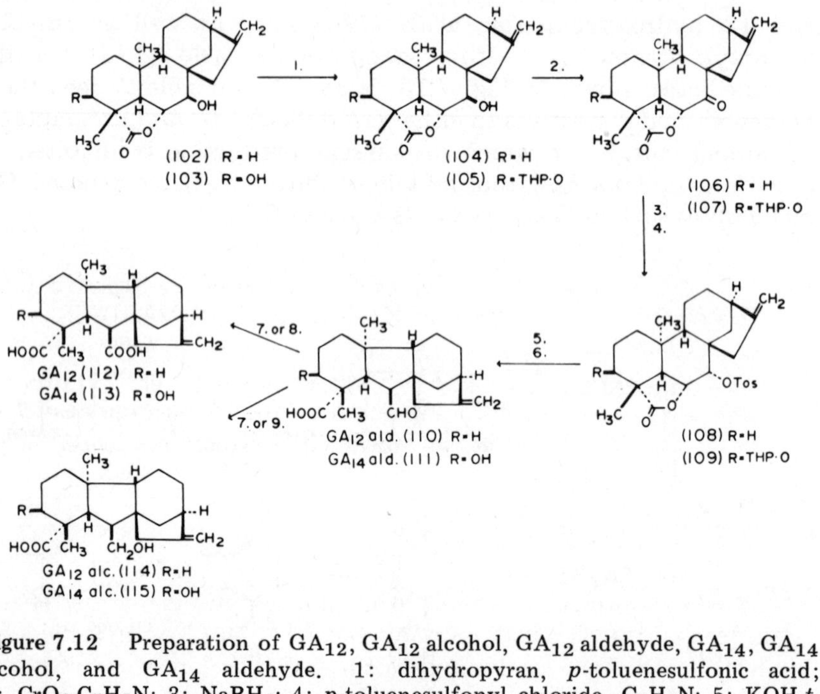

Figure 7.12   Preparation of GA_{12}, GA_{12} alcohol, GA_{12} aldehyde, GA_{14}, GA_{14} alcohol, and GA_{14} aldehyde.   1: dihydropyran, p-toluenesulfonic acid; 2: CrO_3-C_5H_5N; 3: $NaBH_4$; 4: p-toluenesulfonyl chloride, C_5H_5N; 5: KOH-t-BuOH-H_2O; 6: H_3O^+; 7: CH_3ONa under N_2; 8: Jones oxidation (R=H only); 9: $NaBH_4$.

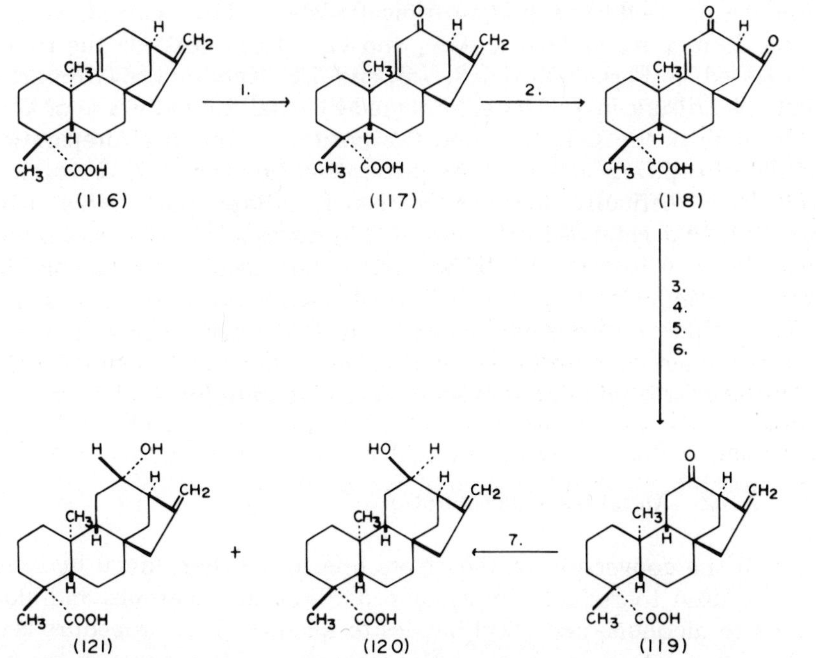

Figure 7.13 Preparation of *ent*-12α- and 12β-hydroxykaurenoic acids from grandiflorenic acid. 1: *t*-BuCrO$_4$; 2: OsO$_4$, NaIO$_4$; 3: H$_2$-10% Pd/CaCO$_3$; 4: [(CH$_3$)$_3$Si]$_2$NH, (CH$_3$)$_3$SiCl; 5: (C$_6$H$_5$)$_3$P=CH$_2$; 6: H$_3$O$^+$; 7: NaBH$_4$.

7.3 CHEMICAL PROCEDURES FOR ISOTOPIC LABELING OF GIBBERELLINS AND *ENT*-KAURANE DERIVATIVES

Most of the reactions described in Section 7.2 can be used to prepare labeled compounds by employing reagents labeled with radioisotopes such as ^{3}H and ^{14}C, or stable isotopes such as ^{2}H and ^{13}C.

7.3.1 Catalytic reduction

Catalytic reduction is used to introduce hydrogen or its ^{2}H and ^{3}H isotopes into either an *ent*-gibberellane or *ent*-kaurane skeleton. Application is, however, limited to the introduction of the isotopes into the olefinic unsaturation in ring A of GAs and ring A or B of *ent*-kauranes. As describes in Section 7.2.1, pyridine is used as a cosolvent to protect the exocyclic methylene group. Catalytic reduction can be used to prepare isotopically labeled GA_1, GA_5, GA_8, and GA_{20} from GA_3 as well as GA_4 and GA_9 from GA_7 via the routes illustrated in Figs. 7.1 to 7.3, 7.5, and 7.8. Kende (1967) reported that the tritium in [^{3}H]GA_1 obtained by catalytic reduction of GA_3 is located at the C-1, C-2, and C-3 positions. Chemical preparation of the isotopically labeled GAs possessing the same ring A structure as GA_3 is difficult. However, Nadeau and Rappaport (1974) have reported that catalytic tritiation of GA_3 affords [^{3}H]GA_3 as a minor product. The location of ^{3}H was not determined. Since labeling by catalytic reduction requires the use of gaseous isotopes, such as ^{3}H$_2$, considerable caution must be exercised. It is worth noting, however, that commercial enterprises such as Amersham International offer carrier-free tritium reduction services at moderate prices.

7.3.2 Metal hydride reduction

In the conversion of GAs from one to another, metal hydrides can be used to reduce simple or α,β-unsaturated ketones and aldehydes to alcohols, and alkyl halides to alkanes. This procedure is of value when introducing ^{2}H or ^{3}H into ring A of GAs through the use of isotopically labeled reagents such as NaB^xH_4, $(n\text{-}Bu)_3Sn^xH$. It has been used to prepare GA_5, GA_{20}, and GA_9 and their labeled counterparts such as [$3\alpha\text{-}^2$H]GA_9, [$3\beta\text{-}^2$H]GA_9, [$1\beta,3\text{-}^2$H$_2$]GA_5, [$1\beta,3\text{-}^3$H$_2$]-GA_5, [$1\beta,3\alpha\text{-}^2$H$_2$]GA_{20}, and [$1\beta,3\alpha\text{-}^3$H$_2$]GA_{20} from GA_3 or GA_7 via the corresponding 3-keto intermediates (*14*), (*19*), and (*40*) as shown in Figs. 7.2, 7.3, and 7.5 (Beale et al., 1980; Hanson, 1980). The location of the isotopes in these GAs has been discussed by Beale and MacMillan (1980). Hanson (1980) has reported the synthesis of [$3,13\text{-}^xH_2$]GA_9 from GA_1 via the procedure illustrated in Fig. 7.2. GA_{12} aldehyde labeled at C-7 with ^{3}H has been prepared by successive reduction of the aldehyde with NaB^3H_4 and oxidation with CrO_3-pyridine (Dockerill, Evans, & Hanson, 1977).

7.3.3 Isotopic exchange

Isotopic exchange is used to introduce ^{2}H or ^{3}H into positions α to a carbonyl group. The reaction is carried out under relatively strong basic conditions such as $CH_3ONa\text{-}CH_3O^xH/^xH_2O$, and clearly both the substrate and the product must be stable in such an environment. The procedure has been used to prepare $[6\text{-}^2H]GA_{12}$ aldehyde, $[6\text{-}^3H]GA_{12}$ aldehyde, and $[6\text{-}^3H]GA_{14}$ aldehyde (Bearder, MacMillan, & Phinney, 1973a; Hedden et al., 1974). Graebe and Hedden (1975) also used an exchange reaction to obtain $ent\text{-}[6\text{-}^3H_2]7\alpha\text{-}hydroxy\text{-}$ kaurenoic acid from $ent\text{-}7\text{-}oxo\text{-}kaurenoic$ acid. Bearder et al. (1974) have reported that treatment of $ent\text{-}kaur\text{-}16\text{-}ene$ with $CF_3COO^xH\text{-}^xH_2O$ affords a mixture of $ent\text{-}kaur\text{-}16\text{-}ene$ and $ent\text{-}kaur\text{-}15\text{-}ene$ labeled at the 15- and 17-positions.

7.3.4 Wittig reaction

The Wittig reaction is used to regenerate the exocyclic methylene group at C-16 from the corresponding nor-keto compounds of $ent\text{-}$ gibberell-16-enes and $ent\text{-}kaurenes$. This reaction can be used to introduce isotopes such as ^2H, ^3H, ^{13}C, and ^{14}C into the exocyclic methylene function. It is currently the only chemical method of introducing ^{13}C or ^{14}C into a GA molecule. The advantage of using C-17 labeled GAs in metabolism studies is that the label is retained until the $ent\text{-}gibberell\text{-}16\text{-}ene$ skeleton is cleaved. A Wittig reaction can be used to label most GAs and $ent\text{-}kaurenes$, although protection of carboxyl and hydroxyl groups is necessary. This can be achieved by methylating carboxyls and converting hydroxyl groups to either acetates or trimethylsilyl ethers (see Section 7.2.5).

Methyltriphenylphosphonium iodide and n-butyl lithium are often used as the ylide and the base in Wittig reactions. However, Nakahara, Mori, and Matsui (1971) and Bearder et al. (1976b) found that in the synthesis of steviol and $[17\text{-}^3H]steviol$ better yields were obtained with methyltriphenylphosphonium bromide and potassium t-butoxide. Sponsel and MacMillan (1977) reported that the tritium is approximately equally distributed at C-15 and C-17 in $[^3H]GA_9$, and at C-14, C-15, and C-17 in $[^3H]GA_{20}$ prepared from the corresponding 17-nor-ketones and methyltriphenylphosphonium bromide-potassium t-butoxide.

There are two ways to cleave the exocyclic methylene of GAs and $ent\text{-}kaur\text{-}16\text{-}enes$ when preparing 17-nor-keto compounds for use as substrates in the Wittig reaction. The first is ozonolysis and the second is periodate oxidation of α-glycol generated by OsO_4 oxidation. Ozonolysis is carried out at $-70°C$ to prevent formation of either the lactone (124) or the keto acid (128). Substrates possessing a 13-hydroxy group must be protected prior to OsO_4-oxidation in order to avoid production of the keto-aldehyde (127) (Fig. 7.14).

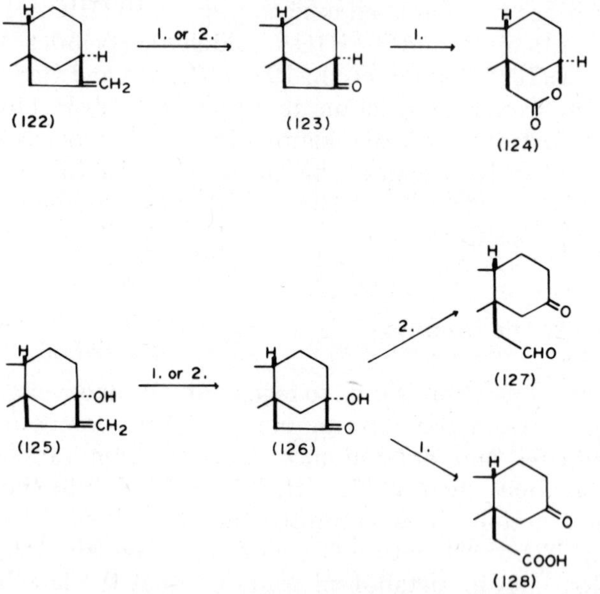

Figure 7.14 Preparation of 17-nor-keto compounds. 1: O_3; 2: OsO_4, $NaIO_4$.

7.4 BIOCHEMICAL AND MICROBIOLOGICAL TECHNIQUES

Mutants and inhibitors have been widely used in investigations of GA biosynthesis in *G. fujikuroi*. GA-deficient test systems are of particular value because various GAs and ent-kaurene analogues can be used as substrates for the production of an array of novel GA metabolites. There are also some *in vitro* systems from higher plants that are useful for the small-scale preparation of isotopically labeled GAs.

7.4.1 Utilization of B1-41a and R-9 mutants of *Gibberella fujikuroi*

Spector and Phinney (1968) reported on two genes in *G. fujikuroi* that control different steps in the GA biosynthetic pathway. In the R-9 mutant, hydroxylation at C-13 is blocked and as a consequence GA_4 and GA_7 are not metabolized to GA_1 and GA_3 (Bearder, MacMillan, & Phinney, 1973b). This mutant is of value in obtaining relatively large amounts of GA_4 and GA_7, which serve as substrates for the chemical synthesis discussed in Section 7.2.2.

A second mutant, B1-41a, which was induced from the wild-type strain GF-1a by irradiation with UV light, is unable to convert *ent*-kaurenal to *ent*-kaurenoic acid and, hence, is deficient in endogenous GAs (Bearder et al., 1974). The enzymes in the GA biosynthesis pathway beyond *ent*-kaurenoic acid do not exhibit absolute substrate specificity. B1-41a, therefore, can metabolize certain exogenous GA and *ent*-kaurenoic acid derivatives to produce a range of GAs, many of which are not endogenous constituents of the parent wild-type strain, GF-1a. Hedden et al. (1974) reported that B1-41a converted GA_{14} aldehyde to GA_3 (15%), GA_7 (17%), GA_1 (5%), GA_4 (19%), GA_{13} (5%), and GA_{14} (9%). GA_{14} alcohol yielded a similar spectrum of metabolites except that GA_1 was absent. Bearder, MacMillan, and Phinney (1975b) found that B1-41a cultured at pH 7.0 converted GA_{12} exclusively to 3-deoxy GAs such as GA_9, GA_{24}, and GA_{25}. The mutant also converts steviol (*129*), the aglycone of stevioside from *Stevia rebaudiana*, to a number of 13-hydroxy products including GA_1, GA_{17}, GA_{18}, GA_{19}, GA_{20}, and GA_{53} (Bearder et al., 1975b). B1-41a cultures have also been used to obtain $[^3H]GA_{20}$ from $[^3H]$steviol acetate (Bearder et al., 1976b) and GA_{45} (*131*) from *ent*-15α-hydrokaurenoic acid (*130*) (Bearder et al., 1975a). GA_9 and its methyl ester are metabolized to GA_{20} (7%), GA_{40} (20%), GA_{10} (2%), etc. and their methyl esters respectively (Bearder et al., 1976a). The production of 12,16-cyclo-GA_9 (*134*) and 12,16-cyclo-GA_{12} (*113*) from trachylobanic acid (*132*) by B1-41a cultures has also been reported (Bearder, MacMillan, & Matsuo, 1979b).

7.4.2 Utilization of wild-type *Gibberella fujikuroi* and inhibitors of Gibberellin biosynthesis

Kende, Ninneman, and Lang, (1963) reported that AMO-1618 (*138*) (see Fig. 7.15) and CCC (*140*) inhibited the biosynthesis of GAs in *G. fujikuroi* without adversely affecting mycelial growth. It is now known that in *G. fujikuroi* cultures and cell-free preparations from higher plants, the retardants block the cyclization of geranylgeranyl pyrophosphate to diterpenes such as *ent*-kaurene without apparently inhibiting subsequent steps in the GA biosynthesis pathway (Adamiec, 1966; Cross & Myers 1969; Dennis, Upper, & West, 1965; Harada & Lang, 1965; Lang, 1970; Robinson & West, 1970a, 1970b; Shechter & West, 1969; see also Chapter 2, this volume). Phosphon D has a similar mode of action in *in vitro G. fujikuroi* test systems, but it does not inhibit *ent*-kaurene synthetase *in vivo* because it is broken down

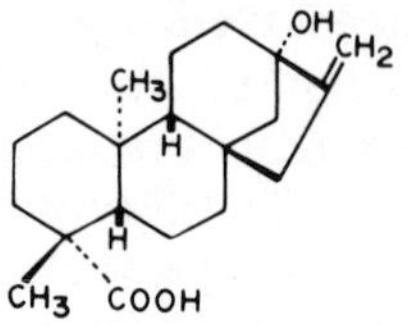

(129)

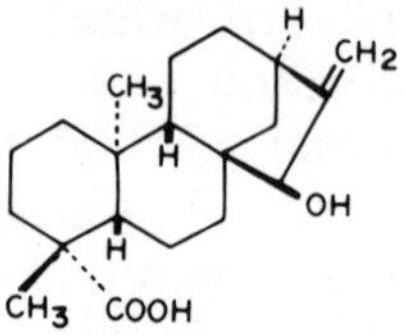

(130)

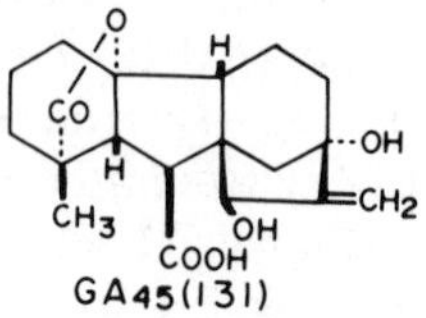

GA45(131)

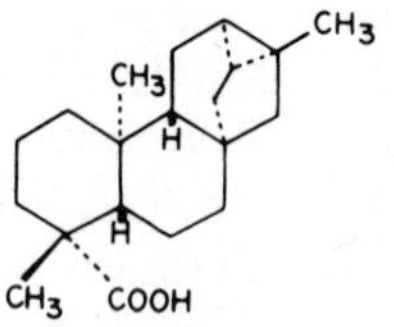

(132)

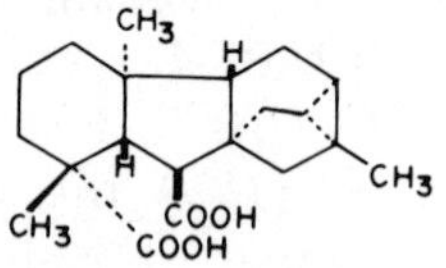

Cyclo GA12 (133)

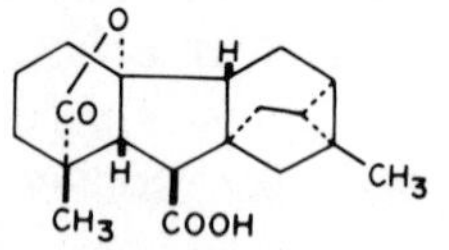

Cyclo GA9 (134)

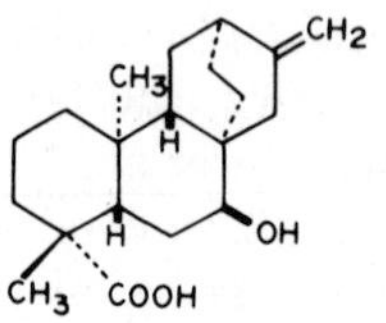

(135)

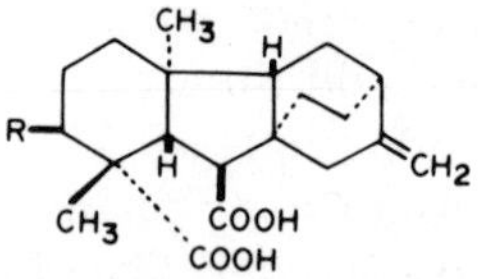

Atiso GA12 (136) R = H
Atiso GA14 (137) R = OH

Structures (129)–(137)

by the fungus and used as a nitrogen source (Harada & Lang 1965; Shechter & West 1969). A CCC- or AMO-1618-treated wild-type culture is thus analogous to the B1-41a mutant in that it is deficient in GAs yet can convert precursors located after the biosynthetic block to a range of GA structures. The system can, therefore, be used to produce GA analogues and isotopically labeled GAs uncontaminated by endogenous GAs. In practice, precultured *G. fujikuroi* is transferred to a medium containing either CCC or AMO-1618 and cultured until the endogenous GA pools are depleted. Hyphae are then transferred to a third medium containing the blocking reagent and the exogenous GA substrate.

Dockerill and Hanson (1978) have reported that *ent*-[^{14}C] kaurene and *ent*-[^{14}C]kaurenoic acid are converted into GA$_3$, GA$_4$, and GA$_7$ by *G. fujikuroi* in the presence of $10^{-4} M$ AMO-1618. GA production is also inhibited by the quarternary ammonium salt (*141*) which was synthesized by Haruta et al. (1974). *G. fujikuroi* cultured in 1 g liter^{-1} of (*141*) converted *ent*-[^{14}C]kaurene to GA_3 with an efficiency of 28% (Hedden et al., 1979). Cho et al. (1979) synthesized (*142*), an analogue of (*141*), which also inhibits GA biosynthesis in *G. fujikuroi*. Murofushi et al. (1979a) used *G. fujikuroi* (G-2) cultures containing 1 g liter^{-1} of (*142*) to prepare GA_1, GA_{18}, GA_{19}, GA_{53}, and various 13-hydroxy kaurenolides from steviol. Wada, Imai, and Suibata (1979) have reported the production of GA_{45} (*131*) and related 15β-hydroxy GAs from *ent*-15α-hydroxykaurenoic acid (*130*) by *G. fujikuroi* (G-2) in the presence of 50 ppm of 1-decylimidazole. This retardant inhibits oxidation step(s) between *ent*-kaurene and *ent*-kaurenoic acid (Wada, 1978). It is of interest to note, however,

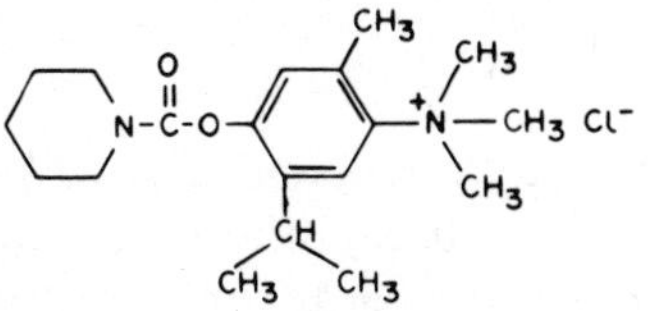

(138)

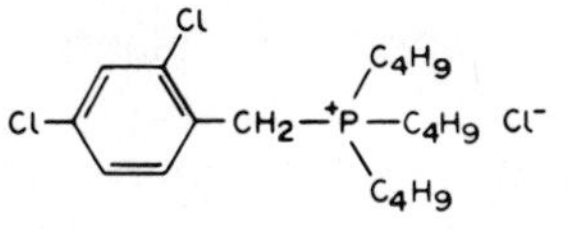

(139)

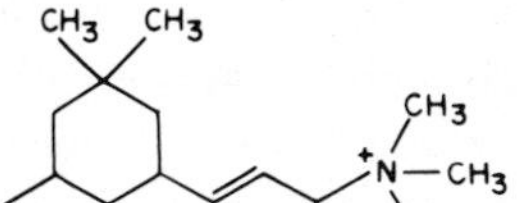

(140)

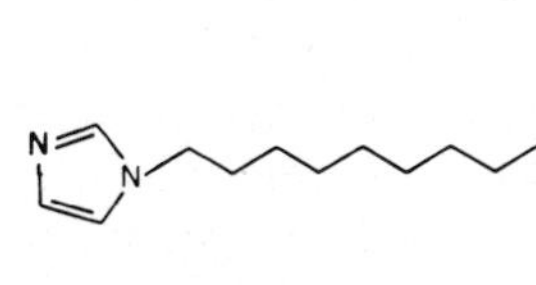

(141)

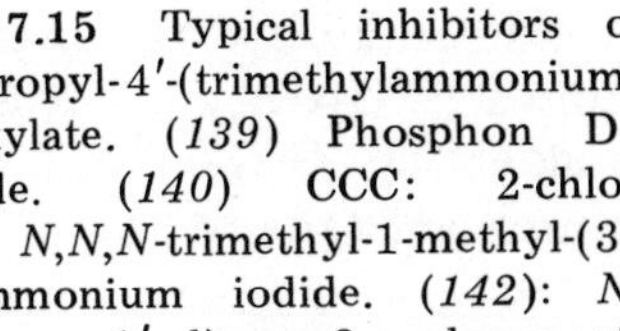

(142)

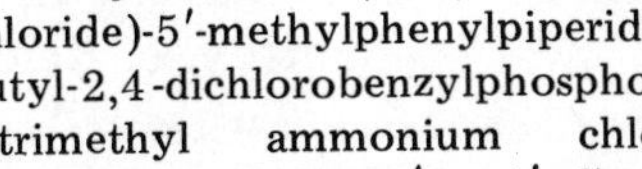

(143)

Figure 7.15 Typical inhibitors of GA biosynthesis. (*138*) AMO-1618: 2′-isopropyl-4′-(trimethylammonium chloride)-5′-methylphenylpiperidine-1-carboxylate. (*139*) Phosphon D: tributyl-2,4-dichlorobenzylphosphonium chloride. (*140*) CCC: 2-chloroethyltrimethyl ammonium chloride. (*141*): N,N,N-trimethyl-1-methyl-(3′,3′,5′-trimethyl-cyclohex-2′-en-1′-yl)prop-2-enylammonium iodide. (*142*): N,N,N-trimethyl-1-methyl-(3′,3′,5′-trimethyl-cyclohexan-1′-yl)prop-2-enylammonium iodide. (*143*): 1-decylimidazole.

that when *G. fujikuroi* (G-2) is cultured in the presence of CCC, *ent*-15α-hydroxykaurenoic acid is converted into *ent*-7α,15α-dihydroxy-kaurenoic acid but seemingly not into 15β-hydroxy GAs (Wada et al., 1979). Hanson et al. (1979) have reported that AMO-1618-treated *G. fujikuroi* cultures convert *ent*-7α-hydroxyatis-16-ene-19-oic acid (*135*) to atisa GA_{12} (*136*) and atisa GA_{14} (*137*). Steric hindrance prevented oxidation of the C-20 methyl group.

Further details on the production of assorted GA and *ent*-kaurene analogues by wild-type and mutant strains of *G. fujikuroi* can be obtained from Chapter 5.

7.4.3 Utilization of cell-free systems from *Gibberella fujikuroi* and higher plants

Cell-free systems derived from *G. fujikuroi* and higher plant species have also been used for studies on GA biosynthesis. Although these *in vitro* systems are not applicable for large-scale preparation of GAs and related compounds, they have nonetheless been used by several research groups to prepare valuable labeled precursors for bio-synthetic investigations. The *in vitro* conversion of *ent*-kaurenoids to *ent*-gibberellanes has so far been successfully demonstrated in *Cucurbita maxima* (Graebe et al., 1972), *Pisum sativum* (Ropers et al., 1978), and *Phaseolus coccineus* (Ceccarelli, Lorenzi, & Alpi, 1981). These systems are described in detail in Chapter 3, and so only representative cases will be touched on here.

Shechter and West (1969) prepared [^{14}C]geranylgeranyl pyrophosphate, [^{14}C]copalyl pyrophosphate, and *ent*-[^{14}C]kaurene from [2-^{14}C]mevalonate using a soluble enzyme system prepared from a cell-free extract of *G. fujikuroi*. Cell-free preparations from the liquid endosperm of *C. maxima* seed have been extensively investigated and can be used to prepare a ubiquitous assortment of isotopically labeled GAs and GA precursors (Graebe et al., 1974; Graebe, 1980). Under appropriate conditions it is possible to obtain a better than 30% conversion of [2-^{14}C]mevalonic acid into *ent*-kaurene, *ent*-kaurenol, *ent*-kaurenal, *ent*-kaurenoic acid, *ent*-7α-hydroxykaurenoic acid, and GA_{12} aldehyde. Reincubation of [^{14}C]GA_{12} aldehyde can yield GA_4, GA_{12}, GA_{13}, GA_{14}, GA_{14} aldehyde, GA_{15}, GA_{24}, GA_{25}, GA_{36}, GA_{37}, GA_{43}, and a range of 12α-hydroxy GAs.

7.5 SUMMARY

An array of chemical, biochemical, and microbiological procedures have been described which can be used to prepare a wide range of rare GAs and their isotopically labeled derivatives from readily available GA and *ent*-kaurane substrates. These procedures are likely to be used with increasing frequency and will be of significant practical value to many investigators in the field of GA biochemistry and physiology.

REFERENCES

Adamiec, A. (1966). The effect of plant growth regulators on strains of *Fusarium moniliforme* Cheld. [*Gibberella fujikuroi* (Saw.) Wr.]. Acta Soc. Bot. Pol. *35*, 487–509.

Bartlett, P. A., Johnson, W. S. (1970). An improved reagent for O-alkyl cleavage of methyl esters by nucleophilic displacement. Tetrahedron Lett. *1970*, 4459–4462.

Beale, M. H., Gaskin, P., Kirkwood, P. S., & MacMillan, J. (1980). Partial synthesis of gibberellin A_9 and $(3\alpha$- and 3β-$^2H_1)$gibberellin A_9; gibberellin A_5 and $(1\beta,3$-$^2H_2$ and -$^3H_2)$gibberellin A_5; and gibberellin A_{20} and $(1\beta,3$-$^2H_2$ and -$^3H_2)$gibberellin A_{20}. J. Chem. Soc., Perkin Trans. I, 885–891.

Beale, M. H., & MacMillan, J. (1980). Mechanism and stereochemistry of conjugate reduction of enones from gibberellins A_3 and A_7. J. Chem. Soc., Perkin Trans. I, 877–884.

Bearder, J. R., Dennis, F. G., MacMillan, J., Martin, G. C., & Phinney, B. O. (1975a). A new gibberellin (A_{45}) from seed of *Pyrus communis* L. Tetrahedron Lett., 669–670.

Bearder, J. R., Frydman, V. M., Gaskin, P., Hatton, I. K., Harvey, W. E., MacMillan, J., & Phinney, B. O. (1976a). Fungal products part XVII. Microbiological hydroxylation of gibberellin A_9 and its methyl ester. J. Chem. Soc., Perkin Trans. I, 178–183.

Bearder, J. R., Frydman, V. M., Gaskin, P., MacMillan, J., & Phinney, B. O. (1976b), Fungal products part XVI. Conversion of isosteviol and steviol acetate in gibberellin analogues by mutant B1-41a of *Gibberella fujikuroi* and the preparation of (^{3}H)gibberellin A_{20}. J. Chem. Soc., Perkin Trans. I, 173-177.

Bearder, J. R., & MacMillan, J. (1973). Fungal products part IX. Gibberellins A_{16}, A_{36}, A_{37}, A_{41}, and A_{42} from *Gibberella fujikuroi*. J. Chem. Soc., Perkin Trans. I, 2824-2830.

Bearder, J. R., MacMillan, J., Lichterfelde, C. C., & Hanson, J., R. (1979a). The removal of C(20) in gibberellins. J. Chem. Soc., Perkin Trans. I, 1918-1921.

Bearder, J. R., MacMillan, J., & Matsuo, A. (1979b). Conversion of trachylobanic acid into novel pentacyclic analogues of gibberellins by *Gibberella fujikuroi*, mutant B1-41a. J. Chem. Soc., Chem. Commun., 649-650.

Bearder, J. R., MacMillan, J., & Phinney, B. O. (1973a). 3-Hydroxylation of gibberellin A_{12}-aldehyde in *Gibberella fujikuroi* strain REC-193A. Phytochem. *12*, 2173-2179.

Bearder, J. R., MacMillan, J., & Phinney, B. O. (1973b). Conversion of gibberellin A_1 into gibberellin A_3 by the mutant R-9 of *Gibberella fujikuroi*. Phytochem. *12*, 2655-2659.

Bearder, J. R., MacMillan, J., & Phinney, B. O. (1975b). Fungal products part XIV. Metabolic pathways from *ent*-kaurenoic acid to the fungal gibberellin in mutant B1-41a of *Gibberella fujikuroi*. J. Chem. Soc., Perkin Trans. I, 721-726.

Bearder, J. R., MacMillan, J., Wels, C. N., Chaffey, M. B., & Phinney, B. O. (1974). Position of the metabolic block for gibberellin biosynthesis in mutant B1-41a of *Gibberella fujikuroi*. Phytochem. *13*, 911-917.

Beeley, L. J., Gaskin, P., & MacMillan, J. (1975). Gibberellin A_{43} and other terpenes in endosperm of *Echinocystis macrocarpa*. Phytochem. *14*, 779-783.

Beeley, L. J., & MacMillan, J. (1976). Partial Synthesis of 2-hydroxygibberellins; Characterization of two new gibberellins, A_{46} and A_{47}. J. Chem. Soc., Perkin Trans. I, 1022-1028.

Bowen, D. H., Cloke, C., Harrison, D. M., & MacMillan, J. (1975). Partial synthesis of gibberellin A_{37} from gibberellin A_{13}. J. Chem. Soc., Perkin Trans. I, 83-88.

Ceccarelli, N., Lorenzi, R., & Alpi, A. (1981). Gibberellin biosynthesis in *Phaseolus coccineus* suspension. Z. Pflanzenphysiol. *102*, 37-44.

Cho, K. Y., Sakurai, A., Kamiya, Y., Takahashi, N., & Tamura, S. (1979). Effects of the new plant growth retardants of quarternary ammonium iodide on gibberellin biosynthesis in *Gibberella fujikuroi*. Plant Cell Physiol. *20*, 75–81.

Corey, E. J., Brennan, T. M., & Carney, R. L. (1971). Stereospecific elaboration of the A ring of gibberellic acid by partial synthesis. J. Amer. Chem. Soc. *93*, 7316–7317.

Cross, B. E., & Myers, P. L. (1969). The effect of plant growth retardants on the biosynthesis of diterpenes by *Gibberella fujikuroi*. Phytochem. *8*, 79–83.

Cross, B. E., Norton, K., & Stewart, J. C., (1968). The biosynthesis of the gibberellins part III. J. Chem. Soc. (C), 1054–1063.

Dennis, D. T., Upper, C. D., & West, C. A. (1965). An enzymic site of inhibition of gibberellin biosynthesis by amo 1618 and other plant growth retardants. Plant Physiol. *40*, 1948–1952.

Dockerill, B., Evans, R., & Hanson, J. R. (1977). Removal of C-20 in gibberellin biosynthesis. J. Chem. Soc., Chem. Commun., 919–921.

Dockerill, B., & Hanson, J. R. (1978). The fate of C-20 in C_{19}-gibberellin biosynthesis. Phytochem. *17*, 701–704.

Galt, R. H. B., & Hanson, J. R. (1965). New metabolites of *Gibberella fujikuroi* part VII. The preparation of some ring B nor-derivatives. J. Chem. Soc. (C), 1565–1572.

Graebe, J. E. (1980). GA-biosynthesis: The development and application of cell-free systems for biosynthetic studies. In: Plant Growth Substances 1979. Proceedings of the 10th International Conference on Plant Growth substances, pp. 180–197, Skoog, F., ed. Springer-Verlag, Berlin, Heidelberg, New York.

Graebe, J. E., Bowen, D. H., & MacMillan, J. (1972). The conversion of mevalonic acid into gibberellin A_{12}-aldehyde in a cell-free system from *Cucurbita pepo*. Planta (Berlin) *102*, 261–271.

Graebe, J. E., & Hedden, P. (1975). The ring contraction step in gibberellin biosynthesis. J. Chem. Soc., Chem. Commun., 161–162.

Graebe, J. E., Hedden, P., Gaskin, P., & MacMillan, J. (1974). Biosynthesis of gibberellins A_{12}, A_{15}, A_{24}, A_{36} and A_{37} by a cell-free system from *Cucurbita maxima*. Phytochem. *13*, 1433.

Hanson, J. R. (1980). The partial synthesis and labeling of some gibberellins. In: Gibberellins—Chemistry, Physiology and Use, Monograph 5, pp. 5–16, Lenton, J. R., ed. British Plant Growth Regulator Group, Wantage.

Hanson, J. R., Hawker, J., & White, A. F. (1972). Studies in terpenoid biosynthesis part IX. The sequence of oxidation on ring B in kaurene-gibberellin biosynthesis. J. Chem. Soc., Perkin Trans. I, 1892-1895.

Hanson, J. R., Sarah, F. Y., Fraga, B. M., & Hernandez, M. G. (1979). The microbiological preparation of two atisagibberellins. Phytochem. *18*, 1875-1876.

Harada, H., & Lang, A. (1965). Effect of some (2-chloroethyl)-trimethyl-ammonium chloride analogs and other growth retardants on gibberellin biosynthesis in *Fusarium moniliforme*. Plant Physiol. *40*, 176-183.

Harrison, D. M., & MacMillan, J. (1971). Two new gibberellins A_{24} and A_{25} from *Gibberella fujikuroi*, their isolation, structure, and correlation with gibberellin A_{13} and A_{15}. J. Chem. Soc. (C), 631-636.

Haruta, H., Yagi, H., Iwata, T., & Tamura, S. (1974). Synthesis and plant growth retardant activities of trimethylammonium compounds containing a terpenoid moiety. Agr. Biol. Chem. *38*, 141-148.

Hedden, P., MacMillan, J., & Grinsted, M. J. (1973). Fungal products part VIII. New kaurenolides from *Gibberella fujikuroi*. J. Chem. Soc., Perkin Trans. I, 2773-2778.

Hedden, P., MacMillan, J., & Phinney, B. O. (1974). Fungal products part XII. Gibberellin A_{14}-aldehyde, an intermediate in gibberellin biosynthesis in *Gibberella fujikuroi*. J. Chem. Soc., Perkin Trans. I, 587-592.

Hedden, P., Phinney, B. O., MacMillan, J., & Sponsel, V. M. (1979). Metabolism of kaurenoids by *Gibberella fujikuroi* in the presence of the plant growth retardant, N,N,N-trimethyl-1-methyl-(2',6',6'-trimethylcyclohex-2'-en-1'-yl)-prop-2-enylammonium iodide. Phytochem. *16*, 1913-1917.

Jones, D. F., & McCloskey, P. (1963). Selective reduction of gibberellic acid. J. Appl. Chem. *13*, 324-328.

Kende, H. (1967). Preparation of radio active gibberellin A_1 and its metabolism in dwarf peas. Plant Physiol. *42*, 1612-1618.

Kende, H., Ninnemann, H., & Lang, A. (1963). Inhibition of gibberellic acid biosynthesis in *Fusarium moniliforme* by Amo 1618 and CCC. Naturwissenschaften *50*, 599-600.

Lang, A. (1970). Gibberellins; Structure and metabolism. Ann. Rev. Plant Physiol. *21*, 537-570.

Lewis, N. J., & MacMillan, J. (1980). Terpenoids part VIII. Partial synthesis of *ent*-11β-, 12α-, and 12β-hydroxykaur-16-en-19-oic acid from grandiflorenic acid. J. Chem. Soc. Perkin Trans. I, 1270-1278.

MacMillan, J. (1978). Gibberellin metabolism. Pure Appl. Chem. *50*, 995–1014.

MacMillan, J. (1980). Partial synthesis of isotopically labelled gibberellins. In: Plant Growth Substance 1979. Proceedings on the 10th International Conference on Plant Growth Substances, pp. 161–169. Skoog, F., ed. Springer-Verlag, Berlin, Heidelberg, New York.

MacMillan, J., & Pryce, R. J. (1967). The solvolysis of gibberelin A_5 methyl ester toluene-*p*-sulfonate—a Bicyclo[3.2.1]octane-bridged derivative. J. Chem. Soc. (C), 550–554.

MacMillan, J., Seaton, J. C., & Suter, P. J. (1960). Plant hormones—I. Isolation of gibberellin A_1 and gibberellin A_5 from *Phaseolus multiflorus*. Tetrahedron *11*, 60.

Murofushi, N., Durley, R. C., & Pharis, R. P. (1977). Preparation of radioactive gibberellins A_{20}, A_5 and A_8. Agr. Biol. Chem. *41*, 1075–1079.

Murofushi, N., Nagura, S., & Takahashi, N. (1979a). Metabolism of steviol by *Gibberella fujikuroi* in the presence of plant growth retardant. Agr. Biol. Chem. *43*, 1159–1161.

Murofushi, N., Sugimoto, M., Itoh, K., & Takahashi, N. (1979b). Three novel gibberellins produced by *Gibberella fujikuroi*. Agr. Biol. Chem. *43*, 2179–2185.

Murofushi, N., Sugimoto, M., Itoh, K., & Takahashi, N. (1980). A novel gibberellin, GA_{57}, produced by *Gibberella fujikuroi*. Agr. Biol. Chem. *44*, 1583–1587.

Murofushi, N., Yamaguchi, I., Ishigooka, H., & Takahashi, N. (1976). Chemical conversion of gibberellin A_{13} to gibberellin A_4. Agr. Biol. Chem. *40*, 2471–2474.

Murofushi, N., Yokota, T., Watanabe, A., & Takahashi, N. (1973). Isolation and characterization of gibberellins in *Calonyction aculeatum* and structures of gibberellins A_{30}, A_{31}, A_{33} and A_{34}. Agr. Biol. Chem. *37*, 1101–1113.

Nadeau, R., & Rappaport, L. (1974). The synthesis of (^{3}H) gibberellin A_3 and (^{3}H) gibberellin A_1 by the paradium-catalized actions of carrier-free tritium of gibberellin A_3. Phytochem. *13*, 1537–1545.

Nakahara, Y., Mori, K., & Matsui, M. (1971). Diterpenoid total synthesis part XVI. Alternative synthetic routes to (±)- steviol and (±)-kaur-16-en-19-oic acid. Agr. Biol. Chem. *35*, 918–928.

Pitel, D. W., & Vining, L. C. (1970). Preparation of gibberellin A_1-3,4-^{3}H. Can. J. Biochem. *48*, 259–263.

Robinson, D. R., & West, C. A. (1970a). Biosynthesis of cyclic diterpenes in extracts from seedlings of *Ricinus communis* L. I. Identification of diterpene hydrocarbons formed from mevalonate. Biochem. *9*. 70–79.

Robinson, D. R., & West, C. A. (1970b). Biosynthesis of cyclic diterpenes in extract from seedlings of *Ricinus communis* L. II. Conversion of geranyl-geranyl pyrophosphate into diterpene hydrocarbon and partial purification of the cyclization enzyme. Biochem. *9*, 80–89.

Roper, H. J., Graebe, J. E., Gaskin, P., & MacMillan, J. (1978). Gibberellin bio-synthesis in a cell-free system from immature seeds of *Pisum sativum*. Biochem. Biophys. Res. Commun. *80*, 690–697.

Schneider, G., Sembdner, G., & Schreiber, K. (1977). Gibberellin-L. Synthesis von $O(3)$- und $O(13)$-glucosylierten gibberellinen. Tetrahedron *33*, 1391–1397.

Shechter, I., & West, C. A. (1969). Biosynthesis of gibberellins. (IV) Biosynthesis of cyclic diterpenes from *trans*-geranylgeranyl pyrophosphate. J. Biol. Chem. *244*, 3200–3209.

Spector, C., & Phinney, B. O. (1968). Gibberellin biosynthesis; Genetic studies in *Gibberella fujikuroi*. Physiol. Plant. *21*, 127–136.

Sponsel, V. M., & MacMillan, J. (1977). Further studies on the metabolism of gibberellins (GAs) A_9, A_{20} and A_{29} in immature seeds of *Pisum sativum* cv. Progress No. 9. Planta *135*, 129–136.

Wada, K. (1978). New gibberellin biosynthesis inhibitors, 1-decyl- and 1-geranyl-imidazole; Inhibitors of (−)-kaurene 19-oxidation. Agr. Biol. Chem. *42*, 2411–2413.

Wada, K., Imai, T., & Suibata, K. (1979). Microbial productions of unnatural gibberellins from (−)-kaurene derivatives in *Gibberella fujikuroi*. Agr. Biol. Chem. *43*, 1157–1158.

Yamaguchi, I., Miyamoto, M., Yamane, H., Murofushi, N., Takahashi, N., & Fujita, K. (1975a). Elucidation of the structure of gibberellin A_{40} from *Gibberella fujikuroi*. J. Chem. Soc., Perkin Trans. I, 996–999.

Yamaguchi, I., Takahashi, N., & Fujita, K. (1975b). Application of ^{13}C nuclear magnetic resonance to the study of gibberellins. J. Chem. Soc., Perkin Trans. I, 992–996.

Yokota, T., Reeve, D. R., & Crozier, A. (1976). The synthesis of ^{3}H-gibberellin A_9 with high specific activity. Agr. Biol. Chem. *40*, 2091–2094.

8

Modern Methods of Analysis of Gibberellins

Alan Crozier & Richard C. Durley

8.1 INTRODUCTION

Analysis of endogenous gibberellins (GAs) is one of the more complex areas of analytical chemistry and at a practical level can be divided into the following sequential steps: (i) extraction and partitioning, (ii) group purification procedures, (iii) separation, and (iv) identification. GAs are comparatively major components in extracts from the fungus *Gibberella fujikuroi*, and thus little purification is necessary before identification can be attempted. In contrast, they are only trace constituents in extracts from higher plants; as a consequence, substantial purification is required before realistic attempts can be made at characterization. This problem is especially severe with vegetative tissues because GA levels are usually several orders of magnitude lower than those encountered in immature seed. In such circumstances, a multistep analytical sequence is required to attain a degree of sample purity that will facilitate accurate analysis.

The application of a standardized set of purification procedures to samples from diverse sources, without prior experimentation, is an approach that sooner or later will result in inaccurate analysis, as not only the GAs but also the nature and the levels of contaminants will vary greatly from one extract to another. The exact combination of procedures to be used is best determined by an on-the-spot assessment. When deciding what techniques to employ it is important to bear in mind two general points. First, in the initial stages of purification, the substantial extract weights that are encountered necessitate the utilization of chromatographic techniques that offer a high sample capacity; second, purification is most effectively achieved if the individual procedures are based on widely different separatory mechanisms.

Sample losses invariably occur during purification, and this adversely affects the accuracy of quantitative analyses. Such errors can be corrected through the use of an appropriate internal marker which is added to the sample at the extraction stage. The most suitable internal markers are isotopically labeled analogues of the particular GA under study. ^{2}H-, ^{3}H-, ^{18}O-, ^{13}C-, and ^{14}C-GAs tend to behave in the same manner as their endogenous counterparts yet can be differentiated by mass spectrometry or radioassay.

8.2 EXTRACTION AND PARTITIONING TECHNIQUES

Over the years an array of GA extraction and partitioning procedures have been used, and this must cause confusion to newcomers to the field. Tissues are usually extracted with methanol or ethanol, although aqueous buffers have also been tried. In ^{3}H-GA metabolism studies, methanol removes over 95% of the radioactivity in *Phaseolus coccineus* seedlings (Reeve & Crozier, 1978). It is, however, an open question as to whether or not endogenous GAs are removed from all cellular sites with equal efficiency. Browning and Saunders (1977) reported that extraction of isolated chloroplasts of *Triticum aestivum* with the detergent Triton-X yielded far higher levels of GA_4 and GA_9 than methanol extracts. Unfortunately, similar results have not been obtained when the experiment has been carried out in other laboratories (Railton & Rechav, 1979). Buffer extracts from pea seedlings contain fewer impurities and more GA-like activity than methanolic extracts (Jones, 1968). However, it does not necessarily follow that buffer is the more effective in removing GAs from plant tissues, since the bioassay data could just as well reflect reduced inhibitor concentrations as increased GA levels. When methanol and buffer extracts are subjected to several purification steps prior to bioassay, the methanolic extract yields higher levels of GA-like activity (Reid & Crozier, 1970).

Figure 8.1 outlines a set of extraction and partitioning procedures that can be used as a first step in the analysis of most GAs. Tissue is macerated and extracted three times with an excess of cold methanol. The combined methanolic extracts are reduced to the aqueous phase *in vacuo* and the aqueous residue diluted at least 2-fold with pH 8.0, 0.5 M phosphate buffer. This stabilizes the pH and ensures a minimum ionic strength during the ensuing partition procedures. At pH 8.0 the aqueous phase is sufficiently basic to retain even the less polar GAs when partitioning against petroleum spirit, yet not so basic as to risk isomerization. An alternative method is to partition against diethyl ether at pH 9.0. Many investigators, however, choose to partition the aqueous phase against either ethyl acetate or diethyl ether at pH 8.0. The GA partition coefficient data of Durley and Pharis (1972), presented in Tables 8.1 and 8.2, clearly show that this results in the removal of significant quantities of nonpolar GAs, especially GA_4, GA_7, GA_9, and GA_{12}. Durley and Pharis (1972) also noted that the distribution of GAs between ethyl acetate and phosphate buffer is dependent upon the molarity of the buffer. This implies that the concentration of an extract can affect the partitioning behavior of GAs between the aqueous and organic phases.

After partitioning at pH 8.0, the buffer phase can be further purified by either slurrying with insoluble polyvinylpyrrolidone (PVP) and polyamide or elution from a short PVP column (Glenn et al., 1972) before acidification to pH 2.5 and extraction with $5 \times 2/5$ volumes of ethyl acetate. At this pH the partition coefficients are such that the bulk of the free GAs are removed by the ethyl acetate. The tetra-hydroxy compound GA_{32} is the only known free GA that will be retained by the buffer to any extent (Yamaguchi et al., 1970). Certain GA conjugates also migrate into the ethyl acetate. For example, the glucosyl esters of GA_4 and GA_7 both partition into ethyl acetate at pH 2.5 (Hiraga et al., 1974). However, other GA conjugates are extracted from the acidified aqueous phase with n-butanol. The analysis of GA conjugates is discussed in detail in Chapter 6.

Macerate tissue and extract 3 times with an excess of cold methanol.
Combine methanolic extracts and reduce to the aqueous phase *in vacuo.*
Add at least an equivalent volume of pH 8.0, 0.5 M phosphate buffer
and if necessary adjust extract to pH 8.0

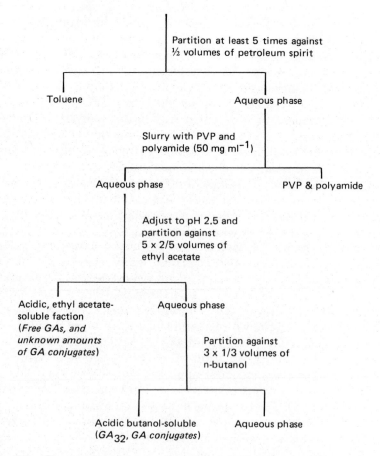

Partition at least 5 times against
½ volumes of petroleum spirit

Toluene

Aqueous phase

Slurry with PVP and
polyamide (50 mg ml^{-1})

Aqueous phase

PVP & polyamide

Adjust to pH 2.5 and
partition against
5 x 2/5 volumes of
ethyl acetate

Acidic, ethyl acetate-
soluble faction
(*Free GAs, and
unknown amounts
of GA conjugates*)

Aqueous phase

Partition against
3 x 1/3 volumes of
n-butanol

Acidic butanol-soluble
(*GA$_{32}$, GA conjugates*)

Aqueous phase

Figure 8.1 Flow diagram of extraction and participating techniques.

Table 8.1. Partition coefficients ($K_d = C_{aq}/C_{org}$) of GAs between ethyl acetate and 1.5 M phosphate buffer solution at five pH values

Gibberellin	pH K_d				
	8.0	6.5	5.0	3.5	2.5
A_1	∞	∞	1.2	0.17	0.11
A_2	∞	7.9	0.97	0.19	0.15
A_3	∞	∞	1.2	0.21	0.17
A_4	2.2	0.29	0.05	0	0
A_5	∞	4.8	0.19	0	0
A_6	∞	5.4	0.49	0.05	0
A_7	3.2	0.56	0.10	0	0
A_8	∞	∞	4.9	0.64	0.45
A_9	0.34	0.06	0	0	0
A_{10}	11.3	1.6	0.33	0	0
A_{12}	0.56	0.04	0	0	0
A_{13}	∞	7.1	0.06	0	0
A_{14}	∞	0.41	0	0	0
A_{16}	∞	3.2	0.16	0	0
A_{17}	∞	∞	0.50	0.04	0
A_{18}	∞	∞	0.42	0	0
A_{19}	∞	4.6	0.81	0.10	0
A_{20}	∞	2.1	0.09	0	0
A_{21}	∞	∞	9.1	0.89	0.08
A_{22}	∞	15.1	1.4	0.51	0.19
A_{23}	∞	∞	19.4	1.0	0.17
A_{24}	∞	0.83	0	0	0
A_{25}	13.1	0.66	0	0	0
A_{26}	∞	∞	3.2	0.44	0.21
A_{27}	∞	1.6	0.18	0.05	0
A_{28}	∞	∞	12.7	0.81	0.07
A_{29}	∞	∞	1.9	0.21	0.15

After Durley & Pharis (1972).
NOTE: Values of K_d below 0.02 are taken as 0; values over 20 are taken as ∞.

Table 8.2. Partition coefficients ($K_d = C_{aq}/C_{org}$) of GA$_3$, GA$_4$, GA$_5$, GA$_9$, and GA$_{13}$ acid between diethyl ether and 0.5 M phosphate buffer at six pH values

Gibberellin	pH K_d					
	9.0	8.0	6.5	5.0	3.5	2.5
A$_3$	∞	∞	∞	18.5	7.0	3.1
A$_4$	∞	∞	4.1	0.72	0.12	0
A$_5$	∞	∞	∞	3.1	0.32	0.12
A$_9$	∞	6.2	0.28	0.07	0	0
A$_{13}$	∞	∞	∞	1.1	0.33	0.15

After Durley & Pharis (1972).
NOTE: Values of K_d below 0.02 are taken as 0; values over 20.0 are taken as ∞.

8.3 GROUP SEPARATORY PROCEDURES

The concentration of the GAs in the acidic, ethyl acetate-soluble fraction is usually very low, and so further purification is required. It is advisable at this stage to use procedures that separate the GAs as a group from other components in the sample; otherwise, unwieldy numbers of subfractions are quickly generated and there will be a marked decrease in the overall speed of analysis.

PVP adsorption chromatography has been widely used as a group purification procedure for GAs. Extracts weighing up to 200 mg are loaded onto a 300×19 mm I.D. column, and the GAs elute between 60 and 160 ml in a 0.1 M, pH 8.0 phosphate buffer mobile phase. Sample weights are reduced substantially, and the recovery of GAs is reported to be more than 90% (Glenn et al., 1972). The technique is flexible in that the support is inexpensive and varying column diameters can be used to accommodate a wide range of sample sizes. Shorter columns than those employed by Glenn et al. (1972) are also effective and have the advantage of enhancing the speed of analysis. Alternatively, as already described in Section 8.2, a PVP slurry technique can be used at the partitioning stage.

Gel permeation or steric exclusion chromatography (SEC) has proved useful as a preparative, group separatory procedure (Reeve & Crozier, 1976, 1978). The system consists of two, $1,000 \times 25$ mm I.D. columns connected in series, packed with Biobeads SX-4, and eluted with tetrahydrofuran (THF) at a flow rate of $2\,\mathrm{ml\,min^{-1}}$. This is the maximum flow rate the support can tolerate without excessive compression of the bed. The gel has an operating range of 0–1,500 daltons, and solutes elute in decreasing order of molecular size. The sample capacity is high and is readily realized because of the excellent solubilizing power of THF, 1.5 ml of which will readily dissolve up to 1.0 gm of an acidic, ethyl acetate-soluble extract. Recoveries, estimated with a range of ^{3}H-GAs, are greater than 90%. The absence of adsorption effects ensures that even with the most impure extracts all solutes will be eluted by a volume of solvent (630 ml) which corresponds to the total volume of the column (V_t). This allows the system to be used repeatedly without fear of sample overlap. [^{3}H]GA$_9$ and [^{3}H]GA$_{43}$, which represent the extremes of the molecular weight range of free GAs, have retention volumes of 550 ml and 470 ml, respectively, in this system with peak widths (w) of 40 ml. Endogenous GAs in extracts can therefore be purified by collecting the 450–570-ml zone. While this technique is well suited for use with large-scale extracts, a high-performance SEC procedure is available which offers a very rapid speed of analysis for the purification of smaller sized samples (Crozier, Zaerr, & Morris, 1980). The procedure involves the use of a 10-μm PL gel support* with a nominal exclusion limit of less than 2,000 daltons. PL gel is a macroporous cross-linked polystyrene divinylbenzene copolymer support that has been specifically designed for high-performance liquid chromatography (HPLC). A 300 $\times$ 8 mm I.D. column eluted with 0.5% acetic acid in THF generates 9,000 theoretical plates and has a sample capacity of more than 100 mg. The exclusion or void volume (V_o) is 5.5 ml, and V_t is 9.5 ml. The V_r of [^{3}H]GA$_{43}$ is 7.0 ml and that of [^{3}H]GA$_9$, 7.6 ml. In both instances, $w = 0.4$ ml. Thus, collection of the 6.8–7.8-ml zone provides a simple means of separating free GAs as a group from many of the extraneous components typically present in plant extracts.

Gräbner, Schneider, and Sembdner (1976) used DEAE Sephadex A-25 anion exchange chromatography to separate abscisic acid, GA$_3$, GA$_7$, and the 3-O-β-D-glucosyl ether of GA$_3$. This procedure is readily adapted for use as a group separatory procedure for free GAs. A 150 $\times$ 40 mm I.D. column of DEAE Sephadex A-25 charged in the acetate form is eluted with four void volumes (600 ml) of methanol-0.2 N acetic acid (1:1 v/v) to remove neutral and weakly acidic impurities. The GAs are then eluted with $ca.$ 90% efficiency with two void volumes of methanol-2.0 N acetic acid (1:1 v/v).

---

*Polymer Laboratories Ltd., Essex Road, Church Stretton, Salop SY6 6AX, England. PL gel supports were originally marketed under the name of μ-Spherogel by Altex Scientific Inc., Berkeley, California, U.S.A.

Purification of GAs in plant extract can also be achieved by exploiting the unusual reverse-phase effects of charcoal adsorption chromatography (Yabuta & Hayashi, 1939). As currently employed, samples are dissolved in 1—2 ml of 20% aqueous acetone and applied to the top of a 120×20 mm I.D. charcoal-celite (1:2 w/w) column. Weakly adsorbed impurities are eluted with 100 ml of 20% aqueous acetone, which is equivalent to four column volumes. The GAs are then removed with 200 ml of acetone. The sample capacity of charcoal is high, and a column of the dimensions described can accommodate extracts weighing up to 500 mg. The method is also readily adaptable for use as a simple slurry procedure, in which case large numbers of extracts can be treated in a matter of minutes. The recovery of GAs from charcoal is usually 75—85%. However, inexplicably high losses do occur from time to time, even with the same batch of charcoal, and as a consequence the procedures should only be used when replicate samples are readily available and internal standards are employed.

A C_{18} Sep-Pak (Waters Associates, Inc) can be used as a simple and rapid reverse-phase group purification procedure for endogenous GAs. A Sep-Pak consists of a small cartridge containing silica gel coated with a chemically bonded octadecylsilane (C_{18} or ODS) stationary phase. As a first step, 10 ml of 70% methanol in pH 3.5, 20 mM ammonium acetate buffer is pushed through the cartridge using a syringe. The extract, dissolved in 5 ml of 70% methanol, is then applied to the Sep-Pak in a similar manner and is followed by a further 5 ml of 70% methanol. Under these circumstances, many of the less polar contaminants are retained, while the GAs pass through the cartridge and can be recovered from the methanolic eluate in yields of over 90%. It is possible to remove polar impurities by adding the sample in 20% methanol and then eluting the cartridge with 20% methanol prior to removing the GAs with 70% methanol. This procedure can, however, result in the early elution of some of the more polar GAs, such as GA_8, in the 20% methanol fraction. It should be noted that the sample capacity of a Sep-Pak is low and overloading will have deleterious effects on both GA recoveries and the degree of purification that is obtained.

A Sephadex G-10 column eluted with 0.1 M, ph 8.0 phosphate buffer will retain GAs by virtue of uncharacterized adsorption phenomena and can be of value as a purification tool (Crozier, Aoki, & Pharis, 1969). However, the procedure is time consuming and is not as effective as other techniques that have been described. Countercurrent distribution has also been used as a preliminary purification procedure for GAs (Crozier et al., 1969), but it is now somewhat outmoded and is likely to do little that cannot be more effectively achieved with other procedures.

8.4 SEPARATORY TECHNIQUES

Extracts that have been subjected to a range of group purification procedures will often require further purification before successful attempts can be made at GA analysis. This is achieved through the use of analytical methods that separate the GAs to some degree. Originally, paper chromatography was the method of choice (see Phinney & West, 1961), but this was superseded by thin-layer chromatography (TLC) (MacMillan & Suter, 1963; Kagawa, Fukinbara, & Sumiki, 1963) which is still widely used especially in conjunction with bioassays. However, liquid-liquid partition column chromatography systems offering a high peak capacity have the ability to simultaneously resolve a large number of components and, as a consequence, provide much better separations than TLC. Several such systems have been used with GAs, and on occasions some truly extraordinary—but sadly unrepeatable—separations have been claimed. In general, good separations have been obtained with techniques utilizing either silica gel or dextran gel supports.

Adequate results can be achieved, without recourse to expensive instrumentation, with a normal-phase silica gel partition column (Powell & Tautvydas, 1967) originally developed to analyze indole-3-acetic acid and other indoles (Powell, 1963). The system involves partitioning a 0–100% gradient of ethyl acetate in hexane against a 40% (v/v) $0.5\,M$ formic acid stationary phase on a Mallinckrodt CC-4 silica gel support. The elution pattern of 12 GAs presented in Table 8.3 indicates that separation is primarily determined by the degree of hydroxylation. Durley et al. (1972) subsequently reported that the technique works well only with certain batches of CC-4 silica gel because columns tend to temporarily "dry out" at an early point in the gradient. They therefore devised a modified procedure that uses a Woelm silica gel support with a 15% water stationary phase and an ethyl acetate-hexane gradient that started at 65% ethyl acetate and dropped to 50% before rising to 100%. The mobile-phase solvents were saturated with $0.5\,M$ formic acid. The retention properties of 33 GAs in this unusual chromatographic system are recorded in Table 8.4. Separation is related to the number and the position of polar groups on the *ent*-gibberellane skeleton. The more hydroxyl groups in the molecule, the later the GA is eluted. The position of

Table 8.3. Retention characteristics of GAs on a straight-phase silica gel partition colum

Fraction number	Gibberellin
2	GA_9
4-5	GA_4, GA_7
6	GA_5
8	GA_6, GA_{20}
11	GA_{13}, GA_{19}
13-14	GA_1, GA_3
18	GA_8
18-19	GA_{23}

After Durley et al. (1972).
NOTE: Column: 13 × 200 mm Mallinckrodt CC-4 silica gel. Stationary phase: 40%, 0.5 M formic acid. Mobile phase: 160-min gradient, 0–100% ethyl acetate in hexane. Flow rate: *ca.* 3 ml min^{-1}. Sample: GAs as indicated. Detector: 25 successive 20-ml fractions collected and GA content determined by gas chromatography.

the hydroxyl group also influences retention properties. In addition, C_{19}-GAs with an α-lactone ring elute earlier than equivalent C_{20}-GAs with a C-20 aldehyde function, while tricarboxylic GAs are more highly retained than their dicarboxylic and monocarboxylic counterparts.

It is now known that the "drying out" phenomenon experienced with the Powell-Tautvydas column is due to out-gassing of the solvents (Reeve et al., 1976). This can be overcome by degassing the hexane and ethyl acetate under reduced pressure immediately prior to use. Readsorption of atmospheric oxygen can be suppressed by entraining a stream of nitrogen or helium over the solvent reservoirs. When these precautions are taken, the procedures of Powell and Tautvydas (1967) are reproducible and batch-to-batch variations in Mallinckrodt CC-4 silica gel are not apparent. Other silica gels work equally well, and in certain instances their performance is far superior.

In the early 1970s dramatic advances took place in liquid chromatography technology, especially the development of efficient microparticulate silica gel supports (see Majors & MacDonald, 1973). This facilitated vast improvements in the performance of the silica gel partition system, which is especially well suited for the separation of GAs in plant extracts, since a 450 × 10 mm I.D. column can accommodate multicomponent samples weighing up to 100 mg. The high sample capacity is a direct consequence of the high stationary-phase loading. The relatively high miscibility of the ethyl acetate mobile phase and formic acid stationary phase does, however, present special problems that must be overcome if high column efficiencies are to be obtained. This can be achieved through the use of a stationary-phase trap in the solvent delivery line and a precolumn, which, along

Table 8.4. Retention characteristics of GAs on a Woelm silica gel partition column

Fraction number	Gibberellin
2	A_9, A_{12}
3	A_{11}, A_{14}, A_{24}, A_{14}, A_{31}
4	A_4, A_5, A_6, A_7, A_{14}, A_{15}, A_{20}, A_{25}, A_{31}
5	A_{10}
6	A_{10}
8	A_{27}, A_{34}
9	A_{27}, A_{34}
10	A_{16}, A_{27}, A_{34}, A_{33}
11	A_{33}
12	A_{30}
13	A_1, A_3, A_{30}
14	A_{19}
15	A_2, A_{19}
16	A_2, A_{13}, A_{22}
17	A_{18}, A_{22}, A_{26}, A_{29}
18	A_{18}, A_{26}, A_{29}
19	A_{17}
20	A_{23}
21	A_{21}, A_{23}
22	
23	A_8, A_{28}
24	A_{28}

After Durley et al. (1972).
NOTE: Column: 200 × 13 mm I.D. Woelm silica gel. Stationary phase: 20% water. Mobile phase: 160-min gradient, 65 → 50 → 100% ethyl acetate in hexane, solvents saturated with 0.5 M formic acid. Flow rate: *ca.* 3 ml min^{-1}. Sample: GAs as indicated. Detector: 25 successive fractions collected and GA content determined by gas chromatography.

with the analytical column, is held at $30 \pm 0.05°C$ to ensure equilibration of the incoming mobile phase with the stationary phase (Reeve et al., 1976; Reeve & Crozier, 1978).

Reeve et al. (1976) assessed the performance of various types of silica gel in this system by chromatographing UV-absorbing phenol under various conditions and calculating column efficiencies by established procedures. Three silica gel supports were used:

Mallinckrodt CC-4: irregularly shaped particles with a wide size range (approximately 60–250 μm). This is the support used by Powell and Tautvydas (1967).

Merckogel SI 200: spherical, 40–63-μm particles with a 200-Å pore diameter.

Partisil 20: irregularly shaped 20-μm particles of closely controlled size distribution with 55–60-Å pore diameters.

The relationship between plate height (H) and linear solvent velocity (v) for columns packed with these gels is presented in Fig. 8.2. In each instance the concentration of ethyl acetate in the mobile phase was adjusted so that phenol eluted with a capacity factor (k') of 1.5. The performance of both the Mallinckrodt CC-4 and Merckogel SI 200 falls off at increased flow rates, the effect being much more marked with CC-4. Much better H values were obtained with Partisil

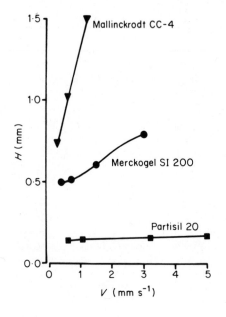

Figure 8.2 The relationship of plate height (H) to linear solvent velocity (v). Column: 450×10 mm I.D., packed with Mallinckrodt CC-4 Silicar, Merckogel SI-200, or Partisil 20. Sample: phenol. Stationary phase: 40%, 0.5 M formic acid. Mobile phase: hexane-ethyl acetate, ratio adjusted to give a k' of 1.5 for phenol. Detector: absorbance monitor at A_{254}. (Reeve et al., 1976.)

20, and no significant fall-off was evident at higher solvent velocities. From the practical viewpoint, Partisil 20 is clearly the superior support as it can generate high efficiencies without sacrificing the speed of analysis. Subsequently, 5-μm and 10-μm silica gel supports became commercially available and they enhance performance even further, although operating pressures are higher. A 450×10 mm I.D. column packed with Partisil 10 and eluted at a flow rate of 5 ml min^{-1}, which corresponds to $v = 1.5$ mm s^{-1}, generates up to 3,800 theoretical plates for a solute with a k' of 1.2. Thus, plate height and speed of analysis can be calculated at 0.12 mm and 1.1 effective plates per second. Depending upon solvent composition, a column inlet pressure of 140–200 p.s.i. is required. Recovery from the column is 95% for a wide range of compounds. These performance figures represent a considerable improvement in both efficiency and speed of analysis, when compared with classical liquid chromatography techniques used for the separation of GAs. The system is some ten times faster and twenty times more efficient than the silica gel partition column of Powell and Tautvydas (1967) from which it was derived. It also has the added advantage that instead of being used once and then discarded, columns can be used repeatedly for many years.

The transition from a silica gel partition to a high-performance system requires the use of more sophisticated equipment as described in detail by Reeve et al. (1976) and Reeve and Crozier (1978). On the other hand, it does introduce an element of flexibility in that manipulation of the hexane-ethyl acetate ratio produces a wide range of mobile-phase solvent strengths which can be used to provide a variety of rapid and effective separations. Figure 8.3 illustrates the use of a gradient designed for the rapid analysis of GAs with a wide range of polarities. As in the Powell-Tautvydas system, the GAs and GA precursors separate according to the degree of hydroxylation. It is evident that with the solvent program shown in Fig. 8.3, the chromatograph has insufficient peak capacity to resolve all 66 GAs that have been characterized to date. The peak capacity can be considerably enhanced by adjusting the gradient to give much larger effective k' values, although this would result in longer analysis times. If, however, the GAs under analysis have a limited polarity span, they can be studied in more detail without unduly sacrificing the speed of analysis by using a restricted gradient which specifically enhances the peak capacity within the relevant polarity zone as shown in Fig. 8.4. In metabolic studies metabolism of the applied GA often involves successive hydroxylations and the products are usually chromatographically distinct from each other and from the precursor molecule. In such cases a considerable saving can be made in the analysis time without the need for high effective k' values. This is shown in Fig. 8.5, where the solvent program has been adjusted to allow the repeated separation of GA_4, GA_1, and GA_8 at 30-min intervals.

Silica gel supports with a range of chemically bonded stationary phases have become increasingly available since the mid-1970s. The great advantage of bonded supports is that they are extremely stable,

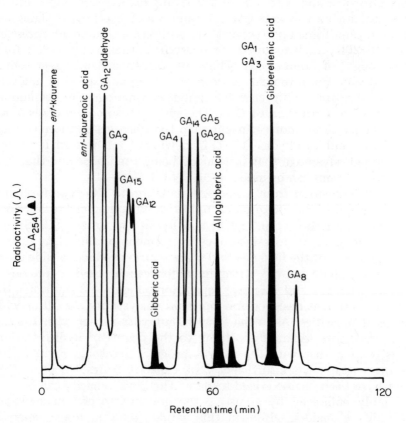

Figure 8.3 Preparative HPLC of radioactive GAs and GA precursors with UV-absorbing internal markers. Column: 450×10 mm I.D. Partisil 20. Stationary phase: 40%, 0.5 M formic acid. Mobile phase: 2-hr gradient 0–100% ethyl acetate in hexane. Flow rate: 5 ml min^{-1}. Sample: *ca.* 24,000 dpm *ent*-[^{14}C]kaurene, 50,000 dpm [^{14}C]GA$_3$, [^{3}H]GA$_5$, [^{14}C]GA$_{12}$, [^{14}C]GA$_{19}$, and [^{3}H]GA$_{20}$; 100,000 dpm *ent*-[^{3}H]kaurenoic and [^{3}H]GA$_1$, [^{3}H]GA$_4$, [^{3}H]GA$_8$, [^{3}H]GA$_9$, [^{3}H]GA$_{12}$ aldehyde, and [^{3}H]GA$_{14}$ and uncalibrated amounts of gibberic acid, allogibberic acid, and gibberellenic acid. Detectors: homogeneous radioactivity monitor, 1,800 cpm full-scale deflection, absorbance monitor at A$_{254}$. (Reeve et al., 1976.)

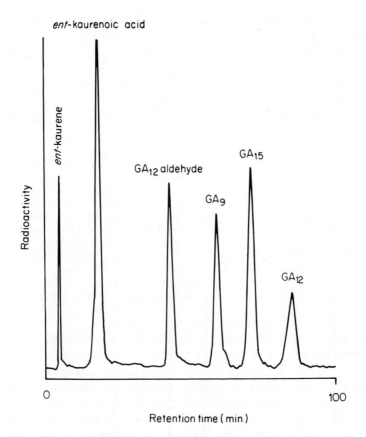

Figure 8.4   Preparative HPLC of radioactive GAs and GA precursors using a
restricted solvent gradient to give increased resolution at the nonpolar end of the
GA polarity spectrum.   Column: 450×10 mm I.D. Partisil 20. Stationary
phase: 40%, 0.5 M formic acid. Mobile phase: 70-min gradient 10–25% ethyl
acetate in hexane. Flow rate: 5 ml $\min^{-1}$. Sample: *ca.* 8,000 dpm *ent*-[^{14}C]-
kaurene, 20,000 dpm [^{14}C]GA$_{12}$ and [^{14}C]GA$_{15}$, and 40,000 dpm *ent*-[^{3}H]-
kaurenoic acid, [^{3}H]GA$_{9}$, and [^{3}H]GA$_{12}$ aldehyde. Detector: homogeneous
radioactivity monitor, 600 cpm full-scale deflection. (Reeve et al., 1976.)

and so the problems of equilibrating the mobile and stationary phases
encountered with the partition system of Reeve et al. (1976) are
avoided.

Reverse-phase liquid chromatography of free GAs on μ-Bondapak
C_{18}, which is a 10-μm silica gel support with a chemically bonded
C_{18} stationary phase, has been reported by Barendse, Van de Werken,
and Takahashi (1980), Jones, Metzger, and Zeevaart (1980), and
Koshioka et al (1983). Details published by Koshioka et al. (1983) of

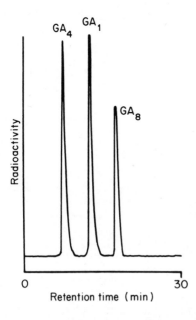

Figure 8.5   Preparative HPLC of GA_1, GA_4, and GA_8 using a restricted solvent for rapid analysis.   Column: 450×10 mm I.D. Partisil 20. Stationary phase: 40%, 0.5 M formic acid. Mobile phase: 20-min gradient 80–100% ethyl acetate in hexane. Flow rate: 5 ml min^{-1}. Sample: *ca.* 20,000 dpm [^{3}H]GA_1, [^{3}H]GA_4, and [^{3}H]GA_8. Detector: homogeneous radioactivity monitor, 600 cpm full-scale deflection. (Reeve et al., 1976.)

the retention characteristics of over 30 GAs are presented in Table 8.5. Some relationships between functional groups and elution order are apparent. As would be anticipated in a reverse-phase system, the more hydroxyl groups on the *ent*-gibberellin skeleton, the less retained the GA. 13-Hydroxy GAs elute earlier than their 2-hydroxy isomers, which, in turn, are less strongly retained than their 3-hydroxy analogues. Methylation of the 7-carboxyl function increases retention times, while $\Delta^{1,2}$ and $\Delta^{2,3}C_{19}$-GAs are less strongly retained than their saturated derivatives. There is no clear cut pattern in the elution of C_{19}- and C_{20}-GAs although the retention properties of GA_{12} and GA_{15}, GA_{14}, and GA_{36}, GA_{14} and GA_{13}, as well as GA_{18} and GA_{23}, implies that 20-CH_3 GAs are more strongly retained than their 20-CHO and 20-COOH equivalents. There are, however, exceptions to this general trend as GA_{53} elutes before GA_{12} and GA_{19}. Adsorption and reverse-phase separations of GAs have also been described by Lin and Heftmann (1981).

The performance of the C_{18} supports mentioned above has been improved by the introduction of 3-μm and 5-μm spherical reverse-phase supports such as ODS-Hypersil, Ultrasphere-ODS, and ODS-Spherisorb. HPLC columns packed with these supports offer much higher efficiencies, and this is an important consideration when attempting to resolve GAs in complex multicomponent samples. It should also be noted that although acidic buffers are routinely used

Table 8.5. Reverse-phase liquid chromatography of GAs

Retention time (min)	Gibberellin
12–13	GA_8
16–20	GA_{29} *
17–18	gibberellenic acid
21–25	GA_{23} *
25–26	epi-GA_1, GA_3, GA_1, GA_{23} catabolite
26–27	$\Delta^{1,10} GA_1$ counterpart
27–28	GA_{18}
28–29	GA_1 methyl ester, $\Delta^{1,10} GA_1$ counterpart 7-methyl ester
30	GA_{43}
30–31	GA_5, GA_{20}
31–32	GA_{44}
32–33	GA_{53}
33–34	GA_{19}, GA_{13}, allogibberic acid
33–35	GA_{34}
34–35	GA_{36} *
35–36	epi-allogibberic acid, GA_{17} *
36–37	GA_{37} *, epi-GA_4
37–38	GA_7, iso-GA_7, GA_4
38–39	GA_9
39–40	GA_{14}, GA_{25} *
40	GA_4 methyl ester
45–47	GA_{12} *
56–57	GA_{12} aldehyde, ent-kaurenoic acid
58–61	ent-kaurene

After Koshioka et al. (1983).

NOTE: Column: 300 × 3.9 mm I.D. μBondapak C_{18}. Mobile phase: 80-min gradient—0–10 min, 10% methanol in 1% aqueous acetic acid; 10–40 min, 10–73% methanol; 40–50 min, 73% methanol; 50–80 min, 73–100% methanol. Flow rate: 2 ml min^{-1}. Samples: GAs as indicated. Detectors: dwarf rice bioassays, absorbance monitor at 254 nm, radioactivity monitor.

*Retention times extrapolated from data of Jones et al. (1980).

to suppress solute ionization in the mobile phase, the chemical microenvironment of a reverse-phase support is such that a significant proportion of the GA molecules ionize on approaching the stationary-phase/silica gel interface. In the chromatographic situation this process is not in equilibrium, and so band broadening occurs. Free GAs thus generate much lower plate counts than the non-ionic text mixtures that are used by manufacturers to assess column performance. For instance, GA separations illustrated by Barendse et al. (1980) indicate efficiencies of 1,000–1,500 theoretical plates on a column that will yield around 7,000 plates when used to chromatograph solutes such as biphenyl, benzophenone, and acetophenone.

High column efficiencies and good GA separations have been achieved with dextran gels as a stationary-phase support. The procedures of Pitel, Vining, and Arsenault (1971) and Vining (1971), using Sephadex G-25, separate the double-bond isomers GA_1/GA_3 and GA_4/GA_7 and some of their closely related derivatives. These columns are, however, of restricted general value as they do not provide adequate resolution of other groups of GAs (Durley et al., 1972). An improved method, devised by MacMillan and Wels (1973), is able to separate a wide range of GAs and GA precursors. A biphasic solvent of light petroleum-ethyl acetate-acetic acid-methanol-water (100: 80:5:40:7) was prepared, and the aqueous phase was used to swell the Sephadex LH-20 support and act as a stationary phase. The gel was packed into a column and eluted with the organic phase. A total of 5,500 theoretical plates was generated on a $1,450 \times 15$ mm I.D. column, and excellent GA separations were obtained (Fig. 8.6). The sample capacity of this column is 100–200 mg, and so most plant extracts can be easily accommodated. The method has the advantage that it is relatively simple and does not require expensive, complex equipment. However, the 30-hr analysis period is a major problem as far as routine analyses are concerned, because it severely limits sample throughout. The speed of analysis, calculated from GA_3 in Fig. 8.6, is only 0.05 effective plates per second. Because of a lack of gel rigidity, it is unlikely that this situation could be improved by either increased solvent velocities or solvent programming (see Bombaugh, 1971). Despite this drawback, the procedure is an attractive proposition in circumstances where only small numbers of samples have to be analyzed.

Whatever method of chromatography is used to separate endogenous GAs, their subsequent detection can be a time-consuming process because of the lack of a specific label. The ^{3}H- and ^{14}C-GAs are, of course, an exception since they can be readily detected with a radioactivity monitor. If a known GA is being analyzed, the appro-

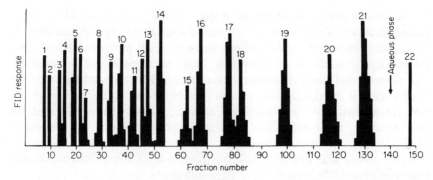

Figure 8.6 Separation of GAs by liquid-liquid partition chromatography on a Sephadex LH-20 support. Column: 1,450 × 15 mm I.D. Sephadex LH-20. Stationary phase: aqueous phase of light petroleum-ethyl acetate-methanol-water (100:80:5:40:7) mixture. Mobile phase: organic phase of above. Flow rate: 50 ml hr^{-1}. Sample: (1) ent-kaurene, (2) ent-kaurenoic acid, (3) GA$_{12}$ aldehyde, (4) GA$_{12}$ alcohol, (5) ent-7α-hydroxykaurenoic acid, GA$_9$, (6) steviol, GA$_{12}$, (7) GA$_{15}$, (8) GA$_{14}$ aldehyde, (9) GA$_{24}$, (10) GA$_4$, (11) GA$_7$, (12) GA$_{25}$, (13) iso-GA$_7$, GA$_{37}$, (14) GA$_{14}$, (15) GA$_5$, (16) mevalonic acid, (17) GA$_{36}$, (18) GA$_{16}$, (19) GA$_2$, (20) GA$_1$, GA$_{13}$, GA$_{17}$, (21) GA$_3$, and (22) GA$_8$ and GA$_{28}$. Detector: 150-ml fractions collected and analyzed by GC with a flame ionization detector. (MacMillan & Wels, 1973.)

priate zone of the chromatogram can be collected and subjected to additional analysis to facilitate identification and quantification. When the identity of the endogenous GAs are unknown, bioassays are commonly used to detect peaks of GA-like activity which are then subjected to physicochemical procedures such as combined gas chromatography-mass spectrometry (GC-MS), in order to characterize the active components. Very often, however, endogenous GAs are not identified in this manner and analyses go no further than bioassay, with the GA content of samples being compared on this basis.

8.5 BIOASSAYS

Historically, bioassays have played an important role in the discovery of GAs. Many GAs were originally detected in plant extracts because of their biological activity, and without such an indicator it is unlikely that the extensive purification that must precede rigorous chemical analysis could have been achieved. Indeed, it is interesting to speculate how much would currently be known about GAs if rice seedlings did not elongate so markedly when infected with *G. fujikuroi*. Over the years, numerous bioassays have been devised. Bailiss and Hill (1971) listed 33 test systems based on processes such as coleoptile, leaf sheath, epicotyl, mesocotyl, and radicle growth, bud dormancy and seed germination, α-amylase synthesis, leaf expansion and senescence, and flower and cone induction. The essential features of some of the more widely used GA assays are reviewed in Table 8.6, and the relative activities of the individual GAs in some of these test systems are listed in Table 8.7. The data are compiled from Crozier et al. (1970), Yokota et al. (1971), Fukui et al. (1972), Yamane et al. (1973), Reeve and Crozier (1975), Hoad et al. (1976), Sponsel, Hoad, and Beeley (1977), and Lenton (unpublished).

When used to detect GAs in plant extracts, bioassays are moderately selective, especially when compared to many physicochemical detectors. However, the selectivity is not absolute, because the bioassay response can be influenced by both inhibitors and synergistic contaminants in the sample. Such interactions will adversely affect the accuracy of the bioassay, a point that has been clearly illustrated in a variety of ways by Reeve and Crozier (1975), Graebe and Ropers (1978), and Letham et al. (1978). In practice, the situation is further compounded because individual GAs exhibit widely varying biological activities (Table 8.7). The threshold doses differ by several orders of magnitude, there is often no parallelism between the slopes of the response curves, and, in addition, the size of the dose required to saturate the response is far from uniform. It can, therefore, be very misleading to express levels of unknown GA-like compounds in terms of micrograms of GA_3 equivalents because the actual amount of GA present may differ by up to several orders of magnitude. If, however, previous work has shown that a particular zone of biological activity is associated with a certain GA, in subsequent experiments it becomes possible to make allowances for the dose-response curve of that GA rather than GA_3. Even this approach is not without its problems as has been demonstrated by testing a purified extract, containing GA_1, from *Phaseolus coccineus* seedlings, in several bioassay systems (Reeve & Crozier, 1980). The observed biological response was related to nanograms of GA_1 by reference to the regression of the response on log-dose GA_1. As this involves a log-normal distribution, the estimates are median rather than mean values and have asymetric confidence limits. Table 8.8 shows the estimates of GA_1 content based on the growth response induced by 1/60 and 1/120 aliquots of the purified sample in the Tanginbozu dwarf rice bioassay. The accuracy of the estimates is dependent upon the validity of the assumption that the dose/response curve of the sample is the same as the GA_1 dose/response curve. Because halving the dose size had no significant effect

Table 8.6. GA bioassay systems

Method	Reference	Minimum detectable level of GA_3	Range of linear response to GA_3
Tanginbozu dwarf rice leaf sheath bioassay	Murakami (1968)	100 pg	100 pg to 1 µg
Progress No. 9 dwarf pea epicotyl bioassay	Köhler & Lang (1963)	1 ng	1 ng to 1 µg
Dwarf maize leaf sheath bioassay	Phinney (1956)	5 ng	5 ng to 5 µg
Cucumber hypocotyl bioassay	Brian, Hemming, & Lowe (1964)	1 ng[a]	1 ng to 1 µg
Lettuce hypocotyl bioassay	Frankland & Wareing (1960)	1 ng ml^{-1} [b]	1 ng ml^{-1} to 1 µg ml^{-1}
Barley half seed bioassay	Nichols & Paleg (1963) Jones & Warner (1967)	100 pg ml^{-1}	100 pg ml^{-1} to 1 µg ml^{-1}
Rumex leaf senescence bioassay	Whyte & Luckwill (1966)	100 pg ml^{-1}	100 pg ml^{-1} to 100 ng ml^{-1}

[a] Test compound GA_4.
[b] Test compound GA_7.

Table 8.7. Relative activities of GAs in five bioassay systems

Gibberellin	Barley aleurone	Dwarf pea	Lettuce hypocotyl	Dwarf rice	Cucumber hypocotyl
GA_1	++++	+++	+++	+++	++
GA_2	++++	++	++	+++	++
GA_3	++++	++++	+++	++++	++
GA_4	+++	+++	++	++	+++
GA_5	++	+++	++	+++	+
GA_6	++	++	++	+++	+
GA_7	+++	+++	++++	+++	++++
GA_8	+	+	+	+	0
GA_9	+	++	+++	++	+++
GA_{10}	+	0	0	+++	++
GA_{11}	+	0	0	+	+
GA_{12}	0	0	0	+	+
GA_{13}	+	0	+	+	0
GA_{14}	0	0	0	+	0
GA_{15}	0	+	++	++	++
GA_{16}	+	++	+	+	+
GA_{17}	0	0	0	+	0
GA_{18}	0	++	+	+++	0
GA_{19}	0	0	+	+++	0
GA_{20}	+	+	+++	+++	0
GA_{21}	0	0	+	0	0
GA_{22}	+++	+++	++	+++	0
GA_{23}	++	++	+	+++	0
GA_{24}	0	+	0	+++	+++
GA_{25}	0	0	0	0	+
GA_{26}	0	0	0	0	0
GA_{27}	0	+	0	+	0
GA_{28}	0	0	0	+	0
GA_{29}	+	0	0	+	0
GA_{30}	+++	+++	+	++	+
GA_{31}	+	++	0	++	0
GA_{32}	+++	+++	++	++++	+++
GA_{33}	+	+	+	+	0
GA_{34}	0	+	0	+	0
GA_{35}	+++	++	+	++	+++
GA_{36}	++	++	+	+++	+++
GA_{37}	++	++	++	+++	+++
GA_{38}	+	+++	0	+	+
GA_{40}		0	+	+	++

Table 8.7. *(Continued)*

Gibberellin	Barley aleurone	Dwarf pea	Lettuce hypocotyl	Dwarf rice	Cucumber hypocotyl
GA_{43}		0	0	0	0
GA_{46}		0	0	0	0
GA_{47}		++	+	+	++
GA_{51}		0	0	0	0
GA_{54}	+++			++	
GA_{55}	+++			++	
GA_{60}	+			+	
GA_{61}	+			++	
GA_{62}	++			++	

NOTE: Relative activities: ++++ very high, +++ high, ++ moderate, + low, 0 very low/inactive.

on the GA_1 estimates, it would seem that this assumption is at least partially valid and that the sample under analysis is acceptably pure. However, interference from extraneous material need not necessarily be revealed by assaying at more than one dose level. The sample was therefore analyzed in other bioassay systems which offer different selectivities. The data in Table 8.8 reveal that lettuce hypocotyl and barley half-seed α-amylase bioassays produced data that yielded GA_1 estimates that were much higher than those based on the dwarf rice bioassay. Without further investigation, it is impossible to establish which figure is the more accurate, and so under the circum-

Table 8.8. Estimated GA_1 content of an extract from 60 light-grown *Phaseolus coccineus* seedlings

Bioassay	Estimated GA_1 levels (ng)	
	Median value	Upper and lower 95% confidence limits
Dwarf rice	18[a]	5–60
	20[b]	4–48
Lettuce hypocotyl	600[c]	130–1,400
Barley half seed	700[c]	300–1,200

[a] 1/60 aliquot assayed.
[b] 1/120 aliquot assayed.
[c] 1/6 aliquot assayed.

stances the best estimate of the GA_1 content of the *Phaseolus* extract is 4–1,400 ng. It should be noted that at no stage has rigorous proof of accuracy been obtained, and thus there is no guarantee that the actual GA_1 content of the sample lies within even this broad range.

The *Phaseolus* analysis cited above is by no means a "worst case" example since the extract was subjected to partitioning procedures and underwent a two-step purification and fractionation on a liquid chromatography column before being tested in three bioassays at various dilutions. In many instances, purification is almost non-existent and estimates of GA content are based on TLC of crude, acidic, ethyl acetate-soluble extracts and a single bioassay at one dilution. The relationship between estimates based on such data and the actual GA content of the sample is likely to range from minimal to non-existent. The data in Table 8.8 also show that because of the inherent variation of plant material and the log-normal dose/response curve of GA bioassays, the confidence limits associated with bioassay-based estimates of GA levels are much wider than is generally appreciated. These observations should not, however, detract from the value of bioassays in qualitative analysis of extracts containing unknown GAs. Here, bioassays play a crucial role in sample purification, detecting zones of GA-like activity in chromatographic eluates prior to characterization by other procedures.

8.6 IMMUNOASSAYS

Although immunological assays are extensively employed in the field of mammalian endocrinology, details of their application to the analysis of GAs have, until recently, been restricted to preliminary reports on relatively insensitive procedures by Fuchs and Fuchs (1969) and Fuchs, Haimovich, and Fuchs (1971). This should not be taken to indicate their general unsuitability in this role; the selectivity, limits of detection, and simplicity of a well-designed immunoassay at least rival—and often exceed—those of many of the techniques currently used to analyze GAs.

Radioimmunoassay is based on the fact that when a saturable amount of antibody is reacted with a fixed quantity of radioactively labeled hormone, any unlabeled hormone will compete with the labeled species for binding sites on the protein. After equilibrium, bound and free hormones are separated and the radioactivity in either fraction is related to the hormone content of the extract via an appropriate standard curve. Typically, such curves are sigmoidal and this restricts the linear range of the assay to less than two orders of magnitude. The limit of detection depends upon a number of experimental variables, including the concentration and antiactivity of the serum and the concentration and specific activity of the hormone. However, it is the magnitude of the association constant of the hormone-protein complex which ultimately limits detection.

The high-selectivity associated with a good radioimmunoassay is very much a feature of the way in which the antibody is prepared. Small molecules such as GAs tend not to be antigenic in their own right and so must be covalently bound to a protein before injection into the host animal. Weiler and Wieczorek (1981) have recently described the preparation of a rabbit antiserum to GA_3-bovine serum albumen in which the GA_3 was coupled to the protein via the 7-carboxyl group. High specific activity $[^{125}I]GA_3$ [N-(p-hydroxybenzyl)putrescine] amide $(2.29 \text{ Ci } \mu\text{mol}^{-1})$ was used as the immunotracer (Fig. 8.7). Inhibition of tracer binding was enhanced at low pH levels, suggesting that the protonated molecule competed more effectively with the tracer for antibody binding sites than the anionic gibberellate. In keeping with this observation, GA_3 methyl ester was found to be an even more effective competitor, and analysis of methylated samples increased the sensitivity of the assay by two orders of magnitude. The association constant was calculated to be 4.7×10^{10} liter mol^{-1}, facilitating quantification of methylated GA_3 in the 10–200 fmol (4–80 pg) range with a detection limit of 5 fmol (Fig. 8.8). Non-specific binding was low at $8 \pm 1\%$. The GA_3 antiserum exhibited high selectivity as sizable cross-reactivity occurred only with GA_7, which is the 13-deoxy derivative of GA_3 (Table 8.9).

The development of this simple yet elegant radioimmunoassay by Weiler and Wieczorek (1981) is an important advance. Recently, radioimmunoassays for GA_1, GA_4, GA_7, GA_9, and GA_{20} have been reported as well as solid-phase enzyme immunoassays for GA_3, GA_4, and GA_7 (Atzorn & Weiler, 1982, 1983a, 1983b). It is now possible that immunoassays may eventually supercede much of the complex methodology that is currently employed for quantitative analysis of endogenous GAs. As well as possessing high sensitivity and selectivity, immunoassays have the advantage that once the antisera and immunotracers are prepared, relatively unskilled hands can be used to analyze over a hundred samples a day. The large number of naturally occurring GAs and their limited availability may hinder the widespread use of immunoassays in GA analysis. However, this problem is not insurmountable—many of the less common GAs can now be synthesized from readily available GA substrates (see Chapter 7), and only a few milligrams of GA are required for the development of an immunoassay.

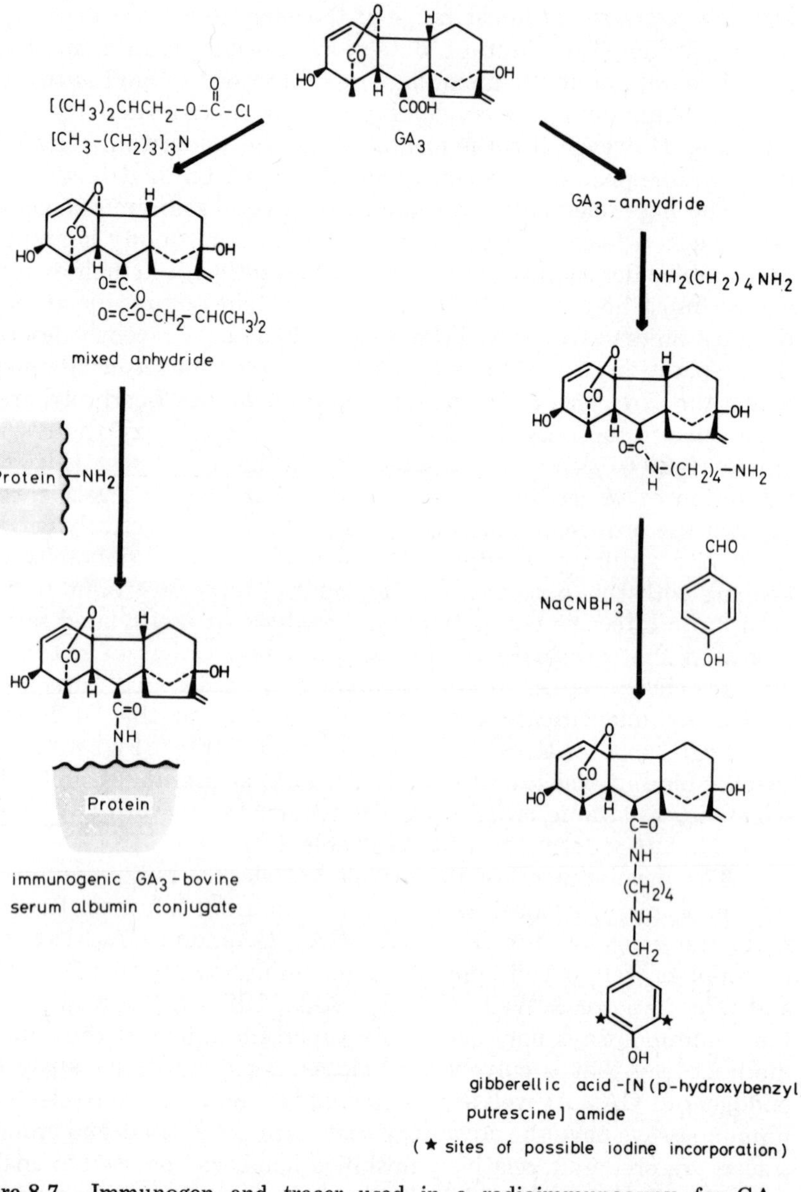

Figure 8.7 Immunogen and tracer used in a radioimmunoassay for GA₃. (Weiler & Wieczorek, 1981.)

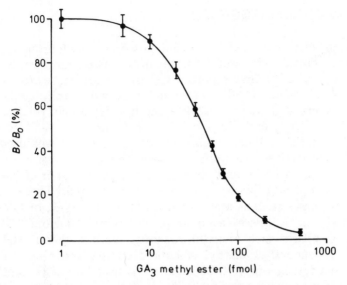

Figure 8.8 Standard curve for GA$_3$ radioimmunoassay. Bars indicate standard deviations of triplicates. B_0: amount of tracer bound in the absence of methylated GA$_3$ standard; B: amount of tracer bound in the presence of methylated GA$_3$ standard. (Weiler & Wieczorek, 1981.)

Table 8.9. Cross-reactivities of GA methyl esters and structurally related compounds with anti-GA$_3$-serum

Compound	Cross-reactivity (%)
GA$_3$ methyl ester	100
GA$_3$-O-β-D-(tetracetyl) glucosyl ester	2.9
GA$_7$ methyl ester	65.5
GA$_1$ methyl ester	12.9
GA$_4$ methyl ester	6.3
GA$_9$ methyl ester	1.9
GA$_8$ methyl ester	0.0
GA$_8$-2-O-β-D-glucosyl ether methyl ester	0.0
GA$_{13}$ methyl ester	0.0
Gibberellethione methyl ester	0.0
Helminthosporic acid methyl ester	0.0

After Weiler & Wieczorek (1981).

8.7 GAS CHROMATOGRAPHY

The high efficiency and resolving power of gas chromatography (GC) has proved very useful for separating GAs. Ikegawa, Kagawa, and Sumiki (1963) first reported GC of the methyl esters of GA_1-GA_9. Cavell et al. (1967) extended this work to include both methyl esters and methyl ester trimethylsilyl ethers (MeTMSi) of GA_1-GA_{15}, GA_{18}, and GA_{19} on 2% QF-1 and 2% SE-33 columns. In both reports a flame ionization detector (FID) was used. The MeTMSi derivatives are preferred over the methyl esters because they are more volatile and chromatograph with higher efficiencies. Silicone stationary phases are most widely used. These include nonpolar phases such as methyl silicone (SE-30, OV-1, or OV-101), moderately polar phases such as phenyl methyl silicone (OV-17), and polar phases such as trifluoropropyl methyl silicone (QF-1 or OV-210), cyanoethyl methyl silicone (XE-60), and phenyl cyano-propyl methyl silicone (OC-225). There is usually a 1–2% coating of the stationary phase on the support.

The retention times of the MeTMSi derivatives of various GAs, GA conjugates, and metabolic products are given in Table 8.10. Certain relationships between molecular structure and retention time can be noted:

(i) On the 2% SE-30 column, the derivatives elute largely according to their volatility, which is a reflection of molecular weight. On the more polar 2% QF-1 and 1% XE-60 columns, separations are also influenced by the presence and position of polar groupings in the molecule. Thus, the number of trimethylsilyl ether $[OSi(CH_3)_3]$ groups in the molecule markedly affects retentions on the 2% SE-30 column: GA_9Me [no $OSi(CH_3)_3$ groups], 4.8 min; $GA_4MeTMSi$ [one $OSi(CH_3)_3$ group], 8.7 min; $GA_1MeTMSi$ [two $OSi(CH_3)_3$ groups], 14.9 min; $GA_8MeTMSi$ [three $OSi(CH_3)_3$, groups], 25.0 min; and $GA_{32}MeTMSi$ [four $OSi(CH_3)_3$ groups], 40.6 min. However, on the 2% QF-1 and 1% XE60 columns, this effect is not nearly so great: $GA_9Me \rightarrow GA_{32}MeTMSi$: $6.7 \rightarrow 28.3$ min and $8.3 \rightarrow 28.8$ min, respectively.

(ii) The MeTMSi derivatives of high molecular weight C_{20}-GAs, such as GA_{13}, GA_{14}, GA_{19}, and GA_{23}, are more retained on 2% SE-30 than on 2% QF-1. Their derivatized γ-lactone C_{19}-GA counterparts, GA_4, GA_{20}, and GA_1, each have similar retention times on both columns.

(iii) The MeTMSi derivatives of the GA pairs, GA_1/GA_3, $GA_4/$

GA_7, and GA_5/GA_{20}, are well separated on 1% XE-60 but not on 2% SE-30. Separations intermediate between these two extremes are obtained on 2% QF-1. The excellent resolution on 1% XE-60 is presumably due to the ability of the support to detect differences in molecular orientation caused by the presence or absence of 1,2- and 2,3-double bonds.

(iv) The position of the $OSi(CH_3)_3$ groups affects retention times. GA derivatives with a 13-$OSi(CH_3)_3$ group have longer retention times than derivatives with either a 2α- or 2β-$OSi(CH_3)_3$ group (compare $GA_1MeTMSi$ to $GA_{34}MeTMSi$ or $GA_{47}MeTMSi$). GA derivatives with a 12α-$OSi(CH_3)_3$ group are more retained than their 13-$OSi(CH_3)_3$ counterparts (compare $GA_{31}MeTMSi$ to $GA_5MeTMSi$, and GA_{30}-MeTMSi to $GA_3MeTMSi$).

(v) The orientation of the $OSi(CH_3)_3$ group also affects retention time. Thus $GA_1MeTMSi$ has shorter retention times than 3-*epi*-GA_1-MeTMSi, and similarly $GA_{47}MeTMSi$ is less retained than GA_{34}-MeTMSi on all columns. In both cases the epimer with the equatorial $OSi(CH_3)_3$ group has the longer retention time.

(vi) The GAs with a δ-lactone function, namely GA_{15}, GA_{27}, GA_{37}, and GA_{38}, have long retention times on all columns.

(vii) The MeTMSi derivatives of GA glucosides and glucosyl esters have very long retention times. Elevated temperatures are needed to elute these high molecular weight conjugates.

The application of GC to the analysis of GAs in plant extracts has not been a great success, because the purity of most samples is such that an FID, which is a nonspecific mass detector, has to cope with high background levels and numerous extraneous peaks. As a consequence, the great advantage of GC—its high peak capacity—is lost as there is no guarantee that the mass peaks being measured are, in fact, attributable to GAs. These criticisms certainly apply to reports by Ross and Bradbeer (1971), Perez and Lachman (1971), Shindy and Smith (1975), Wightman (1979), and Parups (1980). However, GC-based quantitative estimates of GA levels in *Phaseolus coccineus* seed and leaves of *Bryophyllum daigremontianum* by Durley, MacMillan, and Pryce (1971) and Zeevaart (1973) are probably much more reliable as the extracts were purified extensively and GC-MS had been used to identify the GAs involved in preliminary qualitative studies.

The selectivity and low picogram limits of detection of an electron capture detector have proved invaluable in the analysis of abscisic acid by GC (Seeley & Powell, 1970). Unfortunately, GAs do not possess electron-capturing properties. Thus, in order to use

Table 8.10. Gas chromatograph retention times of GA MeTMSi derivatives on three columns

	Retention times (min)		
Gibberellin	2% QF-1	2% SE-30	1% XE-60
GA_1	13.6	14.9	14.9
GA_2	19.4	19.0	20.1
GA_3	16.0	16.5	18.1
GA_4	9.6	8.7	11.3
GA_5	9.7	8.3	13.0
GA_6	16.0	11.0	18.0
GA_7	10.8	9.3	14.4
GA_8	17.0	25.0	18.1
GA_9	6.7	4.8	8.3
GA_{10}	13.8	10.7	16.4
GA_{11}	10.9	6.8	12.7
GA_{12}	2.2	5.0	2.9
GA_{13}	5.2	11.4	8.2
GA_{14}	4.0	8.1	5.8
GA_{15}	23.5	10.8	29.4
GA_{16}	14.3	13.4	15.2
GA_{17}	5.4	10.8	8.8
GA_{18}	5.8	13.4	7.0
GA_{19}	8.2	12.1	11.0
GA_{20}	9.2	8.1	11.4
GA_{21}	25.3	17.2	36.2
GA_{22}	14.1	16.6	17.9
GA_{23}	11.0	18.7	14.4
GA_{24}	6.6	7.9	9.2
GA_{25}	4.9	8.0	7.7
GA_{26}	48.0	24.6	57.0
GA_{27}	34.5	28.6	44.7
GA_{28}	6.6	17.5	9.2
GA_{29}	13.3	16.3	15.7
GA_{30}	20.5	20.6	24.9
GA_{31}	14.1	10.4	17.2
GA_{32}	28.3	40.6	28.8
GA_{33}	18.9	19.6	25.5
GA_{34}	12.2	14.7	12.9
GA_{35}	10.8	13.6	12.4
GA_{36}	8.0	11.8	11.8

Table 8.10. *(Continued)*

Gibberellin	Retention times (min)		
	2% QF-1	2% SE-30	1% XE-60
GA_{37}	29.1	19.4	39.1
GA_{38}	39.4	33.0	53.6
GA_{40}	10.7	9.2	15.5
GA_{42}	6.6	17.7	9.0
GA_{47}	11.2	13.3	13.1
GA_{54}	11.9	11.7	12.1
3-*Epi*-GA_1	25.0	22.4	31.8
Iso-GA_3	12.1	13.5	14.8
3-Oxo-GA_3	22.0	12.0	29.2
Desoxo-GA_5	7.1	4.9	9.3
GA_{29}-catabolite	24.0	14.6	27.4
GA_3 glucoside	21.3[a]	35.7[b]	31.6[c]
GA_8 glucoside	19.7[a]	41.4[b]	27.4[c]
GA_1 glucosyl ester	21.0[a]	32.3[b]	28.2[c]
GA_3 glucosyl ester	18.0[a]	35.2[b]	28.0[c]
GA_4 glucosyl ester	19.0[a]	30.1[b]	26.2[c]

NOTE: Column dimensions: 1.83 m × 2 mm I.D. Support: Gas-chrom Q (80-100 mesh). Helium flow rate: 55 ml min^{-1}. Temp: QF-1, 206°C; SE-30, 203°C; XE-60, 209°C.

[a] Helium flow: 41 ml min^{-1}, temp: 235°C.
[b] Helium flow: 48 ml min^{-1}, temp: 246°C.
[c] Helium flow: 52 ml min^{-1}, temp: 235°C.

GC-ECD, electron-capturing derivatives, such as heptafluorobutyl and trifluoroacetyl ethers of GA methyl esters, have to be prepared (Seeley & Powell, 1974; Küllertz, Eckert, & Schilling, 1978). In these circumstances, the high sensitivity of the ECD is not associated with high selectivity as all the components in an extract with suitable hydroxyl groups will derivatize and so acquire electron-capturing properties. Thus, as far as the analysis of endogenous GAs is concerned, GC-ECD does not alleviate the problems encountered with GC-FID and the technique has not found widespread application.

The analytical situation is greatly simplified when GC is used in metabolism studies because radioactive precursors and metabolites can be selectively monitored with a gas-flow proportional radioactivity monitor. The technique is referred to as gas chromatography-radioactivity counting (GC-RC). Extracts are usually partially purified and radioactive GAs separated by liquid chromatography prior to analysis of the MeTMSi derivatives by GC-RC on the three columns listed in Table 8.10. An example of the data obtained is illustrated in Fig. 8.9. $[1,2\text{-}^3H]GA_{20}$ was applied to immature seed of *Pisum sativum*. After a metabolism period of five days, the seeds were extracted and a partially purified acidic, ethyl acetate-soluble fraction subjected to liquid chromatography on a Woelm silica gel partition column (Durley et al., 1972). Individual zones of radioactivity were then derivatized and examined by GC-RC. The traces in Fig. 8.9 demonstrate that one of the $[1,2\text{-}^3H]GA_{20}$ metabolites has GC-RC retention characteristics similar to those of $[^3H]GA_{29}$ on 2% QF-1, 2% SE-30, and 1% XE-60 columns. In the light of subsequent work, identifications of radioactive GAs, based on GC-RC retention data, have proved very reliable (see Durley & Pharis, 1973; Durley et al., 1974, 1979; Railton, Durley, & Pharis, 1974).

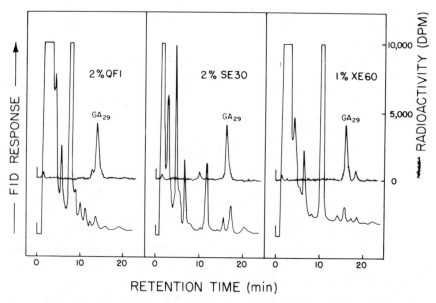

Figure 8.9 GC-RC of partially purified MeTMSi derivatized extract from immature seed of *Pisum sativum* extracted 5 days after the application of $[1,2\text{-}^3H]GA_{20}$. 1,830 × 2 mm I.D. glass columns maintained at 206°C (2% QF-1), 203°C (SE-30), and 209°C (1% Xe-60). Helium carrier gas flow: 55 ml min^{-1}. Detectors: FID and a modified Nuclear-Chicago 4998 gas-flow proportional counter. (Durley et al., 1979.)

8.8 COMBINED GAS CHROMATOGRAPHY-MASS SPECTROMETRY

The problems encountered with GC-FID can be overcome by using GC-MS because the GA MeTMSi derivatives can then be distinguished from each other and from extract impurities on the basis of their mass spectra. When used in this mode, effluent from the GC column passes through a separator, which removes most of the carrier gas, before entering the ion source of the mass spectrometer where, at *ca*. 10^{-6} torr, molecules are ionized and fragmented by bombardment with high-energy electrons. This process is known as electron impact ionization. Instruments offering chemical ionization in a suitable reagent gas, such as methane or ammonia, are increasing in popularity although, as yet, they have not been widely used with GAs. Regardless of the method of ionization, a signal relative to the total ion current (TIC) is derived either by summing the ion current values using a data system or by intercepting a portion of the unresolved ion beam. The TIC gives an indication of the amount of material in the source and is analogous to a FID trace. However, the real analytical power of mass spectrometry lies in the fact that fragment ions can be resolved according to their mass to charge ratio (m/e) by means of a magnetic sector or quadrupole mass analyzer to give positive ion line spectra such as those illustrated in Figure 8.10. It is also possible to obtain negative ion spectra, but these tend to contain fewer fragments and, as a consequence, are less informative. Each line in the spectra shown in Fig. 8.10 represents the ion or ions occurring at a particular atomic mass, and the height of each fragment indicates the percentage abundance relative to the ion used as the base peak (i.e., 100% intensity). Spectra may be recorded on UV-sensitive paper and counted by hand, although in modern instruments the data is processed by computer (Binks et al., 1971).

Electron impact mass spectra of methyl esters and TMSi derivatives of GA_1–GA_{24} have been published by Binks, MacMillan, and Pryce (1969). Since each mass spectrum represents a characteristic fingerprint, identification of GAs can be made by comparison with reference spectra, without the necessity of having GA standards, which are often unavailable. It is beyond the scope of this chapter to fully review details of the mass spectra of the 66 currently characterized GAs and their derivatives. Instead, the discussion will be limited to a consideration of the five most prominent and/or characteristic ions in the spectra of the MeTMSi derivatives of GA_1–GA_{56}, as pre-

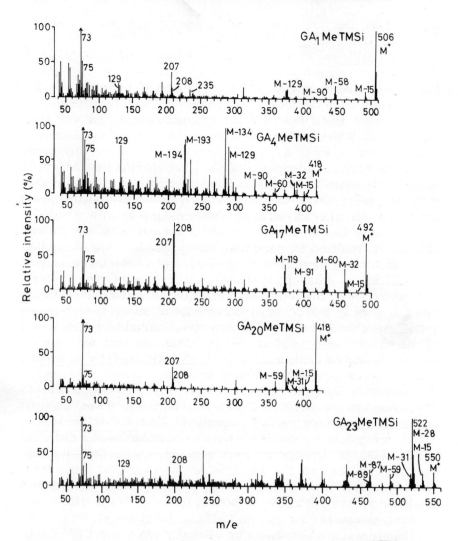

Figure 8.10 Mass spectra of the MeTMSi derivatives of GA_1, GA_4, GA_{17}, GA_{20}, and GA_{23}. (Binks et al., 1969.)

sented in Table 8.11. Readers who are interested in a more thorough discourse on GA mass spectra are referred to MacMillan, Gaskin, and MacNeil (1983).

The majority of the ions listed in Table 8.11 are the most prominent components in the GA spectra above m/e 208. Fragments at m/e values below 208 tend not to be especially characteristic and so have been omitted except where they are very intense, such as m/e 130 in the spectra of GA_2 and GA_{10}.

The intensities of molecular (M^+) and associated fragment ions are useful not only in identifying GAs by comparison with a standard but also in structural diagnosis of an unknown putative GA. The characteristic features of the mass spectra of the MeTMSi derivatives of GAs, listed in Table 8.11 and illustrated in Figs. 8.10 and 8.11, can be briefly summarized as follows:

(i) The M-15 (loss of CH_3 from M^+), M-89/90 [loss of $OSi(CH_3)_3$, $HOSi(CH_3)_3$], m/e 73 [$Si^+(CH_3)_3$], and m/e 75 fragments are associated with the presence of a trimethylsilyl ether grouping. If these ions are present, the GA has at least one hydroxyl group.

(ii) Since all GAs are acids, most spectra contain fragment ions at M-31/32 and M-59/60 associated with the methoxycarbonyl group (see GA_{20}).

(iii) The peaks at M-31/32 and M-59/60 are especially intense in the spectra of C_{20}-GA derivatives with two or three methoxycarbonyl groups (see GA_{17}). Prominent peaks are also observed at M-90/91/92 and M-119/120 due to fragmentation of two methoxycarbonyl groups [M-(31/32 + 59/60) and M-(59/60 + 59/60), respectively]. Peaks at M-149/150 are also common in these GA derivatives if a TMSi ether grouping is also present [M-(59/60) + 89/90)].

(iv) Many spectra have M-28/29 peaks (loss of CO, HCO). The M-28 fragment is especially intense and is larger than M^+ in the spectra of the MeTMSi derivatives of GA_{19} and GA_{23} and the methyl ester of GA_{24}, due to loss of CO from the C-20 aldehyde group. The mass spectrum of GA_{36} MeTMSi, which also contains the C-20 aldehyde function, does not contain an M-28/29 fragment but does exhibit an M-88 peak [M-(60 + 28)]. The M-88 fragment is also prominent in the spectrum of other C-20 aldehydic GAs.

(v) A large M^+ ion indicates either a 13-hydroxylated GA or a 2,3-dihydroxy GA. An exception to this rule occurs with 13-hydroxylated GAs with a 20-aldehyde function (GA_{19}, GA_{23}). In these instances the M-28 peak is intense [see (iv) above].

(vi) Ions at m/e 207/208 indicate a GA with a 13-hydroxyl group. These fragments, which may be very intense (see GA_{17}) or relatively weak (see GA_{20} catabolite; Durley et al., 1979), arise from fragments containing the C/D ring and the 13-$OSi(CH_3)_3$ group.

(vii) The mass spectra of MeTMSi derivatives of GAs with a 3-hydroxyl group and no other hydroxyl group in ring A are characterized by a prominent peak at m/e 129 (see GA_4). This presumably is due to a $(CH_3)_3Si\overset{+}{O}=CH—CH=CH_2$ ion arising from fragmentation of ring A.

Table 8.11. Five prominent and characteristic ions in the mass spectra of the MeTMSi derivatives of GA_1–GA_{56}.

Gibberellin	Mol wt	m/e Characteristic ions (abundance)[a]				
GA_1	506	506(100)	491(13)	448(20)	377(12)	313(17)
GA_2	508	508(27)	493(15)	418(17)	289(19)	130(100)
GA_3	504	504(100)	489(8)	370(9)	347(10)	208(45)
GA_4	418	418(26)	289(70)	284(100)	225(82)	224(76)
GA_5	416	416(100)	401(18)	357(13)	343(11)	299(25)
GA_6	432	432(100)	417(18)	373(17)	302(68)	235(22)
GA_7	416	416(18)	384(50)	356(66)	223(93)	222(100)
GA_8 b	594	594(100)	448(25)	379(20)	375(15)	238(28)
GA_9 b	330	298(100)	270(80)	243(42)	227(51)	226(50)
GA_{10} b	420	420(40)	405(25)	363(13)	331(31)	130(100)
GA_{11} b	344	344(100)	312(97)	284(42)	256(24)	240(26)
GA_{12} b	360	328(22)	300(100)	285(19)	240(30)	239(23)
GA_{13}	492	477(8)	436(10)	400(22)	310(26)	282(22)
GA_{14}	448	433(10)	416(38)	388(22)	298(60)	287(60)
GA_{15}	344	344(23)	312(22)	298(13)	284(50)	239(100)
GA_{16}	506	506(18)	416(20)	390(100)	360(43)	340(43)
GA_{17}	492	492(73)	460(34)	432(37)	401(20)	373(39)
GA_{18}	536	536(47)	521(11)	477(16)	319(29)	238(36)
GA_{19}	462	434(100)	402(35)	375(58)	374(60)	345(24)
GA_{20}	418	418(100)	403(14)	375(45)	359(12)	301(13)
GA_{21}	462	462(100)	447(10)	430(18)	403(41)	345(13)
GA_{22}	504	504(100)	489(28)	401(58)	387(20)	370(20)

	M+					
GA23	550	550(21)	522(100)	463(24)	432(32)	373(38)
GA24 b	374	314(81)	286(79)	285(72)	226(100)	225(78)
GA25 b	404	372(13)	312(82)	284(100)	253(9)	225(44)
GA26	520	520(100)	402(7)	255(8)	217(18)	147(33)
GA27	520	520(71)	430(11)	343(13)	223(25)	217(100)
GA28	580	580(31)	565(13)	371(14)	208(95)	207(100)
GA29	506	506(100)	491(11)	375(15)	303(17)	207(35)
GA30	504	504(9)	369(16)	280(12)	279(12)	221(32)
GA31	416	416(9)	282(35)	223(52)	222(78)	221(56)
GA32	680	680(17)	590(21)	339(9)	325(9)	307(11)
GA33	520	430(45)	383(17)	358(15)	286(19)	237(22)
GA34	506	506(100)	288(9)	229(12)	223(12)	217(22)
GA35	506	506(26)	416(34)	287(27)	282(42)	221(41)
GA36	462	430(50)	402(22)	312(47)	284(94)	211(69)
GA37	432	432(9)	342(15)	310(25)	284(22)	282(15)
GA38	520	520(67)	505(8)	430(8)	238(22)	207(100)
GA39	580	565(23)	488(22)	430(27)	398(36)	370(27)
GA40	418	371(100)	343(83)	299(81)	284(80)	225(57)
GA41	582	567(8)	523(6)	400(8)	283(13)	209(13)

Data taken from Binks et al. (1969) (GA$_1$–GA$_{23}$) and from spectra supplied by Dr. N. Murofushi (Tokyo University) (GA$_{24}$–GA$_{56}$). In data taken from Binks et al. (1969), the intense peak at m/e 73 in the spectra of GA$_1$–GA$_8$, GA$_{10}$, GA$_{14}$, GA$_{16}$, and GA$_{18}$–GA$_{23}$ MeTMSi derivatives was not included when calculating ion abundances.

[a]Peaks at m/e 208 and below are not given since such peaks are often not characteristic of a particular GA. The only exception to this is when the ions are particularly intense compared to fragments above m/e 208.

[b]Methyl ester only.

Table 8.11. *(Continued)*

Gibberellin	Mol wt	m/e Characteristic ions (abundance)[a]				
GA$_{42}$	538	523(26)	416(22)	376(100)	287(92)	259(55)
GA$_{43}$	580	580(9)	431(100)	371(15)	349(22)	217(58)
GA$_{44}$	432	432(63)	417(12)	373(17)	238(41)	207(100)
GA$_{45}$	418	418(100)	403(18)	358(23)	284(14)	225(14)
GA$_{46}$	492	460(41)	400(73)	342(74)	310(41)	282(100)
GA$_{47}$	506	506(100)	459(9)	431(6)	313(11)	217(31)
GA$_{48}$	594	594(58)	504(15)	419(11)	370(12)	191(42)
GA$_{49}$	594	594(71)	504(14)	419(12)	370(13)	191(38)
GA$_{50}$	594	594(52)	504(33)	460(12)	370(19)	309(22)
GA$_{51}$	418	386(25)	328(24)	284(100)	268(67)	225(91)
GA$_{52}$	608	608(67)	518(10)	462(6)	342(9)	217(100)
GA$_{53}$	448	448(47)	389(25)	251(30)	241(18)	235(30)
GA$_{54}$	506	506(34)	416(46)	390(96)	375(43)	300(66)
GA$_{55}$	594	594(100)	553(12)	535(12)	448(22)	375(26)
GA$_{56}$	594	594(100)	522(15)	448(18)	379(11)	375(13)
GA$_{20}$-catabolite[c]	446	446(100)	417(17)	387(31)	327(22)	314(16)

[c] Data taken from Durley et al. (1979).

(viii) The mass spectra of MeTMSi derivatives of GA_{16}, GA_{54}, and GA_{55}, which have 1α- or 1β- and 3β-hydroxyl groups, are characterized by prominent ions at M-116, which probably reflects the loss of a $(CH_3)_3SiOCH=CH_2$ fragment from ring A.

(ix) The mass spectra of MeTMSi derivatives of GAs with hydroxyl groups at both the 2- and 3-positions are characterized by a prominent peak at m/e 147, arising from a rearrangement product of the vicinal $OSi(CH_3)_3$ groups.

(x) The MeTMSi derivatives of GA_4, GA_7, GA_{40}, and GA_{51}, all of which have a single hydroxyl group at either the 2- or 3-positions, have very characteristic mass spectra. All four spectra contain peaks at M-134/138, corresponding to the loss of the C/D ring from the molecular ion, as well as at M-193/4, which can be related to the loss of the methoxylcarbonyl group from the M-134/135.

(xi) GA_{22} is the only currently known GA with an 18-hydroxymethyl group. The M-103 peak in the mass spectrum of GA_{22}MeTMSi is, therefore, likely to arise from the loss of $\dot{C}H_2-O-Si(CH_3)_3$.

(xii) Both GA_2 and GA_{10} have a 16-hydroxyl group. The mass spectra of the MeTMSi derivatives of these GAs contains an intense peak at m/e 130. This fragment presumably corresponds to an ion of the

$$CH_2=\overset{\displaystyle |}{\underset{\displaystyle CH}{\dot{C}}}-\overset{+}{O}-Si(CH_3)_3$$

type and arises from cleavage of ring D.

(xiii) The mass spectra of the MeTMSi derivatives of epimeric GA pairs are similar, differing only slightly in the intensity of selected peaks. The pairs with similar spectra are $GA_1/3\text{-}epi\text{-}GA_1$, GA_8/GA_{56}, GA_{34}/GA_{47}, GA_{48}/GA_{49}, and GA_{55}/GA_{57}. The mass spectra of the MeTMSi derivatives of GA_{34} and GA_{47} are illustrated in Fig. 8.11. It would be very difficult to distinguish between these GAs on the basis of mass spectrometric data, especially if impurity peaks were also present, as would be expected with samples derived from plant extract. Fortunately, GA_{34} and GA_{47}, as well as GA_1 and $3\text{-}epi\text{-}GA_1$, are well separated by GC (Table 8.10). One epimeric pair, GA_{40}/GA_{51}, has similar, but easily distinguishable mass spectra.

Besides methyl ester and MeTMSi derivatives, permethylated GAs have also been examined by GC-MS (Rivier et al., 1981). With free GAs, permethylation offers no improvement over methylation and trimethylsilylation; however, there are distinct advantages with

GA glucosides. Permethylated derivatives are more stable at the high temperatures required for GC of GA conjugates. They also have a lower molecular weight than MeTMSi GA conjugates, which is important because many mass spectrometers are inaccurate in the high mass ranges. In addition, permethylated GA conjugates yield spectra of more diagnostic value (i.e., improved M^+ intensities) than their MeTMSi counterparts.

Approximately 10–100 ng of GA are required to obtain a full-scan mass spectrum. Provided that adequate separation is achieved by GC, acceptable GA spectra can be obtained from relatively impure extracts. When endogenous GAs are qualitatively analyzed by GC-MS, it is advantageous for the mass spectrometer to scan repeatedly throughout the GC run and for the data acquired to be stored in a computer for subsequent retrieval. The value of this approach is illustrated in Fig. 8.12, which contains computer-generated traces from a GC-MS analysis of a semi-purified extract from immature seed of *P. sativum* (P. Hedden, unpublished data). The main peaks in the TIC trace proved to be contaminants. However, when computer-generated mass chromatograms were obtained, using m/e 418, 375, and 301,

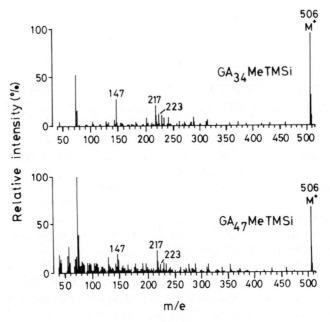

Figure 8.11   Mass spectra of the MeTMSi derivatives of GA_{34} and GA_{47}. (N. Murofushi, unpublished data.)

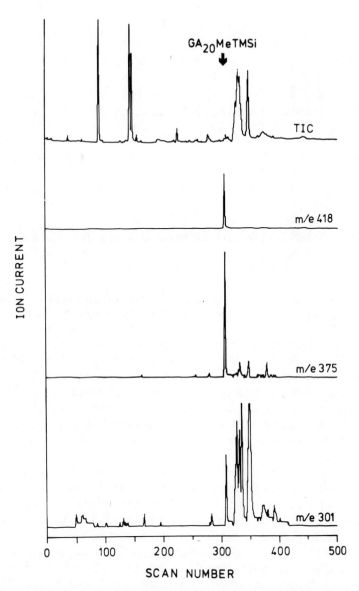

Figure 8.12 Computer-generated traces from a GC-MS analysis of a partially purified MeTMSi derivatized extract from immature seed of *Pisum sativum*. 1-μl sample injected into a $30,000 \times 0.24$ mm I.D. SP-2100 glass capillary WCOT column via a Grob splitless injector. Column maintained at 50°C for 1 min, then programmed to 230°C at 15°C min^{-1} and at 3°C min^{-1} from 230–280°C. Injector temp. 280°C. Helium carrier flow rate: 2 ml min^{-1}. Split: 50:1. GC effluent lead directly to ion source which was held at 290°C. Electron energy: 70 eV. Emission current 0.25 mA. Computer used to collect data from 230°C. (P. Hedden, unpublished data.)

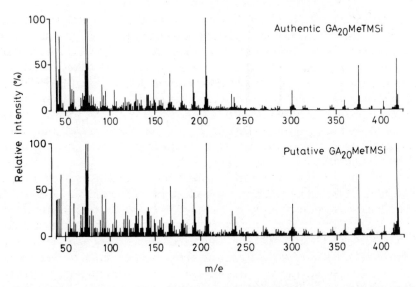

Figure 8.13   Mass spectra of authentic $GA_{20}MeTMSi$ and putative $GA_{20}MeTMSi$ from immature seed of *Pisum sativum*. (P. Hedden, unpublished data.)

which are characteristic ions in the mass spectrum of $GA_{20}MeTMSi$, a GA_{20}-like peak was detected. The computer was then used to produce full-scan spectrum of the putative GA_{20} peak, and the spectrum that was obtained closely matched that of standard $GA_{20}MeTMSi$ (Fig. 8.13). If manual scanning had been employed, it is possible that the GA_{20} may not have been detected as it is a minor component in the *Pisum* extract and does not contribute significantly to the TIC trace.

GC-MS can also be used for selected ion current monitoring (SICM). In this mode, the mass spectrometer is focussed on a limited number of ions that are characteristic fragments in the mass spectrum of the GA derivative under study (see Table 8.11). The mass spectrometer thus operates as a selective GC detector, and even though the TIC profile may contain numerous contaminant peaks, the SICM trace contains far fewer impurities and the GA response is markedly enhanced. When similar m/e values are monitored, SICM offers the same selectivity as mass chromatograms, such as those illustrated in Fig. 8.12. However, because the mass spectrometer is focussed on a limited number of ions, instead of scanning repeatedly, SICM is much more sensitive. If only the most intense of the characteristic ions in the spectrum of the GA derivative of interest is monitored, sensitivity is maximized and with a modern instrument low picogram limits of detection are obtained. However, it is standard practice to monitor at least three ions when analyzing endogenous GAs, and as a consequence there is a corresponding reduction in sensitivity. An example of the use of this method is the identification of GA_{32} in extracts from fruit of *Prunus cerasus* (Bukovac et al., 1979). GC-SICM is a valuable procedure for quantitative analysis of trace quantities of endogenous GAs, although it is necessary to know the identity of the GAs likely to be present in a sample and to have reference compounds available to quantify the detector response and determine GC retention characteristics.

Mass spectrometry can be used to assess the relative isotope content of M^+ and fragment ion clusters. Although the level of ^3H in ^3H-labeled compounds is usually too low to be measured, the $^{13}C/^{12}C$, $^{14}C/^{12}C$, $^{18}O/^{16}O$, and $^2H/^1H$ ratios can often be determined. ^{13}C-, ^{14}C-, ^{-18}O- and ^2H-GAs can therefore be used as internal standards in quantitative analyses as they behave in a similar, if not identical, manner to their endogenous $^{12}C/^{16}O/^1H$ analogues, yet can be distinguished from them by mass spectrometry. The ability of mass spectrometry to measure relative isotope content is also of value in biosynthesis and metabolism studies as it provides a means of simultaneously monitoring the relative levels of $^{14}C/^{16}O/^1H$ endogenous constituents of indeterminate ancestry and metabolites originating from ^{13}C-, ^{14}C-, ^{18}O- and ^2H-labeled precursors. GC-MS has thus been used to investigate (i) the biosynthesis of *ent*-kaurenes and GA_{12} aldehyde from [2-^{14}C] in mevalonic acid in cell-free homogenates of *Cucurbita maxima**** (Bowen, MacMillan, & Graebe, 1972), (ii) the metabolism of [6-$^2H_1$]GA_{12} aldehyde by *G. fujikuroi* (Bearder, MacMillan, & Phinney, 1973), (iii) the conversion of [19-^{18}O]GA_{12} and [19-^{18}O]GA_{12} alcohol to other GAs by *G. fujikuroi* (Bearder & MacMillan, 1976), and (iv) the metabolism of [1β,3α-$^2H_2$][1β,3α-$^3H_2$]-GA_{20}, [2α-$^2H_1$][2α-$^3H_1$]GA_{29}, and [17-^{13}C]GA_{29} in immature seed of *P. sativum* (Sponsel & MacMillan, 1978, 1980).

Because the level of ^3H in ^3H-labeled substrates is usually much too low to be measured by mass spectrometry, the metabolism of ^3H precursors is best analyzed by either GC-RC or HPLC-RC. GC-MS cannot, therefore, be utilized in metabolism studies with tissues containing small endogenous GA pools, as high specific activity ^3H-labeled substrates must be used if the applied dose is to be limited to physiological levels. This is a minor constraint since, overall, GC-MS is an

---

* Formerly known as *Cucurbita pepo*.

extremely powerful and flexible tool that is of great importance to many aspects of GA analysis. In the future, capillary GC-MS is likely to be used with increasing frequency. However, more rigorous sample purification will be essential because the sample capacity of capillary GC is limited.

8.9 HIGH-PERFORMANCE LIQUID CHROMATOGRAPHY

As commonly practiced, HPLC and GC are broadly equivalent in that they display similar efficiencies and speeds of analysis, with the sample capacity of HPLC being at least ten times that of GC. The major difference between the two techniques lies in the thermodynamics of the partitioning process. In liquid-solid and liquid-liquid processes the differences in the free energies of distribution of the solutes $\Delta(\Delta G^{\circ})$ are usually far greater than for gas-solid or gas-liquid systems. Thus, all other factors being equal, HPLC will always give a superior separation to GC. In addition, $\Delta(\Delta G^{\circ})$ is much more dependent upon the properties of the mobile and stationary phase in HPLC than it is in GC, and so HPLC is able to offer a much wider variety of column selectivities.

HPLC has been applied to numerous diverse analytical problems (see Knox, 1979; Snyder & Kirkland, 1979). High column efficiencies and peak capacities, rapid speeds of analysis, the availability of many supports each offering markedly different separatory mechanisms, operation at ambient temperatures, and ease of sample recovery, all contribute to the overall effectiveness of the technique. It should, however, be noted that the high efficiencies ($>$40,000 theoretical plates m^{-1}) are achieved on columns with a 2–5 mm bore. The sample capacity of such columns rarely exceeds 500 μg. Thus, as far as the analysis of endogenous GAs is concerned, the potential of HPLC, like that of GC-MS, can only be fully exploited when applied to extracts of relatively high purity.

Despite its obvious potential, HPLC has, as yet, not been used extensively for the analysis of GAs. Although GAs are readily chromatographed, they exhibit only low UV absorbance and the main dilemma confronting potential users of the technique is the choice of a detector system. Although an absorbance monitor operating at around 210 nm can detect *ca.* 50 ng of GA, the UV cut-off point of most solvents restricts this level of sensitivity to a very limited range of mobile-phase conditions. The problem is compounded when analyzing trace quantities of GAs in multicomponent plant extracts because many of the impurities induce a strong detector response. One course of action is to use bioassays to selectively monitor GA-like activity in HPLC eluates (Barendse et al., 1980; Jones et al., 1980). However, if chromatographic peak capacity is to be maintained, many fractions have to be collected and assayed. This is time consuming, and much of the practicality of HPLC is lost. More often only small numbers of fractions are assayed and, in these circumstances, resolution is clearly being traded for enhanced speed of analysis. This can be an acceptable compromise when HPLC is being used to purify unknown endogenous GA-like compounds prior to analysis by mass spectrometric techniques.

An alternative approach, which involves sacrificing detector selectivity in favor of the resolution and general flexibility of HPLC, is to convert GAs to derivatives that absorb in an accessible region of the UV spectrum. This option is feasible when either radioactive GAs are being analyzed or the identity of the individual GAs in a sample are known or suspected. Reeve and Crozier (1978) made use of GA benzyl esters which were synthesized by esterification of N,N'-dimethylformamide dibenzylacetal and have a λ_{max} of 256 nm. The GA benzyl esters were chromatographed on a silica gel adsorption column which readily separated several isomers because of its ability to distinguish subtle differences in the spatial relationships of the polar groupings of structurally similar molecules. Marked changes in the selectivity of the silica gel column were achieved by using different reagents to modify the mobile phase. The procedures have been used in conjunction with a radioactivity monitor to analyze [^{3}H]GA metabolites from *Phaseolus coccineus* seedlings (Crozier & Reeve, 1977; Reeve & Crozier, 1978; Crozier, 1981) and lettuce hypocotyl sections (Nash, Jones, & Stoddart, 1978). A comprehensive discussion of silica gel adsorption HPLC of GA benzyl esters has also been published (Reeve & Crozier, 1978).

Although GA benzyl esters have proved useful in metabolism studies, it should be noted that the ϵ_{max} of mono derivatives is 205 liter mol^{-1} cm^{-1} and that the limit of detection at 254 nm is only 300 ng. This lack of sensitivity is a serious constraint when it comes to utilizing fully the high resolving power of HPLC to analyze submicrogram quantities of endogenous GAs. Other derivatives do, however, offer much greater potential in this regard. Heftmann, Saunders, and Haddon (1978) prepared *p*-nitrobenzyl GA esters (λ_{max} 265 nm, $\epsilon_{max} > 6,000$) using *o-p*-nitrobenzyl-N,N'-diisopropyl-urea. Unfortunately, when the esters were chromatographed on a preparative silver nitrate-impregnated silica gel column, the speed of analysis was very slow and the performance was poor ($N = 1,500$,

$H = 3.25$ mm). As 50–200-ml peak volumes were obtained, the limit of detection at A_{256} was 100 ng rather than the less than 10 ng that might have been anticipated if conventional HPLC techniques had been used. Morris and Zaerr (1978) used 18-Crown-6 according to the procedures of Durst et al. (1975) to catalyze the conversion of GAs to GA p-bromophenacyl esters (λ_{max} 256 nm, ϵ_{max} 19,100). The limit of detection at A_{254} for mono esters eluting from reverse- and normal-phase HPLC columns was below 5 ng.

A further increase in sensitivity was obtained by using GA methoxycoumaryl esters (GACEs) for HPLC (Crozier, Zaerr, & Morris, 1982). These derivatives were prepared according to the procedures of Dünges (1977) by reacting GAs with an equimolar amount of 4-bromomethyl-7-methoxycoumarin (BMMC), a one-tenth molar equivalent of 18-Crown-6, and a crystal of K_2CO_3 in acetonitrile at 60°C for 2 hr. The reaction is shown in Fig. 8.14. The efficacy of the conversion was determined by analyzing a [^{3}H]GA$_9$ reaction mixture by reverse-phase HPLC-RC. The data obtained are illustrated in Fig. 8.15. The traces show that [^{3}H]GA$_9$ underwent a complete conversion to [^{3}H]GA$_9$CE. Although BMMC breaks down in light, GACEs are stable. They are also strongly fluorescent (λ_{max}^{excit} 320 nm, λ_{max}^{emiss} 400 nm) and can be detected at the low picogram level with a good spectrophotofluorimeter after reverse-phase HPLC. This is shown in Fig. 8.16, where a log-log plot of relative response against simple size gives a line with a slope of 1.0, linear over almost four orders of

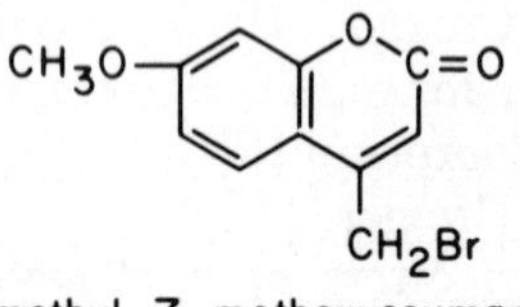

4 - bromomethyl - 7 - methoxycoumarin (BMC)

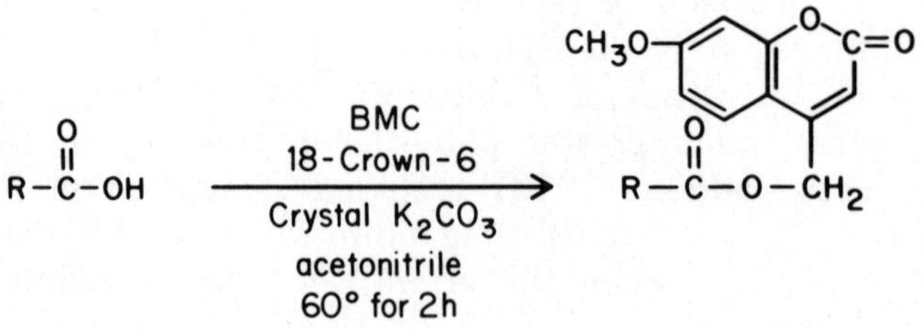

Figure 8.14   Crown ether catalyzed synthesis of methoxycoumaryl esters. (Crozier et al., 1982.)

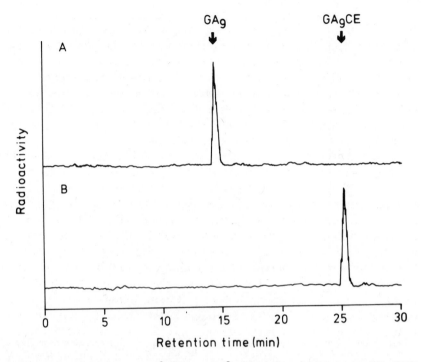

Figure 8.15   Conversion of $[^3H]GA_9$ to $[^3H]GA_9CE$.   250 × 5 mm I.D. ODS Hypersil. Mobile phase: 30-min gradient, 60–100% methanol in 20 mm, pH 3.5 ammonium acetate buffer. Flow rate: 1 ml min^{-1}. Sample: *ca.* 15,000 dpm aliquot of $[^3H]GA_9$-BMMC-18-crown-6 reaction mixture (A) at 0 hr and (B) after 2 hr at 60°C. Detector: homogeneous radioactivity monitor, 600 cpm full-scale deflection. (Crozier et al., 1982.)

magnitude. The limit of detection for GA_3, which forms a mono ester, is *ca.* 1 pg (2.8 fmol) as determined by the point at which the curve intersects the ordinate equivalent to three times the level of background noise. With bis and tris esters this figure is correspondingly lower.

Reverse-phase separations of a range of GACEs obtained by gradient elution from an ODS-Hypersil column are illustrated in Fig. 8.17. The recovery of $[^3H]GA_9CE$ was more than 90%. The system was able to distinguish between closely related GAs. The double-bond isomers GA_1CE/GA_3CE, GA_4CE/GA_7CE, and $GA_5CE/GA_{20}CE$ all separated with baseline resolution, with the $\Delta^{1,2}$ and $\Delta^{2,3}$ derivatives eluting before their saturated analogues. It is of interest to note the effect of solvents on column selectivity. When a methanol-buffer mobile phase was employed, $GA_{13}CE$ and $GA_{14}CE$ co-chromato-

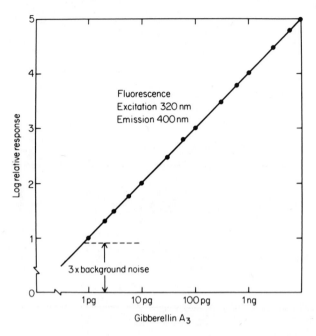

Figure 8.16 Fluorescence detection limit of GA$_3$CE after reverse-phase HPLC. Column: 250 × 5 mm I.D. ODS Hypersil. Mobile phase: 45% ethanol in 20 mM, pH 3.5 ammonium acetate buffer. Flow rate: 1 ml min^{-1}. Sample: GA$_3$CE ($k' = 2.3$) load as indicated. Detector: Perkin-Elmer 650-10 LC spectrophotofluorimeter, excitation 320 nm, emission 400 nm. (Crozier et al., 1982.)

graphed as did GA$_9$CE and GA$_{36}$CE. However, when ethanol was substituted for methanol the compounds were well resolved. In general, increasing the number of free hydroxyl groups decreases retention—13-hydroxylation to a much greater extent than 3β-hydroxylation which, in turn, is more effective than hydroxylation at either the 1α- or 2β-positions. A comparison of the retention characteristics of GA$_4$CE, GA$_{36}$CE, GA$_{14}$CE, and GA$_{13}$CE as well as GA$_9$CE and GA$_{25}$CE shows an elution order of mono>bis>tris esters indicating that increasing the number of methoxycoumaryl functions decreases polarity. The elution of GA$_{14}$CE and GA$_{38}$CE implies that C-20 methyl GACEs are less strongly retained than their C-20 aldehydic counterparts.

Normal-phase separations of GACEs on a CPS-Hypersil column eluted isocratically with 3% ethanol in either 12% or 20% dichloromethane in hexane are illustrated in Fig. 8.18. Recoveries of the ^{3}H-GACEs from the cyanopropyl columns were *ca.* 85%. Although the effects of 3β- and 13-hydroxylation and 1,2- and 2,3-double

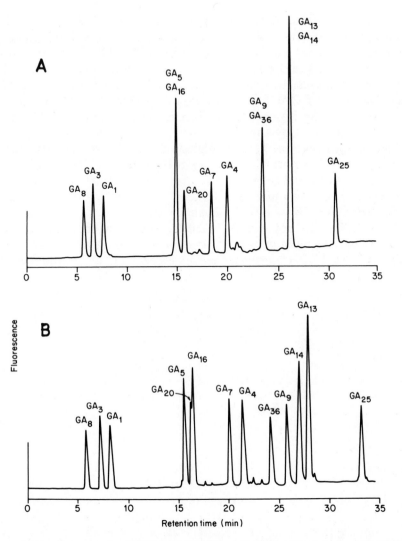

Figure 8.17 Reverse-phase HPLC of GACEs. Column: 250 × 5 mm I.D. ODS Hypersil. Mobile phase: 30-min gradient (A) 60–100% methanol in 20 mM, pH 3.5 ammonium acetate buffer, (B) 40–80% ethanol in 20 mM, pH 3.5 ammonium acetate buffer. Flow rate: 1 ml min^{-1}. Sample: methoxycoumaryl esters of GA_1, GA_3, GA_4, GA_7, GA_8, GA_9, GA_{13}, GA_{14}, GA_{16}, GA_{20}, GA_{25}, and GA_{36}. Detector: Perkin-Elmer 650-10LC spectrophotofluorimeter, excitation 320 nm, emission 400 nm. (Crozier et al., 1982.)

bonds are, as anticipated, the opposite of those observed in reverse-phase analyses, the overall elution pattern is not a mirror image of the ODS-Hypersil profiles in Fig. 8.17. This is due primarily to the behavior of bis and tris GACEs, which exhibit increased k' values with respect to their increased number of methoxycoumaryl groups. The polarity of GACEs, in this regard, is thus the opposite of that observed on the reverse-phase support. The other noticeable difference, as indicated by the elution of $GA_{16}CE$ close to GA_1CE rather than $GA_{20}CE$, is that 1β-hydroxylation has a more marked effect on retention in the normal- than the reverse-phase system. Because of increased levels of background fluorescence and/or quenching in the mobile phase, the limit of detection of mono GACEs eluting from the CPS column was no better than 30 pg.

GACEs are not readily volatile, and as a consequence mass spectra have to be obtained by direct-probe mass spectrometry rather than GC-MS. Direct-probe electron impact and chemical-ionization

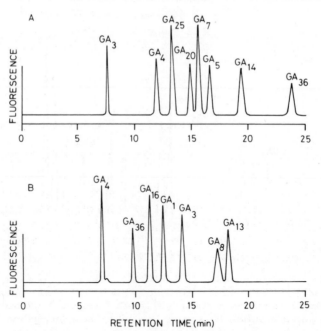

Figure 8.18 Normal phase HPLC of GACEs. Column: 250 × 5 mm I.D. CPS Hypersil. Mobile phase: (A) 3% ethanol in hexane-dichloromethane (88:12) and (B) 3% ethanol in hexane-dichloromethane (80:20). Flow rate: 1 ml min^{-1}. Sample: methoxycoumaryl esters of GA_1, GA_3, GA_4, GA_5, GA_7, GA_8, GA_9, GA_{13}, GA_{14}, GA_{16}, GA_{20}, GA_{25}, and GA_{36}. Detector: Perkin-Elmer 650–10 LC spectrophotofluorimeter, excitation 320 nm, emission 400 nm. (Crozier et al., 1982.)

Table 8.12. Methane chemical-ionization negative-ion mass spectra of GA methoxycoumaryl esters

Compound	Mol wt	Mass spectra
GA_1 CE	536	347—100% (M-189)
GA_3 CE	534	534—100% (M⁻), 345—43% (M-189), 301—7% (M-233), 283—7% (M-251)
GA_4 CE	520	331—100% (M-189)
GA_5 CE	518	329—100% (M-189)
GA_7 CE	518	518—100% (M⁻), 329—34% (M-189), 285—5% (M-233), 267—15% (M-251), 265—5% (M-253)
GA_8 CE	552	363—100% (M-189)
GA_9 CE	504	315—100% (M-189)
GA_{13} CE	942	565—8% (M-377 [-189-188]), 547—100% (M-395 [-189-206]), 359—54% (M-583 [-189-188-206 and/or -189-206-188]), 315—5% (M-627), 314—6% (M-628)
GA_{14} CE	724	535—100% (M-189), 347—18% (M-377 [-189-188])
GA_{16} CE	536	347—100% (M-189), 329—6% (M-207), 303—9% (M-233)
GA_{20} CE	520	331—100% (M-189)
GA_{36} CE	738	549—10% (M-189), 343—100% (M-395 [-189-206])

positive-ion spectra are of no practical value because the dominant fragment in all instances is m/e 191, with no other ions of significant intensity being present. However, chemical-ionization negative-ion spectra are of more diagnostic value (Table 8.12). A strong molecular ion was obtained with the $\Delta^{1,2}$ GACEs, GA_3CE and GA_7CE. M-189 arising from the loss of the methoxycoumaryl moiety was the main fragment in the spectra of the other C_{19}-GA derivatives that were tested. The C_{19}-GA isomers GA_4CE and $GA_{20}CE$ yielded identical spectra. These compounds can, however, be readily distinguished on the basis of their HPLC retention characteristics. Spectra were obtained from the methoxycoumaryl esters of three C_{20}-GAs. M-189 was the strongest ion produced by $GA_{14}CE$, while M-395 was the base peak in the spectra of both $GA_{13}CE$ and $GA_{36}CE$.

The applicability of the HPLC fluorescence procedures to quantitative analysis of endogenous GAs is dependent, in the first instance, on the effectiveness and reliability of the derivatization step.

Initially, variable results were obtained when ^{3}H-GAs were derivatized in the presence of plant extracts. However, the problem was overcome, and greater than 98% conversions were obtained on a routine basis by using acetone rather than acetonetrile as a solvent in the crown ether catalysis. It was also found that 400 nmol of BMMC were required to completely convert 10^6 dpm of $[^3H]GA_9$ (4.6 ng, 14.6 pmol) to $[^3H]GA_9CE$ in the presence of 1 mg of an acidic, ethyl acetate extract from cucumber seedlings. The acid strength even of purified samples will vary somewhat, and so as a general rule of thumb it is advisable to use 1 mmol of BMMC per mg dry weight of extract. $[^3H]GA_9$ in plant extracts is efficiently converted to $[^3H]GA_9CE$ even when present in concentrations as low as 1 part in 10^6 (G. Schneider and A. Crozier, unpublished data). The crown ether derivatization procedure, which is equally effective with mono-, di-, and tricarboxylic GAs, is thus well suited for use with trace quantities of endogenous GAs.

The picogram limits of detection obtained for GACEs with a fluorescence monitor enhance the overall flexibility of HPLC as an analytical tool, especially in the investigation of trace quantities of endogenous GAs in vegetative tissues. It should be noted, however, that the high sensitivity is not accompanied by any degree of detector selectivity. GAs do not exhibit native fluorescence. Thus, when plant extracts are derivatized not only endogenous GAs but also any other carboxylic acids in the sample will form methoxycoumaryl esters and so acquire fluorescent properties. As a consequence, considerable purification is required before homogeneity of detector response can be obtained and an accurate analysis achieved.

The requirement for extensive sample purification was evident when the HPLC fluorescence procedures were used to detect GA_1 in an extract from *Salix pentrandra* shoots (Davies et al., in preparation). Tissues were extracted with methanol, and an acidic, ethyl acetate-soluble fraction was obtained. This was purified by DEAE Sephadex A-25 ion exchange chromatography and preparative SEC and then chromatographed on a high-performance silica gel partition column, fractions from which were tested for biological activity in a range of GA bioassays. An aliquot of a GA_1-like zone of activity was derivatized and re-chromatographed on a silica gel partition column, and the GA_1CE zone was collected and further purified by SEC prior to analysis by gradient elution reverse-phase HPLC which showed that the sample contained several fluorescent peaks, one of which co-chromatographed with GA_1CE (Fig. 8.19). This peak, which was equivalent to 800 ng of GA_1CE, was collected and analyzed isocratically by reverse- and normal-phase HPLC. In both systems, the

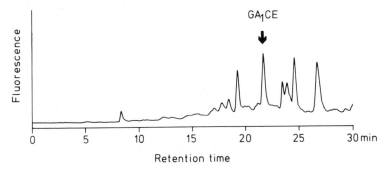

Figure 8.19   Gradient elution reverse-phase HPLC of a partially purified GA_1CE-like component from *Salix pentandra* shoots.   Column: 250×5 mm I.D. ODS Hypersil. Mobile phase: 30-min gradient, 50–80% methanol in 20 mM, pH 3.5 ammonium acetate buffer. Flow rate: 0.75 ml min^{-1}. Sample: 1/800 aliquot of a partially purified GA_1CE-like fraction. Detector: Perkin-Elmer 650-10LC spectrophotofluorimeter, excitation 320 nm, emission 400 nm. (Davies et al., in preparation.)

only component to be detected, other than the void volume solvent peak, was a solute which co-chromatographed with GA_1CE (Fig. 8.20). The data thus provide evidence for the presence of GA_1 in the *Salix* extract. GC-MS analysis of the MeTMSi derivative of an aliquot of the GA_1-like zone of biological activity yielded mass spectra that confirmed the result of the HPLC analysis.

As far as GAs are concerned, HPLC and GC are mutually incompatible techniques. Although GC derivatives such as GA methyl esters can be chromatographed on an HPLC column, they are not readily detected with conventional on-line HPLC monitoring systems. Conversely, GACEs and other derivatives that are suitable for HPLC are far from ideal candidates for GC as they lack the necessary volatility. This means that it is difficult to obtain mass spectra of GACEs by GC-MS. There are three ways around this problem. The first, and the one employed in obtaining the mass spectra presented in Table 8.12, is to use direct-probe mass spectrometry. This practice has limitations with plant extracts because relatively large samples of high purity are required if acceptable spectra are to be obtained. The second approach would be to develop an effective transesterification process to convert fluorescent or UV-absorbing GA esters to a methyl ester that could be silylated and analyzed by GC-MS. One potential problem with this procedure is that GA derivatives amenable to transesterification may well be somewhat unstable and some degree of breakdown could occur during HPLC. The long-term solution is the use of an HPLC directly coupled to a mass spectrometer.

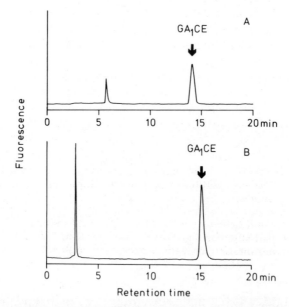

Figure 8.20 Isocratic reverse- (A) and normal-phase HPLC (B) of a $GA_1 CE$-like component from *Salix pentandra* shoots. (A) Column: 250 × 5 mm I.D. ODS Hypersil. Mobile phase: 60% methanol in 20 mM, pH 3.5 ammonium acetate buffer. Flow rate 0.75 ml min^{-1}. (B) Column: 250 × 5 mm I.D. CPS Hypersil. Mobile phase: 3% ethanol in hexane-dichloromethane (80:20). Flow rate: 1 ml min^{-1}. Sample: $GA_1 CE$-like peak from gradient elution reverse-phase HPLC (Fig. 8.19). Detector: Perkin-Elmer 650-10LC spectrophotofluorimeter, excitation 320 nm, emission 400 nm. (Davis et al., in preparation.)

Although commercial interfaces are now being marketed, the technology is still in its infancy and performance, convenience, and reliability have yet to be proven (see McFadden et al., 1977; McFadden, 1979; Karger et al., 1979; Arpino & Guiochon, 1979). When it becomes a practical proposition, HPLC-MS will greatly increase the flexibility of HPLC. It will be possible to obtain spectra with smaller sized samples, and of equal importance as far as the quantitative analysis of endogenous GAs is concerned is that components eluting from HPLC columns could be selectively analyzed by SICM.

8.10 VERIFICATION OF ACCURACY

Theoretical concepts associated with the identification of GAs and other hormones in plant extracts have been proposed by Reeve and Crozier (1980), who point out that to understand fully the nature of the problems encountered in practice it is necessary to take a general view of analytical theory. It is important to realize that the distinction made between "qualitative" and "quantitative" analysis is a semantic convenience rather than a logical reality. Because it is impossible to quantify an unknown in meaningful terms, quantitative analysis is in fact inherently qualitative. The converse also applies, since the statement that "X is GA_1" implies that *all* of sample X has *all* the properties associated with the chemical concept of GA_1. Reeve and Crozier (1980) further argue that quantitative analysis displays all the enigmas of scientific induction. The identification and quantification of a substance can *never* be absolute and, thus, must be considered in association with a complex probability term that defines the chances of making an error when concluding that $Y = x\,\mu g$ of compound Z. Two types of error, namely precision and accuracy, independently contribute to the complex probability term. Precision is a measure of random errors that determine run-to-run variability. Thus, when given a series of estimates of the same sample, it is possible to use statistical methods to calculate the standard deviation (SD) of the data and, with a minimum of assumptions, state that the probability of the precision being no worse than $\pm 2\,SD$ is 0.95. An averaging process can be applied to enhance the precision of an analysis in proportion to the square root of the number of estimates averaged. Accuracy, however, refers to the nonrandom or systematic error of the analysis, and its error and confidence limits are inordinately more difficult to ascertain.

Understanding the distinction between accuracy and precision is critical. It is essential to realise that, regardless of the number of estimates contributing to the final average, there is no guarantee that accurate results will be obtained, because nonrandom error will apply the same bias to each estimate and so be present undiminished in the averaged result. A "target" analogy, such as that illustrated in Fig. 8.21, is a useful means of demonstrating the total independence of the terms accuracy and precision. It is evident from Fig. 8.21 that a rifle must not only be aimed accurately but must also be designed so as to group its shots closely; that it, it must be precise if there is to be a high probability of hitting the "bull's-eye." In the case of plant hormones it is a common mistake to assume that an analysis is accurate because it offers adequate precision. It is apparent from Fig. 8.21 that the degree of repeatability provides no such assurance of accuracy. In practice, verification of accuracy is the single most crucial, yet neglected, factor in the analysis of plant hormones. Reeve and Crozier (1980) have proposed that, in a general context, verification of accuracy consists of (i) defining, in terms of probability limits, the complexity of the analytical problem and (ii) relating this to the

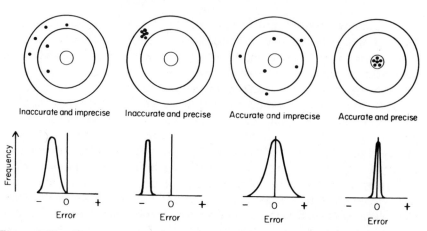

Figure 8.21 Target analogy demonstrating the independence of the error terms, accuracy and precision. Accuracy refers to the nonrandom error, precision to the random error. (Reeve & Crozier, 1980.)

amount of pertinent information obtained during an analysis. At present, practical solutions to this problem require making a number of less-than-ideal assumptions. Even so, figures for accuracy obtained by such methods will be more reliable than those ascertained by conventional procedures where the criteria for verification can range from the whims of technical fashion to standards of intuition that vary enormously from one investigator to another.

Much confusion and dogma surround the entire process of verification of accuracy. Although it is widely believed that only mass spectrometric evidence is acceptable, the method is not foolproof and erroneous identifications do occur, primarily because of subjective interpretations of contaminated, limited scan spectra. In skilled hands a mass spectrometer is an extremely powerful analytical tool, but when used injudiciously it can produce evidence that is barely superior to that generated by less-sophisticated procedures. It is also important to appreciate that when a full-scan mass spectrum is used to characterize a GA, accuracy is dependent upon the discriminatory power of the spectrum relative to the complexity or purity of the sample. However, when identifications are based on SICM, accuracy is only achieved when the procedure offers adequate selectivity for the sample under study.

Reeve and Crozier (1980) have suggested a simple practical test, called a "successive approximation," for detecting situations in which the selectivity of an analysis is inadequate. It relies on the fact that, as the purity of a sample is increased, estimates of GA concentration must show an improvement in accuracy, since even a totally nonselective method will provide accurate results with a perfectly pure sample. The successive approximation works in the following manner. When given a sample purporting to contain a given quantity of a GA_x (E_1), the test for accuracy consists of purifying the sample by a factor of at least 2 and reestimating the GA_x content (E_2). A suitable internal standard should, of course, be used to account for sample losses encountered during purification. If E_1 is accurate, then E_2, taking into account the precision of the method, should not be significantly different. If a difference is found, E_1 must be rejected as inaccurate and E_2 tested by further purification and re-analysis. The process is continued for as long as is necessary to obtain an estimate that does not change on purification. At this point it is possible to conclude that on the basis of the available evidence, there are no grounds for believing the analysis to be inaccurate. Reeve and Crozier (1980) have suggested several statistical methods that can be used to derive the probability of errors arising when making such assumptions.

An alternative and quite independent approach to the verification of accuracy can be taken through the uncertainty associated with sample purity (Reeve & Crozier, 1980). The nature of the problem associated with the analysis of trace quantities of GA in multicomponent extracts is best grasped by viewing extracts as open-ended systems in which an infinite array of organic compounds are *potentially* present. For a variety of reasons, only a limited number of these possible components are likely to occur in amounts that are significant in relation to the quantity of GA present in the sample. Reeve and Crozier (1980) used typical molecular weight distributions of plant extracts to draw probability limits on the number of compounds likely to be encountered. Information theory was then invoked to determine what type of mass spectrum, chromatogram, or other analytical information was necessary to ensure that the discriminating power of the method was sufficient to cope with the number of compounds likely to occur at a given probability level. Their arguments can be summarized as follows.

For any nominal molecular weight n, there will be K_n molecular formulas and the total number of formulas in the range a to b amu is $K_a + K_{a+1} + K_{a+2} + \cdots + K_{b-1} + K_b$. Because any one formula of molecular weight n can be satisfied by as many as H_n isomers, the total number of compounds that occur in the molecular weight range a to b is J, where

$$J = K_a \cdot H_a + K_{a+1} \cdot H_{a+1} + K_{a+2} \cdot H_{a+2} + \cdots$$
$$+ K_{b-1} \cdot H_{b-1} + K_b \cdot H_b \tag{1}$$

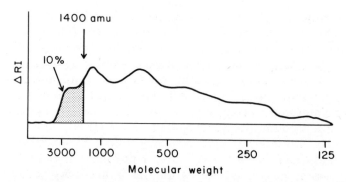

Figure 8.22 Steric exclusion chromatography of an acidic, ethyl acetate extract from *Phaseolus coccineus* seedlings. Column: 2,000 × 25 mm I.D. Biobeads SX-4. Solvent: THF. Detector: differential refractometer. Shaded area represents 10% of the total sample and corresponds to all compounds with molecular weights greater than 1,400 amu. (Reeve & Crozier, 1980.)

This information can be related to a practical situation in the following manner. Figure 8.22 illustrates data that can be used to assess the molecular weight distribution of an acidic, ethyl acetate-soluble extract from *Ph. coccineus* seedlings. The data provide an accurate reflection of the molecular weight distribution of the extract since it was experimentally verified that (i) mechanisms other than steric exclusion do not contribute significantly to the separation, (ii) the differential refractive index response is proportional to the mass of extract eluted, and (ii) the molecular weight calibration is an average for a variety of molecular shapes. Figure 8.22 indicates that 90% of the extract is composed of compounds whose molecular weights range from 100–1,400 amu. Thus, Eq. (1) can be solved for J between the limits $a = 100$ and $b = 1,400$ amu.

By inspection of the molecular formulas presented in the Mass and Abundance Tables of Beynon and Williams (1963), the approximate relationship between K_n and n can be derived as

$$K_n = 0.46n - 18 \tag{2}$$

The relationship between H_n and n is not as easily obtained. No simple method exists for an exact calculation of H_n, although there is general agreement that for a given homologous series, $\log H$ is closely proportional to the carbon number v. Furthermore, the homologous series of different classes of compounds produce parallel $\log H$ versus v curves, which differ from one another by no more than $\log H = \pm 2$ for a given carbon number (see Henze & Blair, 1934; Francis, 1947; Lederberg et al., 1969). Because carbon number, and thus molecular weight, are related to $\log H$ rather than H, the right-hand side of Eq. (1) is dominated by the highest molecular weight term in H and

and can be reduced to

$$J = K_b \cdot H_b \tag{3}$$

Thus, in the case of the *Phaseolus* extract only the values of K and H pertinent to compounds of molecular weight 1,400 need be considered. Using the data of Henze and Blair (1931) for isomeric hydrocarbons of the methane series between C_{10} and C_{40}, the relationship between H and v can be calculated as

$$\log H = 0.41v - 2.6 \tag{4}$$

From this equation it can be predicted that the compound $C_{100}H_{202}$—or, for that matter, almost any compound of molecular weight 1,400—will have approximately 10^{39} potential isomers. However, there is some evidence that, for reasons of symmetry, H will asymptotically approach a limiting value of about 10^{40}. For example, Rouvray (1974) quotes Gouchorov as calculating the number of isomers of $C_{400}H_{802}$ as only 10^{40}. Thus, the estimate of 10^{39} for $C_{100}H_{202}$ is most likely an overestimate rather than an underestimate. Even so, it is of use in that it provides an upper limit for H_b.

Since the number of possible molecular formulas having a nominal molecular weight of 1,400 can be calculated from Eq. (2) as $10^{2.8}$, an approximate upper limit of 10^{42} can be placed on J via Eq. (3). Hence, in the case of the *Phaseolus* extract, an analytical procedure capable of distinguishing between 10^{42} compounds will account for 90% of the mass of the extract. By reference to basic information theory, the amount of information (I) required to describe a system comprised of 10^{42} components can be calculated from the following equation to be some 140 binary digits (bits):

$$I = \frac{\log n}{0.3} \text{ bits} \tag{5}$$

where n is the number of components in the system. Thus, I must exceed 140 bits if the identity of a GA derived from the *Phaseolus* extract is to be defined with a probability of 0.9. It could be argued that in view of the crudity of the approximations, the figure obtained for the minimum information requirement is unlikely to be very meaningful. However, even if J were underestimated by one millionfold, the minimum acceptable value of J would increase by only 20 bits. When this point is considered, along with the fact that some analytical techniques can generate many hundreds of bits of information, it is apparent that practical means are available to establish quantitative accuracy with a high degree of certainty.

When SEC is used as a purification technique, it is possible to select for a narrow range of molecular sizes and thereby greatly restrict the number of compounds *potentially* present in an extract. For example, GA_1 has a molecular weight of 348, and so by selecting only that portion of the eluent from a SEC column that contains GA_1, it is possible to obtain a sample where a large portion of the mass is accounted for by compounds with a molecular weight of less than 400. This represents a great simplification of the analytical situation, since by calculation from Eqs. (2), (3), and (4), J cannot be larger than 10^{11} and thus only 36 bits of information are required to verify accuracy.

The potential information content of analytical techniques can be assessed in terms of bits. A mass spectrum can be thought of as being made up of n individual measurements or information channels, where n is the number of discrete m/e values contained within the spectrum. If the intensity of each m/e value is estimated with a precision of ±3%, application of the equation

$$I = \frac{1.7 - \log q}{0.3} \text{ bits} \tag{6}$$

where q = percentage precision, reveals that the response can be described by 4 bits and hence the information content of the entire spectrum amounts to $4n$ bits. However, ion intensities are commonly recorded as a fraction of the base peak; to allow for this, information equivalent to one m/e value must be subtracted from the total; that is, $I_t = 4(n-1)$ bit. A 30–530-amu mass spectrum thus generates 2,000 bits. However, because of the combined influences of information distribution and correlation, it can be anticipated that the actual information yield is around one quarter of this value, at some 500 bits. Even so, this is well in excess of the minimum requirement of 140 bits for verification of accuracy which can be achieved when the intensities of 141 ions agree with those of the standard to within ±3%.

In practice, the mass spectrum of the sample rarely exactly matches that of the standard, and of the total information I_t contained in the sample spectrum only a portion (I_u) will correspond with the spectral characteristics of the standard. By inspection of Eq. (6), it is apparent that I_u can be calculated according to

$$I_u = \sum_{n=a}^{n=b} \frac{1.7 - \log |L_U - L_u|_n}{0.3} \text{ bits} \tag{7}$$

where $|L_U - L_u|_n$ is the difference, regardless of sign, between the percentage relative intensities of ion peaks having a nominal mass n in the sample and standard spectra, and a and b are the m/e limits of the scans. It should be noted that $|L_U - L_u|_n$ cannot be smaller than the precision with which the percentage relative intensities of the ion peaks are measured, and account should be taken of this fact when employing Eq. (7).

The mass spectrum of the reference compound $GA_{20}MeTMSi$ and the mass spectrum of a substance isolated from a derivatized *P. sativum* extract, as illustrated in Fig. 8.13, can be used to demonstrate the points discussed above. If it is assumed that the precision of the ion intensity measurements is ±3%, it follows from Eq. (6) that 4.1 bits of information are provided for every nominal mass recorded. Thus, the total potential information content of the spectra is $I_t = 4.1 \times (376 - 1) = 1,537$ bits. Of this, the amount of information that correlates the spectrum of the putative $GA_{20}MeTMSi$ with that of the standard can be calculated via Eq. (7) as $I_u = 1,435$ bits (Table 8.13). Hence, it would appear that the putative $GA_{20}MeTMSi$ closely matches that of the reference compound. However, it must be remembered that even quite different substances have many features of their mass spectra in common because of distribution/correlation. As a consequence, as much as 75% I_t bits of the information I_u are likely to be "noninformative" and only $1,435 - (1,537 \times 0.75) = 282$ bits are available to support the conclusion that the *Pisum* sample contains $GA_{20}MeTMSi$. Nevertheless, this is well in excess of the 140-bit minimum proposed by Reeve and Crozier (1980) and is thus sufficient to verify accuracy. It is evident that 140 bits can be furnished by something less than a full scan although there are limits, and if too few ions are monitored and/or too many spurious fragments are present, mass spectrometric evidence will almost certainly fail to provide the necessary verification of accuracy.

The weakest link in the above calculation is the allowance made for information distribution and correlation, and this will be the case until such time as larger and more comprehensive libraries of spectra are available for analysis. However, in spite of this drawback, there is a good case for quoting I_t and I_u in all situations where restrictions on space prevent the publication of full line spectra.

Information theory can also be used to assess analytical data generated by chromatographic techniques. The potential information yield in bits is related, on a one-to-one basis, to the peak capacity (ϕ) of a chromatographic system; thus, $I = \phi$. Giddings (1967) defined peak capacity as the maximum number of components that can be simultaneously separated from each other with unit resolution. This is a particularly useful chromatographic parameter because it is a direct measure of discriminating power. High peak capacities are obtained using chromatographic systems of high efficiency (N) and employing relatively long retention times since for the isocratic/isothermal case

$$\phi = 1 + 0.6N^{0.5} \log (1 + k') \qquad (8)$$

Table 8.13. Calculation of the amount of information in the mass spectrum of a putative GA_{20} MeTMSi sample which matches the spectrum of an authentic sample of GA_{20} MeTMSi. L_U and L_u are the percentage relative ion intensities of the sample and reference spectra, respectively

m/e	L_U (%)	L_u (%)	$\|L_U - L_u\|_n$	Set 3% min	$1.7 - \log\|L_U - L_u\|_n$ / 0.3 (bits)
50	4.5	0	4.5	4.3	3.5
51	0	0	0	3.0	4.1
52	0	0	0	3.0	4.1
53	13.0	5.2	7.8	7.8	2.7
54	4.5	5.2	0.7	3.0	4.1
55	62.3	41.5	20.8	20.8	1.3
56	8.4	7.8	0.6	3.0	4.1
57	8.4	23.3	14.9	14.9	1.8
58	3.9	9.1	5.2	5.2	3.3
59	35.1	20.1	15.0	5.1	3.4
60	3.9	2.6	1.3	3.0	4.1
.	.	.	.	.	.
.	.	.	.	.	.
.	.	.	.	.	.
415	10.4	1.9	8.5	8.5	2.6
416	10.4	3.9	6.5	6.5	3.0
417	14.3	3.9	10.4	10.4	2.3
418	100.0	54.5	55.5	55.5	0.1
419	28.6	15.6	13.0	13.0	2.0
420	7.8	3.9	3.9	3.9	3.7
421	0.6	0	0.6	3.0	4.1
422	0	0	0	3.0	4.1
423	0	0	0	3.0	4.1
424	0	0	0	3.0	4.1
425	0	0	0	3.0	4.1

Total   1,435

where k' is the capacity factor of the last solute to be eluted. Hence, capillary GC has a potential of *ca.* 300 bits, modern HPLC columns up to 100 bits, while classical procedures such as TLC and paper chromatography produce no more than 5 bits.

The theoretical basis for the application of information theory to chromatographic analysis can be considered as follows. In accordance with the basic formal structure presented in Section 3.2.2.3.b of Reeve and Crozier (1980), consider a sample U and standard u analyzed in a chromatographic system with a peak capacity of 10 and which thereby provides ten independent retention time windows A to J (Fig. 8.23). In its simplest form each window provides a single-bit test of the null hypothesis $H_o: U = u$, depending upon whether the signal from the detector is greater or less than, say, twice the background noise. The chromatographic comparison of sample with standard can be summarized as follows:

Test of H_o:	A	B	C	D	E	F	G	H	I	J
pass/fail:	*f*	*p*	*f*	*f*	*p*	*p*	*p*	*f*	*f*	*f*

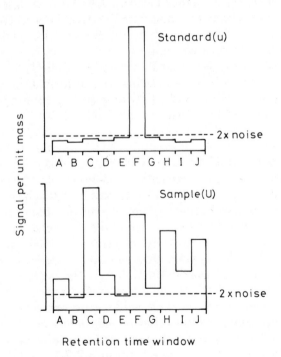

Figure 8.23 Hypothetical analysis of a standard (u) and a sample (U) on a chromatogram with a peak capacity of 10.

The total information accrued is 1 bit per test, making a total of 10 bits, and on the basis of this information it is evident that $U \neq u$. It should be noted that rejection of a null hypothesis can be based on either a lack of response at F, the chromatographic window of the standard, or differences observed between the standard and the sample at windows A–E and G–J, such as occurs in Fig. 8.23. In practical terms, if we assert that a *certain substance isolated from a plant extract* is, say, GA_1 and wish to test that assertion by chromatographic means, we will be interested not only in seeing a peak of the correct retention time for GA_1, *but also* an absence of peaks elsewhere.

It is possible, however, to accept that sample U is impure and to consider only material eluting at retention time window F. In this case, we are dealing with a new sample, U_f, and are interested in testing a new null hypothesis, $H_o: U_f = u$. It follows that the only information pertinent to this new null hypothesis will be the one bit of information given in test F. Thus, in the practical context, if we assert that a peak in a chromatograph of a plant extract is GA_1, we become interested only in a retention window match with the standard, and, as a consequence, the amount of information supporting this assertion falls to 1 bit. In this way chromatographic data are directly analogous to mass spectral information where the information yield decreases as fewer and fewer ions are monitored.

Scott (1982) suggests that Reeve and Crozier (1980) do not use the correct method for calculating the information yielded by chromatographic analysis and as a consequence have overestimated the value of such data. Scott (1982) follows the lead of Massart (1973) and proposes that if a chromatogram is capable of separating, say, 10 components then it has a potential information yield of $\log_2 10$ bits rather than 10 bits. It follows that a retention time match in a chromatogram with a peak capacity ϕ generates $\log_2 \phi$ bits of information. While this conclusion is, in itself, valid, its application to plant extracts which are open-ended systems is a different matter altogether.

Consider a closed system of, say, 10 indivisible objects that can be separated into 10 classes by a particular technique. In such circumstances, it is inarguable that the technique can bring $\log_2 10$ bits of information to bear on the null hypothesis $H_o: U = u$, because in this closed system *both U and u can only be single discrete components*. However, when dealing with plant extracts, the opened nature of the problem dictates that *we can never assume that U is a single discrete component*. All samples, even so-called standards, must ultimately be mixtures—see Section 3.2.2.1., p. 209, in Reeve and Crozier (1980). Consequently, in the chromatographic comparison of unknown against standard illustrated in Fig. 8.23, there can be no justification for the assumption that U_f is a 100% component of unknown nature and, thus, bound to fit only one of the ϕ retention windows available in any chromatographic system. The material eluting at retention window F is a mixture that on the basis of the one relevant bit of information, F, is indistinguishable from the relatively pure mixture, u, which is a standard. It was for this very reason that Reeve and Crozier (1980) formalized the accuracy probability term through definition of sample purity. It is essential to appreciate that accuracy, purity, and identity are inseparable aspects of the same basic concept. Thus, any unspoken assumption as to the purity of a chromatographic peak derived from a plant extract automatically implies knowledge of the identity and quantity of material in that peak. Since the very concept of a sample as an open-end system precludes such assumptions, the Massart equation is inapplicable because it can only be used with simple theoretical systems. In the real world of plant extracts, one can do little better than relate the information content of a chromatogram to its peak capacity on a peak-by-peak, bit-by-bit basis.

Figure 8.24 illustrates a hypothetical chromatogram of an authentic sample of GA_x in which the potential information, by virtue of the peak capacity, is 100 bits. All this information is available for the verification of accuracy in the case of sample A, which produces a trace that closely matches that of the GA_x standard. Sample B, however, contains a large number of impurities, and the correlation is far from perfect. Although GA_x can be quantified on the basis of the appropriate peak area, the amount of information that can be used to verify the accuracy of the estimate is limited to only 1 bit because so few parts of the chromatogram match the authentic GA_x trace. Clearly, sample purity is an important consideration and cannot be ignored, as it is a major factor in determining the amount of information that is accumulated. When analyzing impure samples, the availability of a selective detector is advantageous since traces will yield far more information than equivalent chromatograms obtained with a nonselective detector system. This is where the strength of SICM lies, why GC-RC data have proved reliable in identifying radioactive GAs, and why, in contrast, GC-FID analysis of endogenous GAs has produced many erroneous identifications. In this context, both bioassays and immunoassays can be looked upon as selective GA detectors. Unfortunately, bioassays are very labour-intensive, they lack precision, and the response time can range from 2 to 7 days. Immunoassays are not beset by these problems and are likely to be used with increasing frequency, especially in conjunction with HPLC. It is, however, important to remember that large numbers of fractions must be collected and analyzed if the peak capacity and, hence, the information yield of the chromatogram is to be maintained.

When an analysis is carried out using a number of different chromatographic procedures the information contributed by each step is additive, after due allowance has been made for correlation between the retention characteristics of the individual procedures. The concept of chromatographic correlation is best visualized in terms of two-dimensional TLC or paper chromatography. Figure 8.25 illustrates three hypothetical chromatograms in which the degree of correlation between the first and second solvent ranges from unity (A) to zero (C). It is evident that the discriminating power of the two-dimensional technique is heavily reliant on the ability of

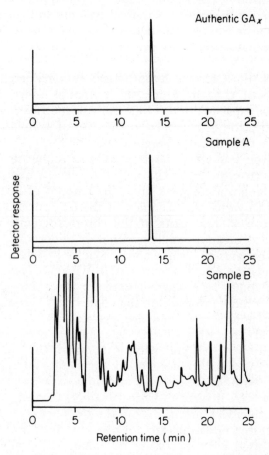

Figure 8.24   Hypothetical analysis of GA_x on a chromatogram with a peak capacity of *ca.* 100. (Crozier, 1981.)

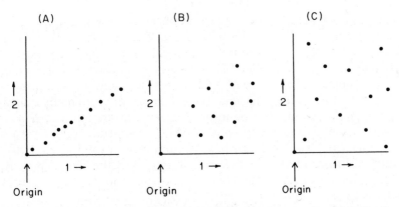

Figure 8.25   Hypothetical two-dimensional chromatograms in which the degree of correlation between solvent systems 1 and 2 varies from unity (A) to zero (C). (Reeve & Crozier, 1980.)

the second solvent system to generate a retention pattern that is completely different from the first. Calculation of the correlation factor between two techniques is not a practical proposition. However, it is safe to assume that the effects of correlation will be minimal when use is made of chromatographic procedures that are mechanistically quite different. Because of the wide range of separatory mechanisms that are available, it is possible to employ up to six HPLC systems without adversely affecting correlation. There is, however, no point in combining more than two or three GC systems, because of the high degree of intersystem correlation.

The HPLC analysis of the *Salix* extract, illustrated in Figs. 8.19 and 8.20, can be used to demonstrate procedures for calculating chromatographic information yield. Because the extract was purified by preparative SEC and the free GA zone collected for further analysis, at least 90% of the components in the sample will have a molecular weight of less than 400. Thus only 36 bits of information are required to verify the accuracy of the GA_1 analysis with a probability of at least 0.9.

The gradient elution reverse-phase HPLC system used to analyze the derivatized GA_1 fraction from *Salix* (Fig. 8.19) has a peak capacity of *ca*. 80. Although the sample contains a fluorescent peak that co-chromatographs with GA_1CE, there are a number of impurities present and as a consequence the null hypothesis $H_o: U = GA_1CE$ must be rejected. The putative GA_1CE peak was collected and further analyzed by isocratic elution from reverse- and normal-phase HPLC columns. In both instances, the only components to be detected were were a void volume solvent peak and a solute that co-chromatographed with GA_1CE (Fig. 8.20). In the reverse-phase system, where $N = 4480$ and $k' = 2.6$, $\phi = 23$ as calculated from Eq. (8). Thus, 22 bits (23 minus 1 bit subtracted because of the void volume peak) support the null hypothesis $H_o : U = GA_1CE$. Similarly, the normal-phase separation on Hypersil CPS, where $N = 6150$, $k' = 6.2$, and $\phi = 41$, provides $41 - 1 = 40$ bits of information. The information supplied by the two systems can be assumed to be additive because intersystem correlation between normal- and reverse-phase HPLC separations is likely to be low. Thus, the HPLC analyses have accumulated a total of $22 + 40 = 62$ bits of information, which is well in excess of the 36 bits required to verify the accuracy of the GA_1CE identification with a probability of at least 0.9.

While the procedures of Reeve and Crozier (1980) provide a means of verifying accuracy, it must be emphasized that they involve a number of less-than-perfect assumptions. However, the semi-objective standards do appear to provide an intuitively acceptable assessment of many of the imponderables that are usually left to subjective judgment.

ACKNOWLEDGMENTS

The authors wish to thank Dr. N. Murofushi (University of Tokyo) for supplying mass spectra of MeTMSi derivatives of GA_{24}-GA_{56}; Dr. P. Hedden (East Malling Research Station) for unpublished GC-MS data; Dr. R. P. Pharis (University of Calgary) for details of reverse-phase HPLC separations of GAs; and Dr. J. R. Lenton (Rothamstead Research Station) for information on the biological activity of GA_{54}, GA_{55}, GA_{60}, GA_{61}, and GA_{62}.

REFERENCES

Arpino, P. J., & Guiochon, G. (1979). LC/MS coupling. Anal. Chem. *51*, 682A–701A.

Atzorn, R., & Weiler, E. W. (1982). The immunoassay of gibberellins. In: Abstracts of XI International Conference on Plant Growth Regulators, pp. 19.

Atzorn, R., & Weiler, E. W., (1983a). The immunoassay of gibberellins, 1. Radioimmunoassay of the gibberellins A_1, A_3, A_4, A_7, A_9 and A_{20}. Planta (in press).

Atzorn, R., & Weiler, E. W. (1983b). The immunoassay of gibberellins, 2. Quantification of GA_3, GA_4 and GA_7 by ultra-sensitive solid-phase enzyme-immunoassay. Planta (in press).

Bailiss, K. W., & Hill, T. A. (1971). Biological assays for gibberellins. Bot. Rev. *37*, 437–479.

Barendse, G. W. M., Van de Werken, P. H., & Takahashi, N. (1980). High performance liquid chromatography of gibberellins. J. Chromatogr. *198*, 449–455.

Bearder, J. R., & MacMillan, J. (1976). Origin of the oxygen atoms in the lactone bridge of C_{19}-gibberellins. J. Chem. Soc., Chem. Commun. 421–422.

Bearder, J. R., MacMillan, J., & Phinney, B. O. (1973). 3-Hydroxylation of gibberellin A_{12}-aldehyde in *Gibberella fujikuroi* strain REC-193A. Phytochem. *12*, 2173–2179.

Beynon, J. E., & Williams, A. E. (1963). Mass and Abundance Tables for Use in Mass Spectrometry. Elsevier, Amsterdam.

Binks, R., Cleaver, R. L., Littler, J. S., & MacMillan, J. (1971). Real-time processing of low resolution mass spectra. Chem. Brit. 7, 8–12.

Binks, R., MacMillan, J., & Pryce, R. J. (1969). Plant hormones VII. Combined gas chromatography-mass spectrometry of the methyl esters of gibberellins A_1 to A_{24} and their trimethylsilyl ethers. Phytochem. *8*, 271–284.

Bombaugh, K. J. (1971). The practice of gel permeation chromatography. In: Modern Practice of Liquid Chromatography, pp. 237–285, Kirkland, J. J., ed. Wiley–Interscience, New York.

Bowen, D. H., MacMillan, J., & Graebe, J. E. (1972). Determination of specific radioactivity of [^{14}C] compounds by mass spectrometry. Phytochem. *11*, 2253–2257.

Brian, P. W., Hemming, H. G., & Lowe, D. (1964). Comparative potency of nine gibberellins. Ann. Bot. *28*, 369–389.

Browning, G., & Saunders, P. F. (1977). Membrane localised gibberellins A_9 and A_4 in wheat chloroplasts. Nature *265*, 375–377.

Bukovac, M. J., Yuda, E., Murofushi, N., & Takahashi, N. (1979). Endogenous plant growth substances in developing fruit of *Prunus cerasus*. Plant Physiol. *63*, 129–132.

Cavell, B. D., MacMillan, J., Pryce, R. J., & Sheppard, A. C. (1967). Thin-layer and gas-liquid chromatography of the gibberellins, direct identification of the gibberellins in a crude plant extract by gas-liquid chromatography. Phytochem. *6*, 867–874.

Crozier, A. (1981). Aspects of the metabolism and physiology of gibberellins. In: Advances in Botanical Research, Vol. 9, pp 33–149, Woolhouse, H. W., ed. Academic Press, London.

Crozier, A., Aoki, H., & Pharis, R. P. (1969). Efficiency of counter-current distribution, Sephadex G-10 and silicic acid partition chromatography in the purification and separation of gibberellin-like substances from plant tissue. J. Exp. Bot. *20*, 786–795.

Crozier, A., Kuo, C. C., Durley, R. C., & Pharis, R. P. (1970). The biological activities of 26 gibberellins in nine plant bioassays. Can. J. Bot. *48*, 867–877.

Crozier, A., & Reeve, D. R. (1977). The application of high performance liquid chromatography to the analysis of plant hormones. In: Plant Growth Regulation, pp. 67–76, Pilet, P. E., ed. Springer-Verlag, Berlin, Heidelberg, New York.

Crozier, A., Zaerr, J. B., & Morris, R. O. (1980). High-performance steric exclusion chromatography of plant hormones. J. Chromatogr. *198*, 57–63.

Crozier, A., Zaerr, J. B., & Morris, R. O. (1982). Reversed- and normal-phase high performance liquid chromatography of gibberellin methoxycoumaryl esters. J. Chromatogr. *238*, 157–166.

Davies, J. K., Jensen, E., Rivier, L., Junttila, O., & Crozier, A. (in preparation). Detection of gibberellins in extracts from *Salix pentandra* shoots by high performance liquid chromatography and combined gas chromatography-mass spectrometry.

Dünges, W. (1977). 4-Bromomethyl-7-methoxycoumarin as a new fluorescent label for fatty acids. Anal. Chem. *49*, 442–445.

Durley, R. C., Crozier, A., Pharis, R. P., & McLaughlin, G. E. (1972). The chromatography of 33 gibberellins on a gradient eluted silica gel partition column. Phytochem. *11*, 3029–3033.

Durley, R. C., MacMillan, J., & Pryce, R. J. (1971). Investigation of gibberellins and other growth substances in seed of *Phaseolus multifloris* and *Phaseolus vulgaris* by gas chromatography and combined gas chromatography-mass spectrometry. Phytochem. *10*, 1891–1908.

Durley, R. C., & Pharis, R. P. (1972). Partition coefficients of 27 gibberellins. Phytochem. *11*, 317–326.

Durley, R. C., & Pharis, R. P. (1973). Interconversion of gibberellin A_4 to gibberellins A_1 and A_{34} by dwarf rice, cultivar Tanginbozu. Planta *109*, 357-361.

Durley, R. C., Railton, I. D., & Pharis, R. P. (1974). The metabolism of gibberellin A_1 and gibberellin A_{14} in seedlings of dwarf *Pisum sativum* In: Plant Growth Substances, 1973, pp. 285-293, Tamura, S., ed. Hirokawa, Tokyo.

Durley, R. C., Sassa, T., & Pharis, R. P. (1979). Metabolism of tritiated gibberellin A_{20} in immature seed of dwarf pea cv. Meteor. Plant Physiol. *64*, 214-219.

Durst, D., Milano, M., Kitka, G. J., Connelly, S. A., & Grushka, E. (1975). Phenacyl esters of fatty acids via crown ether catalysts for enhanced ultraviolet detection in liquid chromatography. Anal. Chem. *47*, 1797-1801.

Francis, A. W. (1947). Numbers of isomeric alkylbenzenes. J. Amer. Chem. Soc. *69*, 1536-1537.

Frankland, B., & Wareing, P. F. (1960). Effect of gibberellic acid on hypocotyl growth of lettuce seedlings. Nature *185*, 255-256.

Fuchs, S., & Fuchs, Y. (1969). Immunological assay for plant hormones using specific antibodies to indoleacetic acid and gibberellic acid. Biochim. Biophys. Acta. *192*, 528-530.

Fuchs, S., Haimovich, J., & Fuchs, Y. (1971). Immunological studies of plant hormones. Detection and estimation by immunological assays. Eur. J. Biochem. *18*, 384-390.

Fukui, H., Ishii, H., Koshimizu, K., Katsumi, M., Ogawa, Y., & Mitsui, T. (1972). The structure of gibberellin A_{23} and the biological properties of 3,13-dihydroxy C_{20}-gibberellins. Agr. Biol. Chem. *36*, 1003-1012.

Giddings, J. C. (1967). Maximum number of components resolvable by gel filtration and other elution chromatographic methods. Anal. Chem. *39*, 1027-1208.

Glenn, J. L., Kuo, C. C., Durley, R. C., & Pharis, R. P. (1972). Use of insoluble polyvinylpyrrolidone for purification of plant extracts and chromatography of plant hormones. Phytochem. *11*, 345-351.

Gräbner, R., Schneider, G., & Sembdner, G. (1976). Gibberelline XLIII. Mitt. Fraktionierung von gibberellinen, gibberellinkonjugaten und anderen phytohormonen durch DEAE-Sephadex-chromatographie. J. Chromatogr. *121*, 110-115.

Graebe, J. E., & Ropers, H. J. (1978). Gibberellins. In: Phytohormones and Related Compounds: A Comprehensive Treatise, Vol. 1, pp. 107-203,

Letham, D. S., Goodwin, P. B., & Higgins, T. J. V., eds. Elsevier/North-Holland, Amsterdam.

Heftmann, E., Saunders, G. A., & Haddon, W. F. (1978). Argenation high-pressure liquid chromatography and mass spectrometry of gibberellin esters. J. Chromatogr. *156*, 71–77.

Henze, H., & Blair, C. (1931). The number of structural isomers of the more important types of aliphatic compound. J. Amer. Chem. Soc. *56*, 157.

Hiraga, K., Yokota, T., & Murofushi, N. (1974). Isolation and characterization of gibberellins in mature seed of *Phaseolus vulgaris*. Agr. Biol. Chem. *38*, 2511–2520.

Hoad, G. V., Pharis, R. P., Railton, I. D., & Durley, R. C. (1976). Activity of the aldehyde and alcohol of gibberellins A_{12} and A_{14}, two derivatives of gibberellin A_{15} and four decomposition products of gibberellin A_3 in 13 plant bioassays. Planta *130*, 113–120.

Ikegawa, N., Kagawa, T., & Sumiki, Y. (1963). Determination of nine gibberellins by gas and thin layer chromatographies. Proc. Japan Acad. *39*, 507–512.

Jones, M. G., Metzger, J. D., & Zeevaart, J. A. D. (1980). Fractionation of gibberellins in plant extracts by reverse phase high performance liquid chromatography. Plant Physiol. *65*, 218–221.

Jones, R. L. (1968). Aqueous extraction of gibberellins in peas. Planta *81*, 97–105.

Jones, R. L., & Varner, J. E. (1967). The bioassay of gibberellins. Planta *72*, 155–161.

Kagawa, T., Fukinbara, T., & Sumiki, Y. (1963). Thin layer chromatography of gibberellins. Agr. Biol. Chem. *27*, 598–599.

Karger, B. L., Kirby, D. P., Vouros, P., Foltz, R. L., & Hidy, B. (1979). On-line reverse phase liquid chromatography-mass spectrometry. Anal. Chem. *51*, 2324–2328.

Knox, J. H. (1979). High Performance Liquid Chromatography. Edinburgh University Press, Edinburgh.

Köhler, D., & Lang, A. (1963). Evidence for substances in higher plants inter-fering with response of dwarf peas to gibberellin. Plant Physiol. *38*, 555–560.

Koshioka, M., Harada, J., Takeno, K., Noma, M., Sassa, T., Ogiyama, K., Taylor, J. S., Rood, S. B., Legge, R. L., & Pharis, R. P. (1983). Reverse phase C_{18}

high performance/pressure liquid chromatography of acidic and conjugated gibberellins. J. Chromatogr. *256*, 101–115.

Küllertz, G., Eckert, H., & Schilling, G. (1978). Quantitative gas chromatographic determination of gibberellic acid (GA_3) in nanogram quantities by electron capture detection. Biochem. Physiol. Pflanz. *173*, 186–187.

Lederberg, J., Sutherland, G. L., Buchanan, B. G., Feigenbaum, E. A., Robertson, A. V., Duffield, A. M., & Djerassi, C. (1969). Application of artificial intelligence for chemical inference. I. The number of possible organic compounds. Acyclic structures containing C, H, O and N. J. Amer. Chem. Soc. *91*, 2973-2976.

Letham, D. S., Higgins, T. J. V., Goodwin, P. B., & Jacobsen, J. V. (1978). Phytohormones in retrospect. In: Phytohormones and Related Compounds: A Comprehensive Treatise, Vol. 1, pp. 1–27, Letham, D. S., Goodwin, P. B., & Higgins, T. J. V., eds. Elsevier/North Holland, Amsterdam.

Lin, J-T., Heftmann, E. (1981). Adsorption and reversed-phase high performance liquid chromatography of gibberellins. J. Chromatogr. *213*, 507–510.

MacMillan, J., Gaskin, P., and MacNeil, K. A. G. (1983). Mass Spectra of Gibberellins, Reference Spectra of Functional Derivatives. Wiley, London, New York, Sydney, Toronto.

MacMillan, J., & Suter, P. J. (1963). Thin layer chromatography of the gibberellins. Nature *197*, 190.

MacMillan, J., & Wels, C. M. (1973). Partition chromatography of gibberellins and related diterpenes on columns of Sephadex LH-20. J. Chromatogr. *87*, 271–276.

McFadden, W. H. (1979). Interfacing chromatography and mass spectrometry. J. Chromatogr. Sci. *17*, 2–17.

McFadden, W. H., Bradford, D. C., Gaines, D. E., & Gower, J. L. (1977). Applications of combined liquid chromatography/mass spectrometry. Int. Laboratory (Oct.), 55–64.

Majors, R. E., & MacDonald, F. R. (1973). Practical implications of modern liquid chromatographic column performance. J. Chromatogr. *83*, 169–179.

Massart, D. L. (1973). The use of information theory for evaluating the quality of thin-layer chromatographic separations. J. Chromatogr. *79*, 157–163.

Morris, R. O., & Zaerr, J. B. (1978). 4-Bromophenacyl esters of gibberellins, useful derivatives for high performance liquid chromatography. Anal. Lett. AII(i), 73–83.

Murakami, Y. (1968). The microdrop method, a new rice seedling test for gibberellins and its use for testing extracts of rice and morning glory. Bot. Mag. 79, 33–43.

Nash, L. J., Jones, R. L., & Stoddart, J. L. (1978). Gibberellin metabolism in excised lettuce hypocotyls: response to GA_9 and conversion of $[^3H]GA_9$. Planta 140, 143–150.

Nicholls, P. B., & Paleg, L. G. (1963). A barley endosperm bioassay for gibberellins. Nature 199, 823–824.

Parups, E. V. (1980). Effect of morphactin on certain plant growth substances in bean roots. Physiol. Plant 49, 281–285.

Perez, A. T., & Lachman, W. H. (1971). Gas-liquid chromatography of endogenous gibberellins in tomato, Lycopersicon esculentum. Phytochem. 10, 2799–2802.

Phinney, B. O. (1956). Growth response of single-gene dwarf mutants in maize to gibberellic acid. Proc. Natl. Acad. Sci. (U.S.A.) 42, 185–189.

Phinney, B. O., & West, C. A. (1961). Gibberellins and plant growth. In: Encyclopedia of Plant Physiology, Vol. 14, pp. 1185–1227, Ruhland, W., ed. Springer-Verlag, Berlin, Heidelberg, New York.

Pitel, D. W., Vining, L. C., & Arsenault, G. P. (1971). Improved methods for preparing pure gibberellins from cultures of Gibberella fujikuroi. Isolation by adsorption or partition chromatography on silicic acid and by partition chromatography on Sephadex columns. Can. J. Biochem. 49, 185–193.

Powell, L. E. (1963). Solvent systems for silica-gel column chromatography of indole derivatives. Nature 200, 79.

Powell, L. E., & Tautvydas, K. J. (1967). Chromatography of gibberellins on silica gel partition columns. Nature 213, 292–293.

Railton, I. D., Durley, R. C., & Pharis, R. P. (1974). Metabolism of tritiated gibberellin A_9 by shoots of dark-grown dwarf pea, cv. Meteor. Plant Physiol. 54, 6–12.

Railton, I. D., & Rechav, M. (1979). Efficiency of extraction of gibberellin-like substances from chloroplasts of Pisum sativum L. Plant Sci. Lett. 14, 75–78.

Reeve, D. R., & Crozier, A. (1975). Gibberellin bioassays. In: Gibberellins and Plant Growth, pp. 35–64, Krishnamoorthy, H. N. ed. Wiley Eastern, New Delhi.

Reeve, D. R., & Crozier, A. (1976). Purification of plant hormone extracts by gel permeation chromatography. Phytochem *15*, 791-793.

Reeve, D. R., & Crozier, A. (1978). The analysis of gibberellins by high performance liquid chromatography. In: Isolation of Plant Growth Substances. Society for Experimental Biology Seminar Series, Vol. 4, pp. 41-77, Hillman, J. R., ed. Cambridge University Press, Cambridge.

Reeve, D. R., & Crozier, A. (1980). Quantitative analysis of plant hormones. In. Hormonal Regulation of Development 1. Molecular Aspects of Plant Hormones. Encyclopedia of Plant Physiology New Series, Vol. 9, pp. 203-280, MacMillan, J., ed. Springer-Verlag, Berlin, Heidelberg, New York.

Reeve, D. R., Yokota, T., Nash, L. J., & Crozier, A. (1976). The development of a high performance liquid chromatograph with a sensitive on-stream radioactive monitor for the analysis of ^{3}H and ^{14}C-labeled gibberellins. J. Exp. Bot. *21*, 1243-1258.

Reid, D. M., & Crozier, A. (1970). CCC-induced increase of gibberellin levels in pea seedlings. Planta *94*, 95-106.

Ross, J. D., & Bradbeer, J. W. (1971). Studies in seed dormancy V. The content of endogenous gibberellins in seeds of *Corylus avellana* L. Planta *100*, 288-302.

Rouvray, D. H. (1974). Isomer enumeration methods. Chem. Soc. Rev. *3*, 355-372.

Rivier, L., Gaskin, P., Albone, K. S., & MacMillan, J. (1981). GC-MS identification of endogenous gibberellins and gibberellin conjugates as their permethylated derivatives. Phytochem. *20*, 607-692.

Scott, I. M. (1982). Information theory and plant growth substance analysis. Plant Cell Environ. *5*, 339-342.

Seeley, S. D., & Powell, L. E. (1970). Electron capture-gas chromatography for sensitive assay of abscisic acid. Anal. Biochem. *35*, 530-533.

Seeley, S. D., & Powell, L. E. (1974). Gas chromatography and detection of microquantities of gibberellins and indoleactic acid as their fluorinated derivatives. Anal. Biochem. *58*, 39-46.

Shindy, W. W., & Smith, O. E. (1975). Identification of plant hormones from cotton ovules. Plant Physiol. *55*, 550-554.

Snyder, L. R., & Kirkland, J. J. (1979). Introduction to Modern Liquid Chromatography, 2nd ed. Wiley-Interscience, London.

Sponsel, V. M., Hoad, G. V., & Beeley, L. J. (1977). The biological activities of some new gibberellins (GAs) in six plant bioassays. Planta 135, 143–147.

Sponsel, V. M., & MacMillan, J. (1978). Metabolism of gibberellin A_{29} in seeds of Pisum sativum cv Progress No. 9; use of $[^{2}H]$ and $[^{3}H]GAs$, and the identification of a new catabolite. Planta 144, 69–78.

Sponsel, V. M., & MacMillan, J. (1980). Metabolism of $[^{13}C]$gibberellin A_{29} to $[^{13}C_1]$gibberellin-catabolite in maturing seeds of Pisum sativum cv. Progress No. 9, Planta 150, 46–52.

Vining, L. C. (1971). Separation of gibberellin A_1 and dihydrogibberellin A_1 by argenation partition chromatography on a Sephadex column. J. Chromatogr. 60, 141–143.

Weiler, E. W., & Wieczorek, U. (1981). Determination of femto-mol quantities of gibberellic acid by radioimmunoassay. Planta 152, 159–167.

Whyte, P., & Luckwill, L. C. (1966). A sensitive bioassay for gibberellins based on retardation of leaf senescence in Rumex obtusifolius (L.). Nature 210, 1360.

Wightman, F. (1979). Modern chromatographic methods for the identification and quantification of plant growth regulators and their application to studies of the changes in hormonal substances in winter wheat during acclimation to cold stress conditions. In: Plant Regulation and World Agriculture, pp. 327–377, Scott, T. K., ed. Plenum, New York.

Yabuta, T., & Hayashi, T. (1939). Biochemical studies on "bakanae" fungus of the rice. Part II. Isolation of "gibberellin," the active principle which makes the rice seedling grow slenderly. J. Agr. Chem. Soc. (Japan) 15, 257–266.

Yamaguchi, I., Yokota, T., Murofushi, N., Ogawa, Y., & Takahashi, N. (1970). Isolation and structure of a new gibberellin from immature seeds of Prunus persica. Agr. Biol. Chem. 34, 1439–1441.

Yamane, H., Yamaguchi, I., Yokota, N., Murofushi, N., Takahashi, N., & Katsumi, K. (1973). Biological activities of new gibberellins A_{30}–A_{35} and A_{35} glucoside. Phytochem. 12, 255–261.

Yokota, T., Murofushi, N., Takahashi, N., & Kalsumi, M. (1971). Biological activities of gibberellins and their glucosides in Pharbitis nil. Phytochem. 10, 2943–2949.

Zeevart, J. A. D. (1973). Gibberellin A_{20} content of Bryophyllum diagremontianum under different photoperiodic conditions as determined by gas-liquid chromatography. Planta 114, 285–288.

INDEX